Student Study Guide/Solutions Manual

to accompany

Principles & Applications of Inorganic, Organic and Biological Chemistry

Prepared by

KATHERINE J. DENNISTON
Towson State University

JOSEPH J. TOPPING
Towson State University

with contributions by

Larry Byrd
Western Kentucky University

Wm. C. Brown Publishers

Dubuque, IA Bogota Boston Buenos Aires Caracas Chicago
Guilford, CT London Madrid Mexico City Sydney Toronto

A Times Mirror Company

ISBN 0-697-25005-9

Printed in the United States of America by Times Mirror Higher Education Group, Inc., 2460 Kerper Boulevard, Dubuque, Iowa, 52001

10 9 8 7 6 5 4 3 2 1

Preface

This student study guide has been written to accompany *Principles and Applications of Inorganic, Organic, and Biological Chemistry* by Caret, Denniston, and Topping. It was designed to complement the text, not to be used in place of the text. Each chapter of the study guide contains the following sections:

1. *Reorganized and expanded set of learning goals.* These have been organized into conceptual goals (new ideas and principles that must be understood), performance goals (types of problems you must be able to solve), and health-related applications (examples of the application of the principles of chemistry to disease states, patient care, and hospital practice).
2. *Concise chapter summary.* Each of the sections in the chapter is discussed briefly to allow you to review key points. The summary follows the organization of the chapter, section by section.
3. *In-chapter solved problems.* Within the chapter summaries you will find a number of solved sample problems. Step-by-step instructions for logically approaching problem solving are presented.
4. *List of key terms.* All the key terms from the main body of the text that are defined in the glossary in the text are listed for quick reference and self-testing. Those from the perspectives have been omitted.
5. *Review problems.* A set of review problems and questions concludes each study guide chapter.

How to Study Chemistry

1. There is no "quick-fix" method of studying chemistry. Understanding the discipline requires study. Problem-solving skills can only be acquired through practice. Above all, it is essential to keep up with the material because the principles and concepts you learn early in the course will be used and applied throughout the course. If you miss a key point early in the course, it will certainly cause you greater and greater problems as the course proceeds.
2. Before you begin to study the chemistry, take a few moments to familiarize yourself with the textbook. In the first pages of the text you will find a Brief Table of Contents and an Expanded Table of Contents. These allow quick reference to major areas of study. There is also a preface that acquaints you with the philosophy and organization of the text.
3. Read the text *before* attending class. This will familiarize you with the terms and concepts to be covered in lecture. If you have particular problems understanding an idea, pay close attention to the instructor's explanation during lecture. If you still have difficulty with the concept, ask your professor or your tutor for additional help. By reading ahead, you can also be prepared with specific questions to ask during the lecture.
4. As you read the text, pay particular attention to the in-chapter examples. These will take you step-by-step through each problem and show you how to work through the logic and mathematics. Often these in-chapter examples are followed by similar in-chapter problems. Test your grasp of the concept by solving a few of these problems.
5. Use your study guide to reinforce your understanding of key concepts. Read the summary of the chapter provided in the study guide. Then work through the solved problems in the study guide.
6. Solve all the problems from the text that are assigned by your instructor. If one type of problem is particularly difficult, select and solve additional problems of that type from among the in-chapter and end-of-chapter problem sets. The answers to all the odd-numbered problems are found at the end of the text.
7. At the end of each chapter in the study guide you will find a number of additional practice problems. Use these to further strengthen your problem-solving skills or as a self-test before an exam.
8. Read through the list of key terms. Be sure that you can write a brief functional definition of each. Note that you don't need to memorize and regurgitate the definitions given in the text. You will achieve a much greater understanding of the term by describing it in your own words.

We are aware that many students experience "exam anxiety." In our experience, confidence in one's own ability to solve problems, to understand basic principles, and to define key terms greatly reduces the extent of such anxiety and generally results in far better performance on exams. Consistent use of this study guide, along with the textbook and class notes, will help to assure success.

Contents

1 *Chemistry: Methods and Measurement*

Learning Goals

1. **Conceptual Goals**
   - Understand the definition of chemistry and know its major subfields.
   - Know the approach to science, the scientific method.
   - Distinguish between the terms hypothesis, theory, and scientific law.
   - Recognize that curiosity, experimentation, and reasoning are at the heart of any scientific study.
   - Know both the differences and relationships between science and technology.
   - Distinguish between data and results.
   - Recognize the need for units of measure.
   - Recall the major units of measure in the metric system.
   - Develop an understanding of significant figures and scientific notation.
   - Develop a general familiarity with the principal experimental quantities and the way in which they are measured.

2. **Performance Goals**
   - Carry out conversion of units within the metric system.
   - Carry out conversions of units between the English and metric systems.
   - Recognize the fundamental experimental basis for significant figures. Practice assigning the proper number of significant digits to a measurement.
   - Generate results with the proper number of significant figures from a calculation.

3. **Health Applications**
   - Know the relationship between mass and energy (calories) in the diet.
   - Describe specific gravity and its importance as a diagnostic tool.

Chapter Overview

The subject matter of chemistry deals with all material substances as well as all the changes that these materials undergo. The basic tools of chemistry will enable you to further your understanding of how chemistry works.

1.1 Chemistry: An Overview

Chemistry is the study of matter and the changes that matter undergoes. **Matter** is anything that has mass and occupies space. The changes that matter undergoes always involve either gain or loss of energy. **Energy** is nonmaterial, and is the ability to do work (to accomplish some change). Thus, a study of chemistry involves matter; energy, and their interrelationship.

1

The major areas of chemistry include **biochemistry**, the study of matter associated with living things; **organic chemistry**, the study of matter principally composed of two elements, carbon and hydrogen; **inorganic chemistry**, the study of matter that consists of all of the other elements and their combinations; **analytical chemistry**, the analysis of matter to determine its composition; and **physical chemistry**, a discipline that attempts to explain the way matter behaves.

1.2 The Methodology

The **scientific method** consists of five interrelated processes:

1. Observation. The description of the properties of a substance is a result of observation. The measurement of the temperature of a liquid or the size or mass of a solid results from observation.
2. Pattern recognition. Discerning a cause-and-effect relationship may give rise to a scientific **law**. A scientific law is a description of the orderly behavior observed in nature.
3. Developing theories. Observation of a phenomenon calls for some explanation. The process of explaining observed behavior begins with a **hypothesis**—an educated guess. If this hypothesis survives extensive testing, it may attain the status of a **theory**. A theory is a hypothesis supported by testing (experimentation) that explains scientific facts and is capable of predicting new facts.
4. Experimentation. The heart of the scientific method is the verification of theories. This verification process results from conducting carefully designed experiments intended to reinforce or refute the model system, the theory, or the hypothesis.
5. Summarizing information. Many phenomena have common causes and explanations. A scientific law summarizes and clarifies large amounts of information.

The development of **science**, using the scientific method, has catalyzed civilization's rapid growth in the last two centuries.

Technology is the conversion of a material from its current form to a more useful form. Technology is applied science: the use of scientific principles to fulfill human needs.

1.3 Data, Results, and Units

A scientific experiment produces **data**. Each piece of data is the result of a single measurement. Examples include the mass of a sample and the time required for a chemical reaction to occur. Mass, length, volume, time, temperature, and energy are the most common types of data obtained from chemical experiments.

Results are the outcome of an experiment. Data and results may be identical, but more often several pieces of data are combined to produce a result.

A **unit** defines the basic quantity of mass, volume, time, and so on. A number not followed by the correct unit conveys no useful information.

1.4 Measurement in Chemistry: English and Metric Units

The English system of measurement has as its most commonly used unit of **weight** the standard pound (lb), its fundamental unit of length, the standard yard (yd), and its basic unit of volume, the standard gallon (gal). The English system is not used in scientific work primarily because of the difficulty involved in converting from one unit to another.

The metric system is a decimal-based system; it is inherently simpler and less ambiguous. In the metric system there are three basic units. **Mass** is represented as the gram, length as the meter, and volume as the liter. Any subunit or multiple unit contains one of these units preceded by a prefix indicating the power of ten by which the base unit is to be multiplied to form the subunit or multiple unit. The most common metric prefixes are shown in Table 1.1 of the textbook and should be memorized.

Unit Conversion: English and Metric Systems

A **conversion factor** or series of conversion factors relates two units when converting from one unit to another. The use of these conversion factors is referred to as the **factor-label method**.

This method is used either to convert from one unit to another within the same system, or convert units from one system to another.

Conversion of Units within the Same System

The factor-label method is a self-indicating system; only if the factor is set up properly will the correct units result. Several commonly used English system conversion factors are included in Table 1.2 of the textbook. Most of these should already be familiar to you.

Conversion of units within the metric system may be accomplished using the factor-label method as well. Unit prefixes, which dictate the conversion factor, make conversion of units easy (refer to the textbook, Table 1.1).

Example 1
Convert 8.62 gallons into pints.

Work

$$8.62 \text{ gallons} \times \frac{4 \text{ quarts}}{1 \text{ gallon}} \times \frac{2 \text{ pints}}{1 \text{ quart}} = 68.96 \text{ pints}$$

Answer
69.0 pints (Three significant figures are needed.)

Example 2
Convert 16.4 kilometers into centimeters.

Work

$$16.4 \ \cancel{\text{kilometers}} \times \frac{1000 \ \cancel{\text{meters}}}{1 \ \cancel{\text{kilometer}}} \times \frac{100 \ \text{centimeters}}{1 \ \cancel{\text{meter}}} = 1.64 \times 10^6 \ \text{cm}$$

Answer

1.64×10^6 centimeters (Three significant figures are needed.)

Example 3
How many kilometers are in 1.568×10^6 centimeters?

Work

$$1.568 \times 10^6 \ \cancel{\text{cm}} \times \frac{1 \ \cancel{\text{m}}}{100 \ \cancel{\text{cm}}} \times \frac{1 \ \text{km}}{1000 \ \cancel{\text{m}}} = \ \text{km}$$

$$1.568 \times 10^6 \ \cancel{\text{cm}} \times \frac{1 \ \cancel{\text{m}}}{1 \times 10^2 \ \cancel{\text{cm}}} \times \frac{1 \ \text{km}}{1 \times 10^3 \ \cancel{\text{m}}} = 1.568 \times 10^1 \ \text{km}$$

Answer

1.568×10^1 km, or 15.68 km (Four significant figures are needed.)

Conversion of Units from One System to Another

The conversion of a quantity expressed in units of one system to an equivalent quantity in the other system (English to metric or metric to English) requires a bridging conversion unit. For example:

Quantity	English		Metric
Mass	1 pound	=	454 grams
	2.2 pounds	=	1 kilogram
Length	inch	=	2.54 centimeters
	1 yard	=	0.91 meters
Volume	1 quart	=	0.946 liters
	1 gallon	=	3.78 liters

The conversion may be represented as a three-step process:

1. Conversion from units stated in the problem to a bridging unit.
2. Conversion to the other system using the bridge.

3.   Conversion within the desired system to units required by the problem.

Example 4

A patient's weight is 129 pounds. What is the patient's weight in kilograms? (Use 454 grams = 1 pound and 1 kilogram = 1000 grams in your set-up.)

Work

$$129 \text{ pounds} \times \frac{454 \text{ grams}}{1 \text{ pound}} \times \frac{1 \text{ kilogram}}{1000 \text{ grams}} = 58.566 \text{ kg}$$

Answer

58.6 kilograms (Three significant figures are needed. See Section 1.6.)

1.5 Error, Accuracy, Precision, and Uncertainty

Error is defined as the difference between the true value and our estimation, or measurement, of the value.

Accuracy is the absence of error, the agreement between the true value and the measured value.

Only discrete objects, for example, the number of jellybeans in a package or the number of problems in this chapter, can be measured with perfect accuracy.

Uncertainty is the degree of doubt in a single measurement. The number of meaningful digits is determined by the measuring device. The presence of some error is a natural consequence of any measurement. Consequently, replicate measurements of the same quantity should be made whenever possible.

Precision is a measure of the agreement of replicate measurements.

It is important to recognize that accuracy and precision are different terms with different meanings. It is possible to have one without the other, as is shown on p. 11 of the textbook, although in scientific measurements taken with proper attention to experimental detail, the two most often go hand in hand.

1.6 Significant Figures and Scientific Notation

Modern pocket calculators often generate more digits than are merited, given the data used in the calculation. It is important that the result of these calculations be reflective of the level of certainty of the measurement. Proper use of significant figures, scientific notation, and rounding-off are essential.

Significant Figures

Significant figures are all digits in a number representing data or results that are known with certainty, *plus the first uncertain digit*.

The number of significant figures associated with a measurement is determined by the measuring device. Conversely, the number of significant figures reported is an indication of the sophistication of the measurement itself.

Recognition of Significant Figures

Only significant digits should be reported as data or results. The six rules enumerated below describe the assignment of significant figures.

Rule 1: All nonzero digits are significant.
Rule 2: The number of significant digits is independent of the position of the decimal point.
Rule 3: Zeros located between nonzero digits are significant.
Rule 4: Zeros at the end of a number (often referred to as trailing zeros) are significant if the number contains a decimal point.
Rule 5: Trailing zeros are insignificant if the number does not contain a decimal point.
Rule 6: Zeros to the left of the first nonzero integer are not significant; they serve only to locate the position of the decimal point.

Scientific Notation

Very large numbers may be represented with the proper number of significant figures using **scientific notation**. Scientific notation, also referred to as exponential notation, involves the representation of a number as a power of ten; the usual convention shows the decimal point in standard position—to the right of the leading digit.

To convert a number greater than 1 to scientific notation, the original decimal point is moved x places to the left, and the resulting number is multiplied by 10^x. The exponent (x) is a positive number equal to the number of places the original decimal point was moved.

To convert a number less than 1 to scientific notation, the original decimal point is moved x places to the right, and the resulting number is multiplied by 10^{-x}. The exponent (–x) is a negative number equal to the number of places the original decimal point was moved.

Example 5
Convert $30,000,000 \times 10^{-4}$ centimeters into correct scientific notation, and use the proper number of significant digits.

Work
$30,000,000 \times 10^{-4} \text{cm} = 3 \times 10^7 \times 10^{-4} \text{cm} = 3 \times 10^3 \text{cm}$
(Notice that only one, the 3, significant digit is present.)

Answer

3×10^3 cm

Significant Figures in Calculation of Results

In the process of addition or subtraction, the position of the decimal point in the quantities being combined determines the number of significant figures in the answer.

In multiplication and division, this is not the case: the decimal point position is irrelevant. What is important is the number of significant figures. Remember that *the answer can be no more precise than the least precise number from which the answer is derived.*

Example 6

Do the following multiplications and divisions using scientific notation. Also round off the answer to the proper number of significant digits.

$$\frac{68.0 \times 10^{-5}}{4.61 \times 10^{-6}} \times \frac{16 \times 10^2}{3.6} = ?$$

Work

1. Start by writing each part in scientific notation form:

$$\frac{6.80 \times 10^1 \times 10^{-5}}{4.61 \times 10^{-6}} \times \frac{1.6 \times 10^1 \times 10^2}{3.6 \times 10^0} = ?$$

$$\frac{6.80 \times 10^{-4}}{4.61 \times 10^{-6}} \times \frac{1.6 \times 10^3}{3.6 \times 10^0} = ?$$

2. Then multiply as below:

$$\frac{(6.80)(1.6) \times 10^{-4} \times 10^3}{(4.61)(3.6) \times 10^{-6} \times 10^0} = ?$$

(Drop the 10^0, and combine exponent parts.)

$$\frac{(6.80)(1.6) \times 10^{-1}}{(4.61)(3.6) \times 10^{-6}} = ?$$

$$\frac{10.88 \times 10^{-1}}{16.596 \times 10^{-6}} = ?$$

$$0.65557 =$$
$$0.66 \times 10^5 = 6.6 \times 10^4$$

(Divide and round off to two significant figures, since 3.6 contains only two significant figures.)

7

$$0.6555795 \text{ x } \frac{10^{-1}}{10^{-6}} = 6.6 \text{ x } 10^{-1} \text{ x } \frac{10^{-1}}{10^{-6}} =$$

$$6.6 \text{ x } 10^{-1} \text{ x } 10^5 = 6.6 \text{ x } 10^4$$

Answer

$6.6 \text{ x } 10^4$

When a number is raised to a power, $n^x = y$, the number of significant figures in the answer (y) is identical to the number contained in the original term (n).

Defined or counted numbers do not determine the number of significant figures. The quantity being converted, not the conversion factor, determines the number of significant figures.

Rounding Off Numbers

A generally accepted rule for rounding off states that if the first digit dropped is 5 or greater, we raise the last significant digit to the next higher number. If the first digit dropped is 4 or less, the last significant digit remains unchanged.

Example 7
Round off 16.00468 grams to two significant figures.

Answer
The first zero past the 6 will not increase the value. Thus, the answer is 16 grams.

1.7 Experimental Quantities

Mass

Mass describes the quantity of matter in an object. The terms weight and mass, in common usage, are often considered synonymous. In fact, they are not. Weight is the manifestation of the force of gravity on an object.

$$\text{weight} = \text{mass} \times \text{acceleration due to gravity}$$

The common conversion units for mass are as follows:

$$1 \text{ gram (g)} = 1 \times 10^{-3} \text{ kilogram (kg)} = 1/454 \text{ pound (lb)}$$

The atomic mass unit is a convenient way to represent the mass of very tiny quantities of matter, such as individual atoms and molecules.

$$1 \text{ atomic mass unit (amu)} = 1.66 \times 10^{-24} \text{ g}$$

Length

The standard metric unit of length is the meter. Large distances are measured conveniently in kilometers, and smaller distances are measured in millimeters or centimeters. Common conversions for length are as follows:

$$1 \text{ meter (m)} = 1 \times 10^2 \text{ centimeters (cm)} = 3.94 \times 10^1 \text{ inches (in)}$$

In the same way that atomic mass units represent mass on an atomic scale, either the nanometer (nm) or the angstrom (Å) is a convenient measure of distance.

$$1 \text{ nm} = 10^{-7} \text{ cm} = 10^{-9} \text{ m}$$

$$1 \text{ Å} = 10^{-8} \text{ cm} = 10^{-10} \text{ m}$$

Volume

The standard metric unit of volume is the liter. A liter is the volume occupied by 1000 grams of water at 4° Celsius. A volume of 1 liter also corresponds to:

$$1 \text{ liter (1)} = 1000 \text{ milliliters (ml)} = 1.06 \text{ quarts (qt)}$$

Time

The standard metric unit of time is the second. The need for accurate measurement of time by chemists is necessary in many applications.

Temperature

Temperature is the degree of "hotness" of an object. Many substances, such as mercury, expand as their temperature increases, and this expansion provides us with a way to measure temperature and temperature changes. The height of the mercury in a thermometer is proportional to the temperature. A mercury thermometer may be calibrated, or scaled, in different units, just like a ruler. Three common temperature scales are Fahrenheit (°F), Celsius (°C) and Kelvin (K). Two convenient reference temperatures used to calibrate a thermometer are the freezing and boiling temperatures of water. Conversion from one temperature scale to another may be accomplished as shown below.

Fahrenheit to Celsius: $°C = \dfrac{(°F - 32)}{1.8}$

Celsius to Fahrenheit: $°F = 1.8 \, °C + 32$

Celsius to Kelvin: $K = °C + 273$

Example 8

What value, in °F, corresponds to 310.0 Kelvin?

Work

$K = °C + 273$
    or
$°C = K - 273$

$°C = 310.0 - 273$

$°C = 37.0$ (The rules of significant figures require that the result be expressed to the nearest tenth of a degree, since 310.0 is to the nearest tenth of a degree.)

$°F = 1.8 \, °C + 32$

$°F = 1.8 \, (37.0) + 32$

$°F = 98.6$

Answer

310.0 Kelvin = 98.6°F

Energy

Energy, the ability to do work, may be categorized as either kinetic energy, the energy of motion, or potential energy, the energy of position. Kinetic energy is energy in process, while potential energy is stored energy. All energy is either kinetic or potential.

Energy may also be classified according to form. The principal forms of energy include light, heat, electrical, mechanical, and chemical energy. Remember the following characteristics of energy:

1. In conventional chemical reactions, energy cannot be created or destroyed.
2. Energy may be converted from one form to another.
3. Energy conversion always occurs with less than 100% efficiency.

4. All chemical reactions involve either a "gain" or "loss" of energy.

Energy absorbed or liberated in chemical reactions is often in the form of heat energy. Heat energy may be represented in units of calories or Joules, their relationship being

$$1 \text{ calorie (cal)} = 4.18 \text{ Joules (J)}$$

One calorie is defined as the amount of heat energy required to change the temperature of 1 gram of water by $1\,°C$.

Density and Specific Gravity

Density, the ratio of mass to volume, $\quad d = \dfrac{mass}{Volume} = \dfrac{m}{V} = \dfrac{gram}{milliliter}$

is independent of the amount of material and is characteristic of a substance; each substance has a unique density.

Values of density are often related to a standard, well-known reference, the density of pure water at $4\,°C$. This "referenced" density is the **specific gravity**.

$$\text{specific gravity} = \frac{\text{density of object (g/mL)}}{\text{density of water}} \text{ (g/mL)}$$

Example 9
A solid that has a mass of 1267.4 grams was found to have the following measurements: length = 9.86 cm; width = 46.6 mm and height = 0.224 m. What is its density?

Work
1. Convert each measurement into centimeters:

    length = 9.86 cm; width 46.6 mm = 4.66 cm; height 0.224 m = 22.4 cm

2. Then use volume equation for a solid:

    $V_{solid} = l \times w \times h$

    $V_{solid} = (9.86 \text{ cm}) (4.66 \text{ cm}) (22.4 \text{ cm}) = 1.0292262 \times 10^3 \text{ cm}^3$
    (Use only three significant figures.)

    $V_{solid} = 1.03 \times 10^3 \text{ cm}^3$

    $V_{solid} = 1.03 \times 10^3 \; \cancel{cm}^3 \times \dfrac{1 \text{ mL}}{1 \; \cancel{cm}^3}$

$$V_{solid} = 1.03 \times 10^3 \text{ mL}$$

3. Then use the density equation:

$$d = \frac{m}{V}$$

$$d = \frac{1267.4 \text{ g}}{1.03 \times 10^3 \text{ mL}}$$

$$d = \frac{1267.4 \times 10^3 \text{ g}}{1.03 \times 10^3 \text{ mL}}$$

$$d = 1.23 \text{ g/mL}$$

Answer
The density equals 1.23 grams/milliliter

Example 10
If the density of carbon tetrachloride is 1.59 grams/mL, what is the mass (in grams) of 2.00 liters of carbon tetrachloride?

Work

$$2.00 \text{ liters} \times \frac{1000 \text{ mL}}{1 \text{ liter}} \times \frac{1.59 \text{ grams}}{1 \text{ mL}} = 3.18 \times 10^3 \text{ grams}$$

Answer
3.18×10^3 grams (Three significant figures are needed.)

Example 11
Calculate the mass in grams of 25 mL of mercury. The density of mercury is 13.5 g/mL.

Work
Use mass = (d) (V), or use the factor-label method as below:

$$25 \text{ mL of mercury} \times \frac{13.5 \text{ g of mercury}}{1 \text{ mL of mercury}} = 337.5 \text{ g}$$

(Round off to two significant figures.)

Answer

3.4×10^2 g

Example 12

Calculate the volume in mL of a liquid (ethyl alcohol) that has a density of 0.789 g/mL and a mass of 5.555 g.

Work

Use $\dfrac{V = m}{d}$, or use the factor label method as below:

$$5.555 \text{ g of liquid } \times \frac{1 \text{ mL of liquid}}{0.789 \text{ g of liquid}} = 7.0405576 \text{ mL}$$

(Round off to three significant figures.)

Answer

7.04 mL

Key Terms

accuracy
analytical chemistry
biochemistry
chemistry
concentration
conversion factor
data
density
energy
error
hypothesis
inorganic chemistry
kinetic energy
law
mass
matter

organic chemistry
physical chemistry
potential energy
precision
results
scientific method
scientific notation
significant figures
specific gravity
technology
temperature
theory
uncertainty
unit
weight

Self Test

1. Calculate the density (in normal unit form) of a liquid with a volume of 348 mL and a mass of 0.3546 kg.

2. $\dfrac{68.0 \times 10^{-3}}{4.61 \times 10^{-1}} = ?$

3. $\dfrac{10.6 \times 10^{-4}}{0.0641 \times 10^{-2}} \times 16.4 \times 10^{-7} \times \dfrac{0.111 \times 10^{4}}{10.0 \times 10^{2}} = ?$

4. 50.0 miles/hour = ? cm/second

5. If one atom of uranium weighs 238.0 atomic mass units, and one atomic mass unit is equal to 1.66×10^{-24} gram, what is the mass, in grams, of one atom of uranium?

6. The speed of light is 186,000 miles per second. What is its speed in cm/second?
   USE: 5280 feet = 1 mile; 12 inches = 1 foot; 2.54 cm = 1 inch

7. $-4.6°F = ? °C$

8. If the density of carbon tetrachloride is 1.59 g/mL, what is the mass of 2.65 liters of carbon tetrachloride in grams?

9. $-40.0°C = ? °F$

10. What is the specific gravity of an object that weighs 13.35 g and has a volume of 25.00 mL? The density of water under the same conditions is 0.980 g/mL.

11. The ability to do work describes what term?

12. A scientific experiment produces what information?

13. Each piece of data is the individual result produced by what process?

14. Which system of measurement is a decimal-based system?

15. In which country was the metric system originally developed?

16. What factors are used in the factor-label method?

17. What is defined as the degree of doubt in a measurement?

18. Which temperature system does not use a degree sign?

19. $98.6°F = ? K$

20. Round off 0.00369865 to two significant figures.

2 *The Structure of the Atom*

Learning Goals

1. **Conceptual Goals**
   - Know the difference between chemical and physical properties.
   - Be aware of the characteristics of the solid, liquid, and gaseous state.
   - Recognize the interrelationship of the structure of matter and its physical and chemical properties.
   - Develop a general overview of modern atomic structure, including the major particles which comprise the atom: protons, neutrons, and electrons.
   - Describe the properties of ions and isotopes.
   - Have a historical view of the development of atomic theory, beginning with Dalton.
   - Recognize the critical role of spectroscopy in the development of atomic theory.
   - Realize the inherent weakness of Bohr's Theory and the need for a more sophisticated "wavemechanical" approach.

2. **Performance Goals**
   - Provide specific examples of physical and chemical properties.
   - Classify matter according to type.
   - Describe the basic postulates of Bohr's Theory.

3. **Health Applications**
   - Describe the role of electromagnetic radiation in medical diagnosis.

Chapter Overview

In this chapter we will explore the fundamental properties of individual bits of matter, atoms. The properties of these tiny units determine the properties of the bulk material.

2.1 Matter and Properties

Matter and Physical Properties

Properties (characteristics) of matter may be classified as either physical or chemical. **Physical properties** enable us to identify different kinds of matter without changing the identity (chemical composition) of the sample. Examples of physical properties include color, odor, taste, melting and boiling temperatures, and compressibility.

Three states of matter exist, and these states are distinguishable by differences in physical properties. They include gases, liquids, and solids.

1. **The gaseous state**. Gases have a very low mass per unit volume, density, because the individual particles that comprise the gas are separated by large distances. Consequently, gases can be compressed (pushed into a smaller volume) or expanded to a larger volume.

2. **The liquid state**. Particles that make up a liquid are much closer together than in the vapor state. Thus, liquids expand and contract only slightly, and the density of liquids is much greater than that of gases.

3. **The solid state**. Solids are characterized by particles that are very close together. Attractive forces between the particles are strong enough to provide a rigid shape or structure to the bulk material. The proximity of particles prevents significant expansion or compression. All common materials, except water, follow this order of decreasing particle separation.

Water is the most common example of a substance that may exist in all three states over a reasonable temperature range. The conversion of ice to liquid water or liquid water to the gaseous state is an example of a **physical change**. A physical change does not alter the composition or identity of the substance undergoing change.

Matter and Chemical Properties

Chemical properties result in a change in composition and can only be observed through chemical reactions. The process of photosynthesis is a common example of chemical change.

Example 1
Which of the following are physical changes?

1. An iron nail rusts.   4. Digestion of foods in the small intestine.
2. A block of ice melts.   5. Gasoline undergoes combustion.
3. Water on the floor evaporates.

Answer
Items 2 and 3

The terms rust, digestion, and combustion all indicate chemical change. The terms melt, boil, freeze, and evaporate all indicate a physical change.

Intensive and Extensive Properties

An **intensive property** is independent of the quantity of the substance. For example, the density of a drop of water is exactly the same as the density of a liter of water. Mass and volume are extensive properties. An **extensive property** depends on the quantity of a substance.

Classification of Matter

All matter can be classified as either a pure substance or a mixture (see Figure 2.3 in the textbook). A **pure substance** is a form of matter that has identical composition and physical and chemical properties throughout. A **mixture** is a combination of two or more pure substances in which the combined substances retain their identity.

A mixture may be either homogeneous or heterogeneous matter. A **homogeneous mixture** has uniform composition. Its particles are well mixed, or thoroughly intermingled. A **heterogeneous mixture** has a nonuniform composition.

Pure substances may also be subcategorized as elements or compounds. An **element** is a pure substance that cannot be converted into a simpler form of matter by any chemical reaction. A **compound** is a substance resulting from the combination of two or more elements in a definite, reproducible fashion.

Example 2
Which of the following are mixtures?

1. soup   3. wine
2 tap water 4. blood

Answer
All, since they contain more than one pure substance.

2.2 Matter and Structure

Atomic Structure

Modern View of Atomic Structure

The basic structural unit of an element is the **atom**, which is the smallest unit of an element that retains the chemical properties of that element. An atom is composed of three primary particles: the **electron**, the **proton** and the **neutron**. These particles are located in one of two distinct regions:

1.   The **nucleus** is a small, dense, positively charged region in the center of the atom. The nucleus is composed of positively charged protons and uncharged neutrons.
2.   Surrounding the nucleus is a diffuse region of negative charge populated by electrons, the source of the negative charge. Electrons are tiny in comparison to the protons and neutrons. The properties of these particles are summarized in Table 2.1 of the textbook.

The number of protons determines the identity of the atom. When the number of protons is equal to the number of electrons, the atom is neutral, because the charges are balanced and effectively cancel.

The **atomic number** (Z) is equal to the number of protons in the atom, and the **mass number**, (A) is equal to the sum of the protons and neutrons (the mass of the electrons is so small as to be insignificant).

Isotopes are atoms of the same element having different masses due to different numbers of neutrons (different atomic mass).

Inspection of the periodic table reveals that the atomic mass of many of the elements is not an integral number. For example the atomic mass of chlorine is actually 35.45 amu, not 35.00 amu. The existence of isotopes accounts for this difference. A natural sample of chlorine is principally composed of two isotopes, $^{35}_{17}Cl$ and $^{37}_{17}Cl$, in approximately a 3:1 ratio, and the tabulated atomic mass is the weighted average of the two isotopes.

Certain isotopes of elements emit particles and energy (**radioactivity**, Chapter 8) that may be useful in tracing the behavior of biochemical systems. These isotopes otherwise behave identically to any other isotope of the same element.

Example 3
Calculate the number of protons, neutrons, and electrons found in carbon–14 ($^{14}_{6}C$)

Work
The atomic number of carbon (from the periodic table) is 6. Thus, there are 6 protons and 6 electrons in one atom of carbon. The 14 in carbon–14 means that this specific isotope of carbon has a mass number of 14. The mass number of an atom is equal to the sum of the protons and neutrons:

mass number = number of protons + number of neutrons

The number of neutrons = mass number – number of protons

Thus, for carbon–14:

number of neutrons = 14 –6 = 8

Answer
Carbon–14 contains 6 protons, 6 electrons, and 8 neutrons.

Example 4
Calculate the number of protons, neutrons, and electrons in the three isotopes of hydrogen:

$$^{1}_{1}H \qquad\qquad ^{2}_{1}H \qquad\qquad ^{3}_{1}H$$

normal hydrogen   deuterium   tritium

Work

Since all are hydrogen atoms, they each contain only 1 proton and 1 electron, because the atomic number of hydrogen is 1.

The mass number of normal hydrogen is 1. Thus, it contains no neutrons.

The mass number of deuterium is 2. Thus, it contains 1 neutron in its nucleus:

number of neutrons = mass number – number of protons

number of neutrons in deuterium = 2 – 1 = 1

The mass number of tritium is 3. Thus, it contains 2 neutrons in its nucleus:

number of neutrons in tritium = 3 – 1 = 2

Answer

The three isotopes each have 1 proton and 1 electron. Normal hydrogen has no neutrons. Deuterium has 1 neutron; tritium has 2 neutrons.

Ions are charged particles that result from a gain of one or more electrons by the parent atom (forming negative ions, or anions) or a loss of one or more electrons from the parent atom (forming positive ions, or cations).

For simplification, the atomic and mass numbers are often omitted. For example, the hydrogen cation would be written as H^+ and the anion as H^-.

Ions are charged particles that result from a grain of one or more electrons by the parent atom (forming negative ions, or **anions**) or a loss of one or more electrons from the parent atom (forming positive ions, or **cations**),

For simplification, the atomic and mass numbers are often omitted. For example, the hydrogen cation would be written as H^+ and the anion as H^-.

Example 5

How many protons and electrons are present in each of the following atoms and ions?

$$H \quad + \quad 1e^- \quad \rightarrow \quad H^-$$

Work

The H with no charge means that it is an atom; atoms have no charge, since they have the same number of protons and electrons. The H^- is an ion that has a negative one charge. This means it has one more electron than it has protons. You can see in the reaction that the hydrogen atom [given as H] has gained one electron to form the ion H⁻.

ATOM				ION
H	+	1e⁻	→	H⁻

+1 = 1 proton   1 proton   = +1
−1 = 1 electron   2 electrons = −2
   0 charge   charge   = −1

Answer

The hydrogen atom contains 1 proton and 1 electron. The −1 hydrogen ion contains 1 proton and 2 electrons.

Remember that if an ion is positive, it was formed by the loss of the required number of electrons from the given atom.

$Mg → 2e^- + Mg^{2+}$ • Two electrons were lost by the atom.
$K → 1e^- + K^+$ • One electron was lost by the atom.
A negative ion, on the other hand, is formed by a gain of one or more electrons by the given atom.
$S + 2e^- → S^{2-}$
$Cl + 1e^- → Cl^-$

2.3 Development of the Atomic Theory

Dalton's Theory

The first experimentally based theory of atomic structure was proposed by John Dalton in the early 1800s. Dalton postulated that:

1. All matter consists of tiny particles called atoms.
2. Atoms cannot be created, divided, destroyed, or converted to any other type of atom.
3. Atoms of a particular element have identical properties.
4. Atoms of different elements have different properties.
5. Atoms combine in simple, whole-number ratios.
6. Chemical change involves joining, separating, or rearranging atoms.

Postulates 1, 4, 5 and 6 are presently regarded as true. The discovery of the processes of nuclear fusion, fission, and radioactivity (Chapter 8) have disproved the postulate that atoms cannot be created or destroyed. Postulate 3, that the atoms of a particular element are identical, was disproved by the later discovery of isotopes.

Electrons, Protons, and Neutrons

Although Dalton pictured atoms as indivisible, various experiments, particularly those of William Crookes and Eugene Goldstein, indicated that the atom is composed of charged (+ and –) particles.

J. J. Thomson demonstrated the electrical and magnetic properties of **cathode rays** (Figures 2.4 and 2.5 in the textbook). Crookes observed rays, which he called cathode rays, emanating from the **cathode** (– charge) of an evacuated (vacuum) tube. Further experiments showed that the ability to produce cathode rays is a characteristic of all materials. In 1897, Thomson announced that cathode rays were streams of negative particles of energy. These particles are electrons.

Similar experiments, conducted by Goldstein, led to the discovery of protons, particles equal in charge to the electron, but opposite in sign.

The neutron has a mass virtually equal to that of the proton, and zero charge. The neutron was first postulated in the early 1920s but it was not until 1932 that James Chadwick demonstrated its existence.

The Nucleus

In the early 1900s, it was believed that protons and electrons were uniformly distributed throughout the atom. However, an experiment by **Hans Geiger** led Ernest Rutherford (in 1911) to propose that the majority of the mass and positive charge of the atom was actually located in a small, dense region, the nucleus, and that the small, negatively charged electrons spread across a much larger, diffuse area outside of the nucleus.

Light and Atomic Structure

When an element is excited (perhaps by the passage of an electrical current) certain wavelengths of light, characteristic of the element under study, are emitted. This range of wavelengths is the **electromagnetic spectrum**. The relationship between the appearance of this spectrum and the structure of the atom was explained by Neils Bohr.

The Bohr Atom

Bohr and his contemporaries were puzzled by the emission spectrum of hydrogen. He developed a hypothesis to explain his observations. This evolved into Bohr's Atomic Theory, summarized below.

1. Atoms can absorb and emit energy via **promotion** of electrons to higher **energy levels** and **relaxation** to lower levels. These levels are referred to as Bohr **orbits**.
2. Energy that is emitted upon relaxation is observed as a single wavelength of light.
3. These spectral lines are a result of electron transitions between "allowed levels" in the atom.
4. The allowed levels are **quantized** energy levels, or orbits.
5. Electrons are found only in these energy levels.

6. The highest-energy orbits are located farthest from the nucleus.
7. Atoms absorb energy by excitation of electrons to higher energy levels producing the **excited state**.
8. Atoms release energy by **relaxation** of electrons to lower energy levels, the **ground state**.
9. Energy differences may be calculated from the wavelengths of light emitted during these electronic transitions.

2.4 Modern Atomic Theory

Development of more sophisticated experimental techniques demonstrated that there were problems with Bohr's Theory, even in the case of hydrogen. Although Bohr's concept of principal energy levels is still valid, the restriction of electrons to fixed orbits is too rigorous. We now speak of the probability of finding an electron in a region of space within the principle energy level, referred to as an atomic orbital. The rapid movement of the electron spreads the charge into a cloud of charge. This cloud is more dense in certain regions, the **electron density** being proportional to the probability of finding the electron at any point in time. Since the atomic orbitals are part of the principal energy levels, they are referred to as sublevels, designated by letters s, p, d, f, and so forth. Orbitals shapes are depicted in Figure 2.6 in the textbook.

Key Terms

anion
atom
anode
atomic mass
atomic number
atomic orbital
cathode
cathode rays
cation
chemical property
chemical reaction
compound
electromagnetic radiation
electromagnetic spectrum
electron
electron density
electronic transitions
element
energy level
excited state
extensive property
gaseous state
ground state
heterogeneous mixture
homogeneous mixture

intensive property
ion
isotope
liquid state
mass number
mixture
natural radioactivity
neutron
nucleus
orbit
physical change
physical property
product
promotion
proton
pure substance
quantization
quantum level
quantum number
reactant
relaxation
solid state
spectroscopy
speed of light
states of matter

Self Test

1. Which of the following are physical properties of an object?
   a. Combustibility
   b. Color
   c. Melting point
   d. Gas phase
   e. Chlorine combines with hydrogen to form HCl.
   f. Density
   g. Volume
   h. $S + O_2 \rightarrow SO_2$
2. Which of the following are mixtures?
   a. NaCl
   b. salt in water
   c. a soft drink
   d. tap water
   e. blood
   f. cake mix
3. How many protons and electrons are found in a magnesium atom?
4. How many protons and electrons are found in a sulfur atom?
5. How many protons and electrons are found in a chloride ion, Cl⁻?
6. Give the number of protons and electrons found in a lithium atom and a lithium ion, Li⁺. rite the reaction showing how a lithium atom can form a lithium ion.
7. What are the three states of matter?
8. Which state of matter will expand to fill any container?
9. Which state of matter has no definite shape or volume?
10. List six examples of physical properties of matter.
11. What is the smallest particle of water that still has all the properties of water?
12. Salt and pepper is an example of what type of mixture?
13. What is the smallest unit of an element that still has all the properties of that element?
14. What is the dense positive center of an atom?
15. Which particle of an atom has the least mass?
16. What number is equal to the number of protons in an atom?
17. The mass number minus the number of protons equals the number of what particle?
18. What do we call atoms of the same element that have different masses?
19. If an atom loses an electron, what charge does the resulting ion have?
20. Symbolically represent the three isotopes of the element hydrogen.

1. a, b, c, d, g, f
2. all except NaCl
3. 12
4. 16
5. 17 p⁺, 18 e⁻
6. Li: 3p 3e⁻   Li⁺ 3p, 2e⁻   Li → 1e⁻ + Li⁺
7. liquid, gas, solid
8. gas
9. gas
10. color, odor, taste, melting point, boiling temp, compressibility
11. H₂O molecule
12. heterogenous mixture
13. atom
14. nucleus
15. electron
16. atomic number
17. neutrons
18. isotopes
19. +
20. $_1^1H$ $_1^2H$ $_1^3H$

23

3 *Elements, Atoms, and the Periodic Table*

Learning Goals

1. **Conceptual Goals**
   - Know the meaning and ramifications of the periodic law.
   - Recognize the important subdivisions of the periodic table, periods, groups (families), metals, and nonmetals.
   - Understand the relationship between the electronic structure of an element and its position in the periodic table.
   - Know the meaning of the octet rule and its predictive usefulness.
   - Understand the meaning of and utility of ionization energies and electron affinities in predicting bond formation.
2. **Performance Goals**
   - Extract information about an element from the periodic table, for example, the mass, number of protons, neutrons, and electrons in an atom of any element.
   - Use the octet rule to predict the charge of common cations and anions.
   - Use the periodic table and its predictive power to estimate relative size of atoms and ions, as well as relative magnitudes of ionization energy and electron affinity.
3. **Health Applications**
   - Recognize the importance of trace metals in the diet and the effect of deficiencies, such as Wilson's disease.

Chapter Overview

The **periodic law** states that properties of elements are periodic functions of their atomic numbers. The periodic table results from this relationship. The periodic table is an organized "map" of the elements that relates their structure to their chemical and physical properties. The chemical and physical properties of elements follow directly from the electronic structure of the atoms that make up these elements. Familiarity with the periodic table allows prediction of the structure and properties of the various elements, and serves as the basis for understanding chemical bonding.

3.1 The Periodic Law and the Periodic Table

Periods and Families

A horizontal row of elements in the periodic table is referred to as a **period**. The periodic table consists of seven periods, six of which contain 2, 8, 8, 18, 18, and 32 elements. The seventh period is still incomplete but potentially holds 32 elements. Note that the lanthanide series is a part of period six and the actinide series is a part of period seven.

The columns of elements in the periodic table are called **groups** or families. The elements of a particular family share many similarities in physical and chemical properties that are related to similarities in electronic structure. The various groups are labeled with Roman numerals, and each is subtitled with the letter A or B.

Group A elements are called **representative elements**, and Group B elements are **transition elements.** Certain families have common names as well as a Roman numeral-letter designation. Group IA elements are also known as the **alkali metals**; Group IIA, as the **alkaline earth metals**; Group VIIA, as the **halogens**; and Group VIIIA, as the **noble gases**.

Metals and Nonmetals

A bold zig-zag line runs from top to bottom of the table beginning to the left of boron (B) and ending between polonium (Po) and astatine (At). This line acts as the boundary between **metals,** to the left, and **nonmetals,** to the right. Elements straddling the boundary such as Ge and As have properties intermediate between metals and nonmetals and are often termed **metalloids**.

Atomic Number and Atomic Mass

The atomic number (Z, number of protons in the nucleus; the nuclear charge) and the atomic mass of each element are readily available from the periodic table. More detailed periodic tables may also provide such information as the electron arrangement, relative sizes of atoms and ions, and most probable ion charges.

Example 1
For each of the following symbols, provide the name of the element, its atomic number (Z), and the family to which it belongs.

1. Li   4. Si   7. I
2. Ra   5. N   8. Cu
3. Al   6. S

Answer
1. Li: lithium; Z = 3; family = IA   5. N: nitrogen; Z = 7; family = VA
2. Ra: radium; Z = 88; family = IIA   6. S: sulfur; Z = 16; family = VIA
3. Al: aluminum; Z = 13, family = IIIA   7. I iodine; Z = 53; family = VIIA
4. Si: silicon; Z = 14; family = IVA   8. Cu: copper; Z = 29; family = IB

3.2 Electron Arrangement and the Periodic Table

The most important factor in chemical bonding is the arrangement of the electrons in the atoms that are combining. The periodic table provides us with a great deal of information about the electron arrangement, or electronic structure, of atoms.

Valence Electrons

Outermost electrons in an atom are **valence electrons**. For representative elements, the number of valence electrons in an atom corresponds to the number of the group or family in which the atom is found. Metals tend to have fewer valence electrons, and nonmetals tend to have more valence electrons.

The energy levels are symbolized by n, with the lowest energy level assigned a value of n =

1. Each energy level may contain up to a fixed maximum number of electrons.
   Two general rules of electron configuration are based upon the periodic law:
2. The number of valence electrons in a neutral atom equals the group number for all representative (A group) elements.
3. The energy level (n = 1, 2, etc.) in which the valence electrons are located corresponds to the period in which the element may be found.

Helium is an exception to rule 1, above. It cannot have eight valence electrons, since all of its electrons are in the n = 1 level which has a maximum capacity of only two electrons.

Energy Levels and Sublevels

The principal energy levels are designated $n = 1, 2, 3$ and so forth. The number of possible sublevels in a principal energy level is also equat to n. When $n = 1$ there can be only one sublevel; $n = 2$ allows two sublevels, and so forth. The total electron capacity of a principal level is $2(n)^2$.

The sublevels, or subshells increase in energy:

$$s < p < d < f$$

Both the principal energy level and type of sublevel are specified when describing the location of an electron. For example: *1s, 2s, 2p*. The first principal energy level (*n*=1) has one possible subshell, *1s*. The second principal energy level (*n*=2) has two possible subshells: *2s* and *2p*. the third principal energy level (*n*=3) has three possible subshells: *3s, 3p* and *3d*. the fourth principal energy level (*n*=4) has four possible subshells: *4s, 4p, 4d*, and *4f*.

An **orbital** is a specific region of a subshell containing a maximum of two electrons. the *s* sublevel contains only one orbital, the *p* sublevel contains three orbitals, the *d* sublevel contains five orbitals, and the *f* sublevel contains seven orbitals. Each orbital may be empty, contain one electron, or be filled, containing two electrons. The *s, p, d*, and *f* sublevels have maximum capacities of 2, 6, 10, and 14 electrons.

Each type of orbital has its own characteristic shape. The *s* orbital is spherically symmetrical; a model appears as a ping-pong ball with its center corresponding to the intersection of imaginary *x, y* and *z* coordinates. There exist three kinds of *p* orbitals, each identical in shape (often modeled as a dumbbell). They differ only in their orientation in space, along the hypothetical *x*-axis, (p$_x$), *y*-axis, (p$_y$), and *z*-axis, (p$_z$). The *d* and *f* orbitals are more complex; the textbook focuses exclusively on *s* and *p* orbitals in subsequent discussion.

Each atomic orbital has a maximum capacity of two electrons. The electrons are perceived to *spin* on an imaginary axis, and the two electrons in the same orbital must have opposite spins,

26

clockwise and counterclockwise. Two electrons in one orbital which possess opposite spins are referred to as *paired* electrons.

Electron Configuration and the Aufbau Principle

The arrangement of electrons in atomic orbitals is referred to as the atom's **electron configuration.** We may represent the electron configuration of atoms of various elements using the Aufbau, or building-up, principle. According to this principle, electrons fill the lowest-energy orbital which is available first. by knowing the order of filling of atomic orbitals, lowest to highest energy, you may write the electron configuration for any element.

Example 2
Name the two element that have electrons only in the first energy level.

Answer
hydrogen and helium

Example 3
List the sublevels in order of lowest energy to highest energy.

Answer
s<p<d<f

Example 4
Write the symbols of all the possible sublevels found in the fourth principal energy level (n=4).

Answer
4*s*, 4*p*, 4*d*, and 4*f*

Example 5
Complete the following statement about an orbital:
Each orbital may be empty, _____, or be _____.

Answer
Each orbital may be empty, contain one electron, or be filled, containing two electrons.

Example 6
What is the maximum number of electrons that may be found in each sublevel?

Answer
$s = 2e^-; p = 6e^-; d = 10e^-; f = 14e^-$

Example 7
Give the electronic configuration for each of the following elements:

1. H
2. He
3. O
4. F
5. Na

6. Al
7. P
8. B
9. S
10. Ar

Answer
1. H: $1s^1$
2. He: $1s^2$
3. O: $1s^2 2s^2 2p^4$
4. F: $1s^2 2s^2 2p^5$
5. Na: $1s^2 2s^2 2p^6 3s^1$

6. Al: $1s^2 2s^2 2p^6 3s^2 3p^1$
7. P: $1s^2 2s^2 2p^6 3s^2 3p^3$
8. B: $1s^2 2s^2 2p^1$
9. S: $1s^2 2s^2 2p^6 3s^2 3p^4$
10. Ar: $1s^2 2s^2 2p^6 3s^2 3p^6$

3.3 The Octet Rule

Elements in the last family, the noble gases, have either two valence electrons (helium) or eight valence electrons (neon, argon, krypton, xenon, and radon). Their most important property is their extreme stability. A full n = 1 energy level (as in helium) or an outer octet of electrons is responsible for this unique stability.

Atoms of elements in other groups are more reactive than the inert gases because they are, in the process of chemical reaction, trying to achieve a more stable "noble gas" configuration by gaining or losing electrons. This is the basis of the **octet rule**. In chemical reactions they will gain, lose, or share the minimum number of electrons necessary to achieve a more stable energy state.

Ion Formation and the Octet Rule

Metallic elements tend to form positively charged ions called **cations**. Positive ions are formed when an atom loses one or more electrons.

These ions are more stable than their corresponding neutral atoms. The ion is isoelectronic (that is, it has a similar number of electrons) with its nearest noble gas neighbor and has an octet of electrons in its outermost energy level.

Nonmetallic elements tend to gain electrons to become isoelectronic with the nearest noble gas element, forming negative ions referred to as **anions**.

Example 8
Show how a sodium atom will form a sodium ion.

Answer
1. Na $\rightarrow$ Na$^+$ + 1e$^-$
   Atom   Ion
2. Na (atom); it has 11e$^-$
3. Na$^+$ (ion); it has 10e$^-$

Example 9
Show how an oxygen atom will form an oxide ion.

Answer
1. O + 2e$^-$ $\rightarrow$ O^{2-}
   oxygen atom   Oxide ion
2. O (atom); it has 8e$^-$
3. O^{2-} (ion); it has 10e$^-$

The transition metals tend to form positive ions by losing electrons, just like the representative metals. However, the transition elements are characterized as "variable valence" elements; depending upon the type of substance they react with, they may form more than one stable ion. For example, iron has two stable ionic forms:
Fe^{2+} and Fe^{3+}

3.4 Trends in the Periodic Table

Atomic Size

The size of the atom will be determined principally by two factors:

1. The energy level (n-level) in which the outermost electron(s) is located increases as we go down a group (recall that the outermost n-level correlates with period number).
2. As the magnitude of the positive charge of the nucleus increases, its "pull" on all of the electrons increases, and the electrons are drawn closer to the nucleus.

Consequently, atomic size increases down a group and decreases across a period.

Ion Size

Three generalizations can be made about the size of ions:

1. Positive ions (cations) are smaller than the parent atom.
2. Negative ions (anions) are larger than the parent atom.
3. Ions with multiple positive charge (such as Cu^{2+}) are even smaller than their corresponding monopositive ion (Cu^+); ions with multiple negative charge (such as O^{2-}) are larger than their corresponding less-negative ion.

Ionization Energy

The energy required to remove an electron from an isolated atom in the gas phase is the **ionization energy**. The magnitude of the ionization energy correlates with the strength of the attractive force between the nucleus and the outermost electron.

1. As we go down a group, the ionization energy decreases, since the atom's size is increasing. The outermost electron is progressively farther from the nuclear charge and hence easier to remove.
2. As we go across a period, atomic size decreases, as the outermost electrons are closer to the nucleus, more tightly held, and more difficult to remove. Therefore, the ionization energy must increase.

A correlation does indeed exist between trends in atomic size and ionization energy. Atomic size decreases from bottom to top of a group and left to right in a period. Ionization energies increase in the same periodic way. Note also that ionization energies are highest for the noble gases; this accounts for the extreme stability and nonreactivity of the noble gases.

Electron Affinity

The energy released when a single electron is added to a neutral atom in the gaseous state is known as the **electron affinity**. Electron affinity is a measure of the ease of forming negative ions. A large value of electron affinity (energy released) indicates that the atom becomes more stable as it becomes a negative ion (through the process of gaining an electron).

Periodic trends for electron affinity are as follows:
1. Electron affinities generally decrease as we go down a group.
2. Electron affinities generally increase as we go across a period.

Be aware that exceptions to these general trends do exist.

Key Terms

actinide series	anion
alkali metal	electron affinity
alkaline earth metals	electron configuration

group	nonmetal
halogen	octet rule
ionization energy	orbital
isoelectronic	period
lanthanide series	periodic law
metalloid	representative element
metal	transition element
noble gas	valence electrons

Self Test

1. What do we call the map used to organize the structure and properties of the different elements?
2. What do we call the outermost electrons of an element, which are involved in chemical bonding?
3. How many valence electrons are in sulfur?
4. Name the only noble gas with two valence electrons.
5. Name the group which contains elements frequently isoelectronic with stable cations and anions.
6. In chemical reactions, elements will gain, lose, or share the minimum number of electrons necessary to achieve what kind of energy state?
7. What do we call an ion that has the same electronic arrangement as its nearest noble gas neighbor?
8. Who were the first to arrange the elements by atomic masses on a table of elements?
9. The modern periodic table is arranged according to what value?
10. What are the Group A elements often called?
11. What are the Group B elements often called?
12. List the values of the principal (main) energy levels of at atom.
13. How many sublevels are present in the second principal energy level?
14. What is the name of a specific region of space of a sublevel where a maximum of two electrons may be found?
15. What orbital has a spherical shape?
16. What is the relationship of the spins of two electrons in the same orbital?
17. What term refers to the arrangement of electrons in atomic orbitals?
18. Write the electron configuration of carbon.
19. Which elements do not readily bond to other elements?
20. All ions in this group always form a + 1 ion.
21. What do we call the ability of an atom to attract electrons to itself?

1. Periodic table
2. valence e⁻s
3. 6
4. He
5. noble gases
6. ground state
7. isoelectric
8. Mendelejev/Lothar Meyer
9. Atomic number
10. representative elements
11. transition
12. N = 1,2,3,4 ...
13. 2
14. orbital
15. S
16. opposite
17. electron configuration
18. 1s² 2s² 2p²
19. noble gases
20. Group 1A alkali metals
21. electronegativity

31

4 Structure and Properties of Ionic and Covalent Compounds

Learning Goals

1. Conceptual Goals
- Describe the essential differences between ionic and covalent compounds.
- Recognize the differences in physical state, melting and boiling points, solid-state structure, and solution chemistry that result from differences in bonding.
- Know the relationship between stability and bond energy.
- Understand the role that molecular geometry plays in determining the solubility and melting and boiling points of compounds.

2. Performance Goals
- Name common inorganic compounds using systematic names, and recognize the common names of frequently used substances.
- Write the formulas of compounds when provided with the name of the compound.
- Draw Lewis structures for covalent compounds and complex inorganic ions.
- Predict the geometry of molecules and ions using the octet rule and Lewis structure.

3. Health Applications
- Describe the way in which the sodium/potassium ratio affects blood pressure.
- Provide examples of low-sodium, high-potassium foods and high-sodium, low-potassium foods.

Chapter Overview

This chapter describes the role of valence electrons in bond formation between atoms. Systems of naming the resultant compounds are discussed, as well as the procedure for writing formulas based on the names of the compounds. The chemical and physical properties of these compounds are related to structure and bonding.

4.1 Chemical Bonding

When two atoms are joined together to make a chemical compound, the force of attraction between the two species is referred to as a **chemical bond**. Interactions involving valence electrons are responsible for the chemical bond.

Lewis Symbols

The **Lewis symbol** is a convenient way of representing atoms singly or in combination. Its principal advantage is that only valence electrons (those which may participate in bonding) are shown. This results in simpler structures and greater clarity. The chemical symbol of the atom is written; this symbol represents the nucleus and all of the lower-energy nonvalence electrons, which do not directly participate in bonding. The valence electrons are indicated by dots (•) or crosses (×) arranged around the atomic symbol.

Types of Chemical Bonds: Ionic and Covalent

Ionic bonding is characterized by an electron transfer process occurring prior to bond formation. In **covalent bonding** electrons are shared between atoms in the bonding process.

The essential features of ionic bonding are as follows:

- Elements with low ionization energy and low electron affinity tend to form positive ions.
- Elements with high ionization energy and high electron affinity tend to form negative ions.
- Ion formation takes place by an electron transfer process.
- The resulting positive and negative ions are held together by the electrostatic force between ions of opposite charge in an ionic bond.
- Reactions between metals and nonmetals (elements far to the left and right, respectively, on the periodic table) tend to result in ionic bonds.

When electrons are shared rather than transferred, the shared electron pair is referred to as a covalent bond. Compounds characterized by covalent bonding are called covalent compounds. Covalent bonds tend to form among atoms with similar tendencies to gain or lose electrons. The most obvious examples are the diatomic molecules H_2 as well as N_2, O_2, F_2 Cl_2, I_2, and Br_2. Bonding in these molecules is totally covalent because there is no net tendency for electron transfer between identical atoms.

Two atoms do not have to be identical in order to form a covalent bond. Compounds such as hydrogen fluoride, water, methane, and ammonia are common examples.

4.2 Naming Compounds and Writing Formulas of Compounds

Proper use of **nomenclature**, the assignment of a correct and unambiguous name to every chemical compound, is fundamental to the study of chemistry. The student must be able to write the name corresponding to a compound and the formula, when provided only with the name.

Ionic Compounds

The "shorthand" symbol for a compound is its **formula**. Examples are: $NaCl$, $MgBr_2$, and $NaIO_3$. The formula identifies the number and type of the various atoms that compose the compounds. The number of like atoms is denoted by a subscript. The presence of one atom is implied when no subscript is present.

Chapter 2 described positive ion formation from elements that

1. are located to the left on the periodic table.

2. are referred to as metals.
3. have low ionization energies and low electron affinities, and hence easily lose electrons.

Elements which form negative ions, on the other hand,

1. are located to the right on the periodic table (but exclude the noble gases).
2. are referred to as nonmetals.
3. have high ionization energies and high electron affinities, and hence easily gain electrons.

Metals and nonmetals react to produce ionic compounds as a result of electron transfer.

Although we refer to ionic compounds as **ion pairs**, in the solid state these ion pairs do not actually exist as individual units. Positive and negative ions arrange themselves in a regular, three-dimensional, repeating array known as a **crystal lattice**.

The names given to ionic compounds are based upon their formulas, with the name of the cation appearing first, followed by the anion name. The positive ion is simply the name of the element, while the negative ion is named as the stem of the elements's name joined to the suffix -ide.

Example 1
Give the correct name for each of the following ionic compounds:

1. $NaCl$	3. $AlCl_3$	5. Ba_3N_2
2. Li_2S	4. CaO	6. MgO

Answer
1. sodium chloride
2. lithium sulfide
3. aluminum chloride
4. calcium oxide
5. barium nitride
6. magnesium oxide

If the cation and anion exist in only one common charged form, there is no ambiguity between formula and name. Sodium chloride must be $NaCl$, and lithium sulfide must be Li_2S, so that the sum of positive and negative charges is zero. With many elements, such as the transition metals, several ions of different charge may exist. Fe^{2+}, Fe^{3+} and Cu^+, Cu^{2+} are a few common examples. Clearly, an ambiguity exists if we use the name iron for both Fe^{2+} and Fe^{3+} or copper for Cu^+ and Cu^{2+}. Two systems have been developed that avoid this problem: the stock system and the common nomenclature system.

In the stock system for naming an ion (the systematic name) a Roman numeral indicates the magnitude of charge of the cation. In the older common nomenclature, the suffix -ous indicates the lower of the ionic charges, and the suffix -ic indicates the higher ionic charge. Systematic names are preferred; they are easier and less ambiguous.

Example 2

Give the correct names for each of the following ionic compounds that contain transition metal ions in their structures.

1. $FeCl_3$ 2. $FeCl_2$

Work

When a transition metal is found in an ionic compound, we must first find the charge on the transition metal before we can name the compound.

1. $FeCl_3$
    a. First use the formula: we find that one Fe ion and three Cl ions are present.
        $Fe^{?+} + 3\ Cl$
    b. From the periodic table, we find that the Cl is in family VIIA. These elements always form a -1 ion.
    c. $Fe^{?+} + 3\ Cl^-$
        We now know that there is a total charge of -3 in the compound. Thus, the iron ion must be Fe^{3+} so it can balance the -3 charge.
        Note: The total + charges must always equal the total $-$ charges in any ionic compound!
        Formula $= FeCl_3$
        Ion-pair form $= Fe^{3+} + 3\ Cl^-$
        Correct name: iron(III) chloride
2. $FeCl_2$
    a. $FeCl_2$ means $Fe^{?+} + 2\ Cl^-$
        Since there are a total of -2, then the $Fe^{?+}$ must be an Fe^{2+} to balance the charges: $+2$ balances -2
    b. Correct name: iron(II) chloride

Answer
1. iron(III) chloride   2. iron(II) chloride

Ions consisting of only a single atom are said to be **monatomic**. In contrast, **polyatomic ions**, such as the hydroxide ion, OH^-, are composed of two or more atoms bonded together. The polyatomic ion has an overall positive or negative charge. Some common polyatomic ions are listed in Table 3.3 of the textbook. Your instructor may suggest that several of the most common polyatomic ions be committed to memory.

It is equally important to write the correct formula when given the compound name. It is essential to be able to predict the charge of monatomic ions and the charge and formula of polyatomic ions. Remember, the relative number of positive and negative ions in the unit must result in a unit (compound) charge of zero.

Example 3
Name the following ionic compounds. Use Roman numerals when needed.

1. $Ca_3(PO_4)_2$ 3. $CsOH$ 5. Na_2SO_4
2. $Mg(C_2H_3O_2)_2$ 4. $FeSO_4$ 6. $Fe_2(CO_3)_3$

Answer
1. calcium phosphate
2. magnesium acetate
3. cesium hydroxide
4. $FeSO_4 = Fe^{2+} + SO_4^{2-}$; iron (II) sulfate
5. sodium sulfate
6. $Fe_2(CO_3)_3 = 2\ Fe + 3\ CO_3^{2-} = 2\ Fe^{3+} + 3\ CO_3^{2-}$
    iron (III) carbonate

Example 4
Write the formulas of the following ionic compounds:

1. sodium chloride   3. iron (III) hydroxide
2. magnesium phosphate   4. ammonium carbonate

Answer
1. sodium chloride
    a. $Na^+ + Cl^-$ IA family $= +1$ charge; VIIA $= -1$ charge
    b. formula $= NaCl$
2. magnesium phosphate
    a. $Mg^{2+} + PO_4^{3-}$ IIA $= +2$ charge; phosphate ion $= -3$
    b. need 3 Mg^{2+} to balance 2 PO_4^{3-}
    +6 balances −6
    c. formula $= Mg_3(PO_4)_2$
3. iron (III) hydroxide
    a. $Fe^{3+} + OH^-$ (Charges are not balanced.)
        need 3 OH^- to balance the +3 charge
        $Fe^{3+} + 3\ OH^-$
        +3 balances −3

b. formula = $Fe(OH)_3$
4. ammonium carbonate.
  a. $NH_4^+ + CO_3^{2-}$ (Charges are not balanced.)
  b. formula = $(NH_4)_2 CO_3$

Covalent Compounds

Covalent compounds are formed by the reaction of nonmetals. Covalently bonded compounds are **molecules**. The existence of this compound unit is a major distinctive feature of covalently bonded substances. The convention used for naming covalent compounds is as follows:

1. The names of the elements are written in the order in which they appear in the formula.
2. A prefix indicating the number of each kind of atom found in the unit is placed before the name of the element.
3. If only one atom of a particular kind is present in the molecule, the prefix mono- is usually omitted.
4. The stem of the name of the last element is used with the suffix -ide.

Common names are often used. For example, H_2O is water, NH_3 is ammonia, C_2H_5OH (ethanol) is alcohol, and $C_6H_{12}O_6$ is glucose. It is useful to be able to correlate both systematic and common names with the corresponding molecular formula and vice versa.

Writing formulas of covalent compounds can only be done from memory when common names are used. You must remember that water is H_2O, ammonia is NH_3, and so forth. This is the major disadvantage of common names; however, they cannot be avoided.

Example 5
Name the following covalent compounds:

1. CO_2 3. SO_2 5. N_2O_4
2. CO 4. SO_3 6. N_2O_3

Work
A prefix must be used to indicate the number of atoms of each nonmetal in the covalent compound.

Answer
1. carbon dioxide
2. carbon monoxide (The prefix mono- is usually not used. However, carbon monoxide has become a common name for CO_2.)
3. sulfur dioxide
4. sulfur trioxide

5. dinitrogen tetraoxide
6. dinitrogen trioxide

4.3 Properties of Ionic and Covalent Compounds

The differences in ionic and covalent bonding account for the different properties of ionic and covalent compounds. Covalently bonded molecules are discrete units, and they have less tendency to form an extended structure in the solid state. Ionic compounds do not have definable units, but form a **crystal lattice** composed of hundreds of billions of positive and negative ions in an extended three-dimensional network.

Major differences in the properties of these compounds are summarized below.

Physical State

All ionic compounds are solids at room temperature, while covalent compounds may be solids (sugar, silicon dioxide), liquids (water, ethyl alcohol), or gases, (carbon monoxide, carbon dioxide).

Melting and Boiling Points

The **melting point** is the temperature at which a solid is converted to a liquid, and the **boiling point** is the temperature at which a liquid is converted to a gas. Considerable energy is needed to break apart an ionic crystal lattice and convert an ionic solid to a liquid or a gas. Therefore, the melting and boiling temperatures for ionic compounds are generally higher than those of covalent compounds, which interact less strongly in the solid state.

Structure of the Compound in the Solid State

Ionic solids are crystalline, characterized by a regular structure, whereas covalent solids may be either crystalline or amorphous (no regular structure).

Solutions of Ionic and Covalent Compounds

Many ionic solids dissolve in solvents, such as water. If soluble, an ionic solid will dissociate in solution to form positive and negative ions (ionization). Because these ions are capable of conducting electric current, these compounds are **electrolytes**, and the solution is termed an **electrolytic solution**. Covalent solids in solution are neutral and are **nonelectrolytes**. The solution is not an electrical conductor.

4.4 Drawing Lewis Structures of Ions and Molecules

Lewis Structure of Molecules

Rule 1: Draw a skeletal structure of the molecule, arranging the atoms in their most probable order.

Rule 2: Determine the number of valence electrons on each atom, and add them together to get the total for the compound.

Rule 3: Distribute the electrons around the atoms (in pairs if possible) in an attempt to satisfy the octet rule: eight electrons around each element (remember, H and He cannot have eight electrons).

Rule 4: After completion of the Lewis structure, the nonbonding electrons are often omitted.

Example 6

Draw the Lewis dot and line structures of carbon dioxide. The carbon is in the middle of two oxygen atoms. The carbon atom is the central atom.

Work

1. O C O (arrangement as in nature)
2. O = 6 valence e^-
    C = 4 valence e^-
    O = 6 valence e^-
    _____
    16 total valence e^- in CO_2
3. Dot structure
    (All must have 8 valence e^- in their final valence shell.)

    $:\ddot{O}::C::\ddot{O}:$

4. Line structure
    $O = C = O$ ($I\bar{O} = C = \bar{O}I$ is also acceptable.)

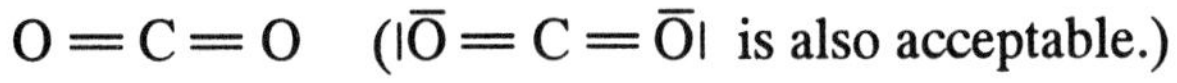

Example 7

Draw the Lewis dot and line structures of SO_3. Sulfur is the central atom in the compound.

Work

1. $\begin{array}{c} O \\ O \quad S \quad O \end{array}$

2. 3 oxygen atoms = 18 valence e^-
    2 sulfur atom = 6 valence e^-
    _____
    24 total valence e^- in SO_3

3. Dot structure (Any one of the following is correct.)

    :Ö: :Ö: :Ö:
    :Ö: S̈ :Ö: or   :Ö::S :Ö: or   :Ö: S :Ö:

4. Line structure (Any one of the following is correct.)

    ```
         O               O                O
         |               |                ‖
    O — S = O    or   O = S — O    or   O — S — O
    ```

Lewis Structure of Polyatomic Ions

The charge on the ion must be accounted for while computing the total number of valence electrons when writing Lewis structures of ions. For negative ions, one valence electron is added for each unit of negative charge and we subtract one valence electron for each unit of positive charge in a positive ion.

Lewis Structure, Stability, Multiple Bonds and Bond Energies

The order of stability often parallels the bond order (the number of bonds between adjacent atoms). Stability is also related to the bond energy, where the **bond energy** is defined as the amount of energy in kilocalories required to break a bond holding two atoms together. The magnitude of the bond energy decreases in the order triple bond > double bond > single bond. The bond length decreases in the order single bond > double bond > triple bond.

Lewis Structure and Resonance

More than one Lewis structure may satisfy the octet rule for a particular compound. When a compound has two or more Lewis structures that contribute to the real structure, the compound displays **resonance**. The contributing Lewis structures are **resonance forms**. The true structure, a hybrid or mixture of the resonance forms, is known as a **resonance hybrid**. The answers for Examples 6 and 7 (above) show a variety of possible structures for SO_3 and CO_3^{2-}. These are resonance forms.

Example 8

Draw the Lewis dot and line structures of the ammonium ion. The nitrogen atom is the central atom in the ion.

Work

1.

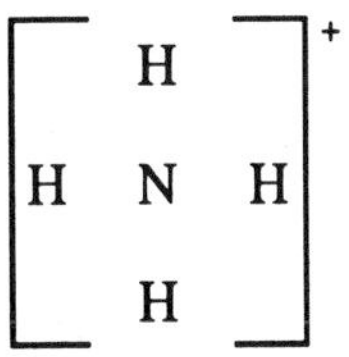

$$\left[\begin{array}{ccc} & H & \\ H & N & H \\ & H & \end{array}\right]^{+}$$

2. N = 5 valence e⁻
    H = 1 valence e⁻
    H = 1 valence e⁻
    H = 1 valence e⁻
    H = 1 valence e⁻
    _____
        9 valence e⁻

Remember however, that since the ion is a +1, one valence electron has been removed. Thus, only eight valence electrons are found in the ammonium ion.

3. 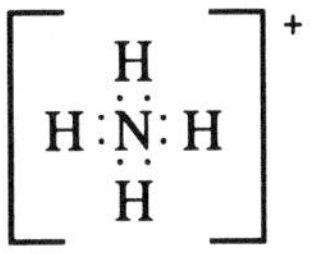
$$\left[\begin{array}{c} H \\ H \!:\! \overset{\cdot\cdot}{N} \!:\! H \\ H \end{array}\right]^{+} \quad \text{is the dot structure.}$$

4. 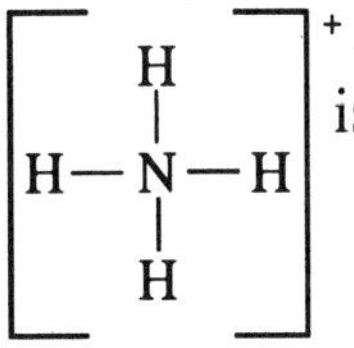
$$\left[\begin{array}{c} H \\ | \\ H\!-\!N\!-\!H \\ | \\ H \end{array}\right]^{+} \quad \text{is the line structure.}$$

Lewis Structures and Molecular Geometry

The shape of a molecule contributes to its properties and reactivity. We may predict the shapes of various molecules by inspecting their Lewis structures for the orientation of their electron pairs. The covalent bond, in which bonding electrons are localized between the atoms, is directional: the bond has a specific orientation in space among the bonded atoms. Electrostatic forces in ionic bonds, in contrast, are nondirectional: they have no specific orientation in space. As a result, the specific orientation of electron pairs in covalent molecules imparts a characteristic shape to the molecules.

Electron pairs around the central atom of the molecule arrange themselves to minimize repulsion; the electrons pairs are as far as possible from each other. This is termed the valence shell electron pair repulsion theory (**VSEPR Theory**). Two electron pairs around the central atom lead to a **linear** arrangement of the attached atoms; three indicate a **trigonal planar** arrangement; and four result in a **tetrahedral** geometry. Molecules with five and six electron pairs also exist. They may have structures which are trigonal bypyramidal or octahedral.

Compounds containing the same central atom will have structures with similar geometries. This is an approximation; it is not always true, but useful in writing reasonable, geometrically accurate structures for a large number of compounds. Consider the following:

BeH_2 (linear)
BeI_2, $BeCl_2$, $Be(OH)_2$ are also linear.
BF_3 (trigonal planar)
BCl_3, BBr_3, BI_3 are also trigonal planar.
CH_4 (tetrahedral)
CF_4, CCl_4, CBr_4 are also tetrahedral.

Furthermore, the periodic similarity of group members is useful in predictions involving bonding. Consider the following:

H_2O (angular)
H_2S, H_2Se are also angular because O, S, and Se are in the same group, VI, 6 valence electrons.

Lewis Structures and Polarity

A molecule is **polar** if its centers of positive and negative charges do not coincide. On the other hand, the positive and negative centers are coincident in non-polar molecules. No dipole exists—the dipole moment is equal to zero.

This property may also be described by considering the equality of electron sharing between the atoms being bonded. The degree of electron sharing can be described using the property of electronegativity. **Electronegativity** is the ability of an atom in a molecule to attract electrons to itself. Electronegativity is represented by a scale derived from the measurement of energies of chemical bonds. Values range from 4.0 (most electronegative element) to 0.7 (least electronegative element). The periodic trends for electronegativity, which decreases from top to bottom and increases from left to right, are similar to both ionization energy and electron affinity.

The atoms of H_2 are identical; their electronegativity is the same. The electrons remain, on average, at the center of the molecule, and the bond is nonpolar. The hydrogen molecule itself is nonpolar as well, with a dipole moment of zero. O_2, N_2, Cl_2, and F_2 are also nonpolar molecules with nonpolar bonds.

Bonds formed from atoms of unequal electronegativity are polar. The more electronegative end of the bond is designated d⁻ (partial negative) and the less electronegative end d⁺ (partial positive). A polar covalent molecule is characterized by **polar covalent bonding**. This implies that the electrons are shared unequally.

A molecule containing all nonpolar bonds must itself be nonpolar. A molecule containing polar bonds may be either polar or nonpolar depending on the relative orientation of the bonds.

4.5 Properties Based on Molecular Geometry

The effects of molecular polarity result from the strength of attractive forces among individual molecules of a compound. These attractions are **intermolecular forces**. The student should not confuse intermolecular forces with intramolecular forces. **Intramolecular forces** are the attractive forces within molecules. It is the inter-molecular forces that determine such properties as solubility, melting point, and boiling point.

Solubility

Solubility is defined as the maximum amount of solute that dissolves in a given amount of solvent. (The solute is present in lesser quantity, and the solvent is present in the greater amount.) Polar molecules are most soluble in polar solvents, and nonpolar molecules are most soluble in nonpolar solvents. This is the rule of "like dissolves like."

Boiling Points of Liquids and Melting Points of Solids

Boiling is the conversion of a liquid to a vapor. Such a process requires energy, overcoming the intermolecular attractive forces in the liquid. The energy required is related to the magnitude of the boiling point and depends upon the strength of the intermolecular attractive forces in the liquid. These attractive forces are directly related to polarity. Molecular size is also an important consideration. The larger or heavier the molecule, and the more polar the molecule, the more difficult it becomes to convert to the gas phase.

The melting points of solids may be described on the basis of intermolecular forces as well.

As a general rule, polar compounds have strong attractive (intermolecular) force, and their boiling and melting points tend to be higher than for nonpolar substances.

4.6 Molecular Orbital Model of Bonding

The bonding rules discussed so far are valid for certain molecules, but inappropriate for others. A more exact theory, atomic and molecular orbitals, provides a more complete picture of bonding.

Review of Atomic Orbitals

Bohr's concept of electrons in fixed orbits was too restrictive and fundamentally incorrect. Although the electron was restricted to a particular energy level, the specific location of the electron can only be described in terms of probability. We referred to regions of high probability of finding an electron as orbitals. These regions were depicted as zones, or "electron clouds."

The principal types of orbitals were designated *s, p, d*, and *f*, denoting different shapes and specific volumes around the nucleus.

Molecular Orbitals

A number of general points concerning molecular orbital theory follow:
1. Valence electrons occupy orbitals that have a particular shape and extend out of the atom in a particular direction. This shape and directionality affects the bonding properties and molecular geometry in much the same way that we pictured using electron pairs and Lewis structures.
2. Bonding can be envisioned as the overlap of atomic orbitals to form **molecular orbitals**.
3. Equal sharing of electron pairs (the nonpolar covalent bond) is represented as a uniformly distributed electron cloud (electrons equally shared) in Figure 4.11.

How Molecular Orbitals Affect Molecular Geometry

The atomic orbital approach to covalent chemical bonding is particularly attractive because it gives us some insight into the shapes of molecules formed from atoms with valence electrons in s, p, d, and f, orbitals. Bond formation results from an overlap of orbitals of atoms participating in the bond producing molecular orbitals. It is therefore reasonable to presume that the size, shape, and orientation of these orbitals will determine the shape of the resulting compound.

Inspect the three-dimensional atomic orbital sketches (pp. 96-97 of the textbook) and try to relate these orbital shapes to the molecular model of the resulting compounds.

Even better agreement between theory and experiment is obtained by combining atomic orbitals to produce new **hybrid orbitals**. Hybrid orbitals exhibit some of the attributes of each atomic orbital that is "mixed" to make the hybrid. Without resorting to hybridization, much useful structural information can be gained from Lewis structures and atomic orbital approaches. The Lewis approach will be used throughout the rest of the textbook.

Key Terms

angular structure	Lewis symbol
boiling point	linear structure
bond energy	lone pair
chemical bond	melting point
covalent bond	molecule
crystal lattice	monatomic ion
dissociation	nomenclature
double bond	nonelectrolyte
electrolyte	polar covalent bonding
electrolytic solution	polar covalent molecule
electronegativity	polyatomic ion
formula	resonance
intermolecular force	resonance form
intramolecular force	resonance hybrid
ionic bonding	single bond
ion pair	solubility

tetrahedral structure
triple bond

Valence shell electron pair repulsion theory

Self Test

1.   The transfer of electrons from one atom to another forms which type of chemical bond?
2.   The sharing of electrons between two atoms forms which type of chemical bond?
3.   What particles are found in ionic compounds?
4.   How many sodium ions are in Na_2S?
5.   On what side of the periodic table are elements found that usually form positive ions?
6.   On what side of the periodic table are elements found that usually form negative ions?
7.   The name of an ionic compound always contains the name of which ion first?
8.   Provide the best name of Na_2O.
9.   Provide the best name of Li_2S.
10. Provide the best name of $AlBr_3$.
11. What is the common name of the iron (II) ion?
12. What is the best name of Fe^{3+}?
13. What does the ending "ic" on the common names of ions mean?
14. Give the common name of $FeSO_4$.
15. Give the systematic name of CuO.
16. Write the formula of sodium sulfate.
17. How many bonding electrons are in H_2O?
18. How many nonbonding electrons are in CO_2?
19. Which end of HCl is slightly more negative?
20. Why is HF a polar molecule?
21. When H_2O is compared to C_2H_4, which has the higher boiling point?
22. When Cl_2 is compared to ICl, which has the higher boiling point?
23. Draw the dot and line structures of ammonia.
24. Draw the dot and line structures of water.
25. Draw the dot and line structures of a hydroxide ion.
26. Draw the dot and line structures of a sulfate ion.
27. Draw the dot and line structures of a hydronium ion. The H_3O^+ ion has oxygen at the center of the ion.
28. Which electrons of an atom are involved in chemical bond formation?
29. What type of bond is formed by the sharing of valence electrons?
30. Write the formula of the ions that result when sodium atoms react with chlorine atoms.
31. Write the formula of the ions that result when calcium atoms react with sulfur atoms.
32. What process results in ion formation in ionic reactions?
33. How many electrons are jointly shared in the hydrogen molecule?
34. How many electrons are shared in H_2O?

5 *Calculations and the Chemical Equation*

Learning Goals

1. Conceptual Goals
- Know the major function served by the chemical equation, the basis for chemical arithmetic.
- Know the relationship between the mole and Avogadro's number.

2. Performance Goals
- Write chemical formulas for common inorganic substances.
- Balance chemical equations given the identity of products and reactants.
- Perform calculations involving conversions from moles to grams or grams to moles.
- Calculate the number of moles of product resulting from a given number of moles of reactants and vice versa.
- Perform computations involving more complex systems that may involve conversion from volume to mass or the reverse, one mass unit to another, and mass to moles or the reverse.

Chapter Overview

You should be able to predict the quantity of a product produced from the reaction of a given amount of reactant. You should also be able to calculate how much of a reactant, or group of reactants, would be required to produce a desired amount of product.

These calculations are based on the **chemical equation**. The chemical equation provides all of the needed information: the combining ratio of elements or compounds (reactants) that are necessary to produce a particular product or products as well as the relationship between the amounts of reactants and products formed.

5.1 The Mole Concept and Atoms

The Mole and Avogadro's Number

Atomic masses have been experimentally determined for each of the elements. Their unit of measurement is the **atomic mass unit**, abbreviated amu.

1 amu = 1.66×10^{-24} grams
A more practical unit for defining a "collection" of atoms is the **mole**.
1 mole of atoms = 6.02×10^{23} atoms of an element

This number is **Avogadro's number**.

The mole and the amu are related. The atomic mass of a given element corresponds to the average mass of a single atom in amu *and* the mass of a mole of atoms in grams.

For example, the average mass of one copper atom is 63.54 amu, while the mass of one *mole* of copper atoms is 63.54 grams. One mole of atoms of *any element* contains the same number, Avogadro's number, of atoms.

Calculating Atoms, Moles, and Mass

Calculations based on the chemical equation, (Section 4.6 of the textbook) utilize the relationship of the number of atoms, number of moles, and mass in grams.

Conversion factors are used to proceed from the information *provided* in the problem to the information *requested* by the problem.

Before you begin any calculations, map out a pattern for the required conversion. You may be given the number of grams and need the number of atoms that corresponds to that mass. Begin by "tracing a path" to the answer:

$$\text{grams} \quad \underset{1}{\overset{\text{Step}}{\rightarrow}} \quad \text{moles} \quad \underset{2}{\overset{\text{Step}}{\rightarrow}} \quad \text{atoms}$$

Two transformations, or conversions are required:

Step 1: grams to moles
Step 2: moles to atoms

The conversion between the three principal measures of quantity of matter, the number of grams (mass), the number of moles, and the number of individual particles (atoms, ions, or molecules) is depicted in Figure 5.2 of the textbook.

Example 1
Calculate the mass of one mole of iron atoms.

Work
From the periodic table, we find that the average mass of one iron atom is 55.85 atomic mass units. Thus, one mole of iron atoms has a mass of 55.85 grams.

5.2 Compounds

The Chemical Formula

Chemical compounds are represented by their **chemical formulas**, a combination of symbols of the various elements that make up the compounds. The chemical formula is based upon the **formula unit,** the smallest collection of atoms that provides the following information:

1. the identity of the atoms present in the compound and
2. the relative numbers of each type of atom.

The term *formula* may be used in reference to any ionic or covalent compound. In Chapter 3 compounds were classified as nonpolar covalent, polar covalent, or ionic, depending upon the type of bonding that holds the unit together.

A covalent compound is composed of discrete molecules, and the term *molecular formula* is most appropriate. Ionic compounds, on the other hand, are composed of ion pairs, not molecules, and *cannot* be described as having a molecular formula.

5.3 The Mole Concept Applied to Compounds

A mole of a compound is based upon the formula mass or **formula weight**. The formula weight is calculated by adding the masses of all the atoms of which the unit is composed. To calculate the formula weight, the formula unit must be known.

Example 2
Calculate the mass of one mole of carbon dioxide (the formula weight of carbon dioxide).

Work
From the periodic table, we find that one atom of carbon has a mass of 12.01 amu, and one atom of oxygen has a mass of 16.00 amu.

CO_2 = 1 carbon atom + 2 oxygen atoms

CO_2 = 1 C + 2 O

CO_2 = 1(12.01 amu) + 2(16.00 amu) = 44.01 amu

Answer
Thus, one molecule of CO_2 has a mass of 44.01 amu, and one mole of carbon dioxide molecules weighs 44.01 grams.

Example 3
How many grams of carbon dioxide are contained in 10.00 moles of carbon dioxide?

Work
Since 1 mole of CO_2 = 44.01 grams, we can use the factor-label method to convert grams into mole units.

We can write the above as two conversion fractions:

Answer
10.00 moles of CO_2 = 44.01 $\times$ 10^2 grams of CO_2

Example 4
Calculate the formula weight of water.

48

Work

Water (H_2O) is composed of two hydrogen atoms plus one oxygen atom. From the periodic table, we find that O = 16.00 amu, and H = 1.008 amu.

H_2O = 2 hydrogen atoms + 1 oxygen atom

H_2O = 2 H + 1 O

H_2O = 2(1.008 amu) + 1(16.00 amu) = 18.01 amu

Answer

The formula weight of water equals 18.01 amu. Notice that amu are used for formula weights. A mole of water, however, is equal to 18.01 grams.

5.4 The Chemical Equation and the Information It Conveys

A Recipe for Chemical Change

The **chemical equation** describes all of the substances that react to produce the product(s). The chemical equation also describes the physical state of the reactants and products as solid, liquid, or vapor. It tells us whether the reaction occurs, and identifies the solvent and experimental conditions. Most important, the relative number of moles of reactants and products appears in the equation. According to the **law of conservation of mass**, matter cannot be either gained or lost in the process of a chemical reaction. The law of conservation of mass states that we must have a *balanced chemical equation.*

Features of a Chemical Equation

1. The identity of products and reactants must be specified.
2. Reactants are written to the left of the reaction arrow → and products to the right.
3. The physical state of reactants and products is shown in parentheses.
4. The symbol Δ over the reaction arrow means that heat energy is necessary for the reaction to occur.
5. The equation must be balanced.

The Experimental Basis of a Chemical Equation

The chemical equation represents a real chemical transformation. Evidence for the reaction may be based on observations such as the formation of a gas, a solid precipitate, the evolution of heat, or a color change in the solution.

5.5 Balancing Chemical Equations

A chemical equation tells us both the identity and state of the reactants and products. **Reactants**, or starting materials, are all substances that undergo change in a chemical reaction, while **products** are substances produced by a chemical reaction.

The chemical equation also shows the *molar quantity* of reactants needed to produce a certain *molar quantity* of products.

The number of moles of each product and reactant is indicated by placing a whole-number *coefficient* before the formula of each substance in the chemical equation.

Many equations are balanced by trial and error. The following steps provide a method for correctly balancing a chemical equation:

Step 1: Count the number of atoms of each element on both product and reactant side.
Step 2: Determine which atoms are not balanced.
Step 3: Balance one atom at a time, using coefficients.
Step 4: After you believe that you have successfully balanced the equation, check, as in Step 1, to be certain that mass conservation has been achieved.

Example 5

Write the balanced reaction of an aqueous solution of hydrochloric acid (HCl) reacting with solid calcium to produce calcium chloride plus hydrogen gas.

Work

1. First write the correct symbols and formulas needed for the reactants and products:
   $$HCl + Ca \rightarrow CaCl_2 + H_2$$

2. Next, count the number of atoms of each element on the right and left side of the reaction arrow:

$$HCl + Ca \rightarrow CaCl_2 + H_2$$

H	Cl	Ca		H	Cl	Ca
1 atom	1 atom	1 atom		2 atoms	2 atoms	1 atom

   Notice that the numbers of atoms on each side are not equal; the equation is **not** balanced.

3. After the correct formulas of all reactants and products are given, you can now use coefficients alone to balance the equation.

4. $$2\,HCl + Ca \rightarrow CaCl_2 + H_2$$

H	Cl	Ca		H	Cl	Ca
2	2	1		2	2	1

   The equation is now balanced.

Answer

$$2\,HCl + Ca \rightarrow CaCl_2 + H_2$$

Note that the equation becomes more complete when the states of each reactant and product are added:

$$2\,HCl(aq) + Ca(s) \rightarrow CaCl_2(aq) + H_2(g)$$

Example 6
Balance the following reaction:

$$FeCl_3(aq) + Cu(s) \rightarrow CuCl_2\,(aq) + Fe(s)$$

Work
1. First, find the number of atoms of each element:

$$FeCl_3 \;\; + \;\; Cu \;\; \rightarrow \;\; CuCl_2 \;\; + \;\; Fe$$

Fe	Cl	Cu		Fe	Cl	Cu
1	3	1		1	2	1

The equation is not balanced.

2. Next, use only coefficients to balance the equation:

$$2\,FeCl_3 \;\; + \;\; 3\,Cu \;\; \rightarrow \;\; 3\,CuCl_2 \;\; + \;\; 2\,Fe$$

Fe	Cl	Cu		Fe	Cl	Cu
2	6	3		2	6	3

The equation is now balanced.

Answer
$$2\,FeCl_3(aq) + 3\,Cu(s) \rightarrow 3\,CuCl_2(aq) + 2\,Fe(s)$$

5.6 The Extent of Chemical Reactions

Not all reactions are complete, those in which reactants are totally consumed. Many reactions, equilibrium reactions, only partly convert reactant to product. When equilibrium is attained, measurable quantities of both reactant and product are observed, and these quantities do not change over time. However, both product formation and the reverse reaction, reactant formation, are continuously taking place; we describe the process as a *dynamic equilibrium*.

Complete reactions are identified by a single reaction arrow ($\rightarrow$), and equilibrium (incomplete) reactions are symbolized by a double reaction arrow ($\rightleftharpoons$).

5.7 Calculations Using the Chemical Equation

General Principles

In doing calculations involving chemical reactions, we apply the following rules:

1. The basis for the calculations is a balanced equation.
2. The calculations are performed in terms of moles.
3. The conservation of mass must be obeyed.

The mole is the basis for calculations. However, masses are generally measured in grams (or kilograms). A facility for interconversion of moles and grams is therefore necessary in calculations involving chemical reactions.

Example 7
In the following balanced equation, how many moles of each element or compound are involved?

$$2\,C_8H_{18} + 25\,O_2 \rightarrow 16\,CO_2 + 18\,H_2O$$

Work
The coefficients in a balanced equation tell us the number of moles of each reactant and product.

Answer
Two moles of octane (C_8H_{18} is found in gasoline) react with 25 moles of oxygen (always O_2 in nature) to form 16 moles of carbon dioxide plus 18 moles of water.

Use of Conversion Factors

1. *Conversion between moles and grams.* Conversion from moles to grams and vice versa requires only the formula weight of the compound of interest.
2. Conversion of moles of reactants and products. Conversion factors, based on the chemical equation, permit us to perform a variety of calculations.

   A general problem-solving strategy is summarized in Figure 4.6 of the textbook.

Example 8
The reaction of propane (C_3H_8 is found in bottled gas) with oxygen produces carbon dioxide and

water. The balanced equation is given below:

$$C_3H_8 + 5 O_2 \rightarrow 3 CO_2 + 4 H_2O$$

Calculate the number of grams of oxygen (O_2, as in nature) needed to completely react with 1 mole of propane.

Work

1. From the balanced equation, we see that 1 mole of propane reacts with 5 moles of O_2 to form 3 moles of CO_2 plus 4 moles of water.
2. One mole of O_2 equals 32.00 grams.
3. Since 5 moles of O_2 are needed to react with 1 mole of propane, then 160.0 grams of oxygen (O_2) are needed.

$$1 \text{ mol } C_3H_8 \times \frac{5 \text{ mol } O_2}{1 \text{ mol } C_3H_8} \times \frac{32.00 \text{ g } O_2}{1 \text{ mol } O_2} = 160.0 \text{ g } O_2$$

Answer

One mole of propane will react with 160.0 grams of oxygen (O_2).

The **theoretical yield** is the *maximum* amount of product that can be produced (in an ideal world). In the "real" world it is difficult to produce the amount calculated as the theoretical yield.

A **percent yield**, the ratio of the actual and theoretical yields multiplied by 10^2, is often used to show the relationship between predicted and experimental quantities. Thus

$$\% \text{ yield} = \frac{\text{actual yield}}{\text{theoretical yield}} \times 100\%$$

Key Terms

atomic mass unit
Avogadro's number
chemical equation
chemical formula
formula unit
formula weight
hydrate
limiting reactant
molar mass

law of conservation of mass
mole
molecular weight
percent yield
product
reactant
theoretical yield

Self Test

1. 6.02×10^{23} molecules of a covalent compound equal how many moles of that compound?
2.   One mole of hydrogen molecules weighs how many grams? Give answer to the nearest tenth of a gram.
3.   How many iron atoms are present in 1 mole of iron?
4.   How many grams of sulfur are found in 0.150 mole of sulfur? Use S = 32.06 amu.
5.   Write the formula of the smallest unit of nitrogen as it is normally found in nature.
6.   The term *molecular formula* is used for what type of compound?
7.   What units are used for the formula weights of atoms and individual molecules?
8.   What is the formula weight of one molecule of H_2O? (nearest hundredth)
     Use H = 1.01; O = 16.00 amu.
9.   The formula weight of a covalent compound may also be called what weight?
10. What term means the starting materials in a chemical reaction?
11. How many moles of hydrogen are needed to react with oxygen to form 2 moles of water?
     $$2\,H_2 + O_2 \rightarrow 2\,H_2O$$
12. How many moles of oxygen are needed to react with hydrogen to form 1 mole of water?
     $$2\,H_2 + O_2 \rightarrow 2\,H_2O$$
13. The number of moles of reactants and products in a balanced chemical equation is given by what kind of numbers?
14. What number will be found in front of HCl when the following equation is balanced?
     $$Ca + HCl \rightarrow CaCl_2 + H_2$$
15. What number will be found in front of H_2O when the following equation is balanced?
     $$Mg(OH)_2 + 2HCl \rightarrow MgCl_2 + 2H_2O$$
16. What number will be found in front of Cl_2 when the following equation is balanced?
     $$2Na + Cl_2 \rightarrow 2NaCl$$
17. How many grams of sodium hydroxide will react with 73.00 grams of hydrochloric acid?
     Use NaOH = 40.00; HCl = 36.45 amu.
     $$NaOH + HCl \rightarrow NaCl + H_2O$$
18. Calculate the number of grams of oxygen that must react with 46.85 grams of C_3H_8 to produce carbon dioxide and water.
     Use C = 12.01, H = 1.01 and O = 16.00 amu.
19. Iron reacts with oxygen to form iron (III) oxide (Fe_2O_3). How many grams of product will be formed from 5.00 grams of Fe?
     Use Fe = 55.85, and 0 = 16.00 amu.
20. How many moles of sulfur are found in 1.81×10^{24} atoms of sulfur?
     Use S = 32.06 amu.
21. What is the name given to the number of atoms in a mole of iron?
22. Write the formula of the smallest unit of oxygen as it is normally found in nature.

6 *States of Matter: Gases, Liquids, and Solids*

Learning Goals

1. **Conceptual Goals**
   - Know the three common states of matter and their general properties.
   - Understand the concept of the various gas laws: Boyle's Law, Charles' Law, Avogadro's Law, the ideal gas law, and Dalton's Law.
   - Recognize the differences between real gases and ideal gases.
   - Understand the properties of the liquid state: compressibility, viscosity, surface tension, and vapor pressure.
   - Describe the processes of melting, boiling, evaporation, and condensation.
   - Know the properties of the various classes of solids: ionic, covalent, molecular, and metallic.
2. **Performance Goals**
   - Perform ideal gas calculations, using each of the gas laws.
3. **Health Applications**
   - Understand the chemical processes involved in the manufacture and use of hot and cold packs.
   - Recognize the essential role of water in the cell.

Chapter Overview

The major differences between solid, liquid, and gaseous material lie in the following properties:

1. The average distance of separation of particles in each state
2. The strength of the attractive forces between the particles
3. The degree of organization of particles

The behavior of the states of matter, their properties, and their interconversion are considered in this chapter.

6.1 The Gaseous State

Boyle's Law

Robert Boyle found that the volume of a gas varies inversely with the pressure exerted by the gas, if the number of moles and the temperature of the gas are held constant. This relationship is known as Boyle's Law.

Mathematically, the product of pressure (P) and volume (V) is a constant:

$$PV = k_1$$

Boyle's law is often used to calculate the volume resulting from a pressure change or vice versa. We consider $P_iV_i = k_1$ the initial condition, the $P_fV_f = k_1$ the final condition. Since (PV), initial or final, is constant and is equal to k_1,

$$P_iV_i = P_fV_f$$

Note that the calculation can be done with volume units of mL or L. It is only important that the units be the same on both sides of the equation. This is also true of pressure; however, the most commonly used unit of pressure measurement is the atmosphere. The relationship between the various pressure units follows: 1 atmosphere (atm) = 76 cm Hg = 760 mm Hg = 760 torr = 30.0 inches Hg = 14.7 psi (pounds per square inch).

Example 1
Convert each of the following into atmospheres.

    1.   77.0 mm of Hg
    2.   30.2 inches of Hg
    3.   16.8 psi (The abbreviation psi means pounds per square inch.)
    4.   800.00 torr

Work
One atmosphere of pressure equals 760 mm of Hg pressure, which also equals 760 torr of pressure, which also equals 14.7 psi, or

$$1 \text{ atmosphere} = 760 \text{ mm Hg} = 760 = 14.7 \frac{lb}{in^2}$$

Answer
1. $77.0 \text{ mm of Hg} \times \dfrac{1 \text{ atm.}}{760 \text{ mm of Hg}} = 1.01 \times 10^{-1} \text{ atm}$

2. $30.2 \text{ in of Hg} \times \dfrac{2.54 \text{ cm}}{1 \text{ inch}} \times \dfrac{1 \text{ meter}}{100 \text{ cm}} \times \dfrac{1000 \text{ mm}}{1 \text{ m}} \times \dfrac{1 \text{ atm.}}{760 \text{ mm of Hg}} = 1.01 \text{ atm}$

3. $16.8 \text{ psi} \times \dfrac{1 \text{ atm.}}{14.7 \text{ psi}} = 1.14 \text{ atm}$

4. $800.0 \text{ torr} \times \dfrac{1 \text{ atm.}}{760 \text{ torr}} = 1.053 \text{ atm}$

Example 2
A given mass of carbon dioxide at 25°C occupies a volume of 500.0 mL at 2.00 atmospheres of pressure. What pressure must be applied to compress the gas to a volume of 50.0 mL, assuming

no temperature change?

Work

1. Use Boyle's Law as given below:

$$P_iV_i = P_fV_f$$

P_i represents the starting pressure; P_f represents the final pressure.
V_i represents the starting volume; V_f represents the final volume.

2. From the given information, we have the following:

$P_i = 2.00$ atmospheres $V_i = 500.0$ mL (the initial volume)

$P_f = ?$ (not known yet) $V_f = 50.0$ mL (the final volume)

3. Rearrange the equation as follows: $P_iV_i = P_fV_f$
$$P_fV_f = P_iV_i$$

$$P_f = \frac{P_iV_i}{V_f}$$

We can thus find the pressure needed to compress the 500.0 mL of gas to 50.0 mL.

4. $$P_f = \frac{P_iV_i}{V_f}$$

$$P_f = \frac{(2.00 \text{ atm.})(500.0 \text{ mL})}{50.0 \text{ mL}}$$

$$P_{final} = 20.0 \text{ atm}$$

Answer

20.0 atmospheres of pressure will be needed.

Charles' law

Jacques Charles found that the volume of a gas varies directly with the absolute temperature (K) if pressure and number of moles of gas are constant. This relationship is Charles's Law.

Mathematically the ratio of volume (V) and temperature (T) is a constant:

$$V/T = k_2$$

In a way analogous to Boyle's Law, and we may use this expression to solve some practical problems.

Example 3
A balloon filled with helium has a volume of 5.00 liters at 0.00°C. What would be the balloon's volume at 25.00°C if the pressure surrounding the balloon remained constant?

Work
1.   Use Charles's Law: $\dfrac{V_i}{T_i} = \dfrac{V_f}{T_f}$

   V_i represents starting volume; V_f means final volume.
   T_i represents starting temperature in Kelvin units.
   T_f represents final temperature in Kelvin units.
2.   Change the °C values into Kelvin values. Use the following equation:

   $K = °C + 273$

   $T_i = 0.00°C$; in K, $T_i = 0.00°C + 273 = 273.00$ K

   $T_f = 25.00°C$; in K, $T_f = 25.00°C + 273 = 298.00$ K

3.   Rearrange the equation: $\dfrac{V_i}{T_i} = \dfrac{V_f}{T_f}$

$$(V_f)(T_i) = (V_i)(T_f)$$

$$V_f = \dfrac{(V_i)\,(T_f)}{T_i}$$

4. $V_{final} = \dfrac{(5.00 \ \text{liters})(298 \ \text{K})}{273 \ \text{K}} = 5.46 \ \text{L}$

Answer
5.46 liters

Combined gas law

Often a sample of gas (a fixed number of moles of gas) undergoes change involving volume, pressure and temperature simultaneously. It would be useful to have one equation that describes such processes.

   A **combined gas law** is such an equation. It can be derived from Boyle's law and Charles's law and takes the form:

$$\frac{P_i V_i}{T_i} = \frac{P_f V_f}{T_f}$$

Avogardro's Law

The relationship between the volume and number of moles of a gas is known as Avogadaro's Law. This law states that equal volumes of a gas contain the same number of moles, if measured under the same conditions of temperature and pressure.

Mathematically, the ratio of volume (V) and number of moles (n) is a constant:

$$V/n = k_3$$

A relationship comparing initial and final conditions, similar to Boyle's and Charles' Laws, may be derived:

$$\frac{V_i}{n_i} = \frac{V_f}{n_f}$$

Molar Volume of a Gas

The volume occupied by 1 mole of any gas is referred to as its molar volume. The basis for this relationship is Avogadro's law. At **standard temperature and pressure** (STP), the molar volume of any gas is 22.4 L. STP conditions are defined as follows:

$$T = 273 \text{ K (or } 0°C)$$
$$P = 1 \text{ atm}$$

Thus 1 mole of N_2 (28g), O_2 (32g), H_2 (2g), and He (4g) all occupy the same volume, 22.4 L at STP.

It is also possible to compute the density of various gases at STP if one recalls that density is the mass/unit volume:

$$d = m/V$$

The Ideal Gas Law

Boyle's law (relating volume and pressure), Charles' law (relating volume and temperature) and Avogadro's law (relating volume to the number of moles) may be combined into a single expression relating all four terms. This expression is the **ideal gas law**:

$$PV = nRT$$

where R is the ideal gas constant.

$$R = 0.0821 \text{ L-atmK}^{-1}\text{mole}^{-1}$$

if the units P (atmospheres), V (liters), n (number of moles), and T (Kelvin) are used.

Example 4
Calculate the number of grams of helium in a 1.00-liter balloon at 27.0°C and under 1.00 atmosphere of pressure.

Work

1. Use the ideal gas law:

   $PV = nRT$

   P is the pressure in atmospheres.

   V is the volume of the gas in liters.

   R is the ideal gas constant, which is equal to

   $$\frac{0.0821\ (L)(atm)}{(K)(mole)}$$

   T is the temperature given in Kelvin units.

   n is the number of moles of gas present.

2. Convert 27.0°C into Kelvin units:

   $K = °C + 273$

   $K = 27.0 + 273 = 300\ K$

3. Rearrange $PV = nRT$ so we can find the number of moles (n in the equation) of gas that is present:

   $$n = \frac{PV}{RT}$$

4. $$n_{helium} = \frac{(1.00\ atm)(1.00\ L)}{\left(\frac{0.0821\ (L)(atm)}{(K)(mole)}\right)(300\ K)}$$

   $n_{helium} = 4.06 \times 10^{-2}$ moles of helium are present.

Answer

4.06×10^{-2} moles of helium are present or, as grams, $4.06 \times 10^{-2}\ mole \times \dfrac{4.00\ grams}{1\ mole\ of\ He} = 0.162\ grams$

Dalton's Law of Partial Pressures

A mixture of gases exerts a pressure that is the sum of the pressure that each gas would exert if it were present alone under similar conditions. This is known as Dalton's Law of partial pressures.

$$P_t = P_1 + P_2 + P_3 + \ldots\ldots P_n$$

where P_t = total pressure and p_1, p_2, p_3, and so on are the partial pressures of the component gases.

Kinetic Molecular Theory of Gases

The fundamental model of particle behavior in the gas phase is the kinetic-molecular theory. This theory describes an ideal gas, in which gas particles exhibit no interactive or repulsive forces, and the volumes of the individual gas particles are assumed to be negligible. The theory may be summarized as follows:

1. A gas consists of particles that are far apart. The volume of the individual particles is assumed to be small in comparison to the average distance of separation of the particles.
2. The gas particles are in continuous, rapid, random motion. Particles change direction only as a result of collisions with other particles or with the wall of a container.
3. Upon collision, there is no net loss of energy; energy may only be transferred from one particle to another.
4. The speed of the particles is directly proportional to the absolute (Kelvin) temperature; as a result, the average kinetic energy is directly proportional to the absolute temperature.

Ideal Gases vs. Real Gases

We have assumed so far, both in theory and calculations, that all gases behave as ideal gases. However, in reality, there is no such thing as an ideal gas. Interactive forces, even between the widely spaced particles of gas, are not totally absent in any sample of gas.

Attractive forces are particularly significant in gases composed of polar molecules. Calculations involving polar gases such as HF, NO, and SO_2, using ideal gas equations (which presume no such interactions), are approximate at best.

Nonpolar molecules exhibit a temporary dipole-dipole interaction known as a London force. The temporary dipole-dipole interaction is weak, and London forces are much weaker than permanent dipole interactions. As a result, the behavior of nonpolar molecules, such as N_2, O_2, and H_2, is explained rather well by the ideal-gas equations.

6.2 The Liquid State

Molecules in the liquid state are close to one another. Attractive forces are large enough to keep the molecules together, in contrast to gases, in which cohesive forces are so low that a gas expands to fill any volume. These attractive forces in a liquid are, however, not large enough to restrict movement, as in solids.

Compressibility

Liquids are almost incompressible. In fact, the spacing between molecules is so small that even the application of many atmospheres of pressure does not significantly decrease the volume of a liquid.

Viscosity

The **viscosity** of a liquid is a measure of its resistance to flow. Viscosity is a function of the attractive forces present between molecules as well as the molecular geometry. Complex molecules, which do not "slide" smoothly past each other, as well as polar molecules, tend to have higher viscosity than less structurally complex, less polar liquids.

Surface Tension

The surface tension of a liquid is a measure of the attractive forces at its surface. Intermolecular attraction is stronger at the surface of a liquid. This increased surface force is responsible for the spherical shape of drops of liquid. Substances known as surfactants may be added to a liquid to decrease surface tension.

Vapor Pressure of a Liquid

According to the kinetic theory, the molecules of a liquid are in continuous motion, with their average kinetic energy proportional to the temperature. Although the average kinetic energy is too small to allow molecules to "escape" from liquid to vapor, a few high-energy molecules possess sufficient energy to escape from the bulk liquid.

At the same time, a fraction of these vaporized molecules lose energy (perhaps by collision with the walls of the container) and return to the liquid state. The process of conversion of liquid to vapor, at a temperature too low to boil, is **evaporation**. The reverse process, conversion of the gas to the liquid state, is **condensation**. After some period of time, the rates of evaporation and condensation become equal, and this constitutes a dynamic equilibrium between liquid and vapor states. The vapor pressure of the liquid is therefore defined as the pressure exerted by the vapor at equilibrium.

The boiling point of a liquid is defined as the temperature at which the vapor pressure of the liquid becomes equal to the prevailing atmospheric pressure. The "normal" atmospheric pressure is 760 torr, or one atmosphere, and the **normal boiling point** is the temperature at which the vapor pressure of the liquid is equal to one atmosphere.

Van der Waals Forces

In 1930 Fritz London demonstrated that he could account for a weak attractive force between any two molecules, whether polar or non-polar. He postulated that the electron distribution in molecules is not fixed; electrons are in continuous motion, relative to the nucleus. So, for a short period of time a non-polar molecule could experience an instantaneous dipole, a short-lived polarity due to a temporary dislocation of the electron cloud. These temporary dipoles could interact with other temporary dipoles, just permanent dipoles interact in polar molecules. We now call these intermolecular forces **London Forces**.

London forces and dipole-dipole interactions are collectively known as **Van der Waals forces**. London forces exist among polar and non-polar molecules because electrons are in constant motion in all molecules. Dipole-dipole attractions occur only among polar molecules. In the next section we will see a third type of intermolecular force, the **hydrogen bond**.

Hydrogen Bonding

Molecules in which a hydrogen atom is bonded to a highly electronegative atom such as nitrogen, oxygen, or fluorine exhibit **hydrogen bonding**. This arrangement of atoms produces a very polar bond, often resulting in a polar molecule with strong intermolecular attractive forces.

Example 5
Illustrate the hydrogen bonding found between a water molecule and its closest three neighboring water molecules.

Work
1. The polar nature of water is caused by the large electronegativity difference between oxygen and hydrogen atoms.
2. One water molecule has the following polar structure:

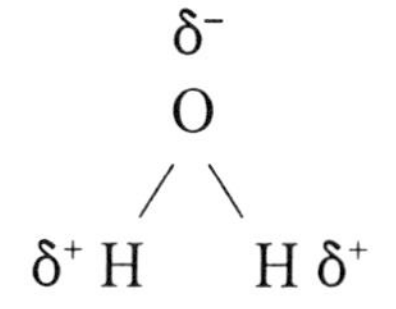

   Notice that the oxygen end is slightly negative, and the two hydrogen ends are slightly positive.
3. The hydrogen bonding in liquid water is due to the attraction of the slightly negative oxygen atom to a slightly positive hydrogen in another water molecule. In similar fashion each slightly positive hydrogen atom attracts an oxygen atom in two other water molecules.
4. Thus, one water molecule has three hydrogen bonds with three other water molecules.

Answer

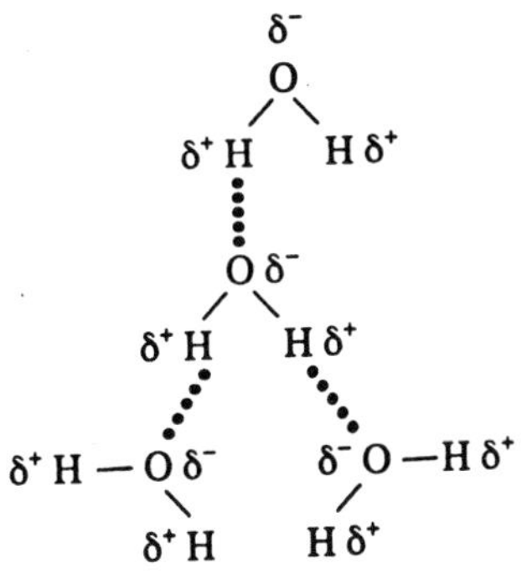

Example 6
Show the hydrogen bonding that will occur between one molecule of methyl alcohol (CH_3OH) and water molecules. The methyl alcohol has the following polar structure:

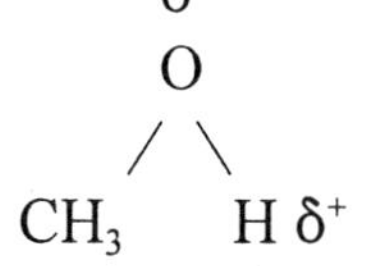

Answer

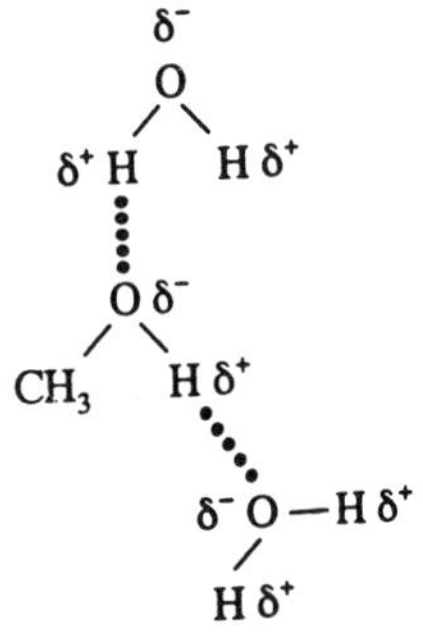

Example 7

Show the hydrogen bonding that will occur between water and the amide structure that is found in proteins;

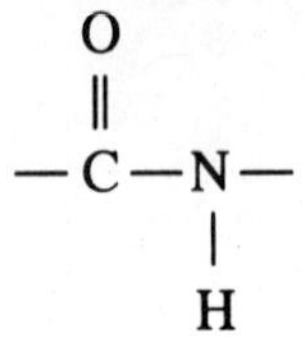

$$
\begin{array}{c}
\text{O} \\
\parallel \\
-\text{C}-\text{N}- \\
| \\
\text{H}
\end{array}
$$

Work

The amide structure has the following polar arrangement:

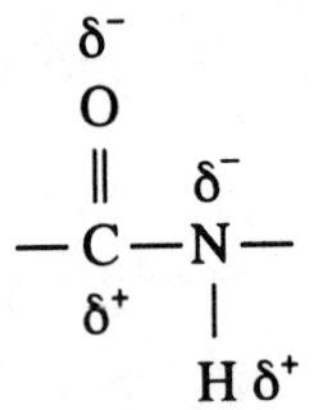

$$
\begin{array}{c}
\delta^- \\
\text{O} \\
\parallel \quad \delta^- \\
-\text{C}-\text{N}- \\
\delta^+ \quad | \\
\text{H } \delta^+
\end{array}
$$

Answer

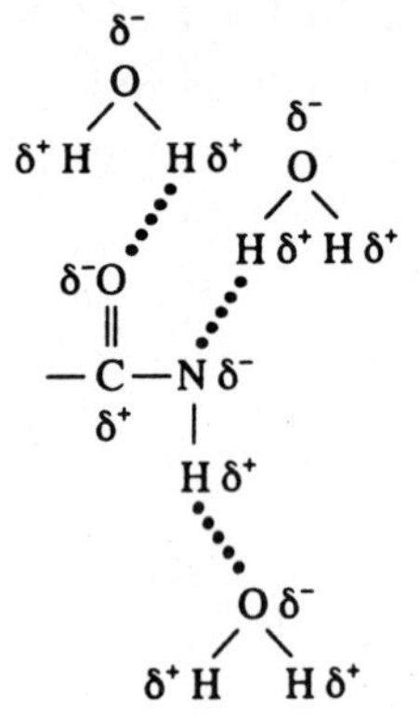

6.3 The Solid State

The close packing of the particles of a solid results from attractive forces that are strong enough to restrict motion. The particles are "locked" together in a defined and highly organized fashion. This results in a fixed shape and volume (recall that gases have no fixed shape or volume, and liquids have a fixed volume but no fixed shape).

Properties of Solids

Solids are incompressible, due to the small distance between particles. Solids may be **crystalline**, having a regular repeating structure, or **amorphous**, having no organized structure.

Types of Crystalline Solids

Crystalline solids may exist in one of four general groups:

1. *Ionic solids*. The units that make up an **ionic solid** are positive and negative ions. Electrostatic forces hold the crystal together.
2. *Covalent solids*. The units that make up a **covalent solid** are atoms held together by covalent bonds.
3. *Molecular solids*. The units composing **molecular solids** are molecules held together by intermolecular forces (dipole-dipole and London).
4. *Metallic solids*. **Metallic solids** are composed of metal atoms held together by metallic bonds. Metallic bonds are formed by the overlap of orbitals of metal atoms, resulting in regions of high electron density surrounding the positive metal centers. Electrons in these regions are extremely mobile, resulting in the high conductivity (ability to carry electrical current) exhibited by many metallic solids.

Key Terms

amorphous solid
Avogadro's Law
barometer
Boyle's Law
Charles's Law
combined gas law
condensation
covalent solid
crystalline solid
Dalton's Law
evaporation
exothermic reaction
hydrogen bonding
ideal gas
ideal gas law
ionic solid

kinetic-molecular theory
London forces
melting point
metallic bond
metallic solid
molar volume of a gas
molecular solid
normal boiling point
partial pressure
pressure
standard temperature and pressure
surface tension
surfactant
van der Waals forces
thermodynamics
vapor pressure of a liquid
viscosity

Self Test

1. _____ is the conversion of a liquid to a gas below the boiling point of the liquid.
2. A(n) _____ gas is one in which gas particles do not interact and the volume of the individual gas particles is assumed to be negligible.
3. Convert 29.96 inches of mercury pressure into mm of Hg pressure.
4. A given mass of gas at 20.0°C occupies a volume of 250.0 mL at 1.00 atmosphere of pressure. What pressure must be applied to compress the gas to a volume of 50.0 mL, assuming no temperature change?

5. A balloon filled with air has a volume of 2.00 liters at 24.0°C. What would be the balloon's volume at 37.0°C if the pressure surrounding the balloon remained constant?

6. Calculate the number of moles of helium in a 0.500-liter balloon at 25.0°C and under 0.960 atmospheres of pressure.

7. Show the hydrogen bonds between hydrogen fluoride molecules. The polar structure of HF is $^{\delta+}H - F^{\delta-}$.

8. The total pressure of our atmosphere is primarily equal to the sum of the pressures of what two gases?

9. What law explains why oxygen is distributed to the cells and the waste product, carbon dioxide, is expelled by the lungs?

10. What is defined as a measure of the attractive forces at the surface of a liquid?

11. What type of substances may be added to a liquid to decrease its surface tension?

12. What is defined as the pressure exerted by the vapor of a liquid at equilibrium?

13. What is defined as the temperature at which a solid is converted into a liquid?

14. Ionic solids have what kind of melting points?

15. If the temperature of 10.0 liters of gas is increased from 273 K to 546 K, what will be the new volume of that gas? Assume that all other conditions are held constant.

16. _____ is the conversion of a gas to a liquid.

17. _____ solids have no organized, regular structure.

18. A(n) _____ is a device for measuring gas pressure.

19. A(n) _____ is a device for measuring heat.

20. Ice is a _____ solid.

7 *Reactions and Solutions*

Learning Goals

1. **Conceptual Goals**
   - Know what is meant by the terms solution, solute, and solvent.
   - Describe the various types of solutions, and give examples of each.
   - Know the importance of solution chemistry in chemical processes.
   - Recognize the relationship between solubility and equilibrium.
   - Describe and explain concentration-dependent solution properties: vapor-pressure lowering, freezing-point depression, boiling point elevation, and osmotic pressure.
   - Describe the chemical and physical properties of water that make it a truly unique solvent.
2. **Performance Goals**
   - Classify chemical reactions by type: combination, decomposition, or replacement.
3. **Health Applications**
   - Understand the relationship of pressure and nitrogen solubility in the blood, which results in a condition known as the bends.

Chapter Overview

Chemical reactions may be classified as:
   - combination (joining reactants)
   - decomposition (breaking up reactants)
   - replacement (exchange reactions)

A **solution** is a homogeneous mixture of two or more substances. A solution is composed of one or more **solutes** dissolved in a **solvent**. When the solvent is water we refer to the homogeneous mixture as an **aqueous solution**.

7.1 Classification of Chemical Reactions

Spontaneous chemical reactions occur for a variety of reasons, linked by the tendency to achieve the lowest (most stable) electronic energy state. Strong electrolytes will react to form weak (less dissociated) electrolytes, if possible. Reactions forming gaseous products are favored. Formation of insoluble solid products favors a chemical reaction.

Chemical reactions involve the combination of reactants to produce products, the decomposition of reactant(s) into products, or the replacement of one or more elements in a compound to yield products.

Combination Reactions

Combination reactions involve the joining of two or more atoms or compounds, producing a product of different composition.

Decomposition Reactions

Decomposition reactions produce two or more products from a single reactant.

Replacement Reactions

Replacement reactions are subcategorized as either single-replacement or double-replacement. Single-replacement reactions occur between atoms and compounds; the atom replaces another in the compound, producing a new compound and the replaced atom. Double-replacement reactions, on the other hand, involve two compounds undergoing a "change of partners."

7.2 Types of Chemical Reactions

Precipitation Reactions

Precipitation reactions involve the conversion of soluble reactants into one or more insoluble products. Information contained in solubility tables enables prediction of possible insoluble products.

Reactions with Oxygen

Reactions involving combination of metals or nonmetals with oxygen are generally exothermic; they are often used specifically for their heat-generating properties. Examples include the reaction of oxygen with carbon/hydrogen-containing compounds (natural gas, gasoline, and oil).

Acid-Base and Oxidation-Reduction Reactions

Another approach to the classification of chemical reactions is based upon a consideration of charge transfer. Acid-base reactions involve the transfer of a hydrogen ion, H^+, from one reactant to another. Another important reaction type, oxidation-reduction, takes place because of the transfer of one or more electrons from one reactant to another.

7.3 Properties of Solutions

General Properties of Liquid Solutions

Liquid solutions are clear and transparent. They may be colored or colorless, depending upon the properties of the solute and solvent.

Solutions may be classified as **electrolytic** or **nonelectrolytic**. Electrolytic solutions are formed from ionic compounds that dissociate in solution to produce ions. Electrolytic solutions are good conductors of electricity. Nonelectrolytic solutions are formed from nondissociating molecular solutes (nonelectrolytes), and these solutions are nonconducting.

Solutions and Colloids

A **colloidal suspension** is not a true solution; the colloid particles are not identical in size nor homogeneously distributed throughout the solution. Particles with diameters between one nanometer and two hundred nanometers are colloids. Smaller particles form a true solution; larger particles are precipitates.

A colloidal suspension is characterized by its light-scattering property, the Tyndall effect.

Degree of Solubility

The degree of solubility, how much solute can dissolve in a given volume of solvent, is difficult to predict, but general trends are based upon the following considerations:

1. The magnitude of difference between polarity of solute and solvent (the rule of "like dissolves like").
2. Temperature. An increase in temperature usually, but no always, increases solubility.
3. Pressure. Pressure has little effect on the solubility of solids and liquids in liquids. However, the solubility of gases in solution is extremely pressure dependent.

When a solution contains all the solute that can be dissolved at a particular temperature, it is **saturated**. Cooling a saturated solution often results in the formation of a **precipitate**. Occasionally, however, on cooling, the excess solute may remain in solution for a period of time. Such a solution is described as a **supersaturated solution**.

Solubility and Equilibrium

When an excess of solute is brought into contact with a solvent, the resulting dissolution establishes a dynamic equilibrium.

Solubility of Gases: Henry's Law

Henry's Law states that the number of moles of a gas dissolved in a liquid at a given temperature is proportional to the partial pressure of the gas. In other words, the gas solubility is directly proportional to the pressure of that gas in the atmosphere that is in contact with the liquid.

7.4 Concentration of Solutions: Percentage

The amount of solute dissolved in a given amount of solution is defined as the solution concentration. The concentration of a solution has a profound effect on the properties of a solution, both physical (melting and boiling points) and chemical (solution reactivity). The more widely used concentration units are considered below.

Weight/Volume Percent

$$\% \ (W/V) = \frac{\text{grams of solute}}{\text{mL of solution}} \times 10^2$$

Example 1

Calculate the %(W/V) of a sodium chloride solution that was made by mixing 15.0 grams of NaCl with enough distilled water to prepare 500.0 mL of solution.

Work

1. Use the equation for $\% \ (W/V)$:

2. The solute is NaCl. $\qquad \% \ (W/V) = \frac{\text{grams of solute}}{\text{mL of solution}} \times 10^2$

3. $\% \ (W/V) = \frac{15.0 \text{ grams}}{500.0 \text{ mL}} \times 10^2$

   $\% \ (W/V) = 3.00\%$

Answer

The solution is a 3.00% (W / V) sodium chloride solution.

Example 2

How many grams of glucose are found in 1.00 liter of 5.00% (W/V) glucose solution?

Work

1. The 5.00% (W/V) glucose solution tells us that there are 5.00 grams of glucose per 100.00 mL of total solution.
2. Use conversion factors to find the number of grams of glucose that are present.

3. $1.00 \text{ liter} \times \dfrac{1000 \text{ mL}}{1 \text{ L}} \times \dfrac{5.00 \text{ grams of the glucose}}{100.00 \text{ mL of solution}} = 50.0 \text{ mL}$

Answer

In 1 liter of 5.00% (W/V) glucose solution there are 50.0 grams of glucose.

Example 3

If a patient is to receive 2.00 grams of a specific drug, and the drug is supplied as a 5.00% (W/V) drug solution, how many milliliters of the drug solution must the patient be given?

Work

1. The 5.00% (W/V) drug solution means that there are 5.00 grams of the drug per 100.00 mL of the drug solution.
2. Use conversion factors to find the number of mL of drug solution needed.

3. $2.00 \text{ grams} \times \dfrac{100.00 \text{ mL of drug solution}}{5.00 \text{ grams of the drug}} = 40.0 \text{ mL}$

Answer

The patient must take 40.0 mL of the 5.00% (W / V) drug solution to receive 2.00 grams of the specific drug.

Volume/Volume Percent

$$\% \ (W/V) = \frac{\text{mL solute}}{\text{mL solution}} \times 10^2$$

Example 4

Calculate the % (V/V) of ethyl alcohol that is found in an alcoholic drink prepared by adding 30.0 mL of ethyl alcohol to enough water to prepare 150.0 mL of total solution.

Work

1. Use the equation for % (V/V):

$$\% \ (V/V) = \frac{\text{volume of solute in mL}}{\text{total volume of the solution in mL}} \times 10^2$$

2. $\% \ (V/V) = \dfrac{30.0 \text{ mL}}{150.0 \text{ mL}} \times 10^2$

$\% \ (V/V) = 20.0\%$

Answer

The solution is a 20.0% (V/V) ethyl alcohol solution.

Weight/Weight Percent

$$\%(W/W) = \frac{\text{grams solute}}{\text{grams solution}} \times 10^2$$

Example 5

Calculate the % (W/W) of gold in a wedding band that has a mass of 12.0g and was found to contain 8.97 g of gold.

71

Work

1. Use the equation for % (W / W):

$$\%(W/W) = \frac{\text{mass of solute in grams}}{\text{mass of solution in grams}} \times 10^2$$

2. $\%(W/W) = \frac{8.97 \text{ g}}{12.0 \text{ g}} \times 10^2$

   $\% (W / W) = 74.8\%$

Answer

The wedding band is 74.8% *(W / W)* gold.

7.5 Concentration of Solutions: Moles and Equivalents

Molarity

Molarity, symbolized M, is defined as the number of moles of solute per liter of solution:

$$M = \frac{\text{moles solute}}{\text{L solution}}$$

Example 6

Calculate the molarity of a sodium chloride solution made by mixing 3.51 grams of solid NaCl in enough water to prepare 500.0 mL of total solution.

Work

1. Use the molarity equation: $\quad \text{Molarity} = \frac{\text{moles solute}}{\text{L solution}}$

2. 500.0 mL of solution is the same as 0.5000 liter
   ***When using molarity, convert all volumes to liters!**

3. The number of moles of solute (NaCl) in the solution is found by first determining the weight of 1 mole of NaCl.

   $$1 \text{ mole of NaCl} = 58.5 \text{ grams}$$

4. Since we have only 3.51 grams of NaCl, we use a conversion factor to find the number of moles of NaCl that is really in our solution:

$$(3.51 \text{ grams of NaCl})\left(\frac{1 \text{ mole of NaCl}}{58.5 \text{ grams of NaCl}} \right) = 0.0600 \text{ moles NaCl}$$

5. We can now use the molarity equation, since we know all the needed values:
The number of moles of NaCl in the solution = 0.0600 moles
The number of liters of solution = 0.5000 liter

6. $\text{Molarity} = \dfrac{\text{moles pf solute}}{\text{liters of solution}}$

$\text{Molarity} = \dfrac{0.0600 \text{ moles}}{0.5000 \text{ liter}}$

Molarity of our solution = 0.120 M

Answer
We have prepared a 0.120 molar NaCl solution.

Example 7
Calculate the number of grams of solid silver nitrate needed to prepare 250.0 mL of a 0.100 molar $AgNO_3$ solution.

Work
1. One mole of $AgNO_3$ is found to be 169.9 grams/mole (from the periodic table).
2. Convert 250.0 mL into liter units:

$250.0 \text{ mL} \times \dfrac{1 \text{ liter}}{10^3 \text{ mL}} = 0.250 \text{ liter solution}$

3. 0.100 molar $AgNO_3$ solution means that there is 0.100 mole of $AgNO_3$ per 1 liter of solution.
4. Using suitable conversion factors:

$0.250 \text{ L} \times \dfrac{0.100 \text{ mole of } AgNO_3}{1 \text{ liter of solution}} \times \dfrac{169.9 \text{ grams of } AgNO_3}{1 \text{ mole of } AgNO_3} = 4.25 \text{ grams of } AgNO_3$

Answer
We need to add 4.25 grams of solid $AgNO_3$ to prepare 250.0 mL of a 0.100 molar $AgNO_3$ solution.

The technique of dilution is often used to prepare less-concentrated solutions. The approach to such a calculation is outlined below:

$$M_1 = \text{molarity prior to dilution.}$$
$$M_2 = \text{molarity after dilution.}$$
$$V_1 = \text{volume prior to dilution.}$$
$$V_2 = \text{volume after dilution.}$$
$$(M_1)(V_1) = (M_2)(V_2)$$

Knowing any three of these terms enables one to calculate the fourth.

The dilution equation is valid with any concentration units, such as % (W / V) or % (V / V), as well as molarity. Be certain to use the same units for both initial and final concentration values. Only in this way will proper unit cancellation occur.

Example 8

How many mL of 12.0 M HCl solution do we need to add to distilled water so we can prepare 500.0 mL of 2.50 M HCl solution?

Work

1. Use $(M_1)(V_1) = M_2)(V_2)$

    V_1 means the starting volume: we do not know this value yet.

    M_1 is the molarity of the solution we have available to us. We will dilute some of it to make our final solution.

    M_2 means the molarity of our final solution. It always will be the lower value.

    V_2 means the final volume of our diluted solution.

2. From our problem:

    V_1 = ? (not known yet)

    $M_1 = 12.0$ M HCl solution (the one we start with)

    $M_2 = 2.50$ M HCl solution (the one we want to make)

    V_2 = 500.0 mL (the volume we want to prepare)

3. Rearrange the equation to find V_1:

    $$(M_1)(V_1) = (M_2)(V_2)$$

    $$V_1 = \frac{(M_2)(V_2)}{M_1}$$

4. $V_1 \; + \; \dfrac{(2.50 \text{ M})(500.0 \text{ mL})}{(12.0 \text{ M})}$

    $$V_1 = 104 \text{ mL}$$

Answer

If we measure out 104 mL of the 12.0 M HCl solution and then add distilled water to it in another container until we have made 500.0 mL of total solution, we will have prepared 500.0 mL of a 2.50 M HCl solution by dilution.

Representation of Concentration of Ions in Solution

The concentration of ions in solution may be represented as moles per liter (molarity) and equivalents per liter.

Molarity emphasizes the number of individual ions. A one molar solution of Na^+ contains Avogadro's number, 6.02×10^{23} Na^+. In contrast, equivalents per liter emphasizes charge; one equivalent of Na^+ contains Avogadro's number of positive charge.

We defined 1 mole as the number of grams of an atom, molecule, or ion corresponding to Avogadro's number of particles. One equivalent of an ion is the number of grams of the ion corresponding to Avogadro's number of electrical charges.

Changing from moles/L to equivalents/L (or the reverse) can be accomplished using conversion factors.

Example 9

Calculate the number of equivalents per liter of Mg^{2+} in a solution that is 1.0×10^{-4} M $MgCl_2$.

Work

A series of conversion factors are required to

1. Convert M $MgCl_2$ to M Mg^{2+}
2. Convert moles of Mg^{2+}/L to moles of charge/L
3. Convert moles of charge/L to equivalents of Mg^{2+}/L
   The setup is as follows:

$$\frac{1.0 \times 10^{-4} \text{ mol } MgCl_2}{1 \text{ L}} \times \frac{1 \text{ mol } Mg^{2+}}{1 \text{ mol } MgCl_2} \times \frac{2 \text{ mol charge}}{1 \text{ mol } Mg^{2+}} \times \frac{1 \text{ eq } Mg^{2+}}{1 \text{ mol charge}} = \frac{2.0 \times 10^{-4} \text{ eq } Mg^{2+}}{\text{liter}}$$

Answer

A 1.0×10^{-4} M $MgCl_2$ solution contains 2.0×10^{-4} eq/L of Mg^{2+}.

7.4 Concentration-Dependent Solution Properties

Solution properties that are dependent upon the concentration of the solute particles, rather than the identity of the solute, are referred to as **colligative properties**. There are four colligative properties of solutions.

Vapor Pressure Lowering

Raoult's law states that when a solute is added to a solvent, the vapor pressure of the solvent decreases in proportion to the concentration of solute particles.

Freezing-Point Depression and Boiling-Point Elevation

When a nonvolatile solid is added to a solvent, the freezing point of the resulting solution decreases (a lower temperature is required to form the solid state), and the boiling point increases (requiring a higher temperature to form the gaseous state).

The magnitude of the freezing-point depression (ΔT_f) is proportional to the solute concentration over a limited range of concentration.

$$\Delta T_f = K_f \text{ (solute concentration)}$$

The boiling point elevation is also proportional to the solute concentration.

$$\Delta T_b = K_b \text{ (solute concentration)}$$

If the value of the proportionality factor is known, the magnitude of the freezing-point depression or boiling-point elevation may be calculated for a solution of known concentration of particles.

We have already worked with one mole-based unit, molarity, and this concentration unit can be used to calculate either the freezing-point depression or boiling-point elevation.

A second mole-based concentration unit is molality, which is more commonly used in these types of situations. **Molality** (symbolized by m) is defined as the number of moles of solute per kilogram of solvent in a solution:

$$m = \frac{\text{moles solute}}{\text{kg solvent}}$$

Molality does not vary with temperature, whereas molarity is temperature-dependent.

Osmotic Pressure

Osmosis is the movement of solvent from a dilute solution to a more concentrated solution through a **semipermeable membrane**. Pressure must be applied to the more concentrated solution to stop this flow, and the magnitude of the pressure required to just stop the flow is termed the **osmotic pressure**. The "driving force" for the osmotic process is the need to establish an equilibrium between the solutions on either side of the membrane.

The osmotic pressure, not unlike the pressure exerted by a gas, may be treated quantitatively. Osmotic pressure, symbolized by p, follows the same form as the ideal gas equation:

	Ideal Gas			**Osmotic Pressure**

$$\text{PV} = n\text{RT} \qquad\qquad \pi\text{V} = n\text{RT}$$

$$\text{or} \quad \text{P} = n\frac{\text{RT}}{\text{V}} \qquad\qquad \text{or} \quad \pi = \frac{n\text{RT}}{\text{V}}$$

$$\text{and, since} \quad \text{M} = \frac{n}{\text{V}} \qquad \text{and, since} \quad \text{M} = \frac{n}{\text{V}}$$

$$\text{then,} \quad \text{P} = \text{MRT} \qquad \text{then,} \quad \pi = \text{MRT}$$

The osmotic pressure may be calculated from the solution concentration at any given temperature. Osmosis is a colligative property; it is dependent on the concentration of solute particles.

By convention, the molarity of particles in solution is termed osmolarity, abbreviated Osm, for osmotic pressure calculations.

Osmosis is important as a transport mechanism to all living organisms. A living cell contains an aqueous solution and material movement in and out of cells is based partly upon osmosis. If the concentration of the fluid surrounding red blood cells is higher than that inside the cell (a **hypertonic solution**), water flows from the cell, causing it to collapse. This process is known as **crenation**. On the other hand, too low a concentration of this fluid relative to the solution within the blood cell (a **hypotonic solution**) will cause water to flow into the cell, causing cell rupture, a process known as **hemolysis**.

Two solutions are **isotonic** if they have identical osmotic pressures. In that case, the osmotic pressure differential across the cell is zero, and no cell disruption occurs.

7.7 Water as a Solvent

The role of water in the solution process deserves special attention because of its many unique characteristics:

1. It is often referred to as the universal solvent.
2. In view of its small size, it is remarkable that it exists as a liquid at room temperature.
3. Contrary to our expectations, liquid water is more dense than its solid form, ice.
4. Water is the principal biological solvent. Approximately 60% of the adult human body is water.

7.8 Electrolyte in Body Fluids

Proper cell function in biological systems is critically dependent on the concentration of electrolytes. The maintenance of proper muscle and nervous system function depends on the sodium/potassium ratio inside and outside of the cell. A stable osmotic pressure in biological fluids depends on the electrolyte concentration inside and outside of the cell.

Key Terms

acid-base reaction
aqueous solution
colligative property
colloidal suspension
combination reaction
concentration
crenation
decomposition reaction
double-replacement reaction
electrolyte
equivalent
hemolysis
Henry's Law
hydrogen bonding
hypertonic solution
hypotonic solution
isotonic solution
molality
molarity

nonelectrolyte
osmolarity
osmosis
osmotic pressure
oxidation-reduction reaction
precipitate
Raoult's Law
saturated solution
semipermeable membrane
single-replacement reaction
solubility
solute
solution
solvent
supersaturated solution
volume/volume percent
weight/volume percent
weight/weight percent

Self Test

1. What is the % *(W / V)* of NaOH in 150.0 mL of a solution that contains 1.65 grams of sodium hydroxide?
2. How many milliliters of a 5.000% *(W / V)* glucose solution would contain 40.00 grams of glucose?
3. How many liters of D-5-W solution [5.00% *(W / V)* glucose solution] would need to be given to a patient by I.V. so that she would receive 250.0 food-calories of energy from the given glucose? 1.00 gram of glucose produces 4.00 food-calories of energy in the human body.
4. How many moles of KCl are present in 100.0 mL of a 0.552 molar potassium chloride solution?
5. How many grams of KCl are present in 50.0 mL of a 0.125 molar potassium chloride solution?
    KCl = 74.60 grams/mole.
6. If 20.80 grams of HCl are added to enough distilled water to form 3.0 liters of solution, what is the molarity of the solution? HCl = 36.46 grams/mole.
7. Calculate the molarity of a solution that contains 5.60 grams of KNO_3 in 300.0 mL of solution.
    KNO_3 = 101.1 grams/mole.
8. How many mL of 12.0 M HCl solution are required to prepare 250.0 mL of a 2.50 M HCl

solution?

9. What term defines the amount of solute in grams or moles dissolved in a given amount of solution?

10. Normal physiological saline solution is a 0.900% (W / V) NaCl solution. How many grams of solid NaCl are needed to prepare 1.50 liters of this solution?

11. What term is defined as the number of moles of solute per liter of solution?

12. What term defines the movement of solvent through a semipermeable membrane from a dilute solution to a more concentrated solution?

13. Is a concentrated solution hypertonic or hypotonic when compared to a dilute solution?

14. What is the term used to describe solutions of exactly the same particle concentration?

15. _____ is the shrinkage of red blood cells due to water loss to the surrounding medium.

16. An(n) _____ is a substance that ionizes in water.

17. A(n) _____ is the more concentrated solution of two separated by a semipermeable membrane.

18. A(n) _____ is the more dilute solution of two separated by a semipermeable membrane.

19. _____ is the number of moles of solute per liter of solution.

20. _____ is the net flow of solvent through a semipermeable membrane.

21. _____ is the pressure required to stop net transfer of solvent across a semipermeable membrane.

22. A(n) _____ is a solution in which undissolved solute is in equilibrium with the solution.

23. The _____ is a component of a solution that is present in lesser quantity than the solvent.

24. A(n) _____ is a solution that is more concentrated than a saturated solution.

8   *Chemical and Physical Change: Energy, Rate, and Equilibrium*

Learning Goals

1. **Conceptual Goals**
   - Recognize the role that energetics (or thermodynamics) plays in chemical and physical change.
   - Know what is meant by the rate of a reaction and the importance of the role of kinetics in chemical and physical change.
   - Know what is meant by the terms activation energy and the activated complex.
   - Develop a "feel" for the way reactant structure, concentration, and temperature affect the rate of a chemical reaction.
   - Know the role of a catalyst in a chemical reaction.
   - Understand what is meant by the term equilibrium.

2. **Performance Goals**
   - State the first and second laws of thermodynamics, and know their implications.
   - Describe experiments that yield thermochemical information.
   - Calculate fuel values based on experimental data.
   - Describe why the physical state of reactants has such a significant effect on reaction rates.
   - State LeChatelier's Principle, and relate this principle to the concept of reversibility.

3. **Health Applications**
   - Understand the chemical processes involved in the manufacture and use of hot and cold packs.

Chapter Overview

Three concepts play an important role in determining the extent and speed of a chemical reaction:

1. Thermodynamics deals with energy changes associated with chemical reactions.
2. Kinetics concerns itself with the rate or speed of a chemical reaction.
3. Equilibrium considerations determine the completeness of a reaction.

    A reaction may be thermodynamically favored but very slow; conversely, a fast reaction may result in very little product being formed. Thus, each is considered independently.

    A system is at equilibrium in a reaction that does not go to completion when the rates of the forward and reverse reactions are equal.

    The concepts of energy, rate, and equilibrium can be applied to changes in state as well as to chemical reactions. The behavior of the states of matter (gas, liquid, and solid), their properties, and their interconversion are considered in this chapter.

8.1 Thermodynamics

Thermodynamics is the study of energy, work, and heat. Thermodynamics may be applied to chemical change or physical change. There exist three basic laws of thermodynamics; only the first two are of concern in Chapter 5.

The Chemical Reaction and Energy

Every chemical reaction involves a change in energy. The first law of thermodynamics, also known as the law of conservation of energy, states that energy cannot be created or destroyed in the course of the reaction. It may only be converted from one form to another or transferred from one component of the system to another.

Our current view of events at the molecular level are consistent with the first law. When substances react:

- molecules and atoms in a reaction mixture are in constant, random motion;
- these molecules and atoms frequently collide with each other;
- only some collisions, those with sufficient energy, will break bonds in molecules; and
- when reactant bonds are broken, new bonds may be formed and products result.

Exothermic and Endothermic Reactions

A reaction that absorbs energy (exhibits a net gain in energy) is termed **endothermic**. The products of the reaction possess more energy than the reactants. On the other hand, a reaction that releases energy (exhibits a net loss of energy) is termed **exothermic**. The products of the reaction possess less energy than the reactants.

Figure 8.1 in the textbook graphically represents energy changes in endothermic and exothermic reactions.

The energy released in exothermic reactions is often in the form of **heat energy**. It may also take the form of light or even electricity. A battery is a common example of a chemical reaction producing electrical energy. The energy absorbed during chemical change is stored as **chemical energy.**

Enthalpy

Enthalpy is the term used to represent heat energy and is symbolized by H^o. In chemical change we are primarily interested in the heat gained or released, and that is the enthalpy change, ΔH^o. For endothermic reactions the enthalpy change is positive; $\Delta H^o = +$. For exothermic reactions the enthalpy change is negative; $\Delta H^o = -$. The enthalpy change is most often reported in units of joules (or kilojoules) or calories (or kilocalories).

Spontaneous and Nonspontaneous Reactions

Spontaneous reactions are just that: they occur without any external energy input.

Nonspontaneous reactions must be persuated; they need an input of energy.

It seems that all exothermic reactions should be spontaneous. After all, an external supply of energy does not appear to be necessary; in fact, energy is a product of the reaction. It also seems that all endothermic reactions should be nonspontaneous: energy is a reactant that we must provide. However, these hypotheses are not supported by experimentation.

Experimental measurement has shown that most, but not all, exothermic reactions are spontaneous; likewise, most, but not all, endothermic reactions are not spontaneous. There must be some factor in addition to enthalpy that will help us to explain the less obvious cases of nonspontaneous exothermic reactions and spontaneous endothermic reactions. This other factor is entropy.

Entropy

The second law of thermodynamics states that a system and its surroundings spontaneously tend toward increasing disorder (randomness). A measure of the randomness of a chemical system is referred to as **entropy**. A random, or disordered system is characterized by high entropy; a well-ordered system is said to have low entropy.

Often, reactions that are exothermic and whose products are more disordered (higher in entropy) occur spontaneously, while endothermic reactions that produce products of lower entropy are not spontaneous. If they occur at all, they will require external energy input.

Disorder or randomness increases as we proceed from the solid to liquid to the vapor state. Solids often have an ordered crystalline structure, and liquids have a loose structure, while gas particles are virtually random in their distribution.

Free Energy

In chemical change, the maximum amount of energy that can be converted to a useful form is called **free energy** (G). Free energy incorporates the two factors discussed above, the energy factor and the entropy factor, into a single expression for predicting the spontaneity of chemical change:

$$\Delta G = \Delta H - T\Delta S$$

In this expression ΔG is the difference in free energy between products and reactants, ΔH is the difference in heat energy (enthalpy) between products and reactants, and ΔS is the difference in entropy between products and reactants. The temperature of the reaction is symbolized by T.

We find that a spontaneous reaction will have a negative ΔG. A positive ΔG is indicative of a nonspontaneous reaction. In fact, a change for which ΔG is positive will occur only if energy is added to the system.

8.2 Experimental Determination of Energy Change in Reactions

The measurement of the energy demand or energy release in a chemical reaction is **calorimetry**. This technique involves the measurement of the change in the temperature of a quantity of water that is in contact with the reaction of interest and also isolated from its surroundings. A device used in these types of measurements is a calorimeter, which measures heat changes in calories (or energy, since the calorie is a unit of heat energy).

For an exothermic reaction, heat is released by the reaction to the surrounding water, which is isolated from its surroundings. The **specific heat** (SH_w) is defined as the number of calories needed to raise the temperature of 1 gram of water 1 degree Celsius. This, coupled with the total number of grams of water (m_w) and the temperature increase (ΔT) allows one to calculate the heat released, Q.

$$Q = m_w \times \Delta T \times SH_w$$

Many chemical reactions that produce heat are combustion reactions. In our bodies many food substances, principally carbohydrates and fats, are oxidized to produce energy. In these cases, the amount of energy per gram of food is referred to as it fuel value. A special type of calorimeter, a bomb calorimeter, is useful for the measurement of the **fuel value** (calories) of foods.

Example 1
When equal moles of hydrochloric acid (HCl) and sodium hydroxide (NaOH) are mixed in a calorimeter, the temperature of 50.00 grams of water surrounding the reaction increased from 25.00°C to 38.00°C. If the specific heat of water is 1.000 calories/(gram) (°C), calculate the quantity of energy in calories involved in this reaction.

Work
1. The change in temperature is $\Delta T = 38.00°C - 25.000°C = 13.00°C$
2. Since heat energy is released from the reaction to the water, the reaction is exothermic.
3. The quantity of heat absorbed or released by a reaction (Q) is the product of the mass of water in the calorimeter (m_w), the specific heat of water (SH_w), and the change in temperature (ΔT) of the water as the reaction proceeds from the initial to final states. The above relationship is given by the following equation:
   $Q = (m_w)(SH_w) (\Delta T)$

4. $Q = 50.00 \text{ g} \times 13.00°C \times \dfrac{1.000 \text{ calories}}{(1 \text{ gram} (1°C))} = 650.0 \text{ calories}$

5. 650.0 calories are released during the reaction.

Example 2
When 1.000 gram of glucose ($C_6H_{12}O_6$) was burned in a bomb calorimeter, the temperature of 100.0 grams of water in the calorimeter increased from 25.00°C to 63.00°C. Calculate the fuel value of one gram of glucose. Remember that the SH_w for water is 1.000 calorie per gram of water per 1°C temperature change.

Work
1. Use the equation $Q = (m_w)(SH_w)(\Delta T)$
2. $\Delta T = 63.00°C - 25.00°C = 38.00°C$

3. $Q = 100.0 \text{ g} \times 38.00°C \times \dfrac{1.000 \text{ calories}}{(1 \text{ gram } (1°C))}$

$Q = 3.800 \times 10^3$ calories

4. In food-calorie books we see the value of 1 gram of carbohydrate (glucose is a pure carbohydrate) to be 4 calories. These calorie values are really kilocalories. Thus, our value of 3800 calories, or 3.8 kilocalories, is close to those given in food-calorie handbooks.

Answer

One gram of glucose (or any carbohydrate) has a fuel value of 3.8×10^3 calories, or 3.8 kilocalories, or just 3.8 nutritional calories.

8.3 Kinetics

Chemical kinetics is the study of the rate (speed) of a chemical reaction.

The Chemical Reaction

If the energy made available by the collision of reacting molecules exceeds the bond energies, the bonds will break, and the resulting atoms will recombine in a lower energy configuration. A collision meeting the above conditions and producing one or more product molecules is referred to as an effective collision. Only effective collisions lead to chemical reaction.

Activation Energy and the Activated Complex

The minimum amount of energy required to produce a chemical reaction is the **activation energy** for the reaction. As implied above, a large component of the activation energy is the bond energy of the reacting molecules.

The chemical reaction may be represented in terms of the changes in potential energy that occur as a function of the time of the reaction. Several important characteristics of this relationship follow:
1. The reaction proceeds from reactants to products through an extremely unstable intermediate state that we term the **activated complex**.
2. Formation of the activated complex requires energy. The difference between the energy of reactants and activated complex is the activation energy.
3. For an exothermic reaction, the overall energy change must be a net release of energy. The net release of energy is the difference in energy between products and reactants.

Experimental Factors Affecting Reaction *Rate*

Five major experimental conditions influence the rate of a chemical reaction.
1. Structure of the reacting species
2. The concentration of reactants

3. The temperature of reactants
4. The physical state of reactants
5. The presence of a catalyst

A catalyst is a substance that increases the rate of a reaction. If added to a reaction mixture, the catalytic substance undergoes no net change, nor does it alter the outcome of the reaction.

Mathematical Representation of Reaction Rate

For a generalized reaction:

$$A + B \rightarrow C + D$$

The rate of the reaction is represented by:

$$rate = k [A]^n [B]^{n^1}$$

In this expression, the brackets, [ ], represent molar concentration and

$$[A] = molar\ concentration\ of\ species\ A$$
$$[B] = molar\ concentration\ of\ species\ B$$

The exponents, n and n^1, are experimentally determined. They cannot be deduced from the chemical equation.

The symbol k represents a constant that is unique to the reaction under study. It is termed the **rate constant**.

Knowing the form of the rate expression enables chemists to design chemical reactions that will produce product in the shortest time. In chemical industry, where the quantity of product produced and sold determines the amount of profit, a thorough understanding of the kinetics of reactions is of paramount importance.

8.4 Equilibrium

Many chemical reactions do not proceed to "completion." After a period of time, determined by the kinetics of the reaction, the concentration of reactants no longer decreases and the concentration of products no longer increases. At this point, a mixture of products and reactants exists, and its composition would remain constant unless the experimental conditions were changed. This mixture is in a state of chemical equilibrium.

Rate and Reversibility of Reactions

Many reactions may proceed in either direction, left to right or right to left. The concentration of the various species is fixed at equilibrium because product is being consumed and formed at the same rate. In other words, the reaction continues indefinitely (dynamic) but the concentrations of products and reactants is fixed (equilibrium). This is a dynamic equilibrium.

Physical Equilibrium

The most common examples of physical equilibrium involve interconversion between the various states of matter. In the environment there may be an equilibrium between the water in a lake and water in the surrounding atmosphere (as humidity).

$$H_2O(l) \rightleftharpoons H_2O(g)$$

When the temperature is 32°F (0°C) we may observe both ice and liquid water on the road (a very dangerous situation!).

$$H_2O(s) \rightleftharpoons H_2O(l)$$

Each situation is represented by a unique equilibrium constant expression.

$$K_{eq} = \frac{[H_2O_g]}{[H_2O_l]}$$

and

$$K_{eq} = \frac{[H_2O_l]}{[H_2O_s]}$$

Chemical Equilibrium

Chemical change, as well as physical change, can attain equilibrium. For example:

$$H_2(g) + I_2(g) \rightleftharpoons 2HI(g)$$

At equilibrium, the rate of *disappearance* of H_2 and I_2 is equal to the rate of *formation* of H_2 and I_2. The rates of the forward and reverse reactions are equal and the concentrations of H_2, I_2 and HI are fixed. The process is dynamic, with a continuous interconversion of products and reactants. We represent this process as:

$$K_{eq} = \frac{[H_2][I_2]}{[HI]^2}$$

The Generalized Equilibrium Constant Expression

The most precise description of an equilibrium process is achieved through the use of the equilibrium constant, K_{eq}. We write the general form of an equilibrium chemical reaction as

$$aA + bB \rightleftharpoons cC + dD$$

where A and B represent reactants, C and D represent products, and *a, b, c,* and *d,* are the co-efficients of the balanced equation. The equilibrium constant expression for this general case is

$$K_{eq} = \frac{[C]^c [D]^d}{[A]^a [B]^b}$$

Note:

1. Products of the overall equilibrium reaction are in the numerator, and reactants are in the denominator.
2. Brackets [ ] represent molar concentration, M.
3. The exponents correspond to the coefficients of the balanced equation.

It does not matter what initial amounts (concentrations) of reactions or products we choose. When the system reaches equilibrium, the calculated value of K_{eq} will not change. The magnitude of K_{eq} can only be altered by changing the temperature; thus, K_{eq} is temperature dependent.

Example 3

Write the equilibrium constant (K_{eq}) for each of the following reactions.

1. $N_2 + 3 H_2 \rightleftharpoons 2 NH_3$
2. $N_2 + O_2 \rightleftharpoons 2 NO$
3. $H_2 + Cl_2 \rightleftharpoons 2 HCl$

Work

The equilibrium constant (K_{eq}) for any equilibrium reaction is equal to the concentration of each product raised to a power equal to its coefficient in the balanced equation divided by all the reactants raised to their coefficients' powers.

Answer

1. $K_{eq} = \dfrac{[NH_3]^2}{[N_2][H_2]^3}$

2. $K_{eq} = \dfrac{[NO]^2}{[N_2][O_2]^3}$

3. $K_{eq} = \dfrac{[HCl]^2}{[H_2][Cl_2]}$

LeChatelier's Principle

LeChatelier's Principle states that if a stress is placed on an equilibrium system, the system will respond by altering the equilibrium in such a way as to minimize the stress.

Product introduced: equilibrium
<------------
shifted

Product introduced: equilibrium
------------>
shifted

Effect of Concentration

Addition of extra product or reactant to a fixed reaction volume is just another way of saying that we have increased the concentration of product or reactant. Removal of material from a fixed volume decreases the concentration. Therefore changing the concentration of one or more components of a reaction mixture is a way to alter the equilibrium composition of an equilibrium mixture.

Effect of Heat

The change in equilibrium composition caused by the addition or removal of heat from an equilibrium mixture can be explained by treating heat as a product or reactant. Adding heat to an exothermic reaction is similar to increasing the amount of product, shifting the equilibrium to the left. Removing heat from an exothermic reaction shifts the equilibrium to the right.

Heat is a reactant in an endothermic reaction; its removal shifts the equilibrium to the left. Adding heat favors product formation.

Effect of Pressure

Only gases are affected significantly by changes in pressure because gases are free to expand and compress in accordance with Boyle's law. However, liquids and solids are not compressible, so their volumes are unaffected by pressure.

Therefore pressure changes will alter equilibrium composition only when they involve a gas or variety of gases as products and/or reactants. If the reaction shifts to conserve volume:

- If the number of moles of gaseous product is greater than the number of moles of gaseous reactant, an increase in pressure shifts the equilibrium to the left.
- If the number of moles of gaseous product is less than the number of moles of gaseous reactant, an increase in pressure shifts the equilibrium to the right.
- If the number of moles of gaseous product and reactant is identical, pressure will have no effect on the equilibrium because there is no volume advantage.

Effect of a Catalyst

A catalyst has no effect on the equilibrium composition. A catalyst increases the rates of both forward and reverse reactions to the same extent. The equilibrium composition and equilibrium concentration do not change when a catalyst is used.

Key Terms

activated complex

activation energy

calorimetry

catalyst

dynamic equilibrium

endothermic reaction

enthalpy

entropy

equilibrium constant

equilibrium reaction

exothermic reaction

free energy

fuel value

kinetics

LeChatelier's Principle

nutritional Calorie

order of the reaction

rate constant

rate of reaction

reversible reaction

specific heat of water

thermodynamics

Self Test

1. When equal moles of HCl and NaOH were mixed in a calorimeter, the temperature of 100.0 grams of water surrounding the reaction increased from 22.5 °C to 36.5 °C. If the specific heat of water is 1.000 calories/(gram)(°C), calculate the quantity of heat energy in calories that was released.

2. Write the equilibrium constant for each of the following:

   a. $2 NO_2 \rightleftharpoons N_2O_4$

   b. $PCl_5 \rightleftharpoons PCl_3 + Cl_2$

   c. $2 NO + 2 H_2 \rightleftharpoons N_2 + 2 H_2O$

3. A random, or disordered, system is characterized by what kind of entropy?

4. Are endothermic reactions producing products of lower entropy usually spontaneous or non-spontaneous?

5. Which of the three states of matter has the highest entropy?

6. Do nonspontaneous reactions have ΔG values which are positive or negative?

7. What device is used to measure heat changes in calories?

8. What is the amount of energy per gram of food substance?

9. What is the name of an extremely unstable intermediate state of a chemical reaction?

10. Many chemical reactions do not proceed to completion. After a period of time, the concentration of reactants no longer decreases, and the concentration of products ceases to increase. At this point, what state describes the chemical reaction?

11 What constant provides the most precise description of an equilibrium process?

12. What French chemist in the nineteenth century discovered that changes in equilibrium were dependent upon stress that might be applied to the system?

13. _____ is the threshold energy that must be overcome to produce a chemical reaction.

14. A(n) _____ speeds up reactions without undergoing change.

15. _____ processes absorb energy in a chemical change.

16. _____ is a tendency toward randomness or disorder.

9 Charge-Transfer Reactions: Acids and Bases and Oxidation-Reduction

Learning Goals

1. **Conceptual Goals**
   - Know how to recognize acids and bases.
   - Describe the role of the solvent in acid-base reactions.
   - Know what is meant by the term pH.
   - Recognize the importance of pH in chemical and biochemical systems.
   - Know the meaning of the term buffer and the application of buffers to chemical and biochemical systems, particularly blood chemistry.
   - Define oxidation and reduction.
2. **Performance Goals**
   - Calculate solution concentration in the commonly used units: weight/volume percent, volume/volume percent, weight/weight percent, and molarity.
   - Calculate pH from concentration data.
   - Calculate hydronium and/or hydroxide ion concentration from pH data.
   - Describe some practical examples of redox processes.
3. **Health Applications**
   - Relate the buffer process to blood pH.

Chapter Overview

A **solution** is a homogeneous (or uniform) mixture of two or more substances. A solution is composed of one or more **solutes**, dissolved in a **solvent**. When the solvent is water, we refer to the homogeneous mixture as an **aqueous solution**.

Two very important classes of reactions involve charge transfer; acid-base reactions (proton transfer) and oxidation-reduction reactions (electron transfer). Each type is described in this chapter and applied in numerous ways throughout the rest of your study of chemistry.

9.1 Acids and Bases

Learning Terminology

Acids have a sour taste, dissolve some metals, and cause vegetable dyes to change color. Release of the hydrogen ion (H) is characteristic of acids. Bases have a bitter taste, slippery feel, and accept hydrogen ions.

Arrhenius Theory of Acids and Bases

An Arrhenius acid dissociates to form hydrogen ions. An Arrhenius base dissociates to form hydroxide ions. The **Arrhenius theory** is less comprehensive than the Brönsted-Lowry theory, discussed below.

Brönsted-Lowry Theory of Acids and Bases

A **Brönsted-Lowry** acid is a proton donor and a Brönsted-Lowry base is a proton acceptor. Examples include:

acid $HCl + H_2O \rightarrow H_3O^+ + Cl^-$

base $NH_3 + H_2O \rightarrow NH_4+ + OH^-$

9.4 Solutions of Acids and Bases

Strength of Acids and Bases

The terms acid or base strength and acid or base concentration are easily confused. Strength is a measure of the degree of dissociation of an acid or base in solution, independent of its concentration. Concentration refers to the amount of acid or base per quantity of solution.

The strength of acids and bases in water is dependent upon the extent to which they react with the solvent, water. Acids and bases are classified as strong when the reaction with water is virtually 100% complete and as weak when the reaction with water is much less than 100% complete.

All strong bases are metal hydroxides. Strong bases completely dissociate, or ionize, in aqueous solution to produce hydroxide ions and metal cations.

Weak acids and weak bases dissolve in water principally in the molecular form. Only a small percentage of the molecules dissociate to form the **hydronium** or hydroxide ion.

The most fundamental chemical difference between strong and weak acids and bases is their equilibrium situation. A strong acid, such as HCl, does not exist in equilibrium with its ions, H_3O^+ and Cl^-. A weak acid, such as acetic acid, establishes a dynamic equilibrium with its ions, H_3O^+ and $C_2H_3O_2^-$. This reaction is represented by the equilibrium constant expression:

$$CH_3COOH + H_2O \rightarrow H_3O^+ + CH_3COO^-$$

where K_a is the equilibrium constant or acid-dissociation constant for acetic acid. The larger the dissociation constant the more dissociated (hence, stronger) the acid will be.

The situation for bases is analogous. The equilibrium expression for the weak base ammonia in water is

$$NH_3 + H_2O \rightarrow NH_4+ + OH^-$$

and the equilibrium constant for this weak base is

Note that in each of the above cases, the solvent, water, is not written as part of the equilibrium constant expression.

The Dissociation of Water

Although pure water is virtually 100% molecular, a small number of water molecules do ionize. This process occurs by the transfer of a proton from one water molecule to another, producing a hydronium and hydroxide ion. This equilibrium is shown below:

$$H_2O + H_2O \rightarrow H_3O^+ + OH^-$$

This process is the **autoionization**, or self-ionization of water. Water is, therefore, a very weak electrolyte and a poor conductor of electricity. Water has both acid and base properties; the dissociation produces both the hydronium and hydroxide ion.

Pure water at room temperature has a hydronium ion concentration of 1.0×10^{-7} M. One hydroxide ion is produced for each hydronium ion; therefore, the hydroxide ion concentration is also 1.0×10^{-7} M.

The product of hydronium and hydroxide ion concentration is referred to as the **ion product for water,**

$$\text{ion product} = [H_3O^+][OH^-] = [1.0 \times 10^{-7}][1.0 \times 10^{-7}] = 1.0 \times 10^{-14} = K_w$$

The ion product is a constant because its value is not dependent upon the nature or concentration of the solute, as long as the temperature does not change. This relationship is the basis for the pH scale.

Example 1
If we have a 0.200 molar solution of sodium hydroxide, what are the hydroxide ion (OH⁻) and hydronium ion (H_3O^+) concentrations in this solution?

Work
1. If we have a 0.200 molar NaOH solution, it means that we have 0.200 mole of Na⁺ ions and 0.200 mole of OH⁻ ions per liter of this solution.
2. Thus, [OH⁻] = 0.200 moles/liter
3. Next use the ion product for water (K_w):
    ion product = $[H_3O^+][OH^-]$
4. Rearrange ion product = $[H_3O^+][OH^-]$ to find $[H_3O^+]$

$$\left[H_3O^+\right] = \frac{\text{ion product}}{\left[OH^-\right]}$$

5. $\left[H_3O^+\right] = \dfrac{\text{ion product}}{\left[OH^-\right]}$

$$\left[H_3O^+\right] = \frac{1.0 \times 10^{-14}}{2.00 \times 10^{-1}}$$

$$[H_3O^+] = 0.50 \times 10^{-13} = 5.0 \times 10^{-14} \text{ M}$$

Answer
In our solution the [OH⁻] = 0.200 moles/L and the $[H_3O^+]$ = 5 × 10⁻¹⁴ moles/L.

The pH Scale

The **pH scale** correlates the hydronium ion concentration with a number, the pH, which serves as a useful indicator of the degree of acidity or basicity of a solution. The pH scale specifies "how acidic" or "how basic" a solution is.

1. Addition of an acid (proton donor) to water increases the $[H_3O^+]$ and decreases the $[OH^-]$.
2. Addition of a base (proton acceptor) to water decreases the $[H_3O^+]$ and increases the $[OH^-]$.
3. $[H_3O^+] = [OH^-]$ when equal amounts of acid and base are present.
4. In all of the above cases, $[H_3O^+][OH^-] = 1.0 \times 10^{-14}$ = ion product for water
   The pH of a solution is defined as the negative logarithm of the molar concentration of the hydronium ion.
   $$pH = -\log [H_3O^+]$$

Example 2
Calculate the pH of a 0.010 molar HCl solution.

Work
1. Since HCl is a strong acid, if 1 mole of HCl dissociates, it produces 1 mole of H_3O^+ ions. Therefore a 1.0×10^{-2} M HCl solution has $[H_3O^+] = 1.0 \times 10^{-2}$ M.
2. Use the pH equation:
   pH = –logarithm of the molar concentration of the hydronium ion
   or
   $pH = -\log [H_3O^+]$([ ] means molar.)
3. $pH = -\log [H_3O^+]$
   $pH = -\log (1.0 \times 10^{-2})$
   $pH = -(\log 1.0 + \log 10^{-2})$ (Note that log 1.0 = zero; log 10^{-2} = –2)
   $pH = -[0 + (-2)]$
   $pH = -(-2)$
   $pH = 2.0$

Answer
The pH of a 0.010 molar HCl solution is pH = 2.0

Example 3
Calculate the pH of a 1.0×10^{-4} molar solution of NaOH.

Work
1. NaOH is a strong base; thus, in our problem $[OH^-] = 1.0 \times 10^{-4}$ molar.
2. Use ion product = $[H_3O^+][OH^-]$ to find $[H_3O^+]$, so we can use the pH equation.
3. $\left[H_3O^+\right] = \dfrac{\text{ion product}}{\left[OH^-\right]}$

   $\left[H_3O^+\right] = \dfrac{1 \times 10^{-14}}{1.0 \times 10^{-4}}$

$[H_3O^+] = 1.0 \times 10^{-10}$

4. Next, use the pH equation:

$$pH = -\log[H_3O^+]$$
$$pH = -\log(1 \times 10^{-10})$$
$$pH = -(\log 1.0 + \log 10^{-10})$$
$$pH = -[0 + (-10)]$$
$$pH = 10.0$$

Answer

The pH of a 1.0×10^{-4} M NaOH solution is pH = 10.0

The Importance of pH and pH Control

Solution pH and pH control play a major role in many facets of our everyday lives:
1. Agriculture
2. Biochemistry
3. Organic chemistry
4. Industry
5. Municipal services
6. Acid rain

Neutralization

The reaction of an acid with a base to produce a salt and water is referred to as **neutralization**. In the strictest sense, neutralization requires equal numbers of moles of H_3O^+ and OH^- to produce a neutral solution (no excess acid or base).

A **neutralization reaction** may be used to determine the concentration of an unknown acid or base solution. The technique of **titration** involves the addition of measured amounts of a standard solution (one whose concentration is known) to neutralize the second, unknown solution. From the volumes of the two solutions and the concentration of the **standard solution**, the concentration of the unknown solution may be determined.

Example 4

A 5.00 g sample of vinegar is titrated with 0.100 M NaOH. If the vinegar requires 40.0 mL of the NaOH for a complete reaction,

    a.   Calculate the M of acetic acid in the vinegar

    b.   Calculate the % (W/V) of acetic acid in the vinegar.

Assume that the density of the vinegar solution is 1.00 g/mL and that the only acidic component of the vinegar solution is acetic acid (CH_3COOH). The reaction is:

$$CH_3COOH(aq) + NaOH(aq)CH_3COONa(aq) + H_2O(l)$$

Work, part a

1. The coefficients of both acetic acid and sodium hydroxide (the reactants) in the equation are 1,

   moles acid = moles base

   at the equivalence point

2. moles acid = $M_{acid} \times V_{liters\ acid}$

   moles base = $M_{base} \times V_{liters\ base}$

3. Therefore,

   $$M_{acid} \times V_{liters\ acid} = M_{base} \times V_{liters\ base}$$

   and,

   $$M_{acid} = M_{base} \times \frac{V_{liters\ base}}{V_{liters\ acid}}$$

4. $M_{base} = 0.100M$

   $$V_{liters\ base} = 40.0\ mL \times \frac{10^{-3}\ L}{1\ mL} = 0.0400\ L$$

   $$V_{liters\ acid} = 5.00\ g \times \boxed{\frac{1\ mL}{1.00\ g}} = 5.00\ mL$$
   <----------- Density conversion

   and

   $$V_{liters\ acid} = 5.00\ mL \times \frac{10^{-3}\ L}{1\ mL} = 0.00500\ L$$

5. Substituting,

   $$M_{acid} = M_{base} \times \frac{V_{liter\ base}}{V_{liter\ acid}}$$

   $$M_{acid} = 0.100\ M \times \frac{0.0400\ L}{0.00500\ L}$$

   $M_{acid} = 0.800\ M$ acetic acid

Work, part b

1. $\%(W/W) = \dfrac{grams\ acetic\ acid}{milliliters\ of\ solution} \times 10^2$

2. grams acetic acid $= \dfrac{0.800 \text{ mL acetic acid}}{1 \text{ L}} \times \dfrac{60.0 \text{ g acetic acid}}{1 \text{ mol acetic acid}} = \dfrac{48 \text{ g acetic acid}}{1 \text{ L}}$

3. Convert L to mL $= \dfrac{48.0 \text{ g acetic acid}}{1 \text{ L} \times \dfrac{1 \text{ mL}}{10^{-3} \text{ L}}} = \dfrac{0.0480 \text{ g}}{1 \text{ mL}}$

4. Substitute into the initial equation for % (W / V)

$$\%(W/W) = \dfrac{0.0480 \text{ g}}{1 \text{ mL}} \times 10^2 = 4.80\%$$

Answer
Part a) The vinegar solution is 0.800 M acetic acid
Part b) The vinegar solution is 4.80(W / V)%

9.3 Acids-Bases Buffers

A **buffer solution** contains components that enable the solution to resist large changes in pH when acids or bases are added.

The Buffer Process

The basis of buffer action is the establishment of an equilibrium between either a weak acid and its salt, or a weak base and its salt.
A buffer solution functions in accordance with **LeChatelier's Principle** (Section 5.5), which states that an equilibrium system, when stressed, will shift its equilibrium to alleviate that stress.

Addition of Base (OH⁻) to a Buffer Solution

For the acetic acid/sodium acetate system:

$$CH_3COOH + H_2O \rightleftharpoons H_3O^+ + CH_3COO^-$$
OH⁻ added, equilibrium shifts to the right
-->

Addition of Acid (H₃O⁺) to a Buffer Solution

For the acetic acid/sodium acetate system:

$$CH_3COOH + H_2O \rightleftharpoons H_3O^+ + CH_3COO^-$$
H₃O⁺ added, equilibrium shifts to the left
<---

Higher-than-normal CO_2 levels shift the above equilibrium to the right increasing $[H_3O^+]$ and lowering the pH. A situation of high blood CO_2 levels and low pH is **acidosis**.
Lower-than-normal CO_2 levels shift the equilibrium to the left, decreasing $[H_3O^+]$ and

making the pH more basic; this condition is termed **alkalosis** (from alkali, implying basic in nature).

9.4 Oxidation - Reduction Processes

Oxidation and Reduction

Oxidation is defined as a loss of electrons. Sodium metal, is, for example, oxidized to a sodium ion, losing one electron:

$$Na \rightarrow Na^+ + e^-$$

Reduction is defined as a gain of electrons. A chlorine atom is reduced to a chloride ion by gaining one electron:

$$Cl + e^- \rightarrow Cl^-$$

Oxidation and reduction are complementary processes. The oxidation half-reaction produces an electron which is the reactant for the reduction half-reaction. The combination of two half reactions produces the complete reaction:

$$Na + Cl \rightarrow Na^+ + Cl^-$$

Note that the electrons cancel; in the electron transfer process no free electrons remain.

In the reaction described above, sodium metal is the **reducing agent**; it releases electrons for the reduction of chlorine. Chlorine is the **oxidizing agent**; it accepts electrons from the sodium, which is oxidized.

The characteristics of oxidizing and reducing agents are summarized below:

Oxidizing Agent	**Reducing Agent**
Is reduced	Is oxidized
Gains electrons	Loses electrons
Causes oxidation	Causes reduction

Oxidation and Reduction Reactions

In one type of corrosion, elemental iron is oxidized to iron oxide:

$$4\ Fe + 3O_2 \rightarrow 2Fe_2O_3$$

The oxidation state of iron changes from zero to +3, and oxygen is reduced from zero to -2.

There are many examples of biological redox reactions. For example, the electron-transport chain of respiration (Chapter 17) involves the reversible oxidation and reduction of iron atoms in cytochrome C:

$$\text{cytochrome C } (Fe^{3+}) + e^- \rightarrow \text{cytochrome C}(Fe^{2+})$$

Key Terms

acid-base reaction
Arrhenius theory
autoionization
Brönsted-Lowry theory
buffer solution
buret
equivalence point
hydronium ion
indicator
ion product for water

neutralization
oxidation
oxidation-reduction reaction
oxidizing agent
pH scale
reducing agent
reduction
standard solution
titration

Self Test

1. Calculate the pH of a solution that contains 6.50×10^{-5} moles of hydronium ion per liter. Choose the best answer from the following:
   a. 5.49
   b. 4.19
   c. 4.00
   d. 5.0
   e. 9.81

2. If 14.8 mL of 0.100 M NaOH solution are needed to completely react with 25.0 mL of an unknown HCl solution, what is the molarity of the unknown HCl solution?

3. If we have a 0.200 molar solution of NaOH, how many H_3O^+ ions are present in the solution?

4. Calculate the pH of a 1.0×10^{-3} molar solution of HCl.

5. $A + B \rightarrow AB$ describes what type of reaction?

6. $2H_2 + O_2 \rightarrow 2H_2O$ describes what type of reaction?

7. $AB \rightarrow A + B$ describes what type of reaction?

8. $CaCO_3 \xrightarrow{heat} CaO + CO_2$ is what type of reaction?

9. $A + BC \rightarrow AC + B$ is what type of reaction?

10. $HCl + NaOH \rightarrow H_2O + NaCl$ is what type of reaction?

11. $Na_2CO_3 + BaCl_2 \rightarrow BaCO_3 + 2\ NaCl$ is what type of reaction?

12. What is the term for the reaction of an acid plus a base to yield a salt plus water?

13. Which theory defines an acid as a proton donor and a base as a proton acceptor?

14. What is the measure of the strength of an acid in water?

15. Which of the following are not strong acids?
   a. HCl
   b. CH_3COOH
   c. H_2SO_4
   d. HNO_3

16. Which of the following are not strong bases?
    a. NaOH
    b. NH_4OH
    c. $Ca(OH)_2$
    d. KOH

10 *Radioactivity and Nuclear Medicine*

Learning Goals

1. **Conceptual Goals**
   - Know what is meant by the term radiochemistry.
   - Know the characteristics of alpha, beta, and gamma radiation.
   - Distinguish between natural and artificial radioactivity.
   - Know what is meant by the term half-life, and understand its importance.
   - Familiarize yourself with common techniques for the detection of radioactivity.
   - Develop an understanding of the common units in which radiation intensity is represented: the curie, the roentgen, the rad, and the rem.
2. **Performance Goals**
   - Explain why only certain isotopes of an element are radioactive.
   - Write balanced equations for common nuclear processes.
   - Describe the various ways in which nuclear energy may be used to generate electricity: fission, fusion, and the breeder reactor.
   - Explain the process of radiocarbon dating.
3. **Health Applications**
   - Cite several examples of the use of radioactive isotopes in medicine.
   - Discuss the biological effects of radiation.
   - Describe the use of ionizing radiation in cancer therapy.
   - List the advantages of MRI over conventional X-ray diagnosis.

Chapter Overview

This chapter considers the nucleus, nuclear properties, and their applications.

10.1 Natural Radioactivity

Radioactivity is the process by which atoms emit high-energy particles or rays. These particles or rays are termed radiation. Nuclear radiation occurs as a result of an alteration in nuclear composition or structure, which happens because the nucleus is unstable and hence radioactive.

Nuclear symbols are analogous to atomic symbols. The nuclear symbols consist of the elemental symbol, the atomic number (equivalent to the number of protons in the nucleus), and the mass number, which is defined as the sum of the neutrons and protons in the nucleus. Note: Be careful not to confuse the **mass number** (neutrons and protons) with the atomic mass, which includes the contribution of electrons, and is a true mass figure.

Only unstable nuclei undergo change and produce radioactivity. Not all atoms of a particular element undergo radioactive decay. When writing the symbols for a nuclear process, it is important to designate the particular isotope involved.

Three types of natural radiation emitted by unstable nuclei are alpha particles, beta particles and gamma rays.

Alpha Particles

Alpha particles (α) contain two protons and two neutrons. An alpha particle is identical to the helium ion (He^{2+}) and is represented as:

$$_2^4He$$

or

$$\alpha$$

Alpha particles emitted by radioisotopes move relatively slowly (approximately 10% of the speed of light), and they are stopped by very little mass.

Beta Particles

The beta particle (β) is a fast-moving electron traveling at approximately 90% of the speed of light as it leaves the nucleus. The beta particle is represented as

$$_{-1}^0e$$

or

$$\beta$$

the subscript -1 is written in the same position as the atomic number and, like the atomic number (number of protons), indicates the charge of the particle.

Beta particles are smaller, faster, and more energetic and penetrating than alpha particles.

Gamma Rays

Gamma rays (γ) are pure energy. Since energy has no mass or charge, the symbol for a gamma ray is simply:

$$\gamma$$

Gamma radiation is highly energetic and is the most penetrating form of nuclear radiation.

Example 1
List the names, symbols, and characteristics of the three basic types of radiation.

Answer
1. The alpha particles have the lowest velocity of the three types of radiation. They have a mass of about 4.0 amu and are composed of two protons and two neutrons. Notice that they have the same structure as the nucleus of a helium atom. Since they have two protons present, they have a charge of +2. The alpha particle may be symbolized as $_2^4He$.

101

2. The beta particle is just a fast-moving electron that has been produced during a nuclear reaction. It has a speed of about 90% the speed of light. Since it is only an electron, it has a charge of -1 and may be symbolized as $_{-1}^{0}e$.

3. The gamma rays are pure energy, in contrast to alpha and beta radiation, which are particles. Gamma radiation is highly energetic and is the most penetrating and deadly to living things. It is represented by the gamma symbol, γ.

Properties of Alpha, Beta, and Gamma Radiation

Important properties of α, β, and γ radiation are summarized in Table 10.1 of the textbook.

10.2 Writing a Balanced Nuclear Equation

A **nuclear equation** represents a nuclear process such as radioactive decay. To write a balanced equation, we must remember the following:

1. The total mass on each side of the reaction arrow must be identical.
2. The sum of the charges of the reactant nuclei must be equal to the sum of the charges of the product nuclei.

Alpha Decay

A reaction involving alpha decay is shown below:

$$_{88}^{226}\text{Ra} \quad \rightarrow \quad _{86}^{222}\text{Rn} \quad + \quad _{2}^{4}\text{He}$$

Radium -226   Radon -222   Helium -4

Beta Decay

Beta decay may be illustrated by the following:

$$_{90}^{234}\text{Th} \quad \rightarrow \quad _{91}^{234}\text{Pa} \quad + \quad _{-1}^{0}e$$

Thorium -234   Protoactinium -234 β particle

Gamma Production

If gamma radiation were the only product of nuclear decay, there would be no change in the mass or identity of the radioactive nuclei, since a gamma ray is pure energy, possessing no mass or charge. The gamma emitter has simply gone to a lower energy state. The decay of the

metastable isotope technetium—99m is shown:

$$^{99m}_{43}Tc \rightarrow ^{99}_{43}Tc + \gamma$$

Predicting Products of Nuclear Decay

It is possible to use a nuclear equation to predict one of the products of a nuclear reaction if the others are known. We know that the mass and charge of the total of all products and reactants must be equal. By difference, we can compute the missing charge and mass that represents the unknown product and deduce its identity.

Example 2
Complete the following nuclear equations:

1. $^{238}_{92}U \rightarrow ^{234}_{90}Th + ?$

2. $^{16}_{7}N \rightarrow ^{16}_{8}O + ?$

3. $^{40}_{19}K \rightarrow ^{0}_{-1}e + ?$

4. $^{197}_{79}Au + ? \rightarrow ^{198}_{79}Au$

Work
Use the following rules:

1. The total mass on each side of the equation arrow must be equal.
2. The sum of the charges of the reactant nuclei must be equal to the sum of the charges of the product nuclei.

Answer
The missing parts are given below:

1. $^{4}_{2}He$

2. $^{0}_{-1}e$

3. $^{40}_{20}Ca$

4. $^{1}_{0}n$

Example 3

Show how molybdenum-98 can be converted by neutron bombardment into technetium-99 plus beta particles. The technetium produced is unstable and can be used as a gamma source for tracer applications.

Work

1. From the periodic table we find that molybdenum has an atomic number of 42, and technetium's is 43.
2. Next we set up the equation:

$$^{98}_{42}\text{Mo} + {}^{1}_{0}\text{n} \rightarrow {}^{99m}_{43}\text{Tc} + {}^{0}_{-1}\text{e}$$

Notice that the "m" following the mass number means that the isotope is metastable.

Answer

$$^{98}_{43}\text{Mo} + {}^{1}_{0}\text{n} \rightarrow {}^{99m}_{43}\text{Tc} + {}^{0}_{-1}\text{e}$$

10.3 Properties of Radioisotopes

Nuclear Structure and Stability

The energy that holds the protons, neutrons, and other particles together in the nucleus is the **binding energy** of the nucleus. When an isotope decays, some of this binding energy is released.

Factors related to nuclear stability include the following:

1. Nuclear stability correlates with the ratio of neutrons to protons in the isotope.
2. Nuclei with large numbers of protons (84 or more) tend to be unstable.
3. Isotopes containing 2, 8, 20, 50, 82, or 126 protons or neutrons are stable. These "**magic numbers**" seem to indicate the presence of stable energy levels in the nucleus.
4. Isotopes with even numbers of protons or neutrons are generally more stable than those with odd numbers of protons or neutrons.

Example 4

Explain why some isotopes are more stable than others.

Answer

Nuclear stability correlates with the ratio of neutrons to protons in a specific isotope. For elements with atomic numbers less than 84, a neutron/proton ratio of 1 usually means a stable isotope of a specific element. Elements that have an atomic number over 83 usually form only unstable isotopes. Isotopes containing the magic numbers of 2, 8, 20, 50, 82, or 126 protons or neutrons are stable. Isotopes with an even number of protons or neutrons are generally more stable than those with odd numbers of protons or neutrons.

Half-Life

The **half-life**, $t_{1/2}$, is the time required for one-half of a given quantity of a substance to undergo change. Each isotope has its own characteristic half-life.

The degree of stability of an isotope is indicated by the isotope's half-life. Isotopes with short half-lives decay rapidly; they are very unstable.

Decay of a radioisotope that has a reasonably short $t_{1/2}$ is experimentally determined by following its activity as a function of time. Graphing the results produces a radioactive decay curve (see Figure 10.1 in the textbook).

Example 5

Radium-226 $\left(^{226}_{88}Ra\right)$ has a half-life of 1620 years. If a watch's hands are coated with paint containing Ra-226 in 1920, why is it still radioactive?

Answer
Since only 72 years (a small fraction of one half-life, $\frac{72y}{1620y}$) have passed, nearly all the radium-226 used in the paint is still radioactive.

Example 6
The following half-lives are known for different isotopes of radium:

>   Ra-223 11 days
>   Ra-224 3.64 days
>   Ra-226 1620 years
>   Ra-228 6.7 years

Why is there such a difference in their half-lives?

Answer
It is primarily due to the proton/neutron ratio in the isotope, but other as yet undiscovered reasons may play a part as well.

Example 7
How much of a 1.00-gram sample of Ra-226 would remain after 3 half-lives? The half-life of Ra-226 is 1620 years.

Work

	first		second		third	
1.00g	---->	0.50 g	---->	0.25 g	---->	0.13 g
	half-life		half-life		half-life	

Answer

0.13 grams of Radium-226 remain.

10.4 Nuclear Power

Energy production

Einstein's equation, relating mass and energy, predicts that a small amount of nuclear mass is converted to a very large amount of energy when the nuclear particle breaks apart.

Einstein's equation is as follows:

$$E = mc^2$$

where

E = energy
m = mass
c = speed of light

Three major modes of generating electrical power through a nuclear reaction involve the processes of fission, fusion, and the use of what is termed a breeder reactor.

Nuclear Fission

Fission (splitting) occurs when a heavy nuclear particle is split into smaller nuclei and large amounts of energy by a smaller nuclear particle (such as a neutron). The fission process intensifies, producing very large amounts of energy This process of intensification is referred to as a chain reaction.

Nuclear Fusion

Fusion (meaning to join together) results from the combination of two small nuclei to form a larger nucleus with the concurrent release of large amounts of energy. The best example of a fusion reactor is the sun. Continuous fusion processes furnish our solar system with light and heat.

$$_1^2H + _1^3H \rightarrow _2^4He + _0^1n + Energy$$

Breeder Reactors

A **breeder reactor** literally manufactures its own fuel. A perceived shortage of fissionable isotopes makes the breeder an attractive alternative to fission reactors. A breeder reactor uses $^{238}_{92}U$, which is abundant but nonfissionable. In a series of steps, the uranium-238 is converted to plutonium-239 which is fissionable and undergoes a fission chain reaction, producing energy.

10.5 Radiocarbon Dating

The approximate age of fossils and other objects of archaeological, anthropological, or historical interest may be established through **radiocarbon dating**. Radiocarbon dating is based on the measurement of the relative amounts (or ratio) of $^{14}_{6}C$ and $^{12}_{6}C$ present in an object. The carbon-14, along with the more abundant carbon-12, is converted into living plant material by the process of photosynthesis. Over time, the amount of carbon-14 slowly decreases, because carbon-14 is radioactive ($t_{1/2}$ = 5760 years); the amount of carbon-12, however, remains constant.

When the fossil is found and studied, the relative amounts of carbon-14 and carbon-12 are determined; using suitable equations involving the $t_{1/2}$ of carbon-14, it is possible to calculate (to within ± a few percent) the time which has elapsed since the formation of the object in question.

10.6 Medical Applications of Radioactivity

The use of radiation in the treatment of various forms of cancer, as well as the newer area of **nuclear medicine**, has become widespread in the past quarter-century.

Cancer Therapy Using Radiation

When high-energy radiation, such as gamma radiation, passes through a cell, it may collide with one of the molecules in the cell and cause it to lose one or more electrons, producing an ion pair. For this reason, such radiation is termed **ionizing radiation**. Ions produced in this way may cause subtle changes in cellular biochemical processes, which may result in diminished or altered cell function, or in extreme cases the death of the cell.

An organ that is cancerous is composed of both healthy cells and malignant cells. Cells undergoing division are particularly sensitive to gamma radiation. Therefore, exposing the tumor area to controlled dosages of high-energy gamma radiation from cobalt-60 (a high-energy gamma ray source) will generally kill a higher percentage of abnormal cells than normal cells.

Nuclear Medicine

The diagnosis of a host of biochemical irregularities or diseases of the human body has been made routine through the use of *radioactive tracers*. Tracers are small amounts of radioactive substances used as probes to study internal organs. Because the isotope is radioactive, its path may be followed, using suitable detection devices. A "picture" of the organ is obtained, often far more detailed than is possible with conventional X-rays. Such techniques are noninvasive; that is, surgery is not required to investigate the condition of the internal organ, eliminating the risk associated with an operation. These techniques are successful because the radioactive isotope of an element has exactly the same chemical behavior as any other isotope of the same element.

Making Isotopes for Medical Applications

The radioactivity produced by unstable isotopes is described as **natural radioactivity**. If a normally stable, nonradioactive nucleus is made radioactive through bombardment with protons, neutrons, or alpha particles, the resulting radioactivity is termed **artificial radioactivity**.

The bombardment process is often accomplished in the core of a **nuclear reactor**, where an abundance of small nuclear particles, particularly neutrons, are available. Alternately, extremely high-velocity charged particles (such as *a* and *b*) may be produced in a particle **accelerator**, such as a cyclotron.

These product isotopes are often used in hospital laboratories as tracers in nuclear medicine.

10.7 Biological Effects of Radiation

Radiation Exposure and Safety

Safety considerations are based on the following:

1. The magnitude of the half-life
2. Shielding
3. Distance from the radioactive source
4. Time of exposure
5. Types of radiation emitted

Virtually all applications of nuclear chemistry create radioactive waste and, along with it, the problems of safe handling and disposal. Most disposal sites at present are considered temporary, until a long-term, safe solution can be found.

10.8 Detection and Measurement of Radiation

The changes that take place when radiation interacts with matter provide the basis of operation for various radiation detection devices.

Nuclear Imaging

An isotope is administered to a patient, and the isotope begins to concentrate in the organ of interest. Photographs of that region of the body are taken at periodic intervals using a special type of film. Upon development of the series of photographs, a record of the organ's uptake of the isotope as a function of time enables the radiologist to assess the condition of the organ.

Computer Imaging

A specialized television camera, sensitive to emitted radiation from a radioactive substance administered to a patient, develops a continuous and instantaneous record of the voyage of the isotope throughout the body.

The Geiger Counter

A **Geiger counter** is an instrument capable of detecting ionizing radiation.

Film Badges

A film badge is merely a piece of photographic film that is sensitive to energies corresponding to radioactive emissions.

10.9 Units of Radiation Measurement

The Curie

The **curie** is a measure of the amount of radioactivity in a radioactive source. The curie is independent of the nature of the radiation (a, b or γ) as well as its effect on biological tissue. A curie is defined as the amount of radioactive material that produces 3.7×10^{10} atomic disintegrations per second.

The Roentgen

The **roentgen** is a measure of ionizing radiation (X ray and gamma ray) only. The roentgen is defined as the amount of radioactive isotope needed to produce 2×10^9 ion pairs when passing through 1 cubic centimeter of air at $0\,^\circ$C. The roentgen is a measure of radiation's interaction with air, and gives no information regarding its effect on biological tissue.

The Rad

The **rad**, or radiation absorbed dosage, provides more meaningful information than either of the previous units of measure. It takes into account the nature of the absorbing material. It is defined as the dosage of radiation able to transfer 2.4×10^{-3} calories of energy to 1 kilogram of matter.

The Rem

The **rem**, or roentgen equivalent for man, describes the biological damage caused by the absorption of different kinds of radiation by the body. The rem is obtained by multiplication of the rad by a factor called the relative biological effect, or **RBE**. The RBE is a function of the type of radiation (α, β, or γ). Although a beta particle is more energetic than an alpha particle, an alpha particle is approximately ten times more damaging to biological tissue. As a result, the RBE is 10 for alpha particles and 1 for beta particles.

An estimated lethal dose, symbolized by **LD$_{50}$**, is 500 rems. The lethal dose, LD$_{50}$, is defined as the dose that would be fatal for 50% of the exposed population within 30 days. Some biological effect, however, is detectable at a level as low as 25 rem.

Key Terms

alpha particle	natural radioactivity
artificial radioactivity	nuclear equation
background radiation	nuclear imaging
beta particle	nuclear medicine
binding energy	nuclear reactor
breeder reactor	particle accelerator
chain reaction	rad
curie	radioactivity
fission	radiocarbon dating
fusion	rem
gamma ray	roentgen
half-life	shielding
ionizing radiation	tracer
lethal dose	
metastable isotope	

Self Test

1. Complete the following nuclear reactions by supplying the missing part:

    a. $^{234}_{90}Th \rightarrow \, ^{234}_{91}Pa \, + \, ?$

    b. $^{27}_{13}Al \, + \, ^{4}_{2}He \rightarrow \, ^{30}_{15}P \, + \, ?$

    c. $^{210}_{82}Pb \rightarrow \, ^{0}_{-1}e \, + \, ?$

2. How many mg of a 100.0 mg sample of Tc-99m will remain after 30 hours? The half-life of Tc-99m is 6 hours.
3. When using the nuclear symbol for an isotope of fluorine, $^{19}_{9}F$, which number is equal to the atomic number?
4. Name the three types of natural radiation emitted by unstable nuclei.
5. Which of the types of radiation is really pure energy?
6. Which of the types of radiation is the least-penetrating form of nuclear radiation?
7. What term means that an isotope is unstable?
8. What is defined as the time period required for one-half of a given quantity of a substance to undergo a nuclear change?
9. Nuclear power plants use what nuclear process to produce energy?
10. What term do we use for cells that undergo unnaturally rapid cell division?
11. Where does iodine tend to concentrate in the human body?
12. What is the term which describes the amount of radiation attributable to our surroundings on a day-to-day basis?

13. The intensity of radiation varies in what way with the square of the distance from the source?

14. What do we call the specialized system that is sensitive to emitted radiation from a radioactive substance administered to a patient and produces a continuous picture of the voyage of the isotope throughout the body?

15. What badge is often worn by staff members that shows their exposure to radiation?

16. Which type of radiation travels at the speed of light?

17. What is defined as the spontaneous decay of a nucleus to produce high-energy particles or rays?

18. Radiocarbon dating involves estimation of the age of objects by measuring the amount of carbon-12 plus which radioactive form of carbon?

19. Which type of radiation is similar to a helium atom?

20. Which radioactive element is found in some homes?

11 *An Introduction to Organic Chemistry: The Saturated Hydrocarbons*

Learning Goals

1. Conceptual Goals
· Define organic chemistry from a historical and modern perspective.
· Compare organic and inorganic compounds.
· Identify the key structural features of the simple alkanes, alkenes, and alkynes.
· Learn the fundamental rules of the I.U.P.A.C. Nomenclature System.

2. Performance Goals
· Name and describe the major classes of hydrocarbons.
· Write condensed and structural formulas for simple hydrocarbons.
· Identify the common functional groups and draw their structures.
· Draw and name alkanes and substituted alkanes.
· Define and provide examples of isomers of simple organic compounds.
· Write equations for combustion and halogenation reactions of alkanes.

3. Health Applications
· Describe several health-related effects of the halogenated hydrocarbons.

Chapter Overview

Until 1828 it was thought that all organic compounds were derived from natural sources. All early attempts to synthesize these compounds in the laboratory failed, and it was proposed that a vital force was necessary for their formation. In 1828, Wöhler synthesized urea (an organic compound) from potassium cyanate and ammonium sulfate (two inorganic compounds). Wöhler's experiment is now recognized as the beginning of modern organic chemistry.

Example 1

Draw the dash structures of ammonium cyanate and urea. Why are these two compounds important to the beginning of modern organic chemistry?

Answer
1. Dash structure of ammonium cyanate:

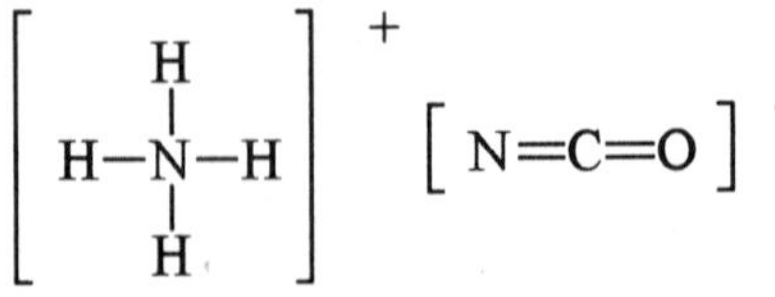

$$\left[\begin{array}{c} H \\ | \\ H{-}N{-}H \\ | \\ H \end{array} \right]^{+} \qquad \left[N{=}C{=}O \right]^{-}$$

2. Dash structure of urea:

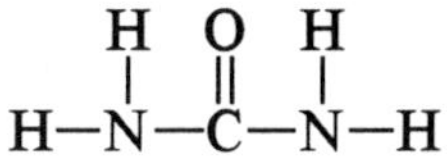

$$\begin{array}{ccc} H & O & H \\ | & \| & | \\ H-N- & C & -N-H \end{array}$$

3. In 1828 the German chemist, Friedrich Wöhler was trying to produce ammonium cyanate, but due to experimental conditions the compound urea was produced. The two compounds have the molecular formula, CON_2H_4, but they have different structures and properties. The "vital force theory" was laid to rest. Up until his discovery, it was believed that only living things could make compounds like urea, which is found as a waste product in the urine of animals.

11.1 The Chemistry of Carbon

There are several reasons for the existence of hundreds of thousands of organic compounds:

1. Carbon atoms are able to form *stable, covalent bonds* with other carbon atoms. Consider the family of alkanes as an example. Each molecule in this family contains only carbon and hydrogen, but each is a different chemical substance with unique chemical and physical properties. There are an infinite number of possible alkanes; alkanes with hundreds or even thousands of carbon and hydrogen atoms may be easily synthesized. Even elemental carbon exists in different forms, allotropic forms, that have different properties.
2. Carbon can form stable bonds with other elements, for example nitrogen, oxygen, sulfur, and the halogens.
3. The number of ways that these elements may combine to form unique structures is practically limitless. Compounds having the same molecular formulas but different structures (hence, different properties), are termed **constitutional isomers.**

Example 2
List the three reasons there are so many organic compounds:

Answer
1.Carbon can form chain, branch, and ring structures with other carbon atoms:
  a. Chain structures with no branches:

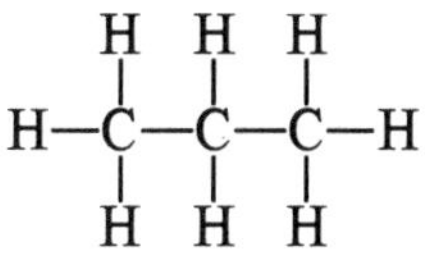

Propane 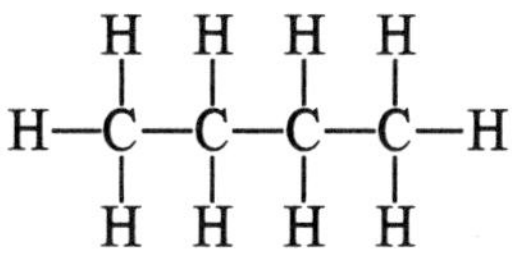   Butane

---

b. Chain structure that also has branch structures:

$$H-\overset{\overset{\displaystyle H}{|}}{\underset{\underset{\displaystyle H}{|}}{C}}-\overset{\overset{\displaystyle H}{|}}{\underset{\underset{\underset{\underset{\displaystyle H}{|}}{\underset{\displaystyle C-H}{|}}}{|}}{C}}-\overset{\overset{\displaystyle H}{|}}{\underset{\underset{\displaystyle H}{|}}{C}}-H$$

Isobutane

c. Ring structures:

$$\begin{array}{cc} H & H \\ | & | \\ H-C-C-H \\ | & | \\ H-C-C-H \\ | & | \\ H & H \end{array}$$

2. Carbon can form compounds that also contain other nonmetals, such as oxygen, sulfur and nitrogen, in their structure:

$$H-\overset{\overset{\displaystyle H}{|}}{\underset{\underset{\displaystyle H}{|}}{C}}-O-H \qquad H-\overset{\overset{\displaystyle H}{|}}{\underset{\underset{\displaystyle H}{|}}{C}}-\overset{\overset{\displaystyle H}{|}}{\underset{\underset{\displaystyle H}{|}}{C}}-\overset{\overset{\displaystyle H}{|}}{N}-H$$

Methyl alcohol   Ethyl amine

3. Carbon can form compounds that have the same molecular formula but a different structure and a different set of properties. Urea and ammonium cyanate are one example of such a set of isomeric compounds.

Important Differences between Organic and Inorganic Molecules

The bonds in organic molecules are almost always covalent bonds, while those found in many inorganic substances are ionic bonds. When comparing organic and inorganic compounds it helps to remember the differences between ionic and covalent bonds.

1. Ionic bonds result from a transfer of one or more electrons. Covalent bonds are formed by a sharing of electrons to form a stable orbital containing two electrons.
2. The ionic bond is electrostatic in nature. It is formed by the attraction of positive and negative ions resulting from the electron transfer process. Charge separation of covalently bonded atoms is much less extreme.

114

3. Ions are arranged in large, three-dimensional crystals, consisting of many positive and negative ions. Covalently bonded substances exist as discrete units: molecules.

4. Ionic compounds often dissociate in solution (electrolytes), whereas covalently bonded molecules retain their identity in solution (nonelectrolytes).

Families of Organic Compounds

Organic compounds are classified according to groups, or families. The two most general headings are **hydrocarbons** and **substituted hydrocarbons**. Compounds that contain only carbon and hydrogen are classified as hydrocarbons: hydrogen (hydro) and carbon (carbon). Hydrocarbons are subdivided into two principal classes: aliphatic and **aromatic hydrocarbons**. **Aliphatic hydrocarbons** are further subdivided into three families, alkanes, alkenes, and alkynes.

The **alkanes** (and cycloalkanes) are termed **saturated hydrocarbons**. They are composed solely of C—C and C—H single bonds. The **unsaturated hydrocarbons**, hydrocarbons that contain at least one (carbon-carbon double bond) or (carbon-carbon triple bond) are referred to as **alkenes** and **alkynes**, respectively. Aromatic compounds contain a benzene ring or a derivative of the benzene ring.

Substitution of one or more functional groups for hydrogens in a hydrocarbon brings about huge changes in properties. A **functional group** is an atom or group of atoms in a molecule principally responsible for the chemical and physical properties of that molecule. All compounds that contain a particular functional group, for example, the —OH (hydroxyl group), are classified as being in the same family (the —OH group is the functional group of the alcohols).

Example 3
To which family of organic compounds do each of the following belong? (Refer to Table 11.1 in the text)

1. CH_3CH_3
2. $CH_3CH=CHCH_3$
3. CH_3OH

$$\underset{\|}{\overset{O}{}}$$

4. CH_3CCH_3
5. CH_3Br

Answer
1. alkane
2. alkene
3. alcohol
4. ketone
5. alkyl halide

11.2 Alkanes

Structure and Properties

Alkanes are hydrocarbons that contain only carbon and hydrogen bonded together through carbon-hydrogen and carbon-carbon single bonds. They have the general formula C_nH_{2n+2}.

Several different types of formulas are used to describe organic molecules. The **structural formula** shows all of the atoms in a molecule and shows all bonds as lines. The **molecular formula** provides the atoms and number of each type of atom in a molecule, but gives no information regarding the bonding pattern involved. The **condensed formula** shows all of the atoms in a molecule and places them in a sequential arrangement that details which atoms are bonded to each other.

A molecular formula tells us the atomic composition of a molecule. However, molecular formulas provide no information about the structure of the molecule. For instance, there are three possible structures for a compound with the molecular formula of C_5H_{12}:

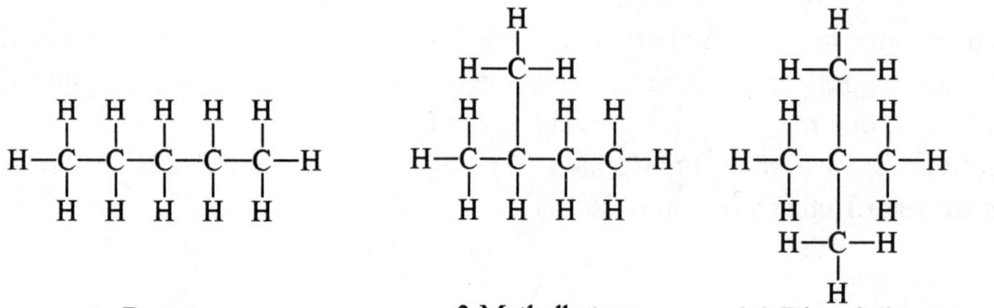

| Pentane | 2-Methylbutane | 2,2-Dimethylpropane |

These are the three possible isomers of pentane. These "line" representations, called structural formulas, provide more information about a molecule.

A suitable compromise between the convenience of the molecular formula and the detail of the structural formula is the condensed formula. The condensed formula for pentane is

$$CH_3CH_2CH_2CH_2CH_3$$

for 2-methylbutane, it is

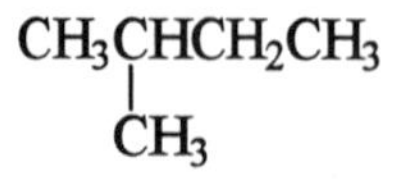

$$CH_3CHCH_2CH_3$$
$$\mspace{40mu}|$$
$$\mspace{30mu}CH_3$$

116

and the condensed formula for 2,2-dimethylpropane is

$$CH_3-\overset{\displaystyle CH_3}{\underset{\displaystyle CH_3}{\overset{|}{\underset{|}{C}}}}-CH_3$$

Example 4

Draw the structural formula of an isomer of ethyl alcohol (CH_3CH_2OH), in which an oxygen atom is placed between two carbon atoms.

Work

1. Hydrogen always forms one bond; carbon forms four bonds and oxygen forms two bonds.

Answer

$$H-\overset{\displaystyle H}{\underset{\displaystyle H}{\overset{|}{\underset{|}{C}}}}-O-\overset{\displaystyle H}{\underset{\displaystyle H}{\overset{|}{\underset{|}{C}}}}-H$$

This compound is dimethyl ether, and it is an isomer of ethyl alcohol.

Example 5

Write the condensed formulas for the following:

1. Propane

$$H-\overset{\displaystyle H}{\underset{\displaystyle H}{\overset{|}{\underset{|}{C}}}}-\overset{\displaystyle H}{\underset{\displaystyle H}{\overset{|}{\underset{|}{C}}}}-\overset{\displaystyle H}{\underset{\displaystyle H}{\overset{|}{\underset{|}{C}}}}-H$$

2. 2-Methylbutane

$$H-\overset{\displaystyle H}{\underset{\displaystyle H}{\overset{|}{\underset{|}{C}}}}-\overset{\displaystyle H}{\underset{\displaystyle H-C-H}{\overset{|}{\underset{|}{C}}}}-\overset{\displaystyle H}{\underset{\displaystyle H}{\overset{|}{\underset{|}{C}}}}-\overset{\displaystyle H}{\underset{\displaystyle H}{\overset{|}{\underset{|}{C}}}}-H$$

3. Methyl alcohol

$$H-\overset{\displaystyle H}{\underset{\displaystyle H}{\overset{|}{\underset{|}{C}}}}-O-H$$

4. Acetone

$$H-\overset{\displaystyle H}{\underset{\displaystyle H}{\overset{|}{\underset{|}{C}}}}-\overset{\displaystyle O}{\underset{}{\overset{\|}{C}}}-\overset{\displaystyle H}{\underset{\displaystyle H}{\overset{|}{\underset{|}{C}}}}-H$$

5. 2-Chloropropane

$$H-\overset{\displaystyle H}{\underset{\displaystyle H}{\overset{|}{\underset{|}{C}}}}-\overset{\displaystyle H}{\underset{\displaystyle Cl}{\overset{|}{\underset{|}{C}}}}-\overset{\displaystyle H}{\underset{\displaystyle H}{\overset{|}{\underset{|}{C}}}}-H$$

Answer

1. $CH_3CH_2CH_3$
2. $CH_3CHCH_3CH_2CH_3$
3. CH_3OH
4. CH_3COCH_3
5. $CH_3CHClCH_3$

Example 6

Write the molecular formulas of the alkanes that contain 4, 8, and 40 carbon atoms.

Work

Use the general formula for finding the molecular formula of any alkane, provided you are given the number of carbon atoms.

C_nH_{2n+2} where n = number of carbon atoms

Answer

4 carbons: C_4H_{10}; 8 carbons: C_8H_{18}; 40 carbons: $C_{40}H_{82}$

Since all the hydrocarbons are composed of nonpolar carbon-carbon and carbon-hydrogen bonds, hydrocarbons are nonpolar molecules. As a result, they are not water-soluble, but are readily soluble in nonpolar solvents. Furthermore, virtually all of the hydrocarbons are less dense than water and have relatively low melting points and boiling points.

Nomenclature

The basic rules used for naming compounds in the I.U.P.A.C. Nomenclature System follow:

1. The name of the compound is defined by the longest continuous carbon chain in the compound. This chain is the **parent compound**. (Refer to Table 11.2 in the text for the names of parent compounds.)

2. Each substituent attached to the parent compound is given a name and a number. The number designates the position of the substituent on the main chain, and the name tells what type of substituent is present at that position.

3. The chain must be numbered from one end to the other in order to provide the lowest position number for each substituent. If more than one substituent is present, number from the end that gives the lowest number of the first substituent encountered regardless of the numbers that result for the other substituents.

4. If the same substituent occurs more than once in the compound, a separate position number is supplied for each substituent, and the prefixes *di-*, *tri-*, *tetra-*, *penta-*, *hexa-*, *hepta-*, and so forth are used.

5. Place the names of the substituents in alphabetical order before the name of the parent compound. Numbers are separated from each other by commas, and numbers are separated from names by hyphens. By convention halogen substituents are placed before alkyl substituents in this priority sequence regardless of the alphabetization.

Example 7

Give the I.U.P.A.C. names of the following branched-chain alkanes:

$$
\begin{array}{cc}
\text{1.} \quad CH_3CHCH_2CH_3 & \text{2.} \quad CH_3CH_2CCH_2CH_2CH_3 \\
\quad\quad\; | & \quad\quad\quad\quad | \\
\quad\quad\; CH_3 & \quad\quad\quad\quad CH_3
\end{array}
$$

(with CH_3 branch above the C in compound 2)

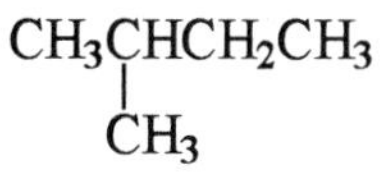

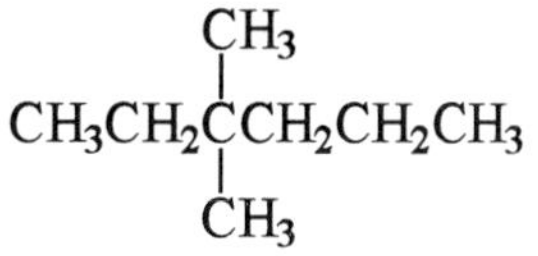

Work

1. Determine the longest continuous carbon chain. This is called the parent compound.
2. Number the parent compound chain so that the branches have the lowest possible numbers.

Answer

1. The parent chain is butane. We will number 1, 2, etc. from left to right. The correct name of this branched alkane is 2-methylbutane.
2. The parent chain is hexane. We will number from left to right. The correct name of this branched alkane is 3,3-dimethylhexane.

Example 8

Give the I.U.P.A.C. name of the following:

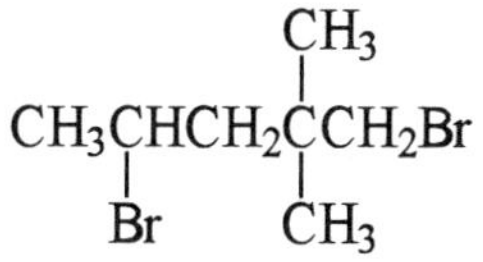

$$
\begin{array}{c}
CH_3 \\
| \\
CH_3CHCH_2CCH_2Br \\
\;\;\;\; | \quad\;\; | \\
\;\;\;\; Br \quad CH_3
\end{array}
$$

Work

1. The parent compound is pentane.
2. Number the parent compound from right to left:

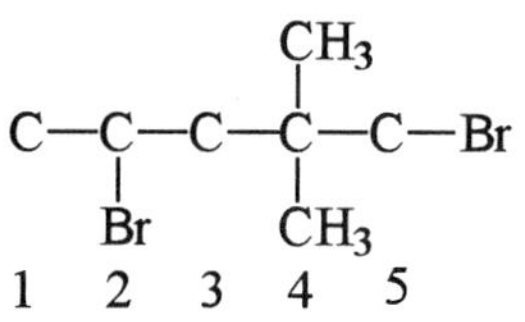

$$
\begin{array}{c}
CH_3 \\
| \\
C-C-C-C-C-Br \\
\;\;\;\;\; | \quad\;\; | \\
\;\;\;\; Br \quad CH_3 \\
1 \;\; 2 \;\; 3 \;\; 4 \;\; 5
\end{array}
$$

Answer

1,4-dibromo-2,2-dimethylpentane

Constitutional Isomers

Two molecules having the same molecular formulas but different structures are called **constitutional isomers**. Isomers having the same molecular formula are unique compounds due to their structural differences. They may have similar physical and chemical properties, but in many cases their properties are quite dissimilar.

Example 9
Draw and give the I.U.P.A.C. names of all the alkane isomers of C_5H_{12}.

Work
1. The first isomer you should always draw is the one that has all the carbons in a continuous chain structure:

$$CH_3CH_2CH_2CH_2CH_3 \qquad \text{I.U.P.A.C. name = pentane.}$$

2. Next draw four carbons in a continuous chain, and connect the fifth carbon as a methyl group to this continuous chain:

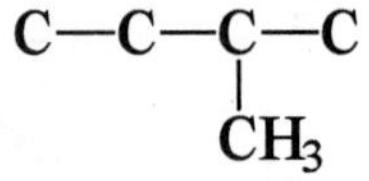

$$
\begin{array}{cccc}
C & C & C & C \\
& & | & \\
& & CH_3 &
\end{array}
$$

Fill in the structure using hydrogen atoms, so that each carbon atom has four bonds.

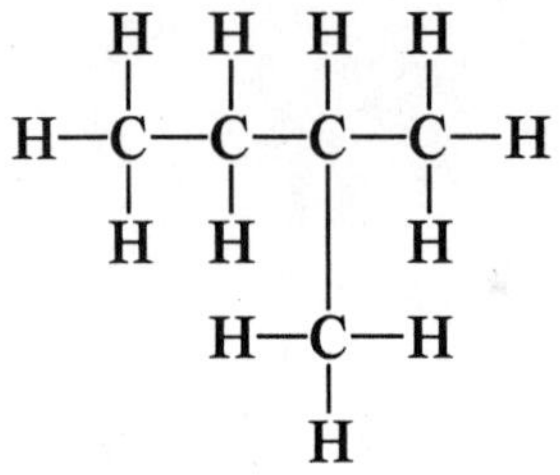

$$
\begin{array}{ccccccc}
& H & H & H & H & \\
& | & | & | & | & \\
H- & C- & C- & C- & C- & H \\
& | & | & | & | & \\
& H & H & | & H & \\
& & & H-C-H & & \\
& & & | & & \\
& & & H & &
\end{array}
$$

Then write the condensed structure:

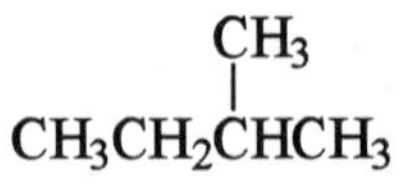

$$
\begin{array}{c}
CH_3 \\
| \\
CH_3CH_2CHCH_3
\end{array}
$$

I.U.P.A.C. name = 2-methylbutane. Note that you can move the methyl group one carbon to the left. However, if you apply the I.U.P.A.C. nomenclature test the product is still 2-methylbutane. Similarly, you can move the methyl group one carbon to the right. The

I.U.P.A.C. nomenclature test reveals that you have drawn pentane, the linear isomer.

3. The last isomer of C_5H_{12} is found by drawing three carbons in a continuous chain and connecting the other two carbons as methyl groups to the middle carbon of the continuous chain:

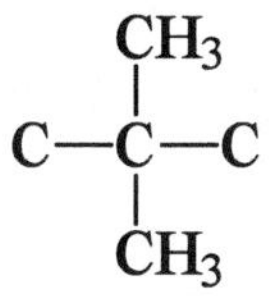

$$
\begin{array}{c}
CH_3 \\
| \\
C—C—C \\
| \\
CH_3
\end{array}
$$

Fill in the structure using hydrogen atoms as before:

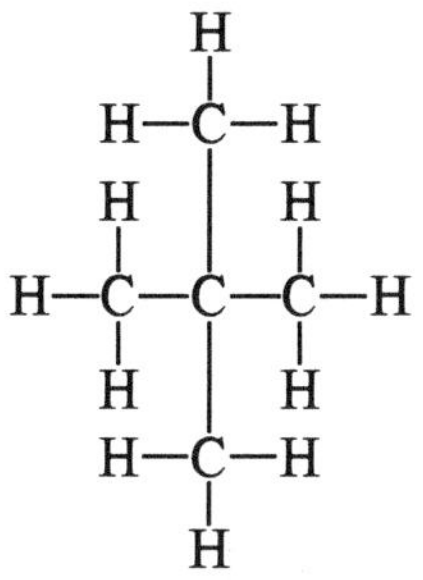

$$
\begin{array}{c}
H \\
| \\
H—C—H \\
H \quad | \quad H \\
| \quad | \quad | \\
H—C—C—C—H \\
H \quad | \quad H \\
H—C—H \\
| \\
H
\end{array}
$$

Then write the condensed structure:

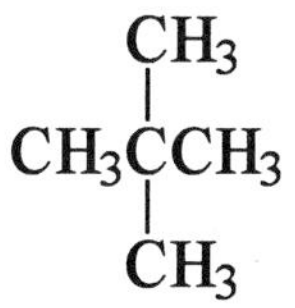

$$
\begin{array}{c}
CH_3 \\
| \\
CH_3CCH_3 \\
| \\
CH_3
\end{array}
$$

I.U.P.A.C. name = 2,2-dimethylpropane.

Remember that all isomers must have different I.U.P.A.C. names. To check whether you have drawn duplicate isomers, name them using the I.U.P.A.C. nomenclature system.

Cycloalkanes

The **cycloalkanes** are another family of hydrocarbons closely related to the alkanes. Cycloalkanes have the general molecular formula C_nH_{2n}. Note that they contain two fewer hydrogens than the corresponding alkane.

Cycloalkanes are named by adding the prefix *cyclo-* before the name of the alkane with the same number of carbon atoms. For example, cyclobutane is a cyclic alkane that has four carbon atoms. Substituted cycloalkanes are named by placing the name of the substituent before the name of the cycloalkane. No number is needed if only a single substituent is present. If more than one substituent is present, then the numbers that result in the lowest possible position numbers for the substituents are used.

11.3 Conformations of Alkanes and Cycloalkanes

There is free rotation around the carbon-carbon single bond, and at room temperature, rotations occur on the order of millions of times per second. Thus, these molecules exist in an infinite variety of arrangements that are rapidly interconverting. These varying arrangements, capable of interconversion by simple rotation about a bond, are called **conformations** or **conformers**.

Newman projections are used to portray conformers. In the **staggered conformation**, the carbon-hydrogen bond of one carbon bisects the hydrogen-carbon bond angle of the second carbon of the compound. The staggered conformation is the most common because it is the most stable structure. The **eclipsed conformation** occurs when the carbon-hydrogen bonds on two carbons line up with one another.

All cycloalkanes, except cyclopropane, also exist in different conformations. The **chair conformation** is the most energetically favored conformer for six-member ring structures. The hydrogen atoms that lie above or below the ring are said to be **axial atoms**; those that lie roughly in the plane of the ring are called **equatorial atoms**.

11.4 *Cis-Trans* Isomerism in Cycloalkanes

Rotation around the bonds in a cyclic structure is limited by the fact that the carbons of the ring are all bonded to other ring carbons. As a result, **geometric isomers**, or *cis-trans isomers*, are produced. In the *cis-* isomer, two identical substituents are on the same side of the ring (either above or below). In the *trans-* isomer, the substituents are on opposite sides of the ring.

11.5 Reactions of Alkanes and Cycloalkanes

Combustion

Alkanes and other hydrocarbons may be oxidized (burned in air), producing carbon dioxide and water; this reaction is called **combustion**. During combustion, a large amount of heat energy is released.

Example 10
Show the complete balanced equation for the combustion of octane.

Work
1. Octane: The -ane ending tells us that the compound is an alkane. The oct- means that it has eight carbon atoms. By using the general molecular formula C_nH_{2n+2} we can derive the

molecular formula of octane. Thus, the molecular formula is C_8H_{18}.

2. The complete combustion of an alkane requires O_2 to react with the alkane to produce $CO_2 + H_2O$.

3. $C_8H_{18} + O_2 \rightarrow CO_2 + H_2O$

The above equation is not balanced.

Answer

The balanced equation for the complete combustion of octane is

$$2\ C_8H_{18} + 25\ O_2 \rightarrow 16\ CO_2 + 18\ H_2O$$

Halogenation

Halogenation of an alkane or cycloalkane is a substitution reaction in which a halogen atom (usually bromine or chlorine) replaces a hydrogen atom.

Typically, when an alkane reacts with certain halogens in the presence of heat and/or light, a substitution reaction results. One of the C—H bonds of the alkane is broken and replaced with a C–X bond in which a halogen atom (X = Br or Cl) has substituted for a hydrogen atom. The products of this reaction are an **alkyl halide** or haloalkane and a hydrogen halide. In more complex alkanes, substitution can occur to some extent at all positions to give a mixture of products.

Example 11

Show the reaction for the stepwise substitution of chlorine atoms for the hydrogen atoms on methane; also, determine the I.U.P.A.C. name for each organic compound produced.

Answer

$$CH_4 + Cl_2 \xrightarrow{\ UV\ } CH_3Cl + HCl$$

$$CH_3Cl + Cl_2 \xrightarrow{\ UV\ } CH_2Cl_2 + HCl$$

$$CH_2Cl_2 + Cl_2 \xrightarrow{\ UV\ } CHCl_3 + HCl$$

$$CHCl_3 + Cl_2 \xrightarrow{\ UV\ } CCl_4 + HCl$$

Key Terms

aliphatic hydrocarbon
alkane
alkyl halide
alkyl group
aromatic hydrocarbon
axial atom
chair conformation
cis-trans isomers
combustion
condensed formula
conformations
conformers
constitutional isomers
cycloalkane
eclipsed conformation
equatorial atom

functional group
geometric isomer
halogenation
hydrocarbon
I.U.P.A.C. Nomenclature System
molecular formula
parent compound
saturated hydrocarbon
structural formula
substituted hydrocarbon
substitution reaction
unsaturated hydrocarbon

Self Test

1. Which of the following are isomers of C_5H_{12}? Give letter(s) as your answers.
    a. $CH_3CH_2CH_2CH_3$
    b. $CH_3CH_2CH_2CH_2CH_3$
    c. $CH_3CHCH_3CH_2CH_3$
    d.
$$CH_3CCH_3 \begin{array}{c} CH_3 \\ | \\ C \\ | \\ CH_3 \end{array}$$

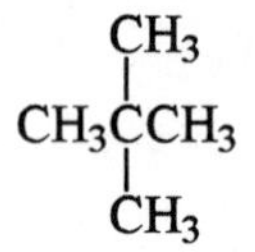

    e. $CH_3CH(CH_3)_2$
2. What is the geometry of a carbon surrounded by four single bonds?
3. What is the angle around carbon surrounded by a double bond and two single bonds?
4. What is the hydrogen-carbon-carbon bond angle in ethyne, HCCH?
5. What is the term used to describe the group in an organic compound that is primarily responsible for the chemical and physical properties of that compound?
6. Name the functional group found in all alcohols.
7. Name the functional group found in all aldehydes.
8. Name the family whose simplest member is benzene.
9. Which of the following is an isomer of 1,2-dibromoethane?
    a. Br_2CHCH_3 c. $BrCH_2CH_2CH_3$
    b. CH_2BrCH_2Br

10. Which of the following is an isomer of C_4H_{10}?
    a. benzene
    b. cyclobutane
    c. 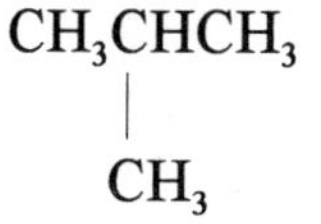 CH_3CHCH_3
       |
       CH_3
    d. $CH_3CHBrCH_2CH_3$
    e. 2,2-dimethylpropane

11. Which of the following is an isomer of CH_3OCH_3?
    a. CH_3CH_2OH
    b. 2-methylbutane
    c. $CH_3CH_2OCH_3$
    d. oxymethylurea
    e. dimethylether

12. Which of the following is the simplest organic compound?
    a. ethyl alcohol
    b. methane
    c. ammonium cyanate
    d. urea
    e. acetylene

13. Which of the following is not a normal property of an inorganic compound?
    a. flammable
    b. high boiling point
    c. soluble in water
    d. reactions are fast
    e. high melting point

14. Which of the following formulas have limited usage in organic chemistry?
    a. space filling formula
    b. condensed formula
    c. molecular formula
    d. structural formula

15. Compounds that contain only carbon and hydrogen atoms are classified in what major subdivision?
    a. alkanes
    b. aliphatics
    c. aromatics
    d. heterocyclic
    e. hydrocarbons

16. What field of study is defined as the chemistry of all the elements except carbon?

17. What organic compound did Freiderich Wöhler synthesize from inorganic reactants?

18. Alkanes contain only carbon and what other element?

19. Name the functional group found in all alkenes.

20. Give the molecular formula of the alkane that contains nine carbon atoms.

21. Give the I.U.P.A.C. names of the following branched-chain alkanes:

a.

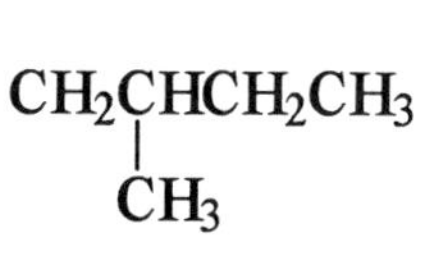

$CH_2CHCH_2CH_3$
    |
    CH_3

b. 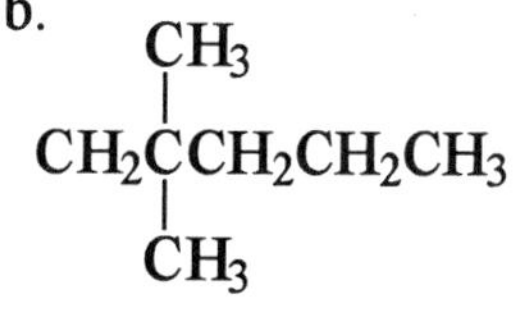
       CH_3
       |
$CH_2CCH_2CH_2CH_3$
       |
       CH_3

c. 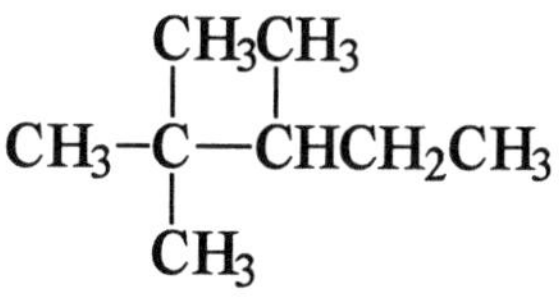
         $CH_3 CH_3$
         | |
$CH_3-C-CHCH_2CH_3$
         |
         CH_3

125

22. Give the I.U.P.A.C. name of the following:

$$CH_3\overset{\overset{\displaystyle H}{|}}{\underset{\underset{\displaystyle Cl}{|}}{C}}CH_2\overset{\overset{\displaystyle CH_3}{|}}{\underset{\underset{\displaystyle H}{|}}{C}}CH_2Br$$

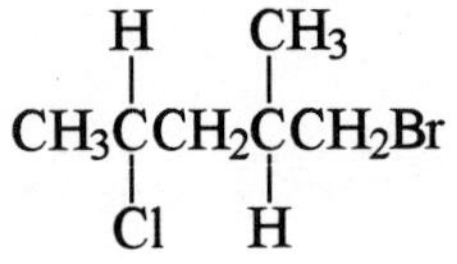

23. Give the I.U.P.A.C. name of the following:

$$CH_3\overset{\overset{\displaystyle CH_3}{|}}{\underset{\underset{\displaystyle Br}{|}}{C}}CH_3$$

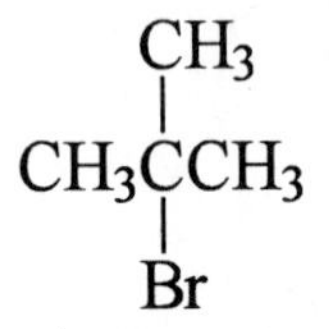

24. Give the I.U.P.A.C. name of the following:

$$CH_3\overset{\overset{\displaystyle CH_2CH_3}{|}}{\underset{\underset{\underset{\underset{\displaystyle Br}{|}}{CH_2CHCH_3}}{|}}{C}}CH_3$$

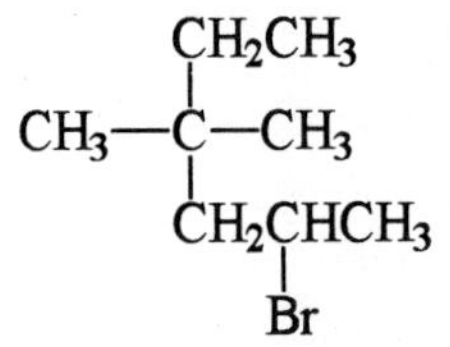

25. Write an equation representing the complete combustion of propane.

26. Write the names of the monosubstitution products that are produced when Cl_2 and propane react under UV light.

27. Describe the solubility of alkanes in water.

28. Are the melting and boiling points of alkanes generally higher or lower than those of other organic compounds?

29. In the I.U.P.A.C. system of naming compounds, the name of the compound is defined by the longest continuous carbon chain in the compound. What do we call this continuous chain?

30. What is the I.U.P.A.C. name of $CHCl_3$?

31. The general formula C_nH_{2n} can be used to find the molecular formula of any member of what alkanelike class of organic compounds?

32. What does the prefix *cyclo-* mean?

33. What is the term for the oxidation of alkanes and other hydrocarbons to carbon dioxide and water?

34. How many moles of water are produced by the complete combustion of one mole of methane?

35. For methane to react with chlorine, what catalyst is needed?

36. What type of organic reaction involves the replacement of one or more atoms in a molecule by new atoms?

12 *The Unsaturated Hydrocarbons: Alkenes, Alkynes, and Aromatics*

Learning Goals

1. Conceptual Goals

· Know the structural features associated with the carbon-carbon double and triple bonds and the benzene ring.

· Be familiar with the physical properties of the alkenes, alkynes, and aromatic hydrocarbons.

· Know the important classes of unsaturated hydrocarbons, aromatic hydrocarbons, and polyhalogenated hydrocarbons; provide examples of each of these classes of compounds.

· Explain Markovnikov's Rule and show examples of its application.

2. Performance Goals

· Draw the structures of and name simple alkenes, alkynes, aromatic hydrocarbons, and aryl halides.

· Predict, write the structures of, and name geometric isomers of alkenes.

· Write equations predicting the products of the simple addition reactions of alkenes: hydrogenation, halogenation, hydration, and hydrohalogenation.

· Discuss the addition reaction mechanism for alkenes, particularly as it pertains to the hydration reaction.

3. Health Applications

· Summarize the theory relating certain aromatic compounds and cancer.

· Describe several health-related effects of the halogenated hydrocarbons.

· List properties of DDT and other halogenated insecticides that relate to environmental and health problems.

Chapter Overview

12.1 Alkenes and Alkynes: Structure and Physical Properties

Alkenes, alkynes, and **aromatic compounds** contain at least one carbon-carbon double or triple bond. As a result of multiple bonding, these compounds contain fewer hydrogens than alkanes with the same number of carbon atoms. They are referred to as unsaturated because they do not contain as many hydrogens as their carbon skeleton will allow.

The structures seen on the next page reveal the structural differences among alkanes, alkenes, and alkynes.

Each of the compounds in the diagram has the same number of carbon atoms, but they differ in the number of hydrogen atoms. Alkynes have the general formula C_nH_{2n-2} and thus contain two fewer hydrogens than the corresponding alkene (general formula C_nH_{2n}). Alkenes have two fewer hydrogens than the corresponding alkane (general formula C_nH_{2n+2}).

Alkanes contain only single bonds; alkenes have at least one carbon-to-carbon double bond; and alkynes contain at least one carbon-to-carbon triple bond. The differences in bond order result in variation in molecular geometry and chemical reactivity among these three families. Alkanes have tetrahedral carbon atoms. When carbon is bonded by one double bond and two single bonds, as in alkenes, the molecule is planar and each bond angle is 120°. When two carbons are bonded by a triple bond, the molecule is linear and the bond angles are 180°.

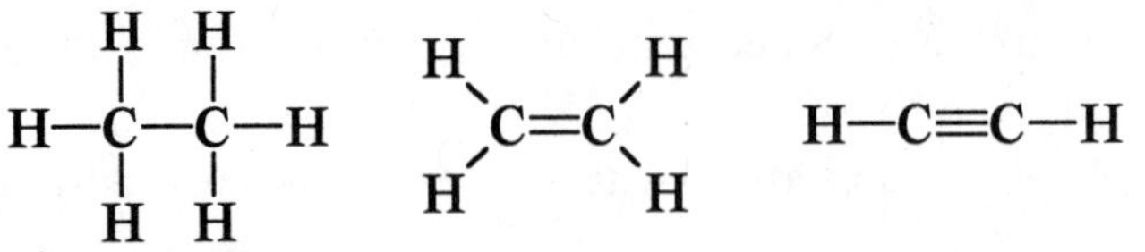

| Ethane (Ethane) | Ethene (Ethylene) | Ethyne (Acetylene) |

Structural formulas

C_2H_6 C_2H_4 C_2H_2

Molecular formulas

CH_3CH_3 $CH_2=CH_2$ $CH\equiv CH$

Condensed formulas

Alkenes and alkynes are nonpolar. As a result they are not water-soluble but are readily soluble in nonpolar solvents and in many low-polarity organic solvents such as ether or chloroform. They are also less dense than water and have relatively low boiling points.

12.2 Alkenes and Alkynes: Nomenclature

The nomenclature of alkenes and alkynes is analogous to that of the alkanes, with the following exceptions. For alkenes, the parent name is derived from the longest continuous carbon chain containing the double bond. Then the *-ane* ending of the alkane is replaced with the *-ene* ending of an alkene. The chain is numbered to give the lowest numbers for the two carbons containing the double bond.

Alkynes are named in the same way as the alkenes, except that the *-ane* ending of the corresponding alkane is replaced with the *-yne* ending of alkynes. The rules used in numbering the alkene chain are also used in alkyne nomenclature.

Example 1
Draw and name the alkane, alkene, and alkyne that contain three carbon atoms.

Work
1. To find the molecular formula of an alkane, use the general formula C_nH_{2n+2}
2. To find the molecular formula of an alkene, use the general formula C_nH_{2n}
3. To find the molecular formula of an alkyne, use the general formula C_nH_{2n-2}

Answer
1. The alkane with three carbon atoms has the molecular formula C_3H_8:

    $CH_3CH_2CH_3$ I.U.P.A.C. name: propane
2. The alkene with three carbon atoms has the molecular formula C_3H_6:

    $CH_3CH=CH_2$ I.U.P.A.C. name: propene
3. The alkyne with three carbon atoms has the molecular formula C_3H_4:

Draw three carbon atoms; and place a triple bond between two carbon atoms, then fill in the structure with hydrogen atoms so each carbon has four bonds.

$$CH_3C\equiv CH$$

I.U.P.A.C. name: propyne

Example 2
Give the I.U.P.A.C. name of the following molecule:

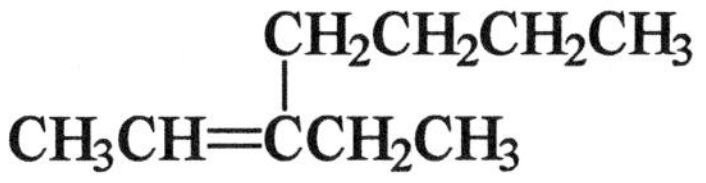

$$\begin{array}{c} CH_2CH_2CH_2CH_3 \\ | \\ CH_3CH=CCH_2CH_3 \end{array}$$

Work

1. First determine the parent compound (longest continuous chain) that contains the double bond. The parent compound is heptene. Since the *hept-* means seven carbons and the *-ene* means double bond, we have

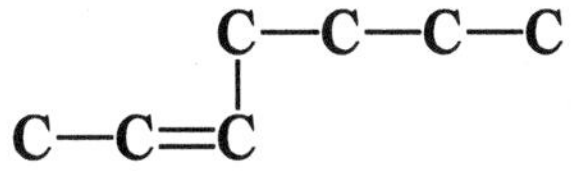

$$\begin{array}{c} C-C-C-C \\ | \\ C-C=C \end{array}$$

129

2. Next, label the parent to give the double bond the lowest possible number:

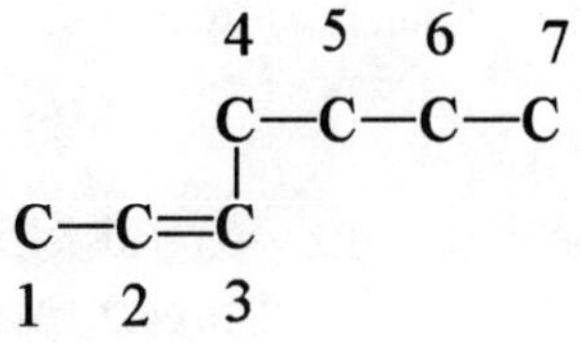

$$
\begin{array}{cccc}
4 & 5 & 6 & 7 \\
\mathrm{C} - \mathrm{C} - \mathrm{C} - \mathrm{C} \\
\end{array}
$$

$$
\begin{array}{ccc}
\mathrm{C} - \mathrm{C} {=} \mathrm{C} \\
1 & 2 & 3
\end{array}
$$

   2-heptene; the 2 tells us that the double bond is between carbons 2 and 3.

3. Finally, list the attached groups, as in naming alkanes.

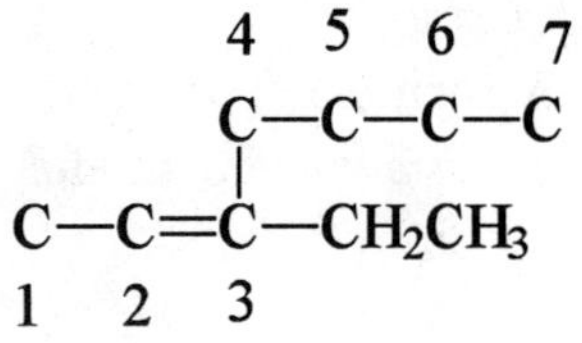

$$
\begin{array}{cccc}
4 & 5 & 6 & 7 \\
\mathrm{C} - \mathrm{C} - \mathrm{C} - \mathrm{C} \\
\end{array}
$$

$$
\begin{array}{ccc}
\mathrm{C} - \mathrm{C} {=} \mathrm{C} - \mathrm{CH_2CH_3} \\
1 & 2 & 3
\end{array}
$$

Answer
3-ethyl-2-heptene

Example 3
Determine the I.U.P.A.C. name of the following molecule:

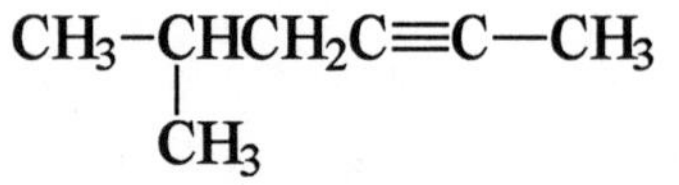

$$CH_3 - \underset{\underset{CH_3}{|}}{CH} CH_2 C {\equiv} C - CH_3$$

Work
1. First determine the parent compound that contains the triple bond. The parent compound is hexyne. The *hex-* means six carbons, and the *-yne* means a triple bond.
2. Label the compound so that the triple bond has the lowest number possible.

$$
\begin{array}{cccccc}
\mathrm{C} - \mathrm{C} - \mathrm{C} - \mathrm{C} {\equiv} \mathrm{C} - \mathrm{C} \\
6 & 5 & 4 & 3 & 2 & 1
\end{array}
$$

   Thus, it is 2-hexyne.

3. Finally, list the attached groups as before:

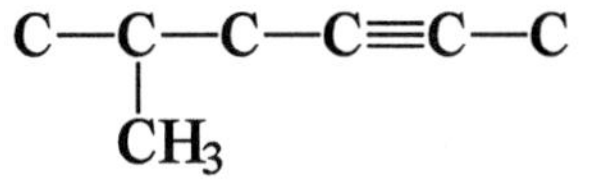

$$C-\underset{\underset{\displaystyle CH_3}{|}}{C}-C-C\equiv C-C$$

Answer

5-methyl-2-hexyne

Example 4

Draw each of the following using structural formulas:

1. 1-Bromo-2-hexyne   2. 2-Butene

Work

1. 1-Bromo-2-hexyne: there are 6 carbons (*hex*-) in the parent compound and a triple bond between carbons 2 and 3 (*2-yne*). Also, there is a bromine (Br–) bonded to carbon 1.

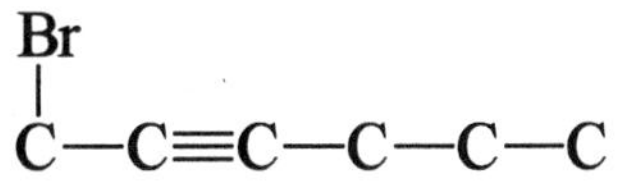

$$\underset{\displaystyle C}{\overset{\displaystyle Br}{\overset{\displaystyle |}{}}}-C\equiv C-C-C-C$$

Next, fill in the other bonds with hydrogen atoms.

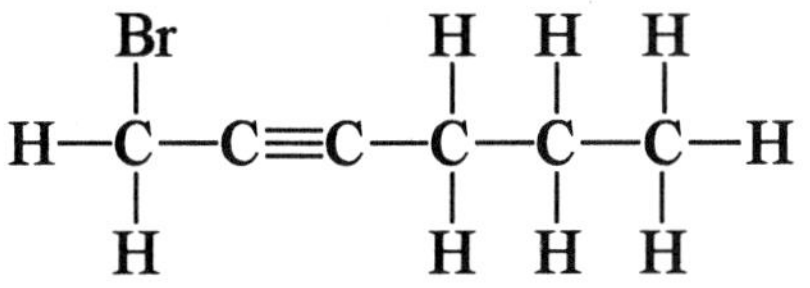

$$H-\underset{\underset{\displaystyle H}{|}}{\overset{\overset{\displaystyle Br}{|}}{C}}-C\equiv C-\underset{\underset{\displaystyle H}{|}}{\overset{\overset{\displaystyle H}{|}}{C}}-\underset{\underset{\displaystyle H}{|}}{\overset{\overset{\displaystyle H}{|}}{C}}-\underset{\underset{\displaystyle H}{|}}{\overset{\overset{\displaystyle H}{|}}{C}}-H \quad \text{or} \quad BrCH_2-C\equiv C-CH_2CH_2CH_3$$

2. 2-Butene: There are four carbons [*but*-] in the parent compound and a double bond between carbons 2 and 3.

$$C-C=C-C$$

Next, fill in the other bonds with hydrogen atoms.

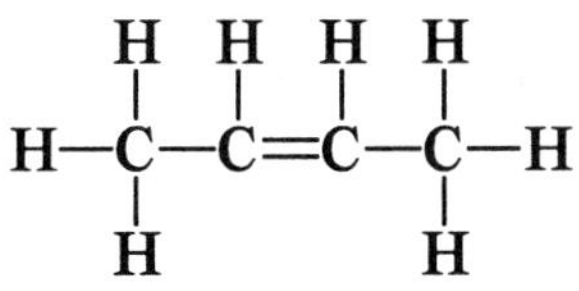

$$H-\underset{\underset{\displaystyle H}{|}}{\overset{\overset{\displaystyle H}{|}}{C}}-\overset{\overset{\displaystyle H}{|}}{C}=\overset{\overset{\displaystyle H}{|}}{C}-\underset{\underset{\displaystyle H}{|}}{\overset{\overset{\displaystyle H}{|}}{C}}-H \quad \text{or} \quad CH_3CH=CHCH_3$$

12.3 Geometric Isomers: A Consequence of Unsaturation

The carbon-carbon double bond is rigid because there is no free rotation around the double bond. The rigidity is caused by the shapes of the orbitals involved in the double bond. The electrons of one carbon-carbon bond lie in a line between the two nuclei. This is a sigma (σ) bond. The second bond, called a pi (π) bond, is formed between two *p* orbitals.   The electrons of the π bond lie above and below the carbon atoms of the double bond. This π bond would have to be broken to allow rotation to occur; thus, the double bond is rigid.

The rigidity of the carbon-carbon double bond in alkenes produces another class of isomers: **geometric isomers**. Geometric isomers are described using the prefixes *cis-* and *trans-*, which provide an easy method for naming and distinguishing between the two isomeric forms.

Geometric isomers are different molecules with different physical and chemical properties. Although their properties may be similar, they are never identical.

The prefixes *cis-* and *trans-* refer to the placement of the substituents attached to the carbon-carbon double bond. When identical groups are on the same side of the double bond, the prefix *cis-* is used; when identical groups are on opposite sides of the double bond, *trans-* is the appropriate prefix.

As we saw in the previous chapter, cycloalkanes can also have geometric isomers. When two substituents are on the same side of the plane of the ring, the compound is the *cis-* isomer. Alternatively, if the two substituents are on opposite sides of the plane of the ring, the compound is the *trans-* isomer.

Example 5
Draw the *cis-* and *trans-* isomers of 2-Butene

Work
1. The carbon-carbon double bond is rigid as a result of the shapes of the orbitals involved in its formation. The specific atoms that are bonded to the two carbon atoms are locked into specific arrangements around the double bond.
2. The *cis-* arrangement means that the two reference groups (say A and A) are on the same side of the double bond. The *trans-* arrangement means the two A groups are across from each other.

cis-isomer	*trans*-isomer

Answer

1. First we draw the 2-butene in a straight form, as before:

$$CH_3CH=CHCH_3$$

2. Next, locate the double bond and place the remaining groups in either the cis- or trans-forms:

cis-2-Butene   trans-2-Butene

12.4 Reactions Involving Alkenes

The major reactions of alkenes involve the addition of atoms or molecules to the carbon-carbon double bond. The principal kinds of **addition reactions** are hydrogenation, halogenation, hydration, and hydrohalogenation.

Hydrogenation: Addition of H_2 to an Alkene

Hydrogenation is the addition of a molecule of hydrogen (H_2) to a carbon-carbon double bond to produce an alkane. Two new C—H single bonds are formed as the double bond is broken.

Hydrogenation generally requires heat and/or pressure. The reaction *always* requires a metal catalyst, such as platinum or nickel, to allow the reaction to occur at a reasonably rapid rate.

The food industry takes advantage of the hydrogenation reaction. Vegetable oils are liquid because the fat molecules have many double bonds. When these oils are hydrogenated, they become solid fats, such as margarine.

Example 6

Write an equation representing the hydrogenation reaction for 2-butene. Indicate the conditions needed, and name all reactants and products.

Work

1. The term *hydrogenation* means the reaction of gaseous H_2 with a specific reactant. A catalyst such as Pt, Pd or Ni is needed to speed up the reaction. Elevated temperature and pressure are also required.

Answer

$$CH_3CH{=}CHCH_3 + H_2 \xrightarrow[\text{pressure}]{\text{Ni, heat}} CH_3CH_2CH_2CH_3$$

2-Butene   Butane

Halogenation: Addition of X_2 to an Alkene

Chlorine (Cl_2) or bromine (Br_2) can be added to a double bond. This reaction, called **halogenation**, proceeds readily and does not require a catalyst.

Example 7

Write an equation representing the halogenation of ethene to produce 1,2-dichloroethane.

Work

1. The term *halogenation* means the addition of a halogen molecule (X_2) across the double bond of an alkene. The double bond is broken and two new C–X bonds are formed.

Answer

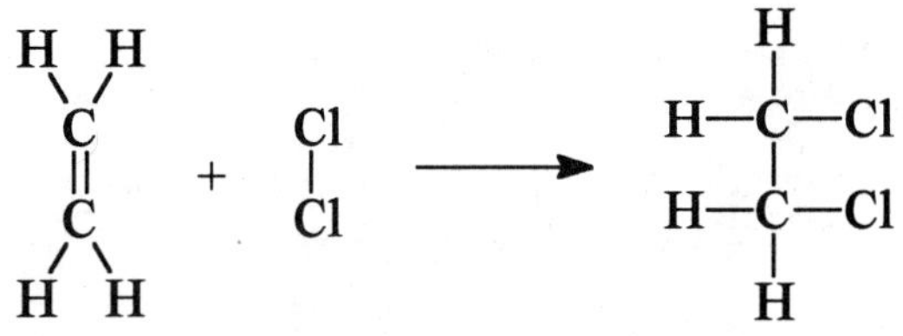

Ethene + Chlorine ⟶ 1,2-Dichloroethane

Halogenation involving bromine (Br_2) can be used to show the presence of double bonds in an organic compound. The reaction mixture is red due to the presence of dissolved bromine. If the red color is lost, it indicates that bromine was consumed, bromination occurred, and the compound must have had a carbon-carbon double bond.

Hydration: Addition of H_2O to an Alkene

A **hydration** reaction is the addition of a water molecule to a carbon-carbon double bond. When an alkene is reacted with water containing a trace of acid, an –OH group bonds to one carbon of the carbon-carbon double bond and an –H atom bonds to the other carbon. The product of the hydration of an alkene is an alcohol.

When an alkene is unsymmetrical (carries two different groups on the double bond carbons) hydration can yield two different products. One is usually favored over the other, as explained by **Markovnikov's Rule** which tells us that the "rich get richer." This means that the carbon of the carbon-carbon double bond that carries the greater number of hydrogen atoms most often will receive the hydrogen atom being added to the double bond. The other carbon becomes bonded to the –OH group.

Example 8

Write an equation representing the hydration of ethene to produce ethanol (ethyl alcohol).

Work

1. The term *hydration* means the addition of a water molecule (HOH) across the double bond of an alkene. The –H bonds to one carbon of the double bond and the –OH attaches to the other carbon of the double bond. The double bond is broken and in its place are two new single bonds.

Answer

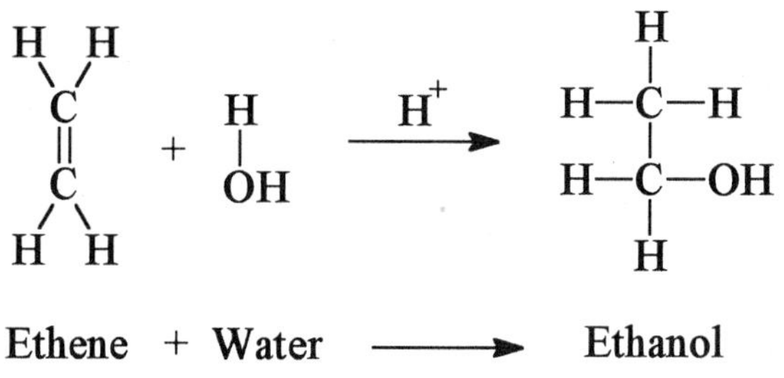

Ethene + Water ⟶ Ethanol

Example 9

Write an equation representing the hydration of propene, and name and draw the major product.

Work

1. The hydration of an unsymmetrical alkene (like propene—having different groups on each side of the double bond) favors one product over the other.

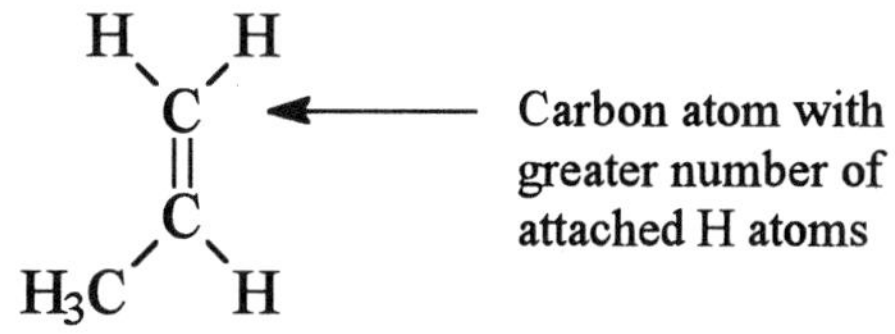

Carbon atom with greater number of attached H atoms

135

The carbon atom of the double bond that has the greater number of directly attached hydrogen atoms most often will receive the hydrogen atom being added to the double bond. The remaining carbon atom bonds to the –OH group. This is known as Markovnikov's Rule.

Notice, that propene is unsymmetrical, since one carbon of the double bond has 2 –H atoms bonded to it while the other carbon of the double bond has only one –H atom and one –CH$_3$ group attached to it.

Answer

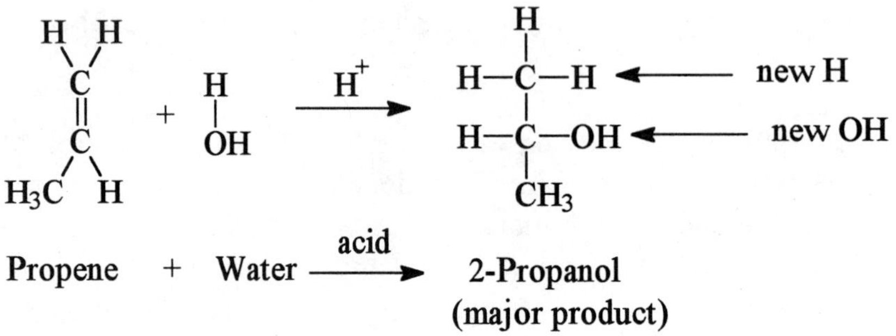

Propene + Water $\xrightarrow{\text{acid}}$ 2-Propanol (major product)

Hydrohalogenation: Addition of HX to an Alkene

A hydrogen halide (H–Br, H–Cl, or H–I) also can be added to an alkene. This addition reaction is called **hydrohalogenation** and the product is an alkyl halide.

The reaction mechanism of hydrohalogenation is the same as that for hydration and the nature of the major and minor alkyl halide products can be predicted by Markovnikov's Rule. This reaction provides an easy, alternative means of preparing alkyl halides.

Example 9
Write an equation representing the reaction of hydrogen bromide with propene. Include the major and minor products. Also, indicate any other needed conditions.

Work
1. The reaction follows Markovnikov's Rule.
2. Since alkenes are insoluble in water, the alkene and the hydrogen bromide (or the HX) are dissolved in CCl$_4$ as the solvent. Alkenes and CCl$_4$ are both nonpolar.

Answer

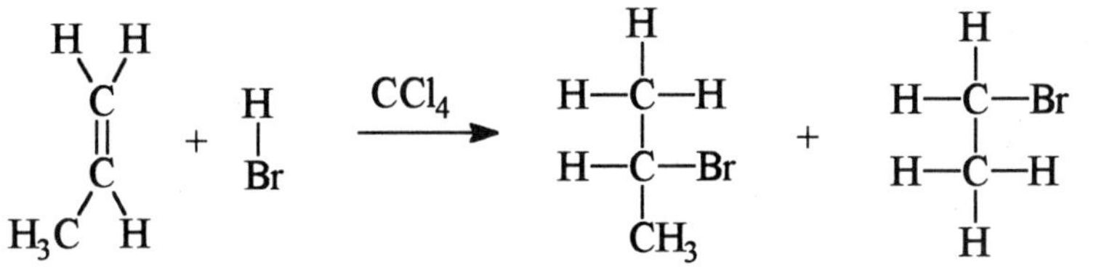

Propene   Hydrogen bromide   2-Bromopropane (major product)   1-Bromopropane (minor product)

Addition Polymers

Addition polymers are produced by the sequential addition of an alkene monomer to produce the **polymer**, which is a macromolecule composed of many repeating structural units (**monomers**). Many useful plastics are addition polymers produced from alkene monomers. Table 12.1 of the text presents a number of common addition polymers.

Example 10

Write an equation representing the reaction that results in the production of the addition polymer polyvinyl chloride (PVC) from vinyl chloride monomers. What is the I.U.P.A.C. name for vinyl chloride?

Work

1. Begin by determining the structure of vinyl chloride (Table 12.1):

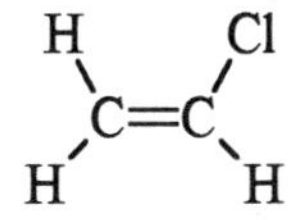

2. Now determine the structure of the addition polymer:

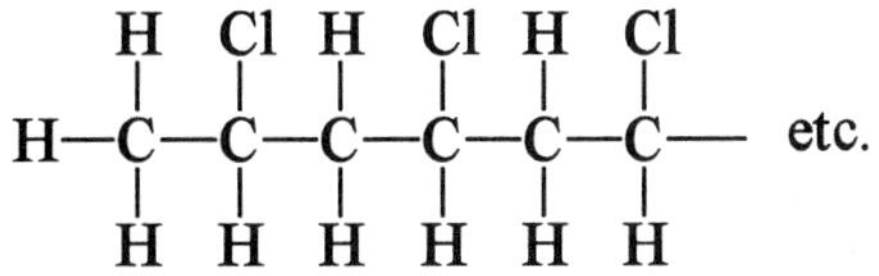

Using the simplified general form for the polymer, write the equation representing the reaction.

Answer

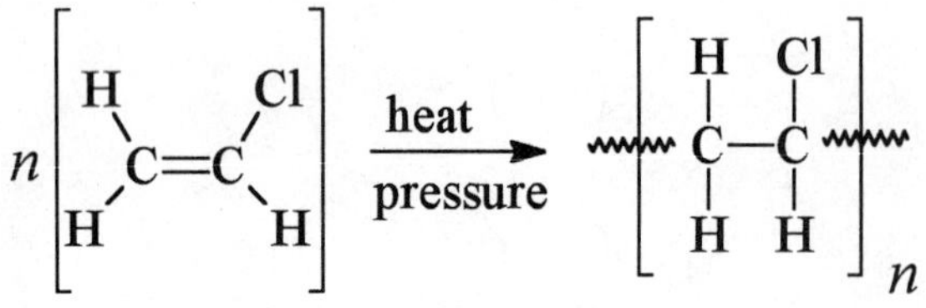

The I.U.P.A.C. name for vinyl chloride is chloropropene.

12.5 The Mechanism of Hydration

A **reaction mechanism** is a detailed, pictorial description of the events that occur as reactants are converted into products. It is based on many, many experimental studies of the structures of the products, the rate of reaction, and the effect of varying concentrations of reactant or temperature.

The electrons of the π bond are involved in all addition reactions because they are more exposed and can readily provide electrons to intermediates that are electron-deficient. An electron-deficient species is called an **electrophile**. An electron-rich species is called a **nucleophile**. A nucleophile $(:Nu^-)$ donates electrons to an electrophile (E^+):

$$E^+ \; + \; :Nu^- \; \rightarrow \; E:Nu$$

The three steps of the reaction mechanism for the hydration of an alkene are shown on the next page and are briefly summarized here:

Step I: The π electrons and H^+ provided by the acid catalyst produce a **carbocation**, a positively charged electron-deficient intermediate in the reaction.

Step II: The carbocation reacts with water to form an oxonium cation. The positive charge is on the oxygen atom.

Step III: In the third step of hydration of an alkene the oxonium ion rapidly reacts to lose a proton and produce an alcohol.

STEP I:

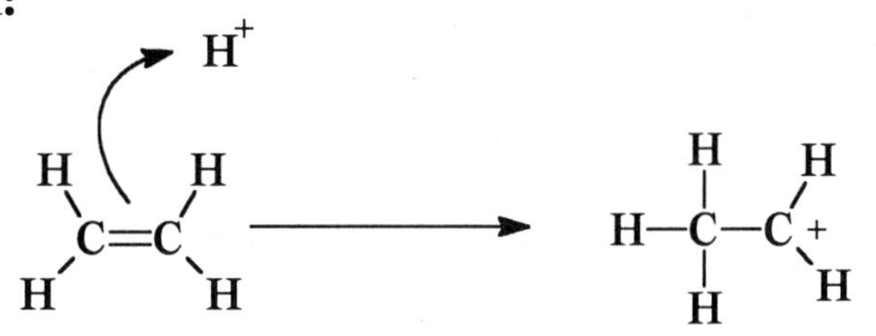

$$H^+$$

$$H_2C=CH_2 \longrightarrow H_3C-CH_2^+$$

Carbocation

STEP II:

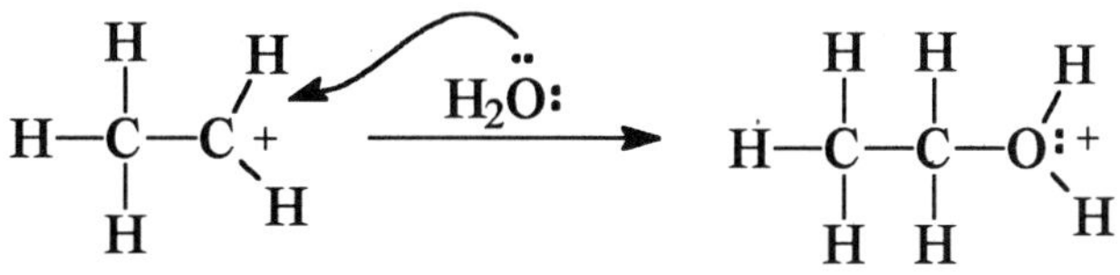

$$H_3C-CH_2^+ + H_2O: \longrightarrow H_3C-CH_2-OH_2^+$$

Oxonium ion

STEP III:

$$H_3C-CH_2-OH_2^+ + H_2O: \longrightarrow H_3C-CH_2-O-H + H_3O^+$$

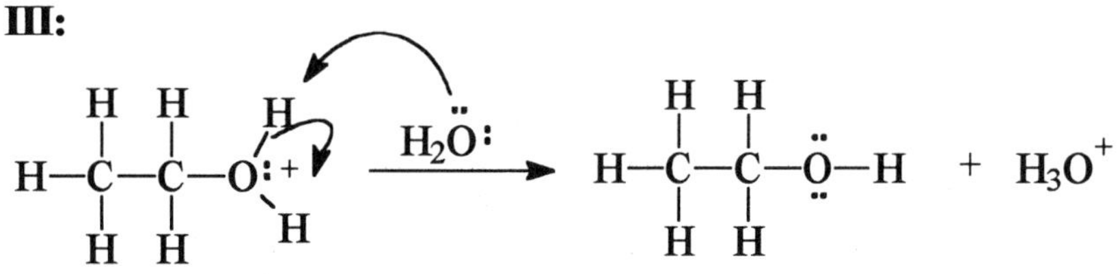

12.6 Aromatic Hydrocarbons

Structure and Physical Properties

Aromatic compounds are all characterized by the presence of an aromaatic ring within the structure. The simplest, and most common, aromatic compound is benzene, which consists of a six carbon ring in a planar hexagonal arrangement. Each carbon is bonded to two other carbon atoms and to a single hydrogen atom.

In 1865 Friedrich Kekulé proposed that single and double bonds alternated around the hexagonal ring. Since benzene did not decolorize bromine, he further proposed that the double and single bonds shifted position rapidly. Today we would show this as a resonance structure.

The current model of benzene structure proposes that each carbon atom is bonded to two other carbon atoms and to a hydrogen atom by sigma (σ) bonds. The remaining six electrons are located in p orbitals that overlap to produce a pi (π) cloud of electrons above and below the ring. Because the electrons are *delocalized* within the π cloud, benzene is unusually stable and resists addition reactions typical of alkenes.

139

Nomenclature

Most simple aromatic compounds are named as derivatives of benzene. Others, for example phenol and toluene, have common historical names that must simply be memorized.

When there are two groups on the ring, there are three possible arrangements. Each arrangement is indicated by a prefix, as follows:

ortho, o: substituents are on adjacent carbon atoms
meta, m: substituents are on carbon atoms separated by one carbon atom
para, p: substituents are on carbon atoms separated by two carbon atoms.

When more than two substituents are attached to the ring, a numerical system must be used in naming the compound.

Example 11
Name the following aromatic compounds:

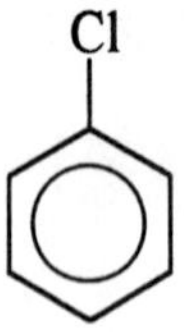

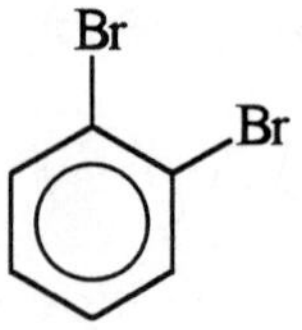

Work
The term aromatic means that there is a benzene ring present.
- If only one group is bonded to the benzene structure, no number is needed in its name.
- If two groups are bonded to the benzene ring, the following naming method is used: (a) if the groups are on adjacent carbons, use *ortho-*; (b) if groups are separated by one carbon, use *meta-*; (c) if two carbons are between groups, use *para-*.
For three groups, it is best to use numbers.

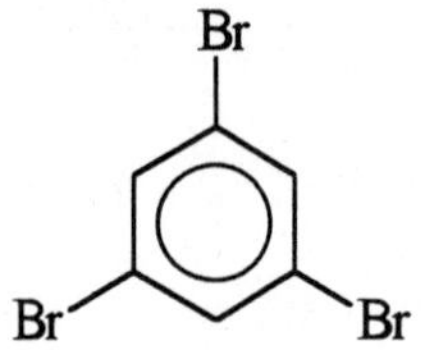

Answer
1. Chlorobenzene
2. 1,4-Dichlorobenzene or *para*-dichlorobenzene
3. 1,3,5-Tribromobenzene
4. 1,2-Dibromobenzene or *ortho*-dibromobenzene

In I.U.P.A.C. nomenclature, the group that results from the removal of a single hydrogen atom from the benzene ring is called the **phenyl group**. Aromatic hydrocarbons having long aliphatic side chains are often named as phenyl-substituted hydrocarbons.

Reactions Involving Benzene

The typical reactions of benzene are **substitution reactions**. In these reactions, a hydrogen atom is replaced by another atom or group of atoms. Benzene can react with chlorine or bromine. The product in these reactions is an aryl halide. Benzene may also undergo sulfonation and nitration.

Key Terms

addition polymer	hydrohalogenation
addition reaction	Markovnikov's Rule
alkene	monomer
alkyne	nucleophile
aromatic hydrocarbon	phenyl group
carbocation	polymer
electrophile	reaction mechanism
geometric isomers	substitution reaction
halogenation	unsaturated hydrocarbon
hydration	
hydrogenation	

Self Test

1. Write the molecular formula for and name the alkane, alkene, and alkyne that contain four carbon atoms. Assume all are in continuous chain forms (no branches).

2. Give the I.U.P.A.C. name of the following compound:

$$CH_3CH\!=\!CCH_2CH_3$$
$$\overset{|}{CH_2CH_3}$$

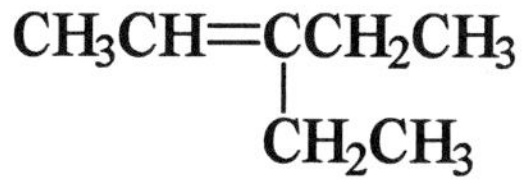

3. Give the I.U.P.A.C. name of the following compound:

$$\overset{Br}{\overset{|}{CH_3CHCHC\!\equiv\!CCH_3}}$$
$$\underset{Br}{\overset{|}{}}$$

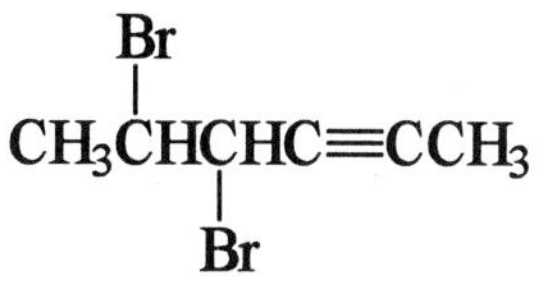

4. Draw the structure of 2-methyl-2-pentene.
5. Draw the structure of *cis*-3-methyl-2-pentene.
6. Name the following compound using the prefixes *cis*- or *trans*-:

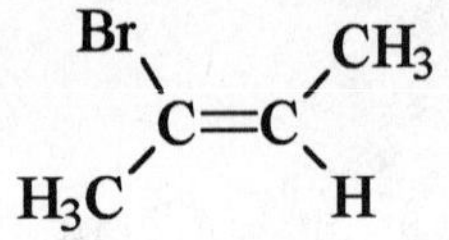

7. Write an equation for the hydrogenation reaction for 1-butene.
8. Write an equation for the hydration of 1-propene. Indicate both the major and minor products.
9. Write an equation for the reaction of hydrogen chloride with 2-methyl-2-pentene.
10. Name the following aromatic compounds:

a.   b. 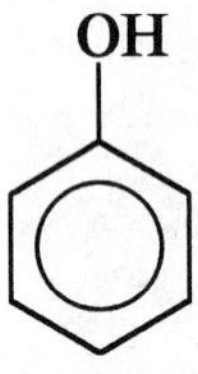   c. 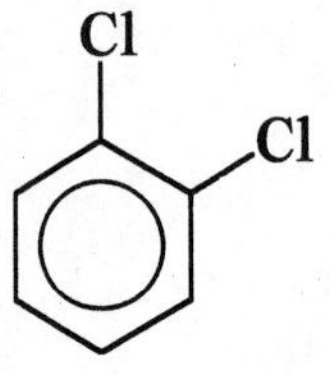   d. 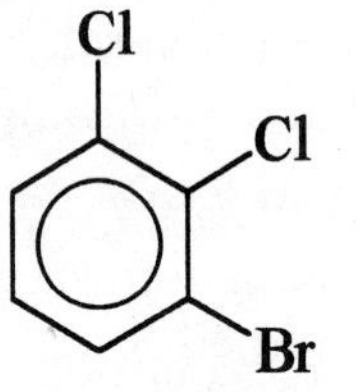   e.

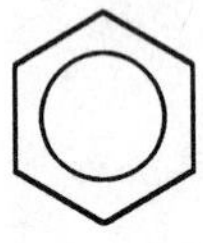

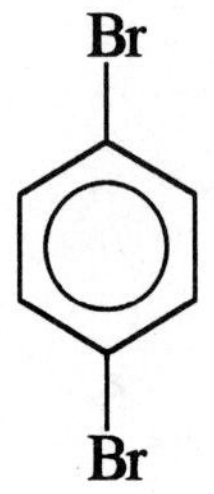

11. What is the common name of ethyne.
12. What is the I.U.P.A.C. name of the following molecule:

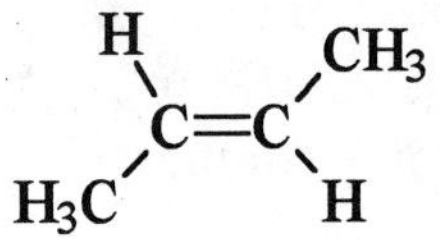

13. What is the I.U.P.A.C. name of the following molecule:

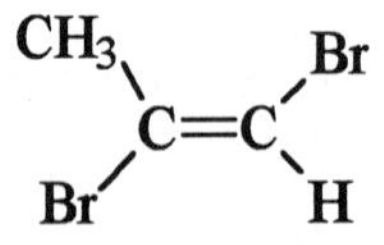

14. What reaction is used to convert alkenes into alkanes?
15. What two catalysts are often used in the hydrogenation of alkenes?

16. Provide the product of the reaction represented by the following equation:

$$CH_2{=}CH_2 \ + \ H_2O \ \xrightarrow{\text{acid}} \ ?$$

17. An aryl unit is similar to what compound?
18. What is the common name of trichloromethane.
19. In ethene the two carbon atoms are joined by what type of bond?
20. Alkenes are characterized by what type of reaction?
21. What does the ending -*ane* mean?
22. What does the ending -*yne* mean?
23. Write an equation showing the reaction of propyne with 2 moles of HBr.
24. Write an equation showing the reaction of 1-butyne with 2 moles of hydrogen gas.
25. Provide the product of the reaction represented by the following equation:

$$CH_3CH_2CH{=}CH_2 \ + \ H_2 \ \xrightarrow{\text{Pd, heat}} \ ?$$

26. What materials are hydrogenated to produce a product such as Crisco?

13 Alcohols, Phenols, Thiols, and Ethers

Learning Goals

1. Conceptual Goals

· Know the names and write structures for the common alcohols, phenols, ethers, and thiols.
· Recognize the members of these four families that are of medical interest.
· Describe the physical properties of each of these four families.
· Classify alcohols as primary, secondary, or tertiary.

2. Performance Goals

· Write equations for the dehydration and oxidation of alcohols.
· Write the mechanism for the dehydration of an alcohol.
· Design simple multistep syntheses involving alcohols and other families.

3. Health Applications

· Appreciate the use of ethanol to treat methanol poisoning.
· Know the chemical reactions that occur in the liver that cause a hangover.
· Describe the fetal alcohol syndrome.
· Summarize the processes that detoxify ethanol.

Chapter Overview

The functional group of the **alcohols** and **phenols** is the **hydroxyl group** (–OH). Alcohols have the general formula R–OH, in which R represents an alkyl group. Thus, the simplest alcohol is methyl alcohol: a methyl group bonded to a hydroxyl group, CH_3OH. Phenols have an aryl group in place of the alkyl group of the alcohols, and thus have the general formula Ar–OH. Phenol, the simplest member of this family, consists of a phenyl group (a benzene ring missing one hydrogen atom) bonded to a hydroxyl group.

     Ethers contain two alkyl or aryl groups attached to an oxygen atom. Thus the functional group of the ethers is R–O–R. **Thiols** are a family of compounds in which the sulfur atom has been substituted for the oxygen atom of an alcohol.

13.1 Alcohols: Structure and Physical Properties

We can think of alcohols (and phenols) as substituted water molecules: an H–O–H molecule in which one hydrogen atom has been replaced by an alkyl or aryl group. The R–O–H portion of an alcohol is planar, as is H–O–H. In addition, the bond angles of R–O–H and H–O–H are both 104.5°.

     Alcohols are polar molecules owing to the hydroxyl group. The oxygen and hydrogen atoms have very different electronegativities (oxygen is 3.5 and hydrogen is 2.1). Consequently, the oxygen atom displays a partial negative charge and the hydrogen atom bears a partial positive charge. Alcohol molecules can form hydrogen bonds with other alcohol molecules and with the molecules of a polar solvent.

Both the polarity of alcohol molecules and the ability to form intermolecular hydrogen bonds exert a strong influence on the physical properties of alcohols. They have higher boiling points than hydrocarbons or ethers of similar molecular weight and show greater solubility in water. In fact, the smallest alcohols (4-5 carbons) are highly soluble in water; those having 5-8 carbons have decreased solubility; and those with more than 8 carbons are insoluble in water.

Example 1
For the following pairs of compounds, indicate the one that has the higher boiling point.

1. CH_4 (methane) or CH_3OH (methanol)
2. $CH_3CH_2CH_2CH_3$ (butane) or $CH_3OCH_2CH_3$ (ethyl methyl ether)
3. ethyl methyl ether or 1-propanol

Work
1. When comparing the boiling points of alkanes, ethers, and alcohols of similar molecular weight, remember that the alcohols have the highest boiling points. The ethers are found to have higher boiling points than the alkanes:

$$alcohol > ethers > alkanes$$

Answer
1. Methanol
2. Ethyl methyl ether
3. 1-Propanol

Example 2
Predict the solubilities of the following alcohols in water solutions:

1. $CH_3CH_2CH_2CH_2CH_2CH_2OH$
2. CH_3CH_2OH
3. $CH_3(CH_2)_8CH_2OH$

Work
1. Since alcohols contain the highly polar hydroxyl group, they are very polar.
2. The smaller members (four to five carbons) of the alcohol family are very soluble in water. However, once the R group (hydrocarbonlike end) has more than five carbon atoms, the alcohol becomes less soluble as the number of its carbon atoms increases. Those with more than eight carbons are insoluble in water.

Answer
1. Not very soluble
2. Very soluble
3. Insoluble in water

13.2 Alcohols: Nomenclature

I.U.P.A.C. Nomenclature

Determine the *parent compound*, in this case the longest continuous carbon chain containing the –OH group. Drop the *-e* ending of the alkane chain and replace it with *-ol*. Number the chain so that the hydroxyl group has the lowest possible number. Add substituents, named and numbered appropriately, as prefixes to the alcohol name.

   If two hydroxyl groups are present, the suffix *-diol* is used; if three hydroxyl groups are present, the suffix used is *-triol*. The position of each of the hydroxyl groups is numbered as usual.

Example 3
Name the following compounds using the I.U.P.A.C. method.

a.   b.   c.   d.

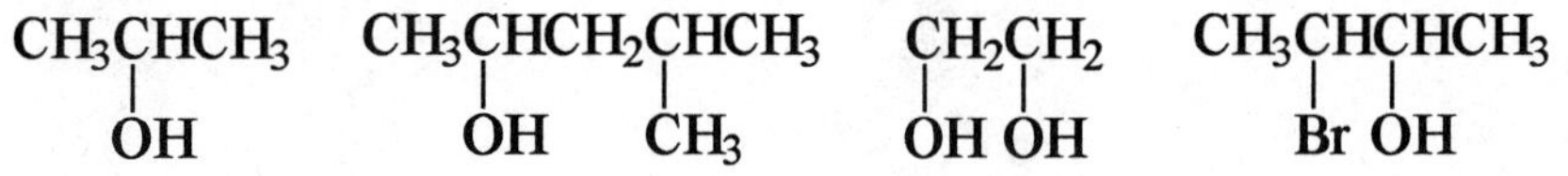

$$\underset{\underset{OH}{|}}{CH_3}CHCH_3 \qquad \underset{\underset{OH}{|}\ \ \underset{CH_3}{|}}{CH_3CHCH_2CHCH_3} \qquad \underset{\underset{OH}{|}\ \underset{OH}{|}}{CH_2CH_2} \qquad \underset{\underset{Br}{|}\ \underset{OH}{|}}{CH_3CHCHCH_3}$$

Work
1. Determine the parent compound that contains the –OH bonded to it.
2. Change the *-ane* ending of the parent alkane's name to *-anol* (drop *e*, add *ol*).
3. Next, number the parent carbon chain so that the –OH has the lowest possible number.
4. If the alcohol has two or more –OH groups, do not drop the e of the *-ane* ending, but add *-diol* or *-triol* to the parent's name.
5. Then use all the other rules we have used previously to finish naming the other groups that may be present.

Answer
a. 2-Propanol
b. 4-Methyl-2-pentanol
c. 1,2-Ethanediol
d. 3-Bromo-2-butanol

Common Names

The common names for the alcohols are derived from the name of the corresponding alkyl group. The name of the alkyl group is followed by the word alcohol.

Example 4

Give the common names of each of the following alcohols:

1. CH_3OH
2. CH_3CH_2OH
3.

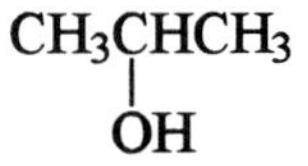

$$CH_3\overset{\underset{\displaystyle |}{OH}}{CH}CH_3$$

4. $CH_3CH_2CH_2CH_2OH$
5.

$$CH_3-\overset{\overset{\displaystyle CH_3}{|}}{\underset{\underset{\displaystyle OH}{|}}{C}}-CH_3$$

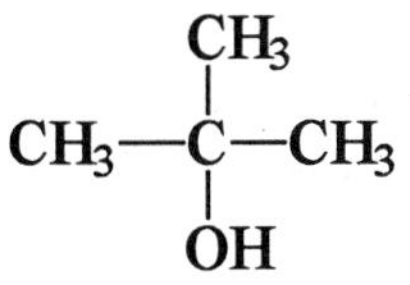

Work

To determine the common names for the alcohols, first find which alkyl group (R–) is bonded to the –OH group. Next, write down the alkyl group name and add the word alcohol to complete the common name of that alcohol.

Answer
1. Methyl alcohol
2. Ethyl alcohol
3. Isopropyl alcohol
4. Butyl alcohol
5. *t*-Butyl alcohol

13.3 Medically Important Alcohols

Several of the small alcohols are important in medicine and industry. The smallest, *methanol* methyl alcohol, wood alcohol) is a common solvent. It is extremely toxic if ingested.

Ethanol (ethyl alcohol, grain alcohol) is a solvent and disinfectant. Ethanol for human consumption is produced by fermentation of sugars by yeast. Denatured alcohol is ethanol to which a denaturing agent has been added to make it undrinkable.

2-Propanol (isopropyl alcohol, rubbing alcohol) is commonly used as a disinfectant and solvent. Like methanol, it is very toxic if ingested.

   1,2-Ethanediol (ethylene glycol) is a common antifreeze for cars.

   1,2,3-Propanetriol (glycerol) is a component of stored fats in the body. It is used in cosmetics and pharmaceuticals.

13.4 Classification of Alcohols

Alcohols are classified as methyl, **primary (1°), secondary (2°), and tertiary (3°)** depending on the number of alkyl groups attached to the **carbinol carbon**, the carbon bearing the hydroxyl group. If there is one alkyl group attached to the carbinol carbon, the alcohol is a primary alcohol. An alcohol with two alkyl groups bonded to the carbinol carbon is a secondary alcohol. A tertiary alcohol has three alkyl groups bonded to the carbinol carbon.

Example 5
Classify the following alcohols as either primary, secondary, or tertiary.

   1. CH_3CH_2OH

   2. 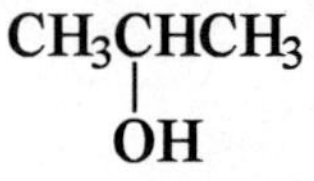 $CH_3\underset{\underset{\displaystyle OH}{|}}{C}HCH_3$

   3. $CH_3CH_2\underset{\underset{\displaystyle OH}{|}}{C}HCH_3$

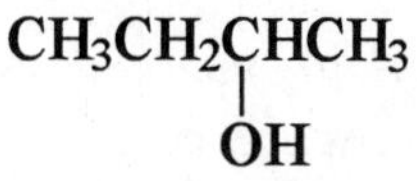

   4. 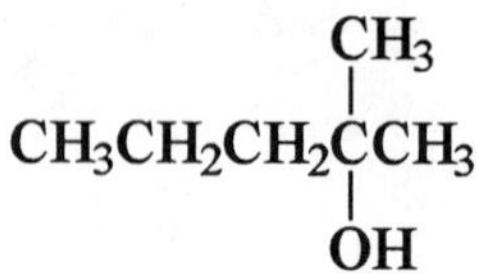 $CH_3CH_2CH_2\overset{\overset{\displaystyle CH_3}{|}}{\underset{\underset{\displaystyle OH}{|}}{C}}CH_3$

Answer
1. Primary
2. Secondary
3. Secondary
4. Tertiary

13.5 Reactions Involving Alcohols

Preparation of Alcohols

In the laboratory, alcohols are prepared by the hydration of alkenes or the reduction of aldehydes and ketones.

Example 6

Show the reduction reactions for the conversions of formaldehyde and acetaldehyde into their corresponding alcohols.

Answer

1.

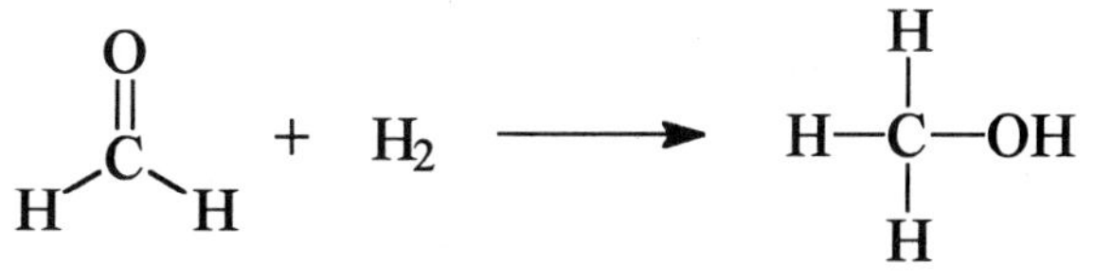

Formaldehyde (methanal)   Methyl alcohol (methanol)

2.

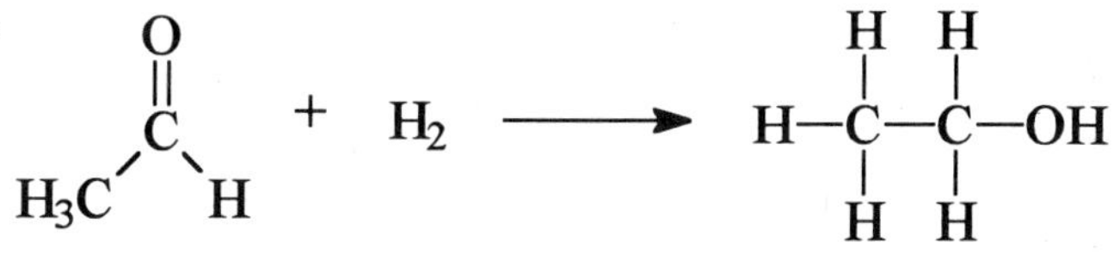

Acetaldehyde (ethanal)   Ethyl alcohol (ethanol)

Dehydration of Alcohols

Alcohols undergo **dehydration** (loss of water) when heated in the presence of concentrated sulfuric or phosphoric acids. The products of dehydration are an alkene and a water molecule. Quite simply, a dehydration reaction is the reverse of the hydration reactions that produce an alcohol from an alkene and water.

The Mechanism of Dehydration

The mechanism of dehydration is the reverse of the mechanism of hydration of an alkene. In the first step the acid catalyst donates a proton that is added to the alcohol to produce an oxonium cation. In step two the oxonium ion loses water to produce a carbocation. Finally the carbocation loses a proton to the solvent, producing an alkene.

Example 7

Write an equation representing the dehydration reactions for the following alcohols:

1. 1-Propanol
2. Ethanol
3. 2-Butanol

Work

The dehydration of an alcohol produces an alkene plus water. Remember that in some cases the dehydration of an alcohol yields a mixture of products. In those instances, it is the more highly branched alkene that is the major product.

Answer

1. $CH_3CH_2CH_2OH \xrightarrow[\text{heat}]{\text{acid}} CH_3CH=CH_2 + H_2O$

     1-Propanol   Propene

2. $CH_3CH_2OH \xrightarrow[\text{heat}]{\text{acid}} CH_2=CH_2 + H_2O$

     Ethanol   Ethene

3.
$$CH_3CH_2\overset{\displaystyle OH}{\underset{\displaystyle H}{\overset{|}{\underset{|}{C}}}}CH_3 \xrightarrow[\text{heat}]{\text{acid}} CH_3CH_2CH=CH_2 + H_2O$$

     2-Butanol   1-Butene

                                       +

$$CH_3CH=CHCH_3 + H_2O$$

                           2-Butene

Two products are formed. The major product is 2-butene.

Oxidation Reactions

Some alcohols can be **oxidized** to produce aldehydes, ketones or carboxylic acids. The most commonly used oxidizing agents are basic potassium permanganate ($KMnO_4/OH^-$) and chromic acid (H_2CrO_4).

Methanol and all primary alcohols produce aldehydes, while secondary alcohols form ketones. Tertiary alcohols cannot be oxidized. This is because the carbon bonded to the hydroxyl group must contain at least one C–H bond in order for oxidation to occur. Since tertiary alcohols contain three C–C bonds to the carbon bonded to the hydroxyl group they cannot be oxidized. Aldehydes can undergo further oxidation to produce carboxylic acids.

Example 8

Write an equation representing the oxidation of methanol or ethanol by the liver. Name all products.

Work

The oxidation of an alcohol first produces an aldehyde or ketone. If an aldehyde is formed, it can be further oxidized to produce a carboxylic acid.

Answer

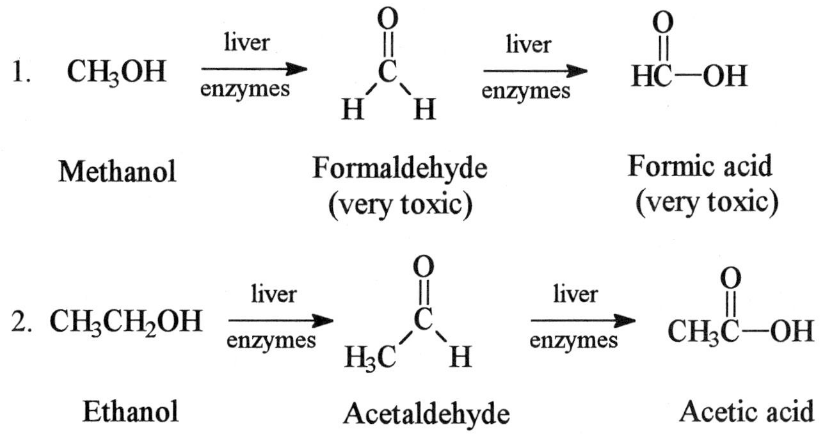

1. CH₃OH → (liver enzymes) Formaldehyde (very toxic) → (liver enzymes) Formic acid (very toxic)

   Methanol

2. CH₃CH₂OH → (liver enzymes) Acetaldehyde → (liver enzymes) Acetic acid

   Ethanol

Example 9

Provide the oxidation products for the following compounds.

1. primary alcohol
2. secondary alcohol
3. tertiary alcohol
4. aldehyde
5. 2-propanol
6. 2-methyl-2-propanol

Answer

1. aldehyde, then to a carboxylic acid

151

2. ketone
3. no reaction
4. carboxylic acid
5. ketone—called acetone
6. no reaction—a tertiary alcohol

13.6 Oxidation and Reduction in Living Systems

In organic and biological chemistry **oxidation** involves a gain of oxygen or a loss of hydrogen from a compound. A **reduction** reaction involves a loss of oxygen or gain of hydrogen. To recognize an oxidation or reduction reaction, simply count the hydrogen and oxygen atoms in the products and reactants and apply the definitions presented above.

Oxidation and reduction reactions are extremely important in the metabolic pathways that harvest energy for use by our bodies. A class of enzymes called *oxidoreductases* catalyze these reactions. *Coenzymes* serve as acceptors, donors, and carriers of hydrogen in these cellular oxidation-reduction reactions. Nicotinamide adenine dinucleotide, NAD^+, is one such coenzyme.

13.7 Phenols

Phenols are compounds in which the hydroxyl group is attached to an aryl group. Owing to the polar hydroxyl group, the phenols are also polar compounds.

The simplest member of this family is known by the common names phenol and carbolic acid. This compound is of interest in the history of medicine. Joseph Lister, a British physician, observed that the incidence of post-surgical infections could be radically decreased if the surgical instruments and the incision were treated with an antimicrobial chemical. The agent he used was carbolic acid. As a result of his observations, the use of antiseptics and disinfectants has become routine medical practice.

13.8 Ethers

Ethers are structurally related to alcohols. However, a quick look at the geometry of the functional group characteristic of the ethers (R–O–R) reveals that these compounds are much less polar than alcohols. Indeed, they are much less water-soluble and have much lower boiling points than the comparable alcohols.

Ethers are chemically inert. Under normal conditions, they will not react with oxidizing agents, reducing agents, or bases.

Common names for ethers are derived from the names of the alkyl groups attached to the ether oxygen. The alkyl group names are listed as prefixes before the word ether and may be ordered by size (small to large) or alphabetically. For instance, the compound $CH_3–O–CH_2CH_3$ can be called methyl ethyl ether (prefixes arranged by size of the alkyl group) or ethyl methyl ether (prefixes ordered alphabetically).

This same compound would be named methoxyethane using the I.U.P.A.C. Nomenclature System. With this system an ether is named as a substituted hydrocarbon, in this case an alkane. In this system, the –OR group is named as an alkoxy group.

Diethyl ether was the first general anesthetic used in medical practice. However, since ethers are highly flammable and can form explosive peroxides upon storage, diethyl ether has largely been replaced by halogenated ethers, such as penthrane and ethrane.

13.9 Thiols

Compounds that contain the –SH group are known as **thiols**. Although similar to alcohols in structure, thiols generally have lower boiling points than corresponding alcohols, though they are higher in molecular weight. Furthermore, thiols, and many other sulfur compounds, have nauseating aromas.

Thiols are also involved in protein structure and conformation. It is the ability of two thiol groups to easily undergo oxidation to a –S–S–(disulfide) bond that helps maintain the correct shape of a protein. The amino acid cysteine has a thiol group and can participate in disulfide bond formation in proteins.

Coenzyme A is a thiol that serves as a "carrier" of acetyl groups and fatty acids in cellular metabolic reactions.

Key Terms

alcohol	oxidation
dehydration	phenol
disulfide	primary (1°) alcohol
ether	reduction
fermentation	secondary (2°) alcohol
hydration	tertiary (3°) alcohol
hydroxyl group	thiol

Self Test

1. From each of the following pairs of compounds, pick the one that has the higher boiling point.
   - a. 1-nonanol or 1-hexanol
   - b. ethane or methanol
   - c. methoxyethane or 1-propanol
   - d. methane or dimethyl ether
2. Which of the following alcohols has the greatest solubility in water?
   - a. 1-propanol
   - b. 1-butanol
   - c. 1-heptanol
   - d. 1-decanol

3. Provide the I.U.P.A.C. name of the following compound:

$$
\begin{array}{ccc}
& \overset{\displaystyle Br}{|} & \\
CH_3CH&CHCH&CH_3 \\
& | \quad\quad | & \\
& OH \quad CH_3 &
\end{array}
$$

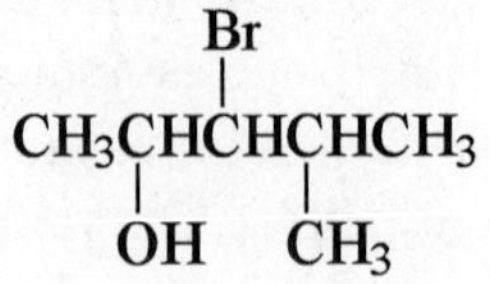

4. Provide the I.U.P.A.C. name of the following compound:

$$
\begin{array}{c}
CH_3CHCH_2CH_2CH_2Cl \\
| \\
OH
\end{array}
$$

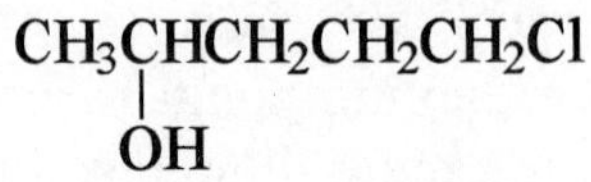

5. Determine the I.U.P.A.C. name of the trialcohol called glycerol: $HOCH_2CHOHCH_2OH$

6. Provide the common names of the following:
   a. $CH_3CH_2CH_2OH$
   b. $CH_3CHOHCH_3$
   c. $C(CH_3)_3OH$

7. Complete the following chemical reactions:

a. 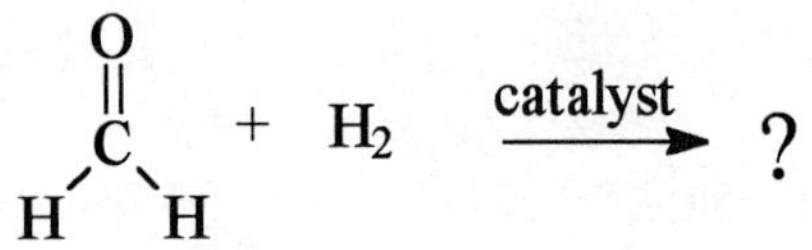
$$
\begin{array}{c}
O \\
\| \\
C \\
H \quad\quad H
\end{array}
\;+\; H_2 \;\xrightarrow{\text{catalyst}}\; ?
$$

b. $CH_2{=}CH_2 \;+\; H_2O \;\xrightarrow[\text{catalyst}]{H^+}\; ?$

8. Complete the oxidations of each of the following alcohols into aldehydes.

a. $CH_3OH \xrightarrow{\text{oxidation}}$

b. $CH_3CH_2OH \xrightarrow{\text{oxidation}}$

9. Write equations representing the dehydration reactions for each of the following alcohols:
   a. ethanol
   b. 1-propanol
   c. 2-butanol

10. Determine the oxidation products of the following alcohols:
    a. CH_3OH c. 2-propanol
    b. ethanol   d. 2-methyl-2-propanol
11. Write equations representing the reactions that result in the conversion of 2-butene into butanone.
12. Provide the common names of each of the following ethers:
    a. CH_3OCH_3 c. $CH_3OCH_2CH_3$
    b. $CH_3CH_2OCH_2CH_3$
13. Which of the following compounds is the most oxidized?
    a. propane   d. acetic acid
    b. 1-propanol   e. propanone
    c. 2-propanol
14. Which of the following compounds is the most reduced?
    a. methane   c. dichloromethane
    b. chloromethane   d. trichloromethane
15. What class of compounds is represented by the general formula, ROH?
16. What class of compounds is represented by the general formula, RSH?
17. Why do alcohols boil at higher temperatures than pure hydrocarbons or ethers of similar molecular weights?
18. What alcohol is produced by the fermentation of sugars and starches? Use the I.U.P.A.C. name.
19. If ethanol has some agent added to it to make it unfit to drink, what do we call that solution?
20. Write the common name of the alcohol that is a viscous, sweet-tasting, non-toxic liquid. (Refer to the text).
21. In a secondary alcohol, how many hydrogen atoms are directly attached to the carbinol carbon atom?
22. Which kind of alcohol will not undergo oxidation under normal conditions?
23. Potassium permanganate and potassium dichromate are used with alcohols in what kind of chemical reaction?
24. Which product of the oxidation of ethanol causes many of the adverse effects of the morning after hangover"?
25. Which class of compounds with a strong odor is found in skunks?
26. Which of the following is the least polar? Give letter as answer.
    a. alcohols
    b. ethers
    c. water
27. A blood level in excess of what percent of ethanol is considered evidence of intoxication?
28. Which of the following cannot hydrogen bond to other molecules?
    a. alcohols
    b. ethers
    c. water

14 *Aldehydes and Ketones*

Learning Goals

1. Conceptual Goals
· Know the names and write structures for the common aldehydes, and ketones.
· Recognize the members of these families that are of medical interest.
· Describe the physical properties of each of these two families

2. Performance Goals
· Write equations for the oxidation and reduction of carbonyl compounds.
· Write equations for the preparation and reactions of hemiacetals, hemiketals, acetals, and ketals.
· Write equations showing aldol condensation reactions.
· Recognize the keto and enol forms of aldehydes and ketones.

3. Health Applications
· Understand the role of vitamin A in vision.
· Be aware of the uses of formaldehyde as a preservative and disinfectant.
· Summarize the processes that detoxify ethanol.

Chapter Overview

The **carbonyl group** (–C = O) is characteristic of many groups of organic compounds, including the **aldehydes** and **ketones**, carboxylic acids and esters. The carbonyl group and the two atoms attached to it are coplanar.

Aldehydes and ketones are the carbonyl-containing compounds that differ from one another in the type of atom or atoms attached to the carbonyl carbon. In ketones, the carbonyl carbon is attached to two carbon atoms, while in aldehydes the carbonyl carbon is attached to at least one hydrogen atom; the second atom attached to the carbonyl carbon in aldehydes may be another hydrogen or a carbon atom.

14.1 Structure and Physical Properties

Owing to the polar carbonyl group, aldehydes and ketones are moderately polar compounds. As a result, they boil at higher temperatures than hydrocarbons or ethers having the same number of carbon atoms, but at temperatures lower than comparable alcohols.

Aldehydes and ketones composed of five or fewer carbon atoms are reasonably soluble in water because of the hydrogen bonding between the carbonyl group and water molecules. Larger members of these carbonyl-containing compounds are less polar, more hydrocarbonlike, and thus more soluble in nonpolar organic solvents.

Example 1

Which member of each of the following pairs of compounds that have similar molecular weights has the higher boiling points?

1. butane or methoxyethane
2. methoxyethane or propanol
3. propanone or methoxyethane

Work

Aldehydes and ketones are moderately polar compounds and boil at higher temperatures than hydrocarbons or ethers of similar molecular weights. The aldehydes and ketones, however, boil at lower temperatures than similar alcohols.

Answer

1. methoxyethane
2. propanol
3. propanone

Example 2

Show the polar attraction that is present between two acetone molecules.

Answer

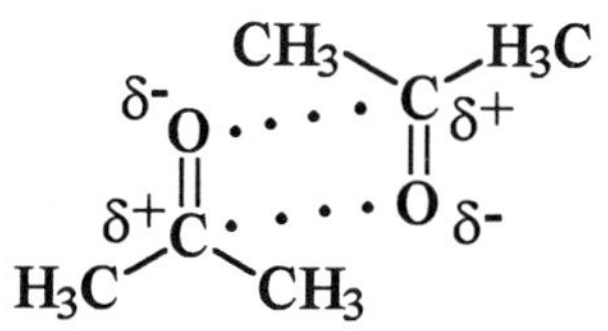

14.2 I.U.P.A.C. Nomenclature and Common Names

Naming Aldehydes

Determine the parent compound, that is, the longest continuous carbon chain containing the carbonyl group. Drop the final *-e* of the parent alkane and replace it with *-al*. The parent chain is always numbered beginning with the carbonyl carbon as carbon-1. All other substituents are named and numbered as usual.

Common names for aldehydes are derived from the same Latin root as the corresponding carboxylic acids (See Table 14.1 and Section 16.1). Substituted aldehydes are named as derivatives of the straight-chain parent compound, using Greek letters to indicate the positions of substituents. The carbon atom bonded to the carbonyl carbon is referred to as the α-carbon.

Naming Ketones

Determine the parent compound. Replace the *–e* ending of the parent alkane with the *-one* suffix of the ketone family. The longest carbon chain is numbered to give the carbonyl carbon the lowest possible number. All other substituents are named and numbered as usual.

The common names for ketones are derived by naming the alkyl groups that are bonded to the carbonyl carbon. These are used as prefixed followed by the work *ketone*. Alkyl groups may be arranged alphabetically or by size.

Example 3
Determine the I.U.P.A.C. and common names of the following aldehydes and ketones:

$$1. \quad \underset{H}{H} - \overset{\overset{\displaystyle O}{\|}}{C} - H$$

$$2. \quad CH_3CH_2\overset{\overset{\displaystyle O}{\|}}{C} - H$$

$$3. \quad CH_3CH_2 - \overset{\overset{\displaystyle O}{\|}}{C} - CH_3$$

$$4. \quad CH_3CH_2CH_2 - \overset{\overset{\displaystyle O}{\|}}{C} - CH_3$$

Work
Aldehydes are named according to the longest continuous carbon chain (parent) that contains the carbonyl group. The final *-e* of the parent alkane is dropped and replaced by *-al* for aldehyde. Numbers are not needed for aldehydes, since they are always the terminal carbon atom. For ketones, the final *-e* of the alkane name is replaced with *-one*, and numbers may be needed to show the position of the carbonyl group in the compound.

Answer
1. methanal (common name: formaldehyde)
2. propanal (common name: propionaldehyde)
3. butanone (common name: ethyl methyl ketone)
4. 2-pentanone (common name: methyl propyl ketone)

Example 4
Write structural formulas for each of the following:

1. 5-bromohexanal
2. 3-methylheptanal
3. 3-bromo-2-pentanone
4. cyclohexanone

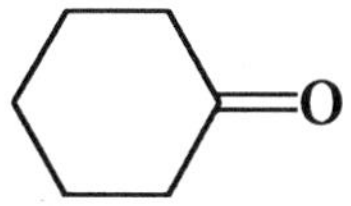

Answer

1. 5-bromohexanal: $CH_3CHCH_2CH_2CH_2$—$\overset{\displaystyle O}{\overset{\|}{C}}$—H
   $\underset{Br}{|}$

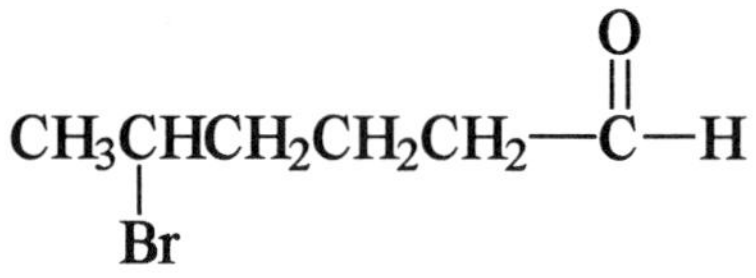

2. 3-methylheptanal: $CH_3CH_2CH_2CH_2CHCH_2$—$\overset{\displaystyle O}{\overset{\|}{C}}$—H
   $\underset{CH_3}{|}$

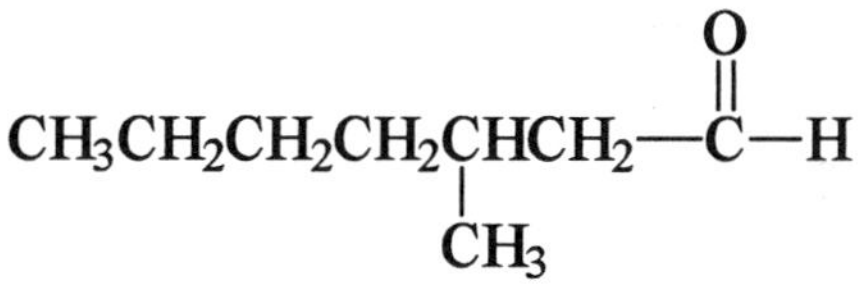

3. 3-bromo-2-pentanone: $CH_3CH_2\overset{\overset{\displaystyle Br}{|}}{C}H$—$\overset{\displaystyle O}{\overset{\|}{C}}$—$CH_3$

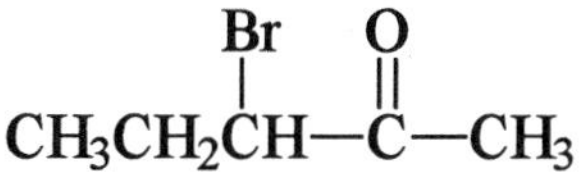

4. cyclohexanone:

14.3 Important Aldehydes and Ketones

Methanal (formaldehyde) is a gas. It is available commercially as an aqueous solution (formalin) that is used to preserve tissue samples.

Ethanal (acetaldehyde) is produced from ethanol in the liver and is responsible for the symptoms of a hangover.

Propanone (acetone) is the simplest ketone. It is an important solvent because it can dissolve organic compounds and is also miscible (mixes) with water. It is found as a solvent in adhesives, paints, and nail polish remover. Many complex members of the ketone family are important in the food industry as food additives. Others are useful as medicinals and agricultural chemicals.

14.4 Reactions Involving Aldehydes and Ketones

Preparation of Aldehydes and Ketones

In the laboratory, aldehydes and ketones are often prepared by the oxidation of the corresponding alcohol. Any aldehyde or ketone can be prepared if the correct alcohol is available.

Oxidation of methyl alcohol gives methanal. Oxidation of a primary alcohol produces an aldehyde; oxidation of a secondary alcohol yields a ketone. Tertiary alcohols do not undergo oxidation. The conclusion we can draw from this information is that the carbonyl carbon must have at least one hydrogen substituent for oxidation to occur.

Example 5

Provide the I.U.P.A.C. names of the oxidation products for the following alcohols:
1. methanol
2. 2-propanol

Answer
a. methanal (or methanoic acid, if methanal is further oxidized)
b. propanone

Oxidation Reactions

Aldehydes are very easily oxidized further to carboxylic acids. In fact, they are so easily oxidized that it is often very difficult to prepare or store them. Oxidation of an aldehyde yields a carboxylic acid:

$$R-\overset{\overset{\displaystyle O}{\|}}{C}-H \xrightarrow{\text{[O]}} R-\overset{\overset{\displaystyle O}{\|}}{C}-OH$$

Aldehyde   Carboxylic acid

Ketones do not undergo further oxidation reactions because a carbon-hydrogen bond to the carbonyl carbon is necessary for the reaction to occur.

Aldehydes and ketones can be distinguished from one another based on their ability to undergo oxidation reactions. The **Tollens' Test**, or Tollens' silver mirror test, is the most common such test. Tollens' reagent consists of a basic solution of $Ag(NH_3)_2^+$. An aldehyde will undergo an oxidation-reduction reaction in which the silver ion (Ag^+) is reduced to silver metal (Ag^0) as the aldehyde is oxidized to a carboxylic acid. The silver metal precipitates from solution and coats the vessel, giving a smooth silver mirror. Because ketones cannot undergo further oxidation they do not react with the Tollens' reagent.

Example 6

Write the major products for each of the following reactions, using Tollens' reagent as the oxidizing agent.

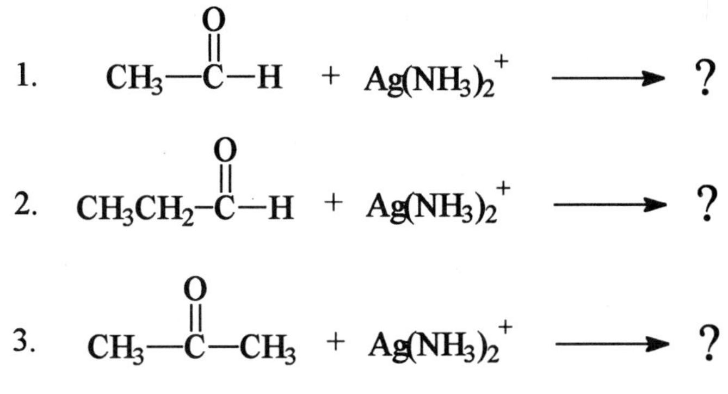

$$1. \quad CH_3{-}\overset{\displaystyle O}{\overset{\|}{C}}{-}H \ + \ Ag(NH_3)_2^{+} \ \longrightarrow \ ?$$

$$2. \quad CH_3CH_2{-}\overset{\displaystyle O}{\overset{\|}{C}}{-}H \ + \ Ag(NH_3)_2^{+} \ \longrightarrow \ ?$$

$$3. \quad CH_3{-}\overset{\displaystyle O}{\overset{\|}{C}}{-}CH_3 \ + \ Ag(NH_3)_2^{+} \ \longrightarrow \ ?$$

Work

Treatment of an aldehyde with Tollens' reagent gives an oxidation-reduction reaction. The aldehyde is oxidized to the carboxylic acid, and the silver ion (Ag^{+}) is reduced to silver metal (Ag^{0}).

Answer

1. $CH_3COOH + Ag^{0}$
2. $CH_3CH_2COOH + Ag^{0}$.
3. No reaction—Tollens' reagent can not oxidize ketones.

Another test used to distinguish between aldehydes and ketones is **Benedict's Test**. In this test Cu^{2+} is reduced to Cu^{+}. Cu^{2+} is soluble and gives a blue solution, while the Cu^{+} precipitates as the red solid Cu_2O. Benedict's test is useful for determining the concentrations of glucose in the urine.

Reduction Reactions

Aldehydes and ketones are easily reduced to the corresponding alcohol by a large number of different reducing agents, designated [H]. The general reduction reaction is shown:

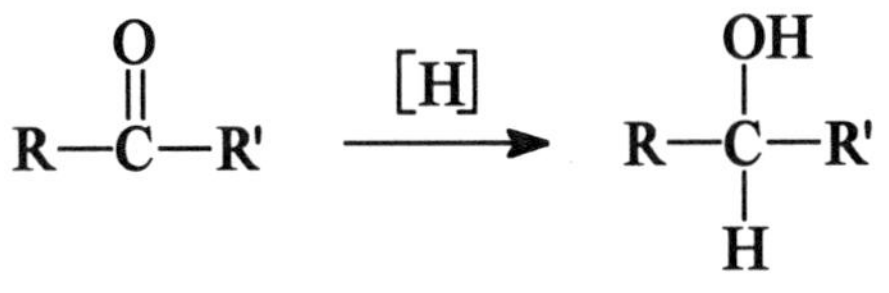

$$R{-}\overset{\displaystyle O}{\overset{\|}{C}}{-}R' \ \xrightarrow{\ [H]\ } \ R{-}\overset{\displaystyle OH}{\underset{\displaystyle H}{\overset{\displaystyle |}{\underset{|}{C}}}}{-}R'$$

$$\text{Aldehyde or ketone} \qquad\qquad \text{Alcohol}$$

The classical reaction for aldehyde and ketone reduction is **hydrogenation**, in which the carbonyl compound is reacted with hydrogen gas. This reaction requires a metal catalyst, pressure and/or heat. The carbon-oxygen double bond (the carbonyl group) is reduced to a carbon-oxygen single bond.

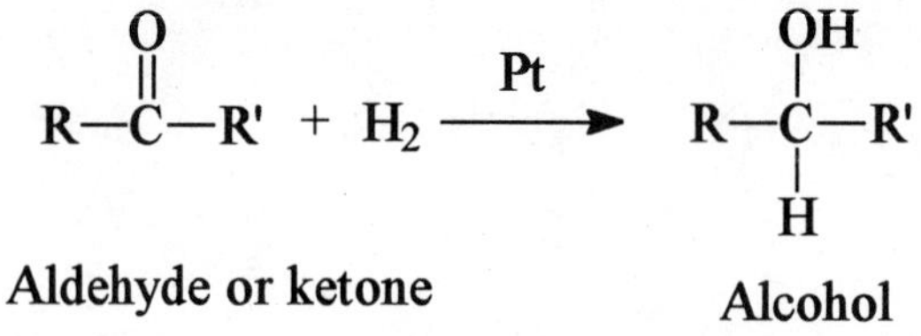

$$R-\overset{\overset{O}{\|}}{C}-R' + H_2 \xrightarrow{Pt} R-\overset{\overset{OH}{|}}{\underset{\underset{H}{|}}{C}}-R'$$

Aldehyde or ketone   Alcohol

Example 7

Label the following as oxidation or reduction reactions:

1. methanol to methanoic acid
2. methanol to formaldehyde
3. benzoic acid to benzaldehyde
4. 2-pentanol to 2-pentanone
5. acetone to 2-propanol

Answer
1. oxidation
2. oxidation
3. reduction
4. oxidation
5. reduction

Addition Reactions

The most common reaction of the carbonyl group is **addition** across the carbon-oxygen double bond. The reaction requires catalytic amounts of acid, represented as H^+ over the reaction arrow.

The addition of one alcohol molecule to an aldehyde produces a **hemiacetal**. Addition of a second alcohol molecule yields an **acetal**.

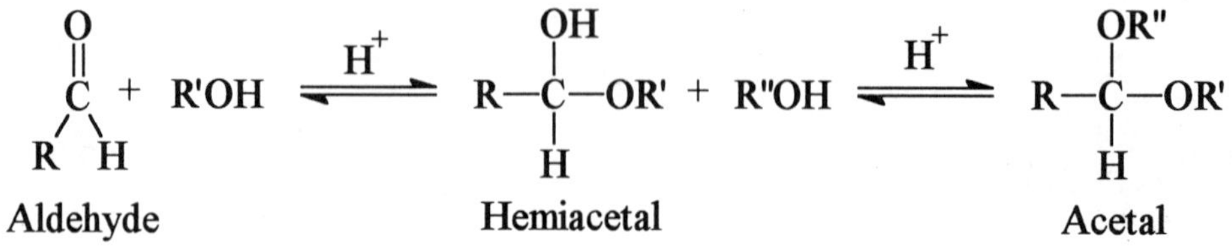

$$\underset{R \quad H}{\overset{\overset{O}{\|}}{C}} + R'OH \underset{}{\overset{H^+}{\rightleftharpoons}} R-\overset{\overset{OH}{|}}{\underset{\underset{H}{|}}{C}}-OR' + R''OH \underset{}{\overset{H^+}{\rightleftharpoons}} R-\overset{\overset{OR''}{|}}{\underset{\underset{H}{|}}{C}}-OR'$$

Aldehyde   Hemiacetal   Acetal

Similarly, the reaction of a ketone with one alcohol molecule produces a **hemiketal**. Addition of a second alcohol molecule yields a **ketal**.

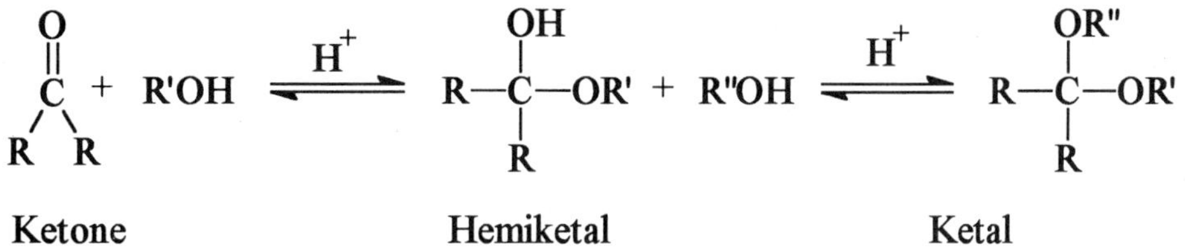

$$\underset{\text{Ketone}}{\underset{R \diagup \diagdown R}{\overset{\overset{O}{\|}}{C}}} + R'OH \;\underset{}{\overset{H^+}{\rightleftharpoons}}\; \underset{\text{Hemiketal}}{R-\overset{\overset{OH}{|}}{\underset{\underset{R}{|}}{C}}-OR'} + R''OH \;\underset{}{\overset{H^+}{\rightleftharpoons}}\; \underset{\text{Ketal}}{R-\overset{\overset{OR''}{|}}{\underset{\underset{R}{|}}{C}}-OR'}$$

Hemiacetals and hemiketals are commonly found in the structures of carbohydrates. Linear sugar molecules have many hydroxyl groups and at least one carbonyl group that undergo an intramolecular reaction in solution to give cyclic hemiacetals or hemiketals.

Example 8
Show the reaction of acetone with methyl alcohol to form a hemiketal compound.

Answer

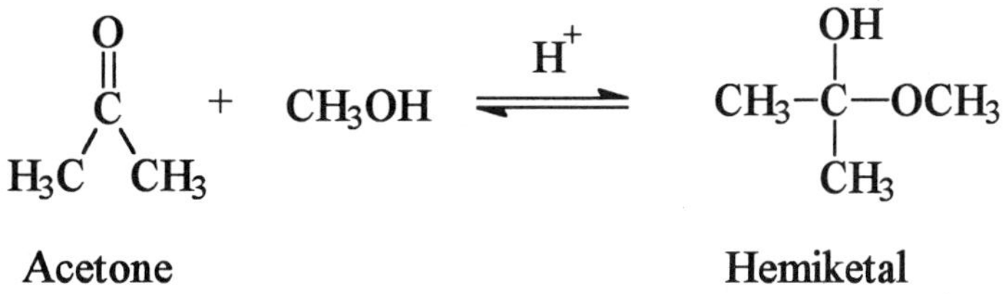

$$\underset{\text{Acetone}}{\underset{H_3C \diagup \diagdown CH_3}{\overset{\overset{O}{\|}}{C}}} + CH_3OH \;\underset{}{\overset{H^+}{\rightleftharpoons}}\; \underset{\text{Hemiketal}}{CH_3-\overset{\overset{OH}{|}}{\underset{\underset{CH_3}{|}}{C}}-OCH_3}$$

Aldol Condensation

The **aldol condensation** is a reaction in which aldehydes or ketones react to form larger molecules. The α-carbon of one aldehyde forms a bond with the carbonyl carbon of a second aldehyde. One biological example of the aldol condensation occurs in the gluconeogenesis pathway. This pathway accomplishes the synthesis of the sugar glucose. The enzyme aldolase catalyzes the condensation of the aldehyde glyceraldehyde-3-phosophate and the ketone dihydroxyacetone phosphate.

Keto-Enol Tautomerism

An aldehyde or a ketone may exist in an equilibrium mixture of two isomers called tautomers. One isomer is the keto form; the other is the enol form. The two forms are shown below:

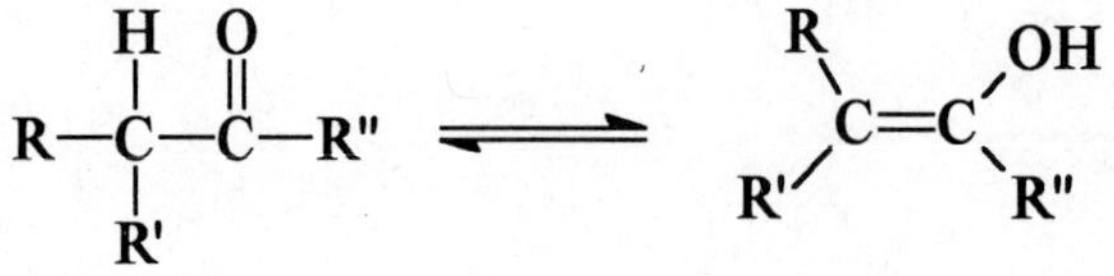

$$R-\underset{\underset{R'}{|}}{\overset{\overset{H}{|}}{C}}-\overset{\overset{O}{\|}}{C}-R'' \rightleftharpoons \underset{R'}{\overset{R}{>}}C=C\underset{R''}{\overset{OH}{<}}$$

Keto form   Enol form

Typically the keto form is more stable than the enol form.

Key Terms

acetal	hemiketal
addition reaction	hydrogenation
aldehyde	ketal
aldol condensation	ketone
Benedict's Test	oxidation
carbonyl group	Tollens' Test
hemiacetal	

Self Test

1. Which of the following compounds of each set has the highest boiling point?
     a. 1-propanol or propane or propanone
     b. pentane or methoxyethane or propanal
2. Draw the polar attraction between two butanone molecules.
3. Give the I.U.P.A.C. names of the following:

a. 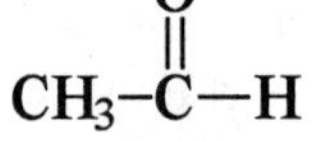 $CH_3-\overset{\overset{O}{\|}}{C}-H$

b. $CH_3-\overset{\overset{O}{\|}}{C}-CH_3$

c. $CH_3CH_2-\overset{\overset{O}{\|}}{C}-H$

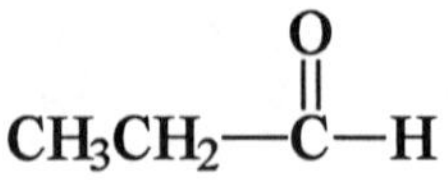

d. 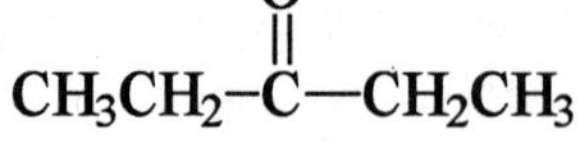 $CH_3CH_2-\overset{\overset{O}{\|}}{C}-CH_2CH_3$

e. 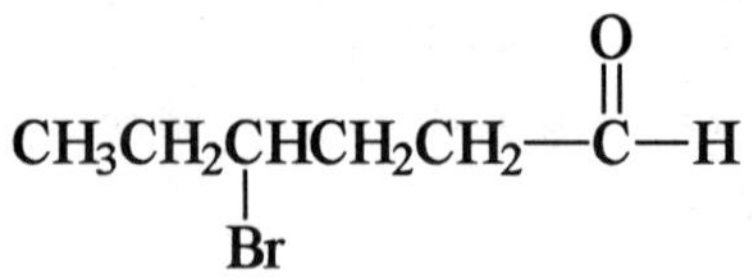 $CH_3CH_2\underset{\underset{Br}{|}}{C}HCH_2CH_2-\overset{\overset{O}{\|}}{C}-H$

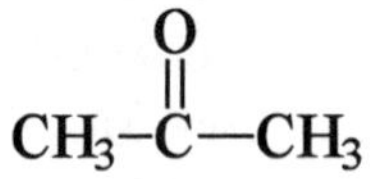

164

4. Why are 2-propanone and 2-butanone not correct I.U.P.A.C. names?
5. Write structural formulas for the following:
   a. 2-bromobutanal
   b. 2-methyl-3-pentanone
6. Show the reactions for the oxidation of methanol and ethanol by the liver.
7. Complete the following oxidation reactions:

a. 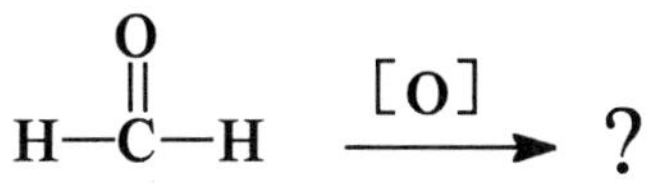 $H-\overset{\displaystyle O}{\overset{\|}{C}}-H \quad \xrightarrow{[O]} \quad ?$

b. $CH_3-\overset{\displaystyle O}{\overset{\|}{C}}-H \quad \xrightarrow{[O]} \quad ?$

c. $CH_3-\overset{\displaystyle O}{\overset{\|}{C}}-CH_3 \quad \xrightarrow{[O]} \quad ?$

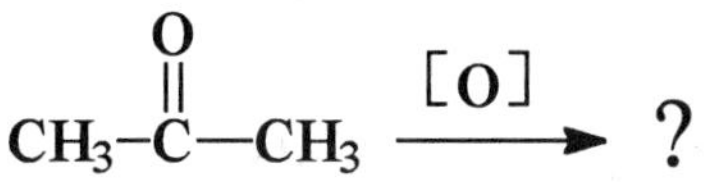

d. 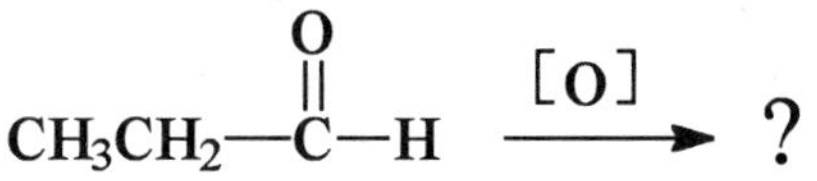 $CH_3CH_2-\overset{\displaystyle O}{\overset{\|}{C}}-H \quad \xrightarrow{[O]} \quad ?$

e. 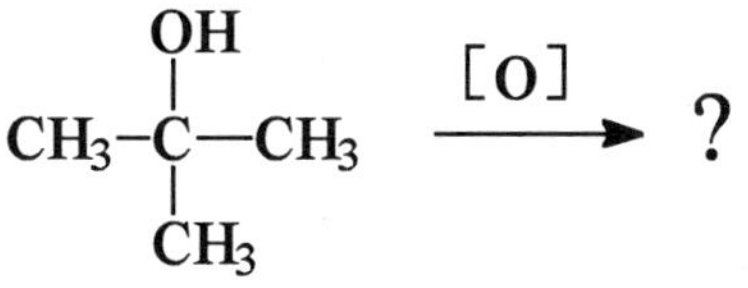 $CH_3-\underset{\underset{\displaystyle CH_3}{|}}{\overset{\overset{\displaystyle OH}{|}}{C}}-CH_3 \quad \xrightarrow{[O]} \quad ?$

8. Give the I.U.P.A.C. names of the oxidation products for the following alcohols:
   a. CH_3CH_2OH
   b. 1-propanol
   c. 2-propanol
   d. 2-methyl-2-propanol
9. Show the major products for each of the following oxidation reactions:

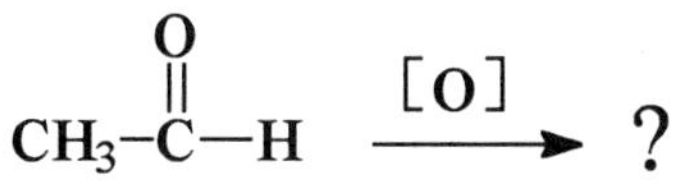

a. 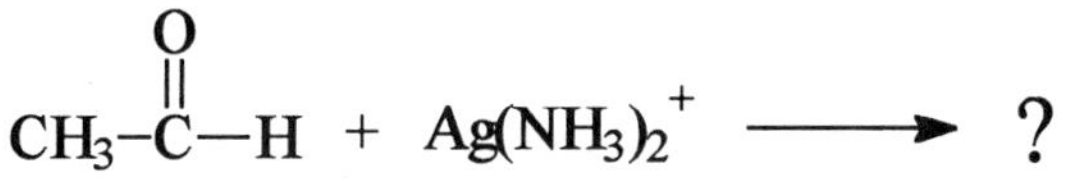 $CH_3-\overset{\displaystyle O}{\overset{\|}{C}}-H \; + \; Ag(NH_3)_2{}^+ \quad \longrightarrow \quad ?$

b. $CH_3-\overset{\displaystyle O}{\overset{\|}{C}}-CH_3 \; + \; Ag(NH_3)_2{}^+ \quad \longrightarrow \quad ?$

c. $CH_3CH_2-\overset{\displaystyle O}{\overset{\|}{C}}-CH_3 \; + \; Ag(NH_3)_2{}^+ \longrightarrow \quad ?$

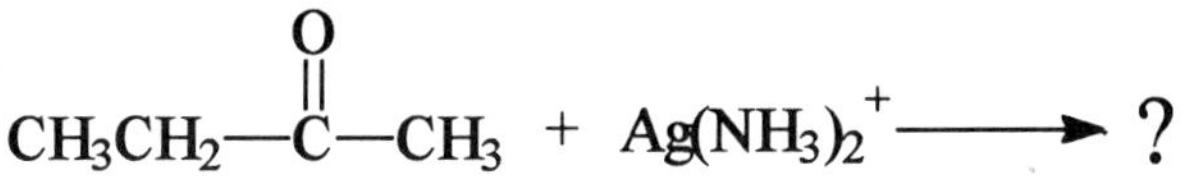

10. Label the following as oxidation or reduction reactions.
    a. conversion of methyl alcohol by the liver into formaldehyde
    b. conversion of acetaldehyde by the liver into acetic acid
    c. conversion of 3-pentanone into 3-pentanol
    d. conversion of ethylene glycol (1,2-ethanediol) to oxalic acid (ethanedioic acid)
11. Complete the following reactions:
    a. $RCHO + ROH \rightleftharpoons$?
    b. $NAD^+ + CH_3OH \rightarrow$?
    c. $NAD^+ + CH_3CH_2OH \rightarrow$?
12. Which class(es) of organic compounds contain a carbonyl functional group?
13. The general formula RCOR represents what class of compounds?
14. Give the common name of the two carbon aldehyde.
15. Formaldehyde is available commercially as an aqueous solution known as what?
16. Aldehydes are easily further oxidized to what compounds?
17. What is the common name of ethanoic acid?
18. The oxidation of benzaldehyde produces what compound?
19. List two common laboratory tests used to distinguish between aldehydes and ketones.
20. Name the product of the following reaction:

$$\begin{array}{c} OH \\ | \\ R-C-O-R \\ | \\ H \end{array} + ROH \xrightarrow{acid} \quad ?$$

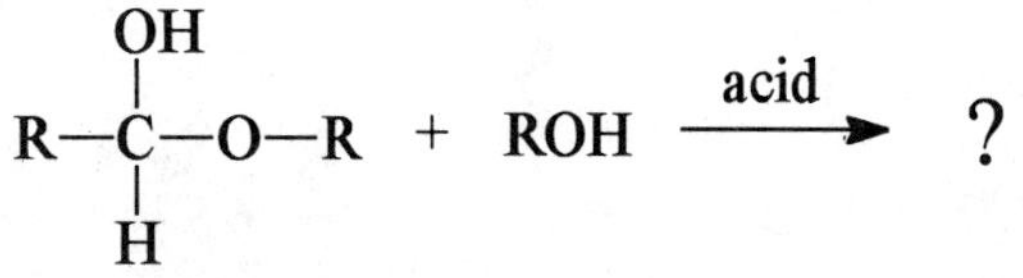

21. When β-carotene is cleaved, what vitamin is produced? (Hint: Review A Human Perspective: The Chemistry of Vision)
22. Why are ketones polar compounds?
23. RCOH is in what class of compounds?
24. What do we call a benzene group that has one hydrogen atom removed from it?
25. The ending -al on the name of a compound means that it is what kind of compound?
26. The ending -one on the name of a compound means that it is what kind of compound?

15 *Carbohydrates*

Learning Goals

1.Conceptual Goals
· Be familiar with the ways in which carbohydrates are classified.
· Understand the concepts of chirality, enantiomers, stereoisomers, and D- and L-configuration.
· Recognize whether a sugar is a reducing or nonreducing sugar.
· Know the general structural features of oligosaccharides and polysaccharides.

2. Performance Goals
· Draw and name the common, simple carbohydrates, using structural formulas and Fischer Projection formulas.
· Using Fischer Projection formulas, be able to differentiate between D- and L- sugars.
· Given the Fischer Projection of a monosaccharide, be able to draw the Haworth Projection of its α- and β-cyclic forms and vice versa.
· Discuss the structural, chemical, and biochemical properties of the monosaccharides, oligosaccharides, and polysaccharides.

3. Health Applications
· Discuss the use of Benedict's reagent to measure the level of glucose in urine.
· Understand the difference between galactosemia and lactose intolerance.
· Recognize the significance of carbohydrate blood group antigens in blood transfusion.
· Understand the mechanism of action of penicillin against bacterial cell wall synthesis.
· Know the difference between simple and complex carbohydrates and the amounts of each recommended in the daily diet.

Chapter Overview

Carbohydrates are produced in plants by photosynthesis. They are the main source of energy for both plants and animals. They are found in many natural sources, such as grains and cereals, breads, fruits, sugar cane, and sugar beets. A healthy diet should include both complex and simple carbohydrates. It is recommended that 58% of the diet should be carbohydrate and that no more than 10% of the daily caloric intake should be sucrose.

15.1 Types of Carbohydrates

Carbohydrates may be categorized by size. **Monosaccharides** are composed of a single (mono-) sugar (saccharide) unit. A **disaccharide** consists of two monosaccharides. Intermediate in size are the **oligosaccharides**, polymers consisting of two to ten monosaccharide units. The largest and most complex carbohydrates are the **polysaccharides**;

polymers consisting of two to ten monosaccharide units. The largest and most complex carbohydrates are the polysaccharides; Oligosaccharides and polysaccharides are chains of monosaccharides held together by **glycosidic bonds** through "bridging" oxygen atoms.

Example 1

List each of the following as either a monosaccharide, disaccharide, or polysaccharide:

1. starch
2. glucose
3. ribose
4. maltose
5. glycogen
6. mannose
7. cellulose
8. glyceraldehyde

Work

Monosaccharides are those carbohydrates that cannot be broken down into any simpler substance by hydrolysis. *Disaccharides* are composed of two monosaccharides joined together by an ether (glycosidic) bond. *Polysaccharides* are composed of many monosaccharides joined together by ether (glycosidic) bonds between each pair.

Answer

1. polysaccharide
2. monosaccharide
3. monosaccharide
4. disaccharide
5. polysaccharide
6. monosaccharide
7. polysaccharide
8. monosaccharide

15.2 Monosaccharides

Nomenclature

If a monosaccharide is a ketone, it is called a **ketose**. If it is an aldehyde, it is called an **aldose**. All carbohydrates contain a large number of hydroxyl groups and are, therefore, *polyhydroxyaldehydes* or *polyhydroxyketones*.

A second system of nomenclature is based on the number of carbon atoms in the main carbohydrate skeleton. A **triose** has three carbons, a **tetrose** has four carbons, and so on. By combining the two systems, a name is derived that provides information about both the structure and composition of a sugar. For instance, a sugar may be an aldotriose, aldohexose, ketotriose, ketotetrose, etc.

In addition, each carbohydrate also has a specific unique name, such as glucose and fructose. Finally, it is important to indicate the isomeric form of the sugar; thus, a D- or L - designation is placed in front of the name. Structures and examples of these names are seen on the next page.

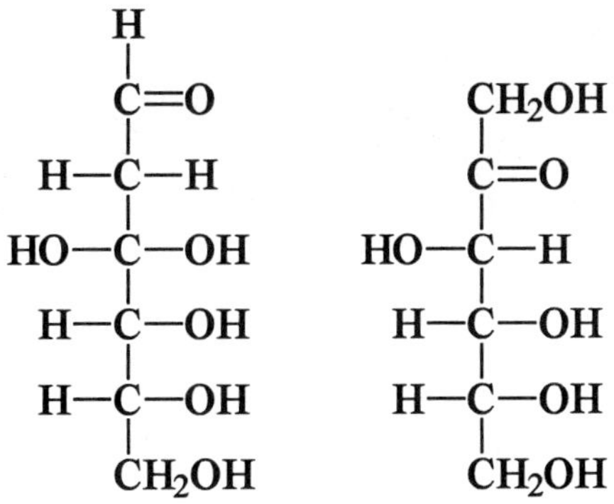

```
        H
        |
        C=O                  CH₂OH
        |                     |
   H—C—H                 C=O
        |                     |
  HO—C—OH             HO—C—H
        |                     |
   H—C—OH               H—C—OH
        |                     |
   H—C—OH               H—C—OH
        |                     |
      CH₂OH                 CH₂OH
```

Aldose	Ketose
Hexose	Hexose
Aldohexose	Ketohexose
D-Glucose	**D-Fructose**

Example 2

Classify each of the following monosaccharides as an aldose or a ketose, and also give the number of carbon atoms found in each.

1. ribose   3. galactose
2. glyceraldehyde   4. fructose

Work

A more detailed system of nomenclature may be used for monosaccharides. They may be classified as an aldose (which means they have an aldehyde structure) or as a ketose (ketone structure). Also, by using suffixes, we can indicate the number of carbon atoms found in each. For example, glucose is classified as an aldohexose, since it has six carbon atoms, and it also contains an aldehyde structure.

Answer

1. aldopentose   3. aldohexose
2. aldotriose   4. ketohexose

Stereoisomers

The prefixes D- and L- are used to distinguish **stereoisomers**. In each member of a pair of stereoisomers, all of the atoms are bonded together using the same bonding pattern; they differ only in the arrangements of their atoms in space. **Stereochemistry** is the study of the spatial arrangement of atoms in a molecule. D- and L-stereoisomers are **enantiomers**, nonsuperimposable mirror images of one another. A carbon atom that has four different

169

groups bonded to it is called an **asymmetric** or **chiral carbon**. A molecule that has a chiral carbon (a chiral molecule) can exist as a pair of enantiomers.

In Figure 15.4 of the text, the stereoisomers of glyceraldehyde are presented. When the hydroxyl group is drawn to the right of the chiral carbon, the isomer is the D-enantiomer. When the hydroxyl group is drawn to the left of the chiral carbon, the molecule is the L-enantiomer.

Example 3
Draw and label the two different stereoisomers of glyceraldehyde.

Answer

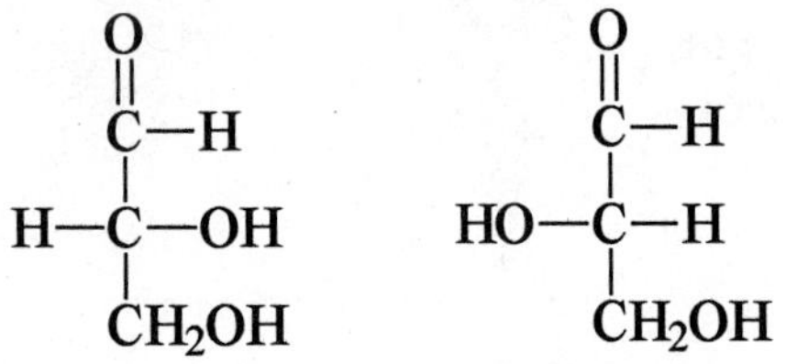

$$
\begin{array}{cc}
\begin{array}{c}
O \\
\parallel \\
C-H \\
| \\
H-C-OH \\
| \\
CH_2OH
\end{array}
&
\begin{array}{c}
O \\
\parallel \\
C-H \\
| \\
HO-C-H \\
| \\
CH_2OH
\end{array}
\end{array}
$$

D-Glyceraldehyde   L-Glyceraldehyde

Example 4
Determine the configuration (D- or L-) for each of the following:

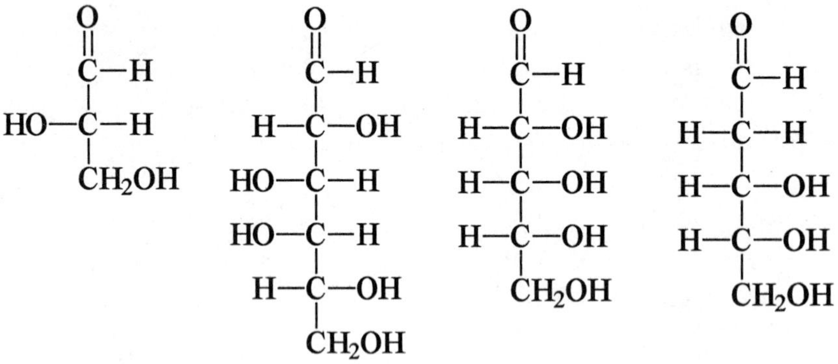

$$
\begin{array}{cccc}
\begin{array}{c}
O \\
\parallel \\
C-H \\
| \\
HO-C-H \\
| \\
CH_2OH
\end{array}
&
\begin{array}{c}
O \\
\parallel \\
C-H \\
| \\
H-C-OH \\
| \\
HO-C-H \\
| \\
HO-C-H \\
| \\
H-C-OH \\
| \\
CH_2OH
\end{array}
&
\begin{array}{c}
O \\
\parallel \\
C-H \\
| \\
H-C-OH \\
| \\
H-C-OH \\
| \\
H-C-OH \\
| \\
CH_2OH
\end{array}
&
\begin{array}{c}
O \\
\parallel \\
C-H \\
| \\
H-C-H \\
| \\
H-C-OH \\
| \\
H-C-OH \\
| \\
CH_2OH
\end{array}
\end{array}
$$

Work
by convention, it is the position of the hydroxyl group on the penultimate (next to the last) carbon that determines whether the compound is a D- or L- enantiomer. The D- isomer is the one in which the −OH on the penultimate carbon is on the right-hand side of the structure. The L- form is just the opposite.

Answer

1. L-glyceraldehyde
2. D-galactose
3. D-ribose
4. D-2-deoxyribose

Some Important Monosaccharides

D-*Glyceraldehyde*

The simplest carbohydrate is the aldotriose glyceraldehyde (Figure 15.4 of the text). When drawing the structure of a monosaccharide, the most oxidized carbon, in this case the carbonyl carbon of the aldehyde, is drawn at the "top." This carbon is numbered C-1. C-2 of glyceraldehyde is chiral; it is covalently bonded to four different groups.

The asymmetric carbon farthest from the oxidized end of the linear monosaccharide determines whether an enantiomer is D- or L-. If the –OH group is on the right, the molecule is in its D- configuration. If the –OH group is on the left, the molecule is in its L-configuration. Most carbohydrates have more than one chiral carbon. The designation D- or L- refers only to the configuration at the penultimate chiral carbon.

Glucose

Glucose (dextrose) found in fruits and honey, is the preferred energy source of many tissues, especially the brain. Thus, the blood glucose concentration must be carefully controlled to allow the body to function optimally. The enantiomer of glucose found in nature is D-glucose.

Generally sugars of five or more carbons exist in cyclic form under physiological conditions. This results from the reaction of the carbonyl group at C-1 of glucose with the hydroxyl group at C-5 to produce a six-membered ring. Recall that the reaction between an aldehyde and an alcohol yields a **hemiacetal**. In the case of the cyclic form of glucose, the product is a cyclic *intramolecular hemiacetal*.

Two isomers are formed in this reaction because cyclization creates a new asymmetric carbon, in this case C-1. These isomers are designated as either α- or β- forms. In the α-isomer, the C-1 hydroxyl group is below the ring. In the β-isomer, it is above the ring.

Haworth projections are used to depict the three-dimensional configuration of cyclic monosaccharide molecules. To draw a Haworth Projection, begin with either the **Fischer projection** or a structural formula of the linear form of the sugar (refer to Example 15.2 in the text). Chemical groups to the left of the carbon chain are placed above the ring in the Haworth projection. Chemical groups to the right of the carbon chain are drawn beneath the carbon ring in the Haworth Projection.

Fructose

Fructose, the sweetest of all sugars, is found in honey and fruits. It is also called levulose and fruit sugar. Fructose is a ketohexose.

Cyclization of D-fructose produces α- and β-D-fructose. In this case the product is an intramolecular **hemiketal**. The most common ring structure formed by D-fructose consists of a five-carbon skeleton.

Galactose

Galactose is a hexose (six-carbon sugar) found most commonly as a component of the disaccharide lactose. **Lactose**, or milk sugar, is the most abundant sugar found in milk. β-D-galactose and a modified form, *N*-acetyl-β-D-galactosamine, are found in the oligosaccharides on the surface of red blood cells. These oligosaccharides are referred to as the *blood group antigens*. They determine whether an individual has Type A, B, AB, or O blood.

Ribose and Deoxyribose: Five-Carbon Sugars

Ribose, a **pentose**, is a component of ribonucleic acids (RNA) and of several coenzymes, compounds required for the action of some enzymes. The molecule that carries the genetic information in the cell is deoxyribonucleic acid (DNA). DNA contains a modified form of D-ribose in which the –OH group at C-2 has been replaced by a hydrogen. This sugar is called *2-deoxyribose*.

Reducing Sugars

The aldehyde group of aldoses is easily oxidized. Thus, **Benedict's reagent** can be used to detect the presence of aldoses, or **reducing sugars**. The aldehyde group is oxidized to a carboxylate group, and Cu^{2+} is reduced to Cu^+.

Although ketones are not readily oxidized, ketoses, such as fructose, react with the Benedict's reagent because they are easily converted to aldoses through the *enediol reaction*.

Benedict's reagent can be used to detect the presence of glucose in the urine.

Example 5
Write an equation representing the oxidation of D-glucose by Benedict's solution.

Answer

$$
\begin{array}{c}
\underset{\text{CH}_2\text{OH}}{\overset{\displaystyle \overset{\text{O}}{\|}}{\underset{\displaystyle}{\text{C}-\text{H}}}}
\end{array}
$$

\[
\begin{array}{c}
\text{O} \\
\|\\
\text{C}-\text{H} \\
\text{H}-\text{C}-\text{OH} \\
\text{HO}-\text{C}-\text{H} \quad + \ 2\,\text{Cu}^{2+} \ + \ 5\,\text{OH}^- \\
\text{H}-\text{C}-\text{OH} \\
\text{H}-\text{C}-\text{OH} \\
\text{CH}_2\text{OH}
\end{array}
\longrightarrow
\begin{array}{c}
\text{O} \\
\|\\
\text{C}-\text{O}^- \\
\text{H}-\text{C}-\text{OH} \\
\text{HO}-\text{C}-\text{H} \quad + \ \text{Cu}_2\text{O} \ + \ 3\,\text{H}_2\text{O} \\
\text{H}-\text{C}-\text{OH} \\
\text{H}-\text{C}-\text{OH} \\
\text{CH}_2\text{OH}
\end{array}
\]

Example 6
Write an equation representing the oxidation of L-glyceraldehyde by Benedict's reagent.

Answer

\[
\begin{array}{c}
\text{O} \\
\|\\
\text{C}-\text{H} \\
\text{HO}-\text{C}-\text{H} \quad + \ 2\,\text{Cu}^{2+} \ + \ 5\,\text{OH}^- \\
\text{CH}_2\text{OH}
\end{array}
\longrightarrow
\begin{array}{c}
\text{O} \\
\|\\
\text{C}-\text{H} \\
\text{HO}-\text{C}-\text{H} \quad + \ \text{Cu}_2\text{O} \ + \ 3\,\text{H}_2\text{O} \\
\text{CH}_2\text{OH}
\end{array}
\]

15.3 Oligosaccharides

The α- or β-hydroxyl group of a cyclic monosaccharide can react with a hydroxyl group of another sugar to form an "oxygen bridge" (ether bond) and produce a disaccharide. This reaction is illustrated in Figure 15.10 in the text for two molecules of glucose.

If the disaccharide has an unreacted hemiacetal –OH group at C-1, it is a reducing sugar, because the cyclic structure can break open, forming a free aldehyde. When drawing the structure of a disaccharide, the nonreducing end is written to the left. Most disaccharides are reducing sugars. The single exception is sucrose, which is a **nonreducing sugar**.

Maltose

Maltose is a disaccharide composed of two glucose molecules. When the reducing end of maltose has a β-hydroxyl group, this disaccharide is called β-maltose. Since the C-1 hydroxyl group of one glucose molecule is attached to C-4 of another glucose molecule, the bond between the two monosaccharides is called an α (1→ 4) glycosidic bond.

Lactose

Milk sugar, or **lactose**, is a dimer of β-D-galactose and either α- or β-D-glucose. In lactose the C-1 hydroxyl group of β-D-galactose is bonded to the C-4 hydroxyl group of glucose. Thus the bond between these two monosaccharides is called a β (1→ 4) glycosidic bond.

Lactose, the principal sugar in milk, must be broken down to glucose and galactose before it can be used by the body as an energy source. Glucose can be used directly in energy-harvesting metabolic reactions. However, galactose must be converted into a phosphorylated form of glucose before it can enter the energy harvesting glycolysis pathway. **Galactosemia** is a genetic disease caused by the absence of an enzyme needed for this conversion. Dulcitol, a reduced form of galactose, accumulates to toxic levels, causing severe mental retardation, cataracts, and early death. Galactosemic infants must be provided a diet that does not contain any milk or milk products.

The inability to digest lactose, called **lactose intolerance**, is caused by the absence of the enzyme lactase. The symptoms include intestinal cramping, diarrhea, and dehydration. Because some intestinal bacteria metabolize lactose and release organic acids and CO_2 gas, the individual suffers further discomfort. Avoiding milk and milk products will eliminate these unpleasant symptoms. Lactase is available in tablet form and can be ingested along with dairy products to avoid the symptoms of lactose intolerance.

Sucrose

Sucrose, common table sugar, is a disaccharide of α-D-glucose joined to β-D-fructose. Examine the structure of sucrose in Figure 15.13 in the text. The glycosidic linkage between α-D-glucose and β-D-fructose involves both of the carbons that were previously part of the hemiacetal or hemiketal. This is called an α,β glycosidic linkage.

Because both carbons that were previously part of a hemiacetal or a hemiketal have reacted to form this linkage, the ring structure cannot open up. No aldehyde or ketone group can be formed; thus, sucrose is not a reducing sugar.

15.4 Polysaccharides

Most carbohydrates found in nature are **polysaccharides**, high molecular weight polymers of glucose.

174

Starch

Plants use the energy of sunlight to produce glucose from CO_2 and H_2O. Most plants store glucose in the form of the polysaccharide starch.

Starch is composed of the glucose polymers **amylose** (80%) and **amylopectin** (20%). Amylose is a linear polymer of up to 4,000 α-D-glucose molecules connected by α $(1\rightarrow 4)$ glycosidic bonds. Amylose exists as a helix that coils at every sixth glucose unit.

During digestion, amylose is degraded by two enzymes. The enzyme α-amylase cleaves the glycosidic bonds of amylose chains at random along the chain, and β-amylase sequentially removes maltose molecules from the reducing end of the amylose chain.

Amylopectin is a highly branched form of amylose. The main chain consists of α $(1\rightarrow 4)$ glycosidic bonds, and the branches of 20–25 glucose molecules are attached to the C-6 hydroxyl groups by α $(1\rightarrow 6)$ glycosidic bonds.

Glycogen

Glycogen is the principal glucose storage form of animals. It is stored in granules in liver and muscle cells. The structure of glycogen is like that of amylopectin, except that glycogen has more branches, and they are shorter.

The liver regulates the level of blood glucose by the formation and degradation of glycogen. When blood glucose levels are high, liver cells take up glucose from the blood and convert it to glycogen. When the blood glucose levels are too low, the liver breaks down glycogen and releases the glucose into the blood stream.

Cellulose

Cellulose, a polymer of about 3,000 β-D-glucose units linked by β $(1\rightarrow 4)$ glycosidic bonds, is the most abundant organic molecule in the world. Cellulose is an unbranched polymer that forms long fibrils. These long, straight, parallel chains of cellulose form a rigid cage that serves as a structural component of plant cell walls.

We cannot digest cellulose because we lack the enzyme *cellulase*. Cellulose serves as a source of dietary fiber, a necessary component of a healthful diet.

Example 7
Write the "word-reactions" for the hydrolysis of the following carbohydrates:

1. starch
2. maltose
3. sucrose
4. glycogen
5. lactose

Answer

1. starch + many H_2O molecules → many glucose molecules
2. maltose + H_2O → 2 glucose molecules
3. sucrose + H_2O → glucose + fructose
4. glycogen + many H_2O molecules → many glucose molecules
5. lactose + H_2O → glucose + galactose

For each of the above reactions, acid conditions are required.

Example 8

Why are termites, cows, and goats able to digest cellulose, while humans are unable to do so?

Answer

The termites, cows, and goats have microorganisms within their digestive systems that produce the enzyme *cellulase*, which allows them to break down cellulose. Humans do not have the cellulase enzyme needed for the hydrolysis of cellulose.

Key Terms

aldose	galactose	lactose	stereochemistry
amylopectin	galactosemia	lactose intolerance	stereoisomers
amylose	glucose	maltose	tetrose
asymmetric carbon	glyceraldehyde	monosaccharide	triose
Benedict's reagent	glycogen	nonreducing sugar	
cellulose	glycosidic bond	oligosaccharide	
chiral	Haworth Projection	pentose	
disaccharide	hemiacetal	polysaccharide	
enantiomers	hemiketal	reducing sugar	
Fischer Projection	hexose	ribose	
fructose	ketose	saccharide	

Self Test

1. Classify each of the following as either a monosaccharide, disaccharide or polysaccharide:
   a. deoxyribose   d. starch
   b. galactose   e. sucrose
   c. glucose   f. fructose

2. Classify each of the following as an aldose or a ketose. Also, include in the classification the number of carbon atoms in each.
   a. glucose
   b. fructose
   c. mannose
   d. ribose
   e. deoxyribose
   f. glyceraldehyde

3. What is the term for a pair of stereoisomers that are nonsuperimposable mirror images of one another?

4. Which polysaccharide serves as a structural element in plant cell walls?

5. Which enantiomeric form of glucose is found in humans?

6. Name the simplest carbohydrate.

7. What two enzymes are involved in the degradation of amylose?

8. Name the bond that holds two monosaccharides together in a disaccharide?

9. What is the structural difference between an aldose and a ketose?

10. What is the difference between a triose and a pentose?

11. Determine the D- or L- configurations for each of the following:

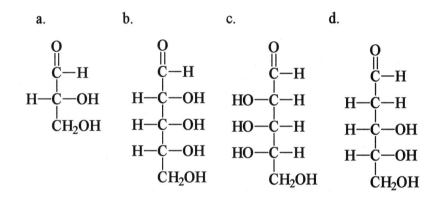

a.   b.   c.   d.

12. List three common names for glucose.

13. Which monosaccharide is the sweetest of all carbohydrates?

14. Which polysaccharide is the major glucose storage form of animals?

15. Give word-reactions for the hydrolysis of the following carbohydrates:
   a. glucose
   b. lactose
   c. sucrose
   d. starch
   e. cellulose

16. Which of the disaccharides will not react with Benedict's solution?

17. What substances are described as either polyhydroxyaldehydes or polyhydroxyketones?

18. Why are humans unable to digest cellulose?

19. What is the term for the condition that results from an inability to digest lactose?

20. Which carbohydrate is found in blood in concentrations up to 0.1% (W/V)?

21. Which form of cyclic D-glucose has the –OH at carbon-1 below the ring?
22. What compound is formed in the reaction between an alcohol and a ketone?
23. What monosaccharide makes up one-third of RNA?
24. Give the molecular formula of ribose.
25. What is the name of the human genetic disease in which galactose cannot be converted into a form that can be used in cellular metabolic reactions?
26. What is the term for carbohydrates that can reduce metal ions?
27. What is the most common type of oligosaccharide?
28. What are the major sites of glycogen synthesis in the human body?
29. What test might be used to distinguish between a monosaccharide and polysaccharide?
30. What test might be used to distinguish between a reducing sugar and a nonreducing sugar?

16   *Carboxylic Acids and Carboxylic Acid Derivatives*

Learning Goals

1. Conceptual Goals
· Know the carboxylic acids and esters that are of natural, medical, or environmental importance.
· Be familiar with the physical properties of these families of organic compounds.
· Understand the methods of preparation of carboxylic acids and esters.

2. Performance Goals
· Draw and name the common carboxylic acids and esters.
· Write equations for the hydrolysis of esters.
· Write equations for the synthesis of a carboxylic acid by oxidation of primary alcohols or aldehydes.
· Write equations for the synthesis of esters from carboxylic acids and alcohols.

3. Health Applications
· Understand the significance of thioesters and phosphoesters in biological systems.
· Define the term saponification and know how a soap works in the emulsification of grease and oils.

Chapter Overview

Carboxylic acids are characterized by the **carboxyl group**, and have the following general structure:

$$Ar-\underset{\underset{O}{\|}}{C}-OH \qquad R-\underset{\underset{O}{\|}}{C}-OH$$

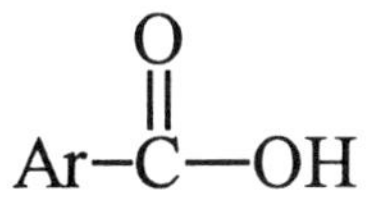

Aromatic carboxylic acid 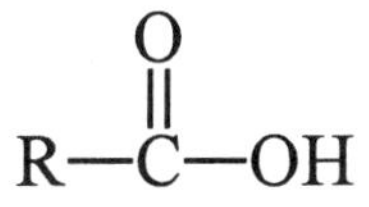   Aliphatic carboxylic acid

The term carboxylic acid tells us that the carboxyl group is derived from a carbonyl group and hydroxyl group. It further tells us that these molecules are acids.

   Esters are produced in the reaction between a carboxylic acid and an alcohol and have the following general structure:

$$R-\underset{\underset{O}{\|}}{C}-O-R \qquad Ar-\underset{\underset{O}{\|}}{C}-O-Ar \qquad Ar-\underset{\underset{O}{\|}}{C}-O-R$$

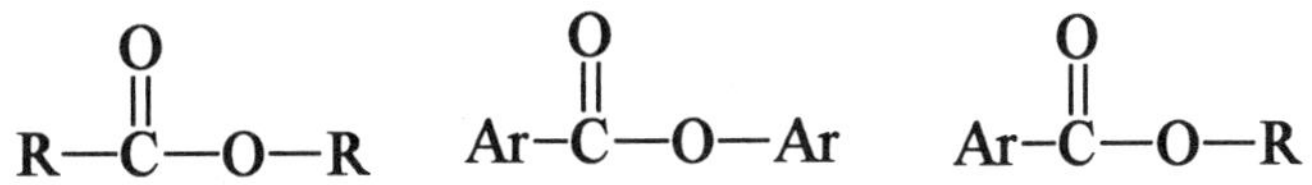

Examples of Aliphatic and Aromatic Esters

179

The following structure is called the **acyl group**:

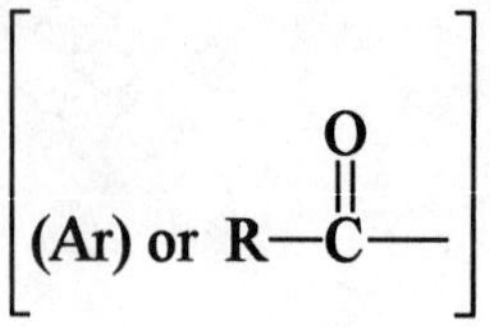

$$\left[(Ar)\ or\ R-\overset{\displaystyle O}{\overset{\displaystyle \|}{C}}- \right]$$

The acyl group is the functional group of the carboxylic acid derivatives, including the esters and amides.

Example 1
The following general formulas are used to represent different classes of organic compounds. Name each of the appropriate families.

1. ROH   4. ROR
2. RCHO   5. RCOOH
3. RCOR   6. RCOOR

Answer
1. alcohol   4. ether
2. aldehyde   5. carboxylic acid
3. ketone   6. ester

16.1 Carboxylic Acids

Structure and Physical Properties

The carboxyl group consists of two very polar functional groups, the carbonyl group and the hydroxyl group. Thus, carboxylic acids are very polar compounds. In addition, carboxylic acids can hydrogen bond to one another. As a result, they boil at higher temperatures than aldehydes, ketones or even alcohols of comparable molecular weight.

Carboxylic acids can form intermolecular hydrogen bonds with water molecules. Thus, small carboxylic acids are water-soluble. However, solubility falls off dramatically as the carbon chain length increases.

The lower molecular weight carboxylic acids have sharp, sour tastes and unpleasant aromas. The longer chain carboxylic acids are called fatty acids and are components of many biologically important lipids.

Example 2

Which member of each of the following sets of compounds has the highest boiling point?

1. ethane or ethanol or ethanoic acid
2. propanal or propanone or 1-propanol
3. methanol or methanal or methanoic acid

Work

The boiling points of most alkanes, alcohols, aldehydes, and carboxylic acids that have similar molecular weights obey the following relationship:

alkane < aldehyde or ketone < alcohol < carboxylic acid

lowest   highest

Answer

1. ethanoic acid   2. 1-propanol   3. methanoic acid

Nomenclature

In the I.U.P.A.C. System, carboxylic acids are named by replacing the *-e* ending of the parent alkane with the suffix *-oic acid*. If there are two carboxyl groups, the suffix *-dioic acid* is used. The parent chain is numbered so that the carboxyl carbon is carbon-1. Other groups are named and numbered in the usual way.

The acyl group of a carboxylic is named by replacing the *-oic acid* suffix with *-yl*. For instance, the acyl group of acetic acid is the acetyl group.

The carboxylic acid derivatives of cycloalkanes are named by adding the suffix *carboxylic acid* to the name of the cycloalkane or substituted cycloalkane. The carboxyl group is defined to be on carbon-1.

Common names of carboxylic acids are frequently used. Table 16.1 in the text shows the I.U.P.A.C. and common names of several carboxylic acids, as well as their sources and the Latin or Greek words that gave rise to the common names.

Aromatic carboxylic acids are usually named as derivatives of benzoic acid. The *-oic acid* or *-ic acid* suffix is attached to the appropriate prefix. However, common names of substituted benzoic acids are frequently used.

Example 3

Give the I.U.P.A.C. names for the following carboxylic acids:

1. CH_3COOH 2. CH_3CH_2COOH 3. $CH_3CHBrCH_2COOH$

Work

1. Determine the longest continuous chain (parent) of carbon atoms that includes the carboxyl group. The carbon atom of the carboxyl group will be labeled as 1 and will not be included in the name.
2. Then drop the -e of the parent alkane name and add -oic acid.

Answer

1. ethanoic acid   2. propanoic acid   3. 3-bromobutanoic acid

---

Example 4

Give the common name of each of the following carboxylic acids:

1. CH_3COOH 3. $HCOOH$
2. $CH_3CH_2CH_2COOH$ 4. $C_{17}H_{35}COOH$

Answer

1. acetic acid   3. formic acid
2. butyric acid   4. stearic acid

Some Important Carboxylic Acids

Many carboxylic acids found in nature are listed in Table 16.1 of the text. Fatty acids can be isolated from a variety of fats and oils. More complex carboxylic acids are also found in a variety of foodstuffs. Citric acid is found in citrus fruits and is added to foods to give a sharp taste (sour candies) or as a preservative and anti-oxidant. Adipic acid adds tartness to soft drinks and helps to retard spoilage. Bacteria in milk produce lactic acid as a product of fermentation of sugars. It contributes a tangy flavor to yogurt and buttermilk and acts as a food preservative. Lactic acid is also produced by muscles during strenuous exercise.

Reactions Involving Carboxylic Acids

Preparation of Carboxylic Acids

Many of the small carboxylic acids are prepared by the oxidation of the corresponding primary alcohol or aldehyde. A variety of oxidizing agents can be used in these reactions and a catalyst is often required. The general reaction is:

$$R-CH_2-OH \xrightarrow{[O]} R-\overset{\overset{\displaystyle O}{\|}}{C}-H \xrightarrow{[O]} R-\overset{\overset{\displaystyle O}{\|}}{C}-OH$$

Primary Alcohol   Aldehyde   Carboxylic acid

182

Acid-Base Reactions

The carboxylic acids behave as weak acids because they are proton donors. They are weak acids (typically less than 5% dissociation) that produce a carboxylate ion and a hydrogen ion in water, as seen in the following example:

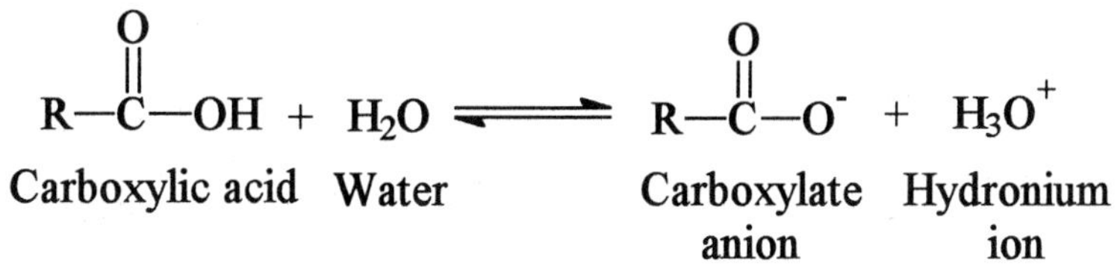

$$R-\overset{\overset{\displaystyle O}{\|}}{C}-OH + H_2O \rightleftharpoons R-\overset{\overset{\displaystyle O}{\|}}{C}-O^- + H_3O^+$$

Carboxylic acid   Water   Carboxylate   Hydronium
                                    anion   ion

When strong bases are added to a carboxylic acid, neutralization occurs:

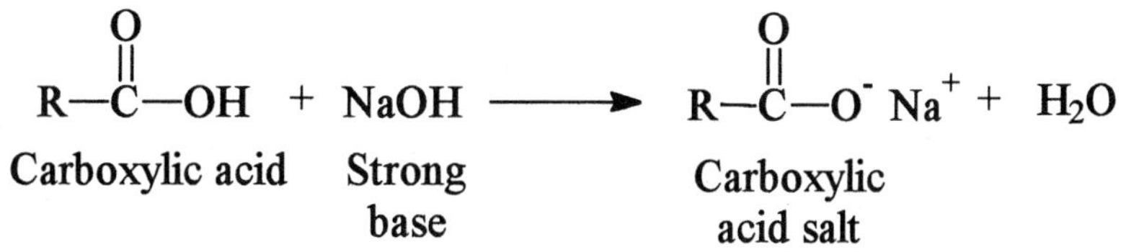

$$R-\overset{\overset{\displaystyle O}{\|}}{C}-OH + NaOH \longrightarrow R-\overset{\overset{\displaystyle O}{\|}}{C}-O^-\,Na^+ + H_2O$$

Carboxylic acid   Strong   Carboxylic
                  base   acid salt

The salt of a carboxylic acid is named by replacing the *-ic acid* suffix with *-ate*. This name is preceded by the name of the appropriate cation, for instance, sodium.

The salts are ionic substances and hence quite soluble in water. The long-chain carboxylic acid salts (fatty acid salts) are good **soaps**. Soaps are made from water, strong base, and natural fats and oils obtained from animals or plants. Soaps form micelles around grease and oil. The hydrophobic alkane tail of the soap dissolves the oils and grease and the hydrophilic carboxylate end of the molecule remains associated with water. This produces a micelle, a tiny sphere with the alkyl group tails of the soap molecules, grease, and oils in the center and the carboxylate on the outside of the sphere, dissolved in the water.

Example 5

Show the reactions of the following carboxylic acids with water:

1. CH_3COOH
2. $HCOOH$
3. $CH_3CH_2CH_2COOH$

Answer

1. Acetic acid dissociates to produce the acetate ion and hydronium ion.

$$CH_3COOH + H_2O \rightleftarrows CH_3COO^- + H_3O^+$$

2. Formic acid dissociates to produce the formate ion and hydronium ion.

$$HCOOH + H_2O \rightleftarrows HCOO^- + H_3O^+$$

3. Butyric acid dissociates to produce the butyrate ion and hydronium ion.

$$CH_3CH_2CH_2COOH + H_2O \rightleftarrows CH_3CH_2CH_2COO^- + H_3O^+$$

Esterification

Carboxylic acids react with alcohols to form esters and water according to the following general reaction:

$$\underset{\text{Carboxylic acid}}{R-\overset{\overset{\displaystyle O}{\|}}{C}-OH} + \underset{\text{Alcohol}}{R'-OH} \underset{\text{heat}}{\overset{H^+}{\rightleftarrows}} \underset{\text{Ester}}{R-\overset{\overset{\displaystyle O}{\|}}{C}-OR'} + \underset{\text{Water}}{H_2O}$$

16.2 Esters

Structure and Physical Properties

Esters are also mildly polar and have pleasant aromas. Many esters are found in natural foodstuffs. Esters boil at approximately the same temperature as aldehydes or ketones of comparable molecular weight. The simpler ones are reasonably soluble and nonreactive in water.

Nomenclature

Esters are formed in the reaction of a carboxylic acid with an alcohol and both of these families are reflected in the naming of the ester. The *alkyl* or *aryl* portion of the alcohol name is used as the prefix and the *-ic acid* ending of the name of the carboxylic acid is replaced with *-ate*.

Example 6

Show the general reaction of an alcohol (R' OH) with a carboxylic acid (RCOOH) to produce an ester.

Answer

$$\underset{\text{Carboxylic acid}}{R-\overset{\displaystyle O}{\overset{\|}{C}}-OH} + \underset{\text{Alcohol}}{R'-OH} \underset{\text{heat}}{\overset{H^+}{\rightleftharpoons}} \underset{\text{Ester}}{R-\overset{\displaystyle O}{\overset{\|}{C}}-OR'} + \underset{\text{Water}}{H_2O}$$

Example 7

Give the common names of the following esters:

1. $CH_3-O-\overset{\displaystyle O}{\overset{\|}{C}}-H$

2. $CH_3CH_2-O-\overset{\displaystyle O}{\overset{\|}{C}}-CH_3$

3. ⬡$-\overset{\displaystyle O}{\overset{\|}{C}}-O-CH_3$

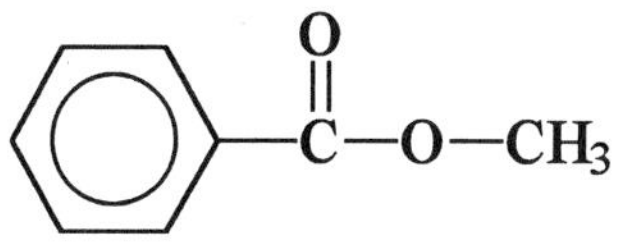

4. $CH_3CH_2-O-\overset{\displaystyle O}{\overset{\|}{C}}-CH_2CH_2CH_3$

Work

Esters are formed from the reaction of a carboxylic acid with an alcohol. The alkyl or aryl portion of the alcohol name is used as a prefix and is followed by the name of the carboxylic acid in which the *-ic acid* (common or I.U.P.A.C.) is replaced by *-ate*.

Answer
1. methyl formate
2. ethyl acetate
3. methyl benzoate
4. ethyl butyrate

Reactions Involving Esters

Preparation of Esters

The conversion of a carboxylic acid to an ester requires heat and is catalyzed by a trace of acid (H^+). The general equation representing this reaction was shown above.

Hydrolysis of Esters

Esters undergo hydrolysis reactions in water. This reaction requires heat and a small amount of acid (H^+) or base (OH^-) to catalyze the reaction, as seen in the following general equations:

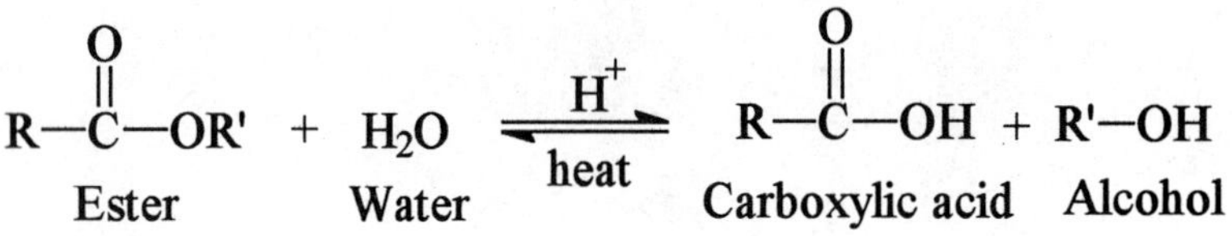

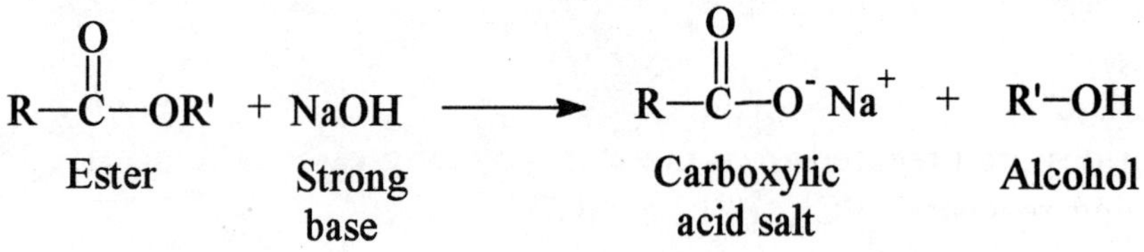

$$\underset{\text{Ester}}{R-\overset{\displaystyle O}{\overset{\|}{C}}-OR'} + \underset{\text{Water}}{H_2O} \underset{\text{heat}}{\overset{H^+}{\rightleftharpoons}} \underset{\text{Carboxylic acid}}{R-\overset{\displaystyle O}{\overset{\|}{C}}-OH} + \underset{\text{Alcohol}}{R'-OH}$$

$$\underset{\text{Ester}}{R-\overset{\displaystyle O}{\overset{\|}{C}}-OR'} + \underset{\substack{\text{Strong} \\ \text{base}}}{NaOH} \longrightarrow \underset{\substack{\text{Carboxylic} \\ \text{acid salt}}}{R-\overset{\displaystyle O}{\overset{\|}{C}}-O^- Na^+} + \underset{\text{Alcohol}}{R'-OH}$$

The base catalyzed hydrolysis of an ester is called saponification, which produces a carboxylic acid salt. The carboxylic acid is formed when the reaction mixture is neutralized with an acid. Saponification is used to hydrolyze fats and oils, which are esters, to the salts of long chain fatty acids—*soaps*.

16.3 Acid Chlorides and Acid Anhydrides

Acid Chlorides

Acid chlorides are named by dropping the *-ic acid* ending of the common or the *-oic acid* ending of the I.U.P.A.C. name of the carboxylic acid and replacing it with *-oyl chloride*.

Example 8
Name the following acid chlorides:

1. 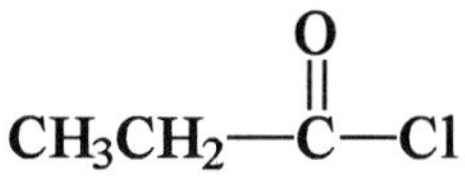 CH₃CH₂—C(=O)—Cl

$$1.\ CH_3CH_2-\overset{\overset{\displaystyle O}{\|}}{C}-Cl$$

$$2.\ BrCH_2-\overset{\overset{\displaystyle O}{\|}}{C}-Cl$$

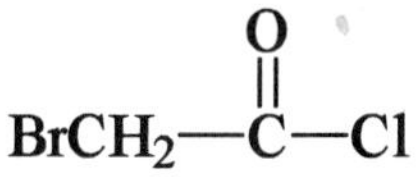

Work
Acid chlorides are named by dropping the *-ic acid* ending of the common name or the *-oic acid* ending of the I.U.P.A.C. name and replacing it with *-oyl chloride*.

Answer
1. propanoyl chloride   2. 2-bromoethanoyl chloride

     Acid chlorides are noxious, irritating chemicals. They are slightly polar and boil at about the same temperature as the corresponding aldehyde or ketone. They cannot be dissolved in water because they react violently with it. The reaction is a hydrolysis reaction that releases HCl and the carboxylic acid.

     Acid chlorides are prepared from the corresponding carboxylic acid by reaction with one of the following reagents: PCl_3, PCl_5, or $SOCl_2$.

Acid Anhydrides

Acid anhydrides are named by replacing the *acid* ending of the carboxylic acid name with the term *anhydride*.

Example 9
Draw the structures of the following acid anhydrides:

1. acetic anhydride
2. propanoic anhydride
3. 3-bromobutanoic anhydride

Work
Anhydrides are formed by the combination of two carboxylic acids, with the concomitant loss of water.

Answer

1. $CH_3-\overset{\overset{\displaystyle O}{\|}}{C}-O-\overset{\overset{\displaystyle O}{\|}}{C}-CH_3$

2. $CH_3CH_2-\overset{\overset{\displaystyle O}{\|}}{C}-O-\overset{\overset{\displaystyle O}{\|}}{C}-CH_2CH_3$

3. $CH_3CHBrCH_2-\overset{\overset{\displaystyle O}{\|}}{C}-O-\overset{\overset{\displaystyle O}{\|}}{C}-CH_2CHBrCH_3$

Anhydrides are formed when two carboxylic acids combine and a water molecule is lost. They boil at much lower temperatures than the corresponding carboxylic acids. Generally they are hydrolyzed by water. This reaction is speeded up by the addition of acid.

16.4 Nature's High Energy Compounds: Phosphoesters and Thioesters

An alcohol can react with phosphoric acid to produce a phosphate ester, or **phosphoester**, as seen here:

$$R-OH \; + \; HO-\overset{\overset{\displaystyle O}{\|}}{\underset{\underset{\displaystyle OH}{|}}{P}}-OH \longrightarrow R-O-\overset{\overset{\displaystyle O}{\|}}{\underset{\underset{\displaystyle OH}{|}}{P}}-OH \; + \; H_2O$$

  Alcohol   Phosphoric acid   Phosphate ester   Water

Phosphate esters of monosaccharides are very important in energy harvesting reactions in the cell.

Adenosine triphosphate (ATP) is the universal energy currency for all living organisms. It consists of a nitrogenous base (adenine) and a phosphate ester of the sugar ribose. There are three phosphoryl groups attached to C-5 of ribose (a triphosphate group). The combination of two phosphate groups results in the loss of a water molecule. Because water is lost, the resulting bond is called a **phosphoric anhydride bond** or *phosphoanhydride bond.*

188

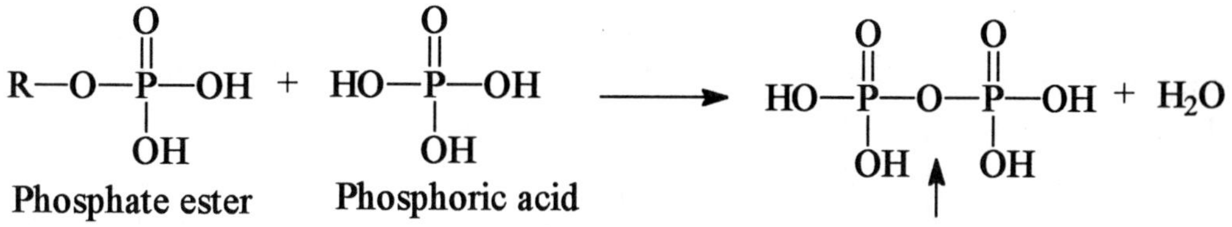

$$\underset{\substack{| \\ \text{OH} \\ \text{Phosphate ester}}}{\overset{\overset{\displaystyle O}{\parallel}}{R-O-P-OH}} + \underset{\substack{| \\ \text{OH} \\ \text{Phosphoric acid}}}{\overset{\overset{\displaystyle O}{\parallel}}{HO-P-OH}} \longrightarrow \underset{\substack{| \\ \text{OH} \quad \text{OH} \\ \text{Phosphoanhydride} \\ \text{bond}}}{\overset{\overset{\displaystyle O \quad O}{\parallel \quad \parallel}}{HO-P-O-P-OH}} + H_2O$$

The energy of ATP is made available through hydrolysis of either of the two phosphoric anhydride bonds. Energy is released in this exothermic reaction. When the phosphoryl group is transferred to a sugar molecule some of that energy resides in the phosphorylated sugar, thereby "energizing" it.

Cellular enzymes can carry out a reaction between a carboxylic acid and a thiol to produce a **thioester**:

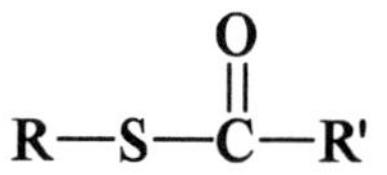

$$\underset{\text{Thioester}}{\overset{\overset{\displaystyle O}{\parallel}}{R-S-C-R'}}$$

This reaction is essential in energy generating pathways as a means of "activating" acyl groups for subsequent breakdown reactions. The complex thiol Coenzyme A is the most important acyl group activator in the cell. The most common thioester is the acetyl ester, called **acetyl coenzyme A** (acetyl CoA). Acetyl CoA transfers the acetyl group from glycolysis to an intermediate of the citric acid cycle. Coenzyme A also serves to activate the acyl group of fatty acids during β-oxidation, the pathway by which fatty acids are oxidized to produce ATP.

Key Terms

acetyl coenzyme A	hydrolysis
acyl group	oxidation
adenosine triphosphate (ATP)	phosphoester
carboxyl group	phosphoric anhydride
carboxylic acid	saponification
carboxylic acid derivative	soap
ester	thioester
fatty acid	

Self Test

1. Give the general formula for each of the following classes of organic compounds:
   - a. alcohol
   - b. ether
   - c. ester
   - d. aldehyde
   - e. carboxylic acid
   - f. ketone
2. Draw the general structure of an acyl group.
3. Give the I.U.P.A.C. names of the following:
   - a. HCOOH
   - b. $CH_3(CH_2)_4COOH$
   - c. CH_3COOH
   - d. $CH_3CH(CH_3)CH_2COOH$
4. Draw the following carboxylic acids:
   - a. acetic acid
   - b. formic acid
   - c. benzoic acid
   - d. lauric acid
5. Provide the common names of each of the following carboxylic acids

$$\text{a. } CH_3(CH_2)_2\overset{\overset{\textstyle O}{\|}}{C}{-}OH$$

b. CH_3CH_2COOH

$$\text{c. } CH_3\underset{\underset{\textstyle CH_3}{|}}{CH}{-}\overset{\overset{\textstyle O}{\|}}{C}{-}OH$$

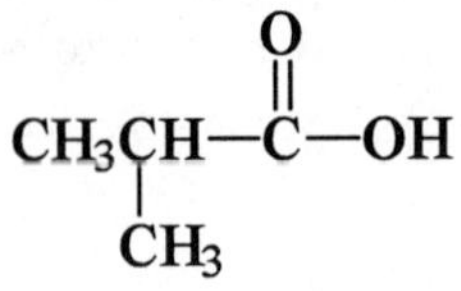

6. Give the common name of the following esters:

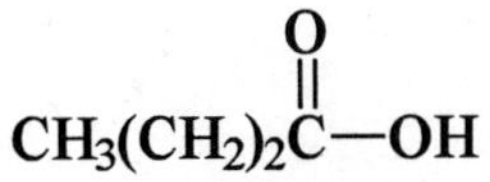

$$\text{a. } H{-}\overset{\overset{\textstyle O}{\|}}{C}{-}O{-}CH_2CH_3$$

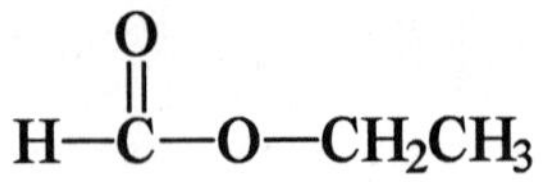

$$\text{b. } CH_3CH_2{-}O{-}\overset{\overset{\textstyle O}{\|}}{C}{-}CH_3$$

$$\text{c. } CH_3CH_2CH_2CH_2CH_2{-}O{-}\overset{\overset{\textstyle O}{\|}}{C}{-}CH_3$$

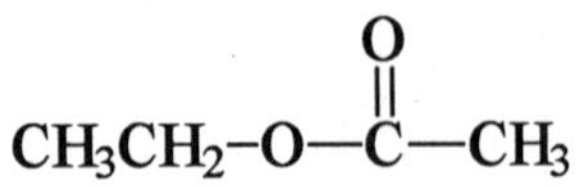

7. Give the I.U.P.A.C. names of the following esters:
   a. methyl acetate

   b. $CH_3CH_2-\overset{\overset{\textstyle O}{\|}}{C}-O-CH_2CH_3$

8. Which of the following compounds that have nearly the same molecular weight has the highest boiling point?
   a. acetic acid   c. propanone
   b. propanal   d. 1-propanol
9. Complete the following reactions:
   a. $CH_3COOH + H_2O \rightleftarrows$?
   b. $CH_3COOH + NaOH \rightarrow$?
   c. $CH_3COOH + CH_3OH$ (acid as a catalyst) $\rightarrow$?
10. What are the hydrolysis products of an ester?
11. What two ions are formed when a soluble carboxylic acid reacts with a molecule of water?
12. For the following reactions, name the products that are formed:
    a. butyric acid + NaOH
    b. acetic acid + $Ca(OH)_2$
13. The long-chain carboxylic acid salts (fatty acid salts) are good _____.
14. What is the term for the large, nonpolar hydrocarbon end of a soap?
15. What two compounds must be reacted to form methyl propanoate?
16. What molecule is the universal energy currency for all cells?
17. What is the name of the base-catalyzed hydrolysis of an ester?
18. Name the functional group found in all carboxylic acids.
19. Which carboxylic acid is found in fire-ants? Give the common name.
20. What is the term for the product of a reaction between a thiol and a carboxylic acid?
21. Provide the I.U.P.A.C. name for the product of the reaction between ethanoic acid and methanol in the presence of heat and a trace of acid.
22. A primary alcohol is mixed with potassium permanganate solution. What reaction occurs and what is the function of the $KMnO_4$?
23. Which class of compound is a weak acid in a water solution.
24. What is the name of the molecule that carries 2-carbon acetyl groups to the citric acid cycle?
25. Aqueous bases, such as a mixture of potassium carbonate and potassium hydroxide, are obtained by leaching wood ashes with what substance?
26. Is the carboxylate part of soap hydrophobic or hydrophilic?
27. What is the term for the "units" formed when particles of oil or grease are surrounded by soap molecules?

28. Provide the missing reactant in the following equation:

$$CH_3CH_2CH_2COOH \; + \; ? \; \longrightarrow \; CH_3CH_2CH_2-\overset{\overset{\textstyle O}{\|}}{C}-O-CH_3$$

29. What is the product of the reaction between an alcohol and phosphoric acid?
30. What is the term for the bond formed between two phosphate groups?

17 Lipids and Their Functions in Biochemical Systems

Learning Goals

1. Conceptual Goals

· Be familiar with the general physical and chemical properties of each of the families of lipids.

· Understand the biological function and origin of the common members of each of the lipid families.

· Be familiar with the structure and function of cell membranes; describe these characteristics as a function of the structure of the lipid molecules that make up the membrane.

2. Performance Goals

· List some examples of the common classes of lipids; describe the principal structural features that differentiate the various classes from each other.

· Write the structures of simple examples of each of the classes of lipids discussed.

· Name the common lipids

· Know the method of synthesizing glycerides and the reactions of fatty acids and glycerides: esterification, hydrolysis, saponification, and hydrogenation.

3. Health Applications

· Be aware of the many functions of prostaglandins in physiological processes.

· Know how aspirin reduces pain.

· Be familiar with the steroid hormones and the physiological results of the abuse of anabolic steroids.

· Understand the role of the lipoproteins in lipid and cholesterol transport in the body.

· Appreciate the role of HDL, LDL, and cholesterol in heart disease.

Chapter Overview

Lipids are a diverse collection of organic molecules grouped together on the basis of their solubility in nonpolar solvents. The four groups of lipids that will be considered in this chapter are fatty acids, glycerides, nonglyceride lipids, and complex lipids.

Example 1

List two properties that all lipids have in common.

Answer

Most are insoluble in water but very soluble in nonpolar solvents.

17.1 Biological Functions of Lipids

Lipids are involved in a variety of biological processes. They are structural components of cell membranes. They are an energy storage form in the body, stored in adipocytes (fat cells). Some lipids, like the steroids, are hormones. Others are vitamins, required for processes such as blood clotting, proper bone development, and vision.

Example 2
List four important functions of the class of compounds called the lipids.

Answer
1. Cell membranes are composed of lipids. The cell membrane creates a barrier between the cell and its environment.
2. Most of the available stored energy in animals is in the form of lipids known as triglycerides.
3. Several of the lipids have hormonelike properties.
4. Others are lipid-soluble vitamins.

17.2 Fatty Acids

Structure and Properties

Fatty acids are long-chain (12–24 carbons) monocarboxylic acids. The general formula for a **saturated fatty acid** is

$$CH_3(CH_2)_nCOOH$$

where n is an even integer between 10 and 22. In a saturated fatty acid, each carbon in the chain is bonded to the maximum number of hydrogen atoms.

In the case of **unsaturated fatty acids** there is at least one carbon-carbon double bond, as seen in this example:

$$CH_3(CH_2)_7CH = CH(CH_2)_7COOH$$

As we observed with alkanes, the melting points of saturated fatty acids increase with increasing carbon number. This general trend is also seen for unsaturated fatty acids, except that the melting point also decreases markedly as the number of carbon-to-carbon double bonds increases.

Example 3
List the properties that most fatty acids have in common.

Answer
1. They are long-chain (12–24) monocarboxylic acids.
2. They often contain an even number of carbon atoms.
3. They form continuous chains of carbon atoms with no branches.
4. They are either saturated, with only single bonds in the R group, or unsaturated, with some double bonds in the R group.

Example 4
What physical property clearly distinguishes a saturated and unsaturated fatty acid with the same number of carbon atoms?

Answer
The melting points of saturated fatty acids are much higher. The greater the number of double bonds in the unsaturated fatty acid, the lower its melting point.

Example 5
Is the fatty acid $C_{13}H_{25}COOH$ saturated or unsaturated?

Answer
It is unsaturated, since all saturated fatty acids (similar to alkanes) will fit the general formula of $C_nH_{2n+1}-COOH$. If it were saturated, its formula would have to be $C_{13}H_{27}COOH$; thus, the $C_{13}H_{25}COOH$ has two fewer hydrogen atoms than the saturated one. This means that the $C_{13}H_{25}COOH$ contains one double bond.

Chemical Reactions of Fatty Acids

The reactions of fatty acids are similar to those of short-chain carboxylic acids.

Esterification

Esterification is the reaction of a fatty acid with an alcohol to form an ester and water.

Example 6

Complete the following esterification reaction:

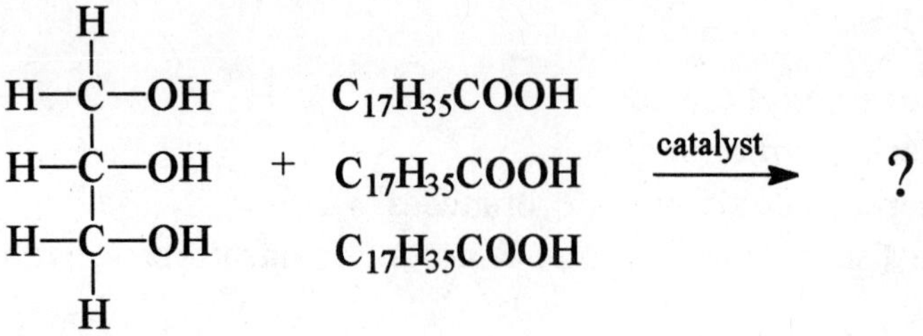

$$
\begin{array}{l}
\text{H} \\
\text{H---C---OH} \\
\text{H---C---OH} \quad + \quad
\begin{array}{l}
\text{C}_{17}\text{H}_{35}\text{COOH} \\
\text{C}_{17}\text{H}_{35}\text{COOH} \\
\text{C}_{17}\text{H}_{35}\text{COOH}
\end{array}
\quad \xrightarrow{\text{catalyst}} \quad ? \\
\text{H---C---OH} \\
\text{H}
\end{array}
$$

Work

A triglyceride is produced by the reaction of three fatty acids and glycerol. Line up the carboxyl group of each fatty acid with the hydroxyl groups (–OH) of the glycerol molecule. Then remove a water molecule between each to form an ester structure at each of the three positions. The resulting triglyceride will always have the following general structure:

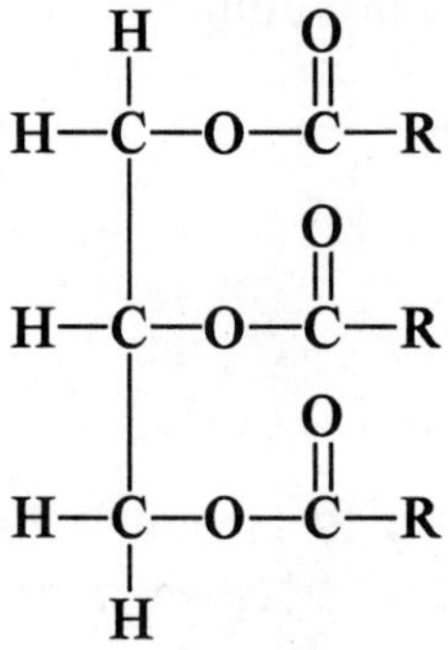

$$
\begin{array}{c}
\quad\quad \text{H} \quad\quad \text{O} \\
\text{H---C---O---C---R} \\
\quad\quad\quad\quad \text{O} \\
\text{H---C---O---C---R} \\
\quad\quad\quad\quad \text{O} \\
\text{H---C---O---C---R} \\
\quad\quad \text{H}
\end{array}
$$

Answer

The R groups will always be the R groups that were found in each specific fatty acid. Notice that there are no hydroxyl or carboxyl groups present; only three ester groups are now holding the glycerol to the three fatty acid. The following triglyceride would be formed, along with three molecules of water:

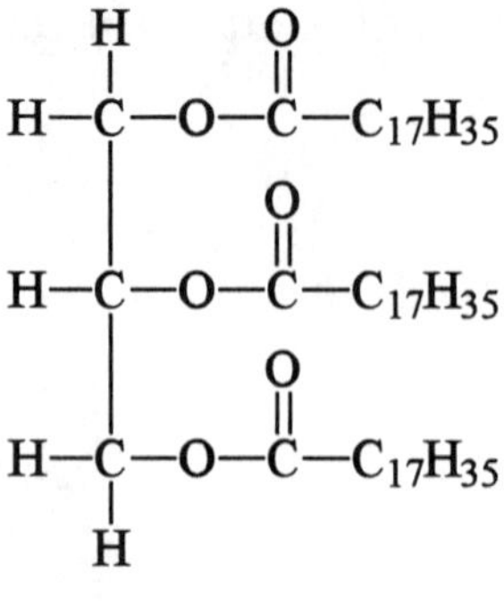

$$
\begin{array}{c}
\quad\quad \text{H} \quad\quad \text{O} \\
\text{H---C---O---C---C}_{17}\text{H}_{35} \\
\quad\quad\quad\quad \text{O} \\
\text{H---C---O---C---C}_{17}\text{H}_{35} \\
\quad\quad\quad\quad \text{O} \\
\text{H---C---O---C---C}_{17}\text{H}_{35} \\
\quad\quad \text{H}
\end{array}
$$

Hydrolysis

Hydrolysis is the addition of water to a fatty acid ester to produce a fatty acid and an alcohol.

Example 7
Give the complete hydrolysis of the following triglyceride:

$$\begin{array}{c}
\text{H} \qquad\quad \text{O} \\
\text{H}-\overset{\displaystyle |}{\underset{\displaystyle |}{\text{C}}}-\text{O}-\overset{\displaystyle \|}{\text{C}}-(\text{CH}_2)_{12}\text{CH}_3 \\
\overset{\displaystyle |}{\phantom{\text{C}}} \qquad\quad \overset{\displaystyle \text{O}}{} \\
\text{H}-\overset{\displaystyle |}{\underset{\displaystyle |}{\text{C}}}-\text{O}-\overset{\displaystyle \|}{\text{C}}-(\text{CH}_2)_{16}\text{CH}_3 \\
\overset{\displaystyle |}{\phantom{\text{C}}} \qquad\quad \overset{\displaystyle \text{O}}{} \\
\text{H}-\overset{\displaystyle |}{\underset{\displaystyle |}{\text{C}}}-\text{O}-\overset{\displaystyle \|}{\text{C}}-(\text{CH}_2)_{14}\text{CH}_3 \\
\text{H}
\end{array}$$

Glyceryl myristostearopalmitate

Work

The hydrolysis of a triglyceride requires three water molecules and a catalyst. The products formed are glycerol and one molecule of each of the fatty acids used to prepare the triglyceride. The name of the above triglyceride tells us that glycerol plus myristic acid, stearic acid, and palmitic acid were used to make it.

Answer

$$\begin{array}{c}
\text{H} \qquad\quad \text{O} \\
\text{H}-\overset{\displaystyle |}{\underset{\displaystyle |}{\text{C}}}-\text{O}-\overset{\displaystyle \|}{\text{C}}-(\text{CH}_2)_{12}\text{CH}_3 \\
\overset{\displaystyle |}{\phantom{\text{C}}} \qquad\quad \overset{\displaystyle \text{O}}{} \\
\text{H}-\overset{\displaystyle |}{\underset{\displaystyle |}{\text{C}}}-\text{O}-\overset{\displaystyle \|}{\text{C}}-(\text{CH}_2)_{16}\text{CH}_3 \quad + \quad 3\text{H}_2\text{O} \xrightarrow{\text{catalyst}} \\
\overset{\displaystyle |}{\phantom{\text{C}}} \qquad\quad \overset{\displaystyle \text{O}}{} \\
\text{H}-\overset{\displaystyle |}{\underset{\displaystyle |}{\text{C}}}-\text{O}-\overset{\displaystyle \|}{\text{C}}-(\text{CH}_2)_{14}\text{CH}_3 \\
\text{H}
\end{array}$$

$$\begin{array}{l}
\text{CH}_2\text{OH} \;+\; \text{CH}_3(\text{CH}_2)_{12}\text{COOH} \\
\overset{\displaystyle |}{\text{CHOH}} \;+\; \text{CH}_3(\text{CH}_2)_{16}\text{COOH} \\
\overset{\displaystyle |}{\text{CH}_2\text{OH}} \;+\; \text{CH}_3(\text{CH}_2)_{14}\text{COOH}
\end{array}$$

Example 8

What are the hydrolysis products for each of the following:

1. glyceryl tristearate + 3H$_2$O
2. glyceryl laurostearopalmitate + 3H$_2$O
3. phosphatidylcholine + 4H$_2$O

Answer

1. glycerol + three molecules of stearic acid
2. glycerol + lauric acid + stearic acid + palmitic acid
3. glycerol + 2 fatty acids + H$_3$PO$_4$ + choline

Saponification

Saponification is the base-catalyzed hydrolysis of an ester. The product of this reaction, an ionized salt, is a soap.

Reaction at the Double Bond

Hydrogenation of an unsaturated fatty acid is the addition of hydrogen to a double bond. This is an example of an addition reaction. Hydrogenation is used in the food industry to convert polyunsaturated vegetable oils into solid fats.

Example 9

Show how linoleic acid can be converted in the laboratory into stearic acid.

Answer

$$CH_3(CH_2)_4CH{=}CHCH_2CH{=}CH(CH_2)_7COOH + 2\,H_2$$

heat
pressure
Ni

$$CH_3(CH_2)_{16}COOH$$
Stearic acid

Eicosanoids: Prostaglandins, Leukotrienes, and Thromboxanes

Linolenic acid and linoleic acid are **essential fatty acids**. They must be obtained in the diet because they cannot be synthesized by the body. Linoleic acid is required for the biosynthesis of **arachidonic acid**, the precursor of a class of 20-carbon, hormonelike molecules known as eicosanoids. The **eicosanoids** include *prostaglandins*, *leukotrienes*, and *thromboxanes*.

The **prostaglandins** are extremely potent biological molecules with hormonelike activity. They are carboxylic acids with a five-carbon ring. All are composed of the basic C_{20} carbon skeleton of prostanoic acid and are grouped under the designations of A, B, E, F, and so on, which indicate the basic carbon skeletal arrangement. Each prostaglandin also has a number designation that indicates the number of carbon-carbon double bonds in the compound. Figure 17.2 in the text provides examples of prostaglandin structure and nomenclature.

Prostaglandins are made in all tissues and exert biological effects on the cells that produce them and on neighboring cells. The amazing array of prostaglandin functions is briefly summarized below.

1. *Blood clotting:* Thromboxane A_2, produced by blood platelets, enhances blood clotting. PGI_2 (prostacyclin), produced by the cells lining the blood vessels, inhibits the clotting process. Working together, these molecules ensure that blood clots form only when necessary.

2. *The inflammatory response:* The inflammatory response is "turned on" when tissue is damaged or invaded by microorganisms. The purpose is to minimize damage and prevent infection. Prostaglandins promote the pain and fever associated with the inflammatory response.

3. *Reproductive system:* PGE_2 stimulates smooth muscle contraction, especially uterine contractions. Dysmenorrhea (painful menstruation) is the result of an excess of two prostaglandins. Drugs that inhibit prostaglandin synthesis provide relief from these symptoms.

4. *Gastrointestinal tract:* Prostaglandins inhibit the secretion of stomach acid and increase the secretion of a protective mucous layer into the stomach. Aspirin inhibits prostaglandin synthesis; thus, prolonged use of this drug may contribute to stomach ulcers.

5. *Kidneys:* In the kidneys, prostaglandins increase the excretion of water and electrolytes.

6. *Respiratory tract:* Leukotrienes promote the constriction of the bronchi associated with asthma. Other prostaglandins have the opposite effect, broncho-dilation.

Aspirin relieves pain by inhibiting prostaglandin synthesis. It works by inhibiting the enzyme cyclooxygenase, which catalyzes the first step in prostaglandin synthesis. The acetyl group from aspirin is covalently bound to the enzyme, inhibiting its activity.

17.3 Glycerides

Glycerides are lipids that contain the alcohol glycerol. They may be subdivided into two classes: **neutral glycerides** and **phosphoglycerides**.

Neutral Glycerides

The esterification of glycerol with one, two, or three fatty acids produces a **mono-**, **di-**, or **triglyceride**. These are also referred to as *mono-*, *di-*, or *triacylglycerols*. Although mono- and diglycerides are present in nature, the most common is the triglyceride, the major storage form of lipids found in fat cells. A glyceride is named by combining the "backbone" name, glyceryl (from glycerol) with the fatty acid name(s).

Neutral glycerides do not dissociate into charged species, because bonding throughout the molecule is covalent and nonpolar. Consequently, they readily stack with one another and are easily stored in adipocytes. In fact, their major function in the body is energy storage. If more nutrients are consumed than are needed for daily metabolic processes, the excess is converted to triglycerides and stored in adipocytes, which form *adipose tissue*. If energy is needed, the triglycerides are broken down, and their stored energy is released for use by the body.

Phosphoglycerides

Phospholipids contain a phosphoryl group (PO_4^{3-}). The presence of the phosphoryl group produces a molecule with a polar head (the phosphoryl group) and a nonpolar tail (the hydrocarbon chain of the fatty acid).

Phosphoglycerides contain acyl groups derived from long-chain fatty acids esterified at C-1 and C-2 of glycerol-3-phosphate. The simplest, **phosphatidate**, contains a free phosphoryl group. More complex phosphoglycerides are formed when the phosphoryl group is bonded to another hydrophilic molecule.

Phosphatidycholine (lecithin), and *phosphatidylethanolamine (cephalin)* are commonly found in cell membranes. The structures of these two phospholipids are shown in Figure 17.5 of the text. Both possess a polar "head" and a nonpolar "tail." The ionic "head" is hydrophilic and interacts with water molecules, while the nonpolar "tail" is hydrophobic and interacts with nonpolar molecules. This bipolar nature is essential to the structure and properties of biological membranes.

Lecithin is also found in egg yolks and soybeans and is commonly used as an **emulsifying agent** in ice cream. An emulsifying agent aids in the suspension of fats in water.

Example 10

Draw a phosphatidylcholine that contains the R groups from linolenic acid (position 1) and linoleic acid (position 2), and then label the hydrophobic and hydrophilic ends.

Work

The general structure of a phosphatidylcholine is as follows. The R groups are from the specific fatty acids that were used to form the ester groups at positions 1 and 2 on the glycerol molecule.

General Structure   Answer

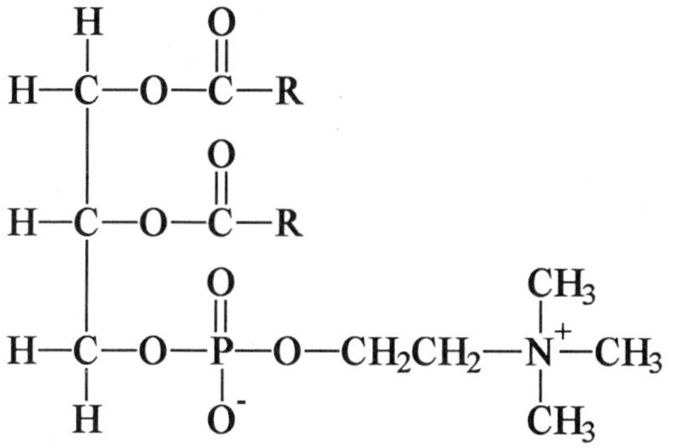

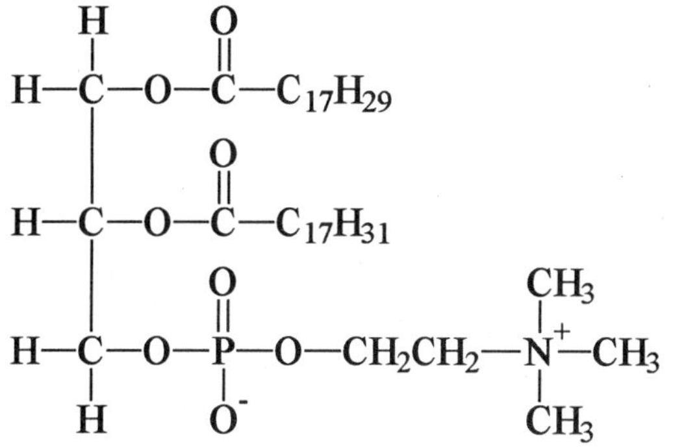

$$
\begin{array}{c}
\text{H} \qquad \text{O} \\
| \qquad\quad \| \\
\text{H}-\text{C}-\text{O}-\text{C}-\text{R} \\
| \qquad\quad \text{O} \\
\qquad\qquad \| \\
\text{H}-\text{C}-\text{O}-\text{C}-\text{R} \\
| \qquad\quad \text{O} \qquad\qquad\qquad \text{CH}_3 \\
\qquad\qquad \| \qquad\qquad\qquad\qquad | \ + \\
\text{H}-\text{C}-\text{O}-\text{P}-\text{O}-\text{CH}_2\text{CH}_2-\text{N}-\text{CH}_3 \\
| \qquad\quad \text{O}^- \qquad\qquad\qquad \text{CH}_3
\end{array}
$$

General structure on the left with R groups. Answer on the right with $\text{C}_{17}\text{H}_{29}$ at position 1 and $\text{C}_{17}\text{H}_{31}$ at position 2.

The hydrophobic regions include all those parts that are not charged. The charged groups are considered to be the hydrophilic ends.

17.4 Nonglyceride Lipids

Sphingolipids

Sphingolipids are lipids that are not derived from glycerol. They are phospholipids derived from the amino alcohol sphingosine.

    Sphingomyelin, a sphingolipid, is abundant in the myelin sheath that surrounds and insulates cells of the central nervous system and is essential to proper nerve transmission.

Steroids

Steroids are an important family of lipids derived from cholesterol. The steroids include the bile salts that aid in the emulsification and digestion of lipids, and the sex hormones testerone and estrone. They are members of a large, diverse collection of lipids called *isoprenoids*, all of which are derived from the five-carbon *isoprene* unit, seen here:

$$CH_2=\underset{\underset{CH_3}{|}}{C}-CH=CH_2$$

All steroids are structured around the steroid nucleus (steroid carbon skeleton), which consists of four fused rings. Two fused rings share one or more common bonds as part of their ring backbones.

Cholesterol is found in most cell membranes, where it functions to regulate the fluidity of the membrane. A high serum cholesterol concentration is associated with heart disease, especially **atherosclerosis** or hardening of the arteries. Cholesterol and other substances coat the inside of the arteries. This causes the arteries to become narrower, and more pressure is needed to cause blood to flow through them. This results in elevated blood pressure (hypertension), which is also linked to heart disease.

Many steroids play roles in the reproductive cycle. *Progesterone*, the most important hormone associated with pregnancy, is synthesized from cholesterol. Produced in the ovaries and placenta, it prepares the uterine lining to accept the fertilized egg. Progesterone is also involved in fetal development and suppression of further ovulation during pregnancy.

Progesterone is the precursor of *testosterone*, a male sex hormone, and estrone, a female sex hormone. Both these hormones are involved in the development of secondary sexual characteristics.

Development of birth control agents has involved application of steroid chemistry. One of the first effective synthetic birth control agents was 19-norprogesterone. Unfortunately, it had to be taken by injection. Norlutin (17-ethynyl-19-nortestosterone) is equally effective and can be taken orally. These compounds all act by inducing a false pregnancy which prevents ovulation.

Cortisone is involved in carbohydrate metabolism and is an important drug used to treat rheumatoid arthritis, asthma, gastrointestinal, and skin disorders. Care must be taken in the use of cortisone because of the possible side effects.

Waxes

The chemical composition of **waxes** is highly variable. All are insoluble in water and are solid at room temperature. Paraffin wax is a mixture of solid straight-chain hydrocarbons. Carbowax is a synthetic polyether. The natural waxes are composed of a long-chain fatty acid esterified to a long-chain alcohol.

Examples of naturally occurring waxes include beeswax; lanolin, used in skin creams; carnauba wax, used in automobile polish; and whale oil, once used as a fuel and for candles.

202

17.5 Complex Lipids

Complex lipids are lipids that are bonded to other types of molecules. **Plasma lipoproteins** are complex lipids that transport other lipids through the blood stream. Because lipids are only slightly soluble in water, their movement from organ to organ requires such a transport system. Lipoprotein particles consist of a core of hydrophobic lipids surrounded by a shell of polar lipids and proteins.

There are four major classes of human plasma lipoproteins. **Chylomicrons** carry triglycerides from the intestine to other tissues. **Very low-density lipoproteins (VLDL)** carry triglycerides synthesized in the liver to other tissues for storage. **Low-density lipoproteins (LDL)** carry cholesterol to peripheral tissues and help regulate cholesterol levels. **High-density lipoproteins (HDL)** transport cholesterol from peripheral tissues to the liver.

The path of lipid transport begins with dietary fat, which is emulsified in the small intestine. Triglycerides are hydrolyzed by lipase, releasing fatty acids and monoglycerides that are absorbed by intestinal cells and reassembled into triglycerides. Chylomicrons are produced, which eventually enter the bloodstream for transport to cells throughout the body. Triglycerides and cholesterol synthesized in the liver are transported in lipoproteins. Triglycerides are carried in VLDL particles, and cholesterol is transported in LDL particles.

LDL particles bind to specific protein receptors (LDL receptors) on cell membranes. The LDL-receptor complex is taken into the cell by a process called *receptor-mediated endocytosis*. In this process, the membrane is pulled into the cell, bringing the entire LDL molecule into the cytoplasm in a membrane-bound sac called an *endosome*. *Lysosomes*, membrane-bound sacs containing digestive enzymes, fuse with the endosome, and the LDL particles are digested by the lysosomal enzymes. Cholesterol is released into the cell cytoplasm, where it inhibits its own biosynthesis, activates an enzyme for its own storage, and inhibits the synthesis of LDL receptors.

A genetic defect in the gene for the LDL receptor causes an accumulation of LDL-cholesterol in the plasma, resulting in atherosclerosis in the sufferer. While high concentrations of LDL in the blood are associated with hardening of the arteries, high levels of HDL in the blood appear to reduce the incidence of atherosclerosis.

17.6 The Structure of Biological Membranes

Biological membranes are *lipid bilayers* consisting of phospholipids and cholesterol. The hydrophobic, hydrocarbon tails stack in the center of the bilayer, and the ionic head groups are exposed on the surfaces.

Fluid Mosaic Structure of Biological Membranes

Membranes are fluid, having the consistency of olive oil. The degree of fluidity is determined by the amounts of saturated and unsaturated fatty acids and by the length of the fatty acid tails. Shorter, more unsaturated fatty acid tails produce a more fluid membrane. Floating in the sea of phospholipids are many proteins that are critical to normal cellular function. When viewed by electron microscopy, these proteins look like a mosaic. Because of the fluid consistency of the membrane and the presence of numerous proteins, our concept of membrane structure is called the **fluid mosaic theory**.

Peripheral proteins are found only on the surfaces of the membrane. **Transmembrane proteins** are embedded within the membrane and extend completely through it.

Membrane Transport

The cell membrane is responsible for the controlled passage of molecules into and out of the cell. Transport of some molecules across the membrane is regulated by integral membrane transport proteins, while other molecules pass through the membranes unassisted, by the **passive transport**.

Passive Diffusion: The Simplest Form of Membrane Transport

Small uncharged molecules, such as O_2, CO_2, and H_2O, pass through the membrane by simple **diffusion**, the *net* movement of a substance from an area of high concentration to an area of low concentration. In this case, diffusion involves the movement of solutes across a cellular membrane. Diffusion is a form of passive transport because it requires no energy expenditure on the part of the cell.

Large or highly charged molecules cannot pass through the lipid bilayer directly because the membrane is **selectively permeable**. That is, the membrane allows the diffusion of some molecules, but not others.

Facilitated Diffusion: Specificity of Molecular Transport

Most molecules diffuse across membranes through specific protein carriers called **permeases**. This process, called **facilitated diffusion**, is also a means of passive transport because the direction of movement will depend on the concentration of the molecule on each side of the membrane. Facilitated diffusion occurs through pores within the permease. Each pore has a shape complementary to the shape of the molecule to be transported. Only molecules having the right shape can pass through the pore and the rate of diffusion is limited by the number of permease molecules in the membrane as well as the concentration of solute on either side of the membrane.

Osmosis: Passive Movement of a Solvent Across a Membrane.

Membranes permeable to solvent but not to solute are called **semi-permeable membranes**. **Osmosis** is the diffusion of a solvent (water in biological systems) through a semi-permeable membrane in response to a water concentration gradient.

The osmotic strength of a solution is referred to as the **osmotic concentration or osmolarity**. It depends only on the ratio of the number of solute particles to the number of solvent particles. Thus, only the concentration of a solution, and not the chemical nature and size of the solute are important. Blood plasma has an osmolarity equivalent to a 0.30 M glucose solution or a 0.15 M NaCl solution (remember that NaCl dissociates into Na^+ and Cl^- in solution). It is important when infusing solutions into the body that the osmolarity of the solutions is the same as that within the cells of the body. Such a solution is said to be **isotonic**. Since the osmolarity is the same inside and outside, there will be no net movement of water into or out of the red blood cells and they will remain the same size.

If red blood cells are exposed to a **hypotonic** solution (one having a lower osmolarity than the cell, the cell will swell and burst. If red blood cells are exposed to a **hypertonic** solution (one with a greater osmolarity than the cell), water will pass out of the cells and they will dramatically shrink in size.

The principle of osmolarity is very important when considering the delivery of intravenous (IV) solutions into a patient. Any fluids infused must be isotonic with the cells of the body.

Energy Requirements for Transport

To survive, it is often necessary for cells to accumulate high concentrations of nutrients that are in low concentration in the environment. In this circumstance, the cell must expend energy to move nutrients from an area of low concentration to an area of higher concentration. This process is called **active transport**. Most ions and food molecules are imported by active transport through the cell membrane. An example of active transport is the Na^+-K^+ pump that maintains a high concentration of Na^+ outside the cell and a high concentration of K^+ inside the cell.

Key Terms

arachidonic acid

atherosclerosis

cholesterol

chylomicron

complex lipid

diglyceride

emulsifying agent

essential fatty acid

esterification

fatty acid

fluid mosaic model

glyceride

high-density lipoprotein (HDL)

hydrogenation

hydrolysis (of triglycerides)

lipid

low-density lipoprotein (LDL)

monoglyceride

neutral glyceride

peripheral protein

phosphatidate

phosphoglyceride

phospholipid

plasma lipoprotein

prostaglandins

saponification

saturated fatty acid

sphingolipid

sphingomyelin

steroid

terpene

transmembrane protein

triglyceride

unsaturated fatty acid

very low-density lipoprotein (VLDL)

wax

Self Test

1. What is the simplest type of lipid?

2. Why are most lipids insoluble in water?

3. What is the main energy storage compound in animals.

4. List two properties of all fatty acids.

5. A triglyceride is called glyceryl stearolauropalmitate. Name the compounds that would be needed to prepare this triglyceride.

6. What products are formed by hydrolysis of a natural wax?

7. List the four major classes of human plasma lipoproteins.

8. What products are released by the complete digestion of a triglyceride in the small intestine?

9. Evidence indicates that high levels of which plasma lipoprotein help reduce the incidence of atherosclerosis?

10. Patients whose diets are high in saturated fat tend to have high levels of what substance in their blood?

11. In the lipid bilayer of cell membranes, which part of the phospholipid is found in the center of the bilayer?

12. What substance dissolved in the hydrophobic region of a biological membrane helps to regulate membrane fluidity?

13. What features of the fatty acid tails of membrane phospholipids determine the degree of fluidity of a biological membrane?

14. Write an equation for the reaction of lauric acid with sodium hydroxide.

15. What structure is found in all steroids?

16. Name foods in your diet that have a high concentration of cholesterol.

17. Which steroid is responsible for both the successful initiation and completion of pregnancy?

18. Which class of biological compounds has members of varying chemical composition, all of which are grouped together on the basis of their solubility?

19. Saturated fatty acids containing 10 or more carbon atoms are found in what physical state at room temperature?

20. What products are formed in the reaction of a fatty acid with an alcohol?

21. List two ions found in hard water.

22. Hydrogenation is used in the food industry to convert polyunsaturated vegetable oils into what?

23. List the three different kinds of eicosanoids.

24. Prostaglandins produced in the kidneys cause what effect on renal blood vessels?

25. What are the properties of neutral glycerides?

26. What two regions characterize phosphoglycerides?

27. Which lipid is found in egg yolks and soybeans and is used as an emulsifying agent in ice cream?

28. Which type of lipoprotein is bound to plasma cholesterol and transports it from the peripheral tissues to the liver?

29. Which steroid is used to suppress the inflammatory response in the treatment of rheumatoid arthritis?

30. What is the purpose of the orally ingested synthetic steroid hormone called norlutin?

Amines and Amides

Learning Goals

1. Conceptual Goals
· Know the common members of the amine and amide families that are of natural, commercial, health, environmental, and industrial interest.
· Be familiar with the physical properties of each of these two families.
· Discuss the basicity of amines relative to other bases we have encountered.
· Recognize that the amide bond is the central feature of the structure of proteins.

2. Performance Goals
· Draw and name the common amines and amides.
· Draw and discuss the structure of the amide bond.
· Write equations for the preparation and hydrolysis of amides.
· Write equations showing the basicity and neutralization of amines.
· Classify amines as primary, secondary, or tertiary.

3. Health Applications
· Appreciate that amphetamines, barbiturates, analgesics, anesthetics, decongestants, and antibiotics number among the medicinal compounds that are amines or amine derivatives.
· Recognize that many carcinogens are amines or amine derivatives.
· Know the functions of serotonin and acetylcholine.
· Understand the relationship between dopamine and Parkinson's Disease.

Chapter Overview

Amines are organic molecules that contain the **amino group,**$-NH_2$, or substituted amino group. They may be aromatic or aliphatic and have the general formula (Ar–) $R-NH_2$. You can think of the amines as substituted ammonia molecules in which one or more of the hydrogens has been substituted by an organic group.

Amides are carboxylic acid derivatives. The amide group is made up of two portions: one portion from a carboxyl group and the other from an amine.

The amino group is found in many important biological molecules, particularly the proteins and nucleic acids (DNA and RNA), and is important in understanding the properties of these large, complex molecules.

18.1 Amines

Structure and Physical Properties

Like ammonia, amines are pyramidal molecules. The nitrogen atom is attached to three groups and has a nonbonding pair of electrons. They are classified by the number of hydrocarbon groups attached to the nitrogen. **Primary (1°) amines** have one R group; **secondary (2°) amines** have two R groups, and **tertiary (3°) amines** have three R groups.

Example 1
Classify the following amines as either primary, secondary, or tertiary.

1. CH_3NH_2
2. CH_3NHCH_3
3. $CH_3N(CH_3)_2$

Work
Amines are classified according to the number of alkyl or aryl groups that are directly attached to the nitrogen atom.

Answer

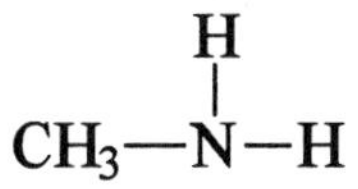

$$CH_3-\underset{\underset{H}{|}}{\overset{\overset{H}{|}}{N}}-H$$

one directly attached carbon atom: primary

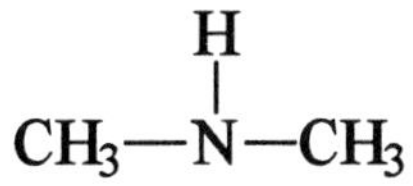

$$CH_3-\underset{\underset{H}{|}}{N}-CH_3$$

two attached carbon atoms: secondary

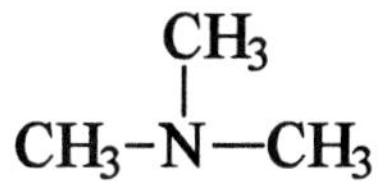

$$CH_3-\underset{\underset{CH_3}{|}}{N}-CH_3$$

three attached carbon atoms: tertiary

Example 2
Write the general formulas for the following;

1. primary amine
2. secondary amine
3. tertiary amine

Answer
1. RNH_2
2. R_2NH
3. R_3N

The N–H bond is polar, and therefore hydrogen bonding occurs between amine molecules. This feature determines the physical properties of the amines, such as boiling point and water solubility. The –NH group is less polar than the –OH group. As a result, the boiling points of amines are lower than comparable alcohols, but higher than comparable ethers or alkanes. The smaller amines are readily soluble in water, but as the size of the hydrocarbon groups increases, their solubility in water decreases.

Example 3

Which member of each of the following sets of compounds with similar molecular weights has the highest boiling point?

1. propane or dimethyl ether or ethanamine
2. ethanol or dimetyl ether or ethanamine

Work

The normal boiling points of alkanes, ethers, primary amines, and primary alcohols obey the following relationship:

$$\text{alkane} < \text{ether} < \text{primary amine} < \text{primary alcohol}$$

Answer

1. ethanamine
2. ethanol

Example 4

Show the hydrogen bonding of methylamine with water.

Work

Since nitrogen is more electronegative than hydrogen, the N–H group is polar.

Answer

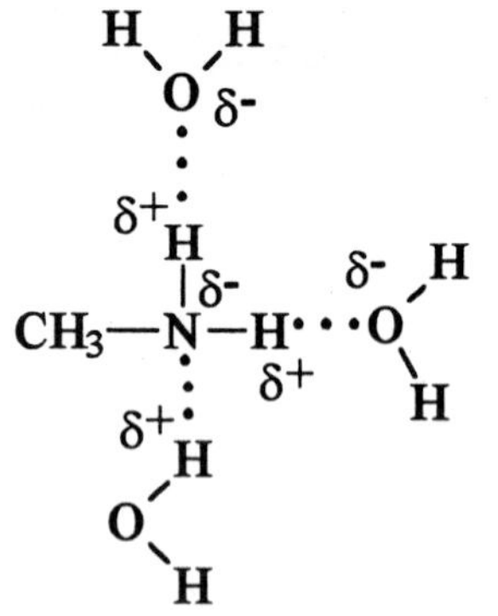

Example 5
The boiling points of methylamine, dimethylamine and trimethylamine are –6.3°, 7°C and 3.5°C respectively. What causes these differences?

Answer
There is a molecular weight difference between each of these molecules (15 grams/mole) due to differences in the number of $-CH_3$ units. The boiling point of trimethylamine is lower than that of dimethylamine because trimethylamine is less polar owing to its symmetrical structure.

Nomenclature

The nomenclature system often used for amines is the *Chemical Abstracts* or CA system. In the CA system, the final *-e* of the name of the parent compound is dropped and the suffix *-amine* is added. For secondary or tertiary amines, the prefix *N*-alkyl is added to the name of the parent compound. Many aromatic amines have special names. An example of this is aniline, a benzene molecule with a substituent amino group.

Common names are also used, especially for the simple amines. The common names of the alkyl groups bonded to the amine nitrogen are followed by the suffix *-amine*. Each alkyl group is listed alphabetically as one continuous word followed by the ending *-amine*.

Example 6
Determine the *Chemical Abstracts* system names of the following amine compounds:

1. CH_3NH_2

2. 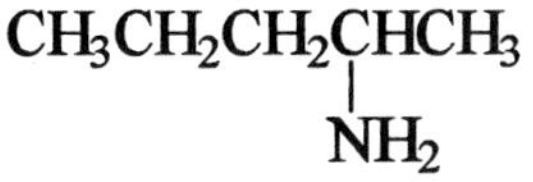 $CH_3CH_2CH_2\underset{\underset{NH_2}{|}}{C}HCH_3$

3. $CH_3-\underset{\underset{NH_2}{|}}{\overset{\overset{H}{|}}{C}}-CH_3$

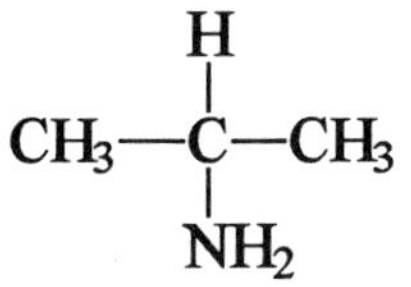

4. $CH_3CH_2CH_2NHCH_3$

Work
1. The parent compound of an amine is the longest continuous chain of carbon atoms to which the amino group is bonded.
2. Drop the final *-e* of the parent alkane and add the suffix *-amine*.
3. If a substituent is present on the nitrogen (other carbon atoms are directly bonded to the nitrogen), it is designated by the prefix *N*-.

Answer
1. methanamine
2. 2-pentanamine
3. 2-propanamine
4. *N*-methyl-1-propanamine

Example 7

Determine the common names for each of the following amines

1. CH_3NH_2

2. $(CH_3)_2NH$

3.
$$CH_3-\underset{\underset{H}{|}}{\overset{\overset{H}{|}}{N}}-\underset{\underset{H}{|}}{\overset{\overset{CH_3}{|}}{C}}-CH_2CH_3$$

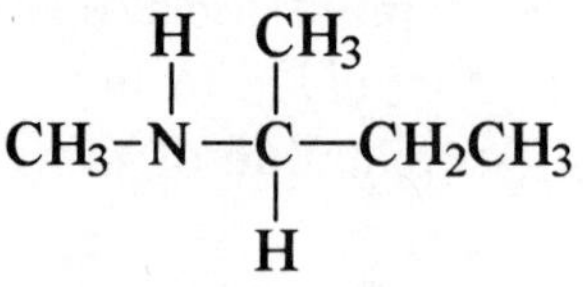

4. $(CH_3CH_2)_2NH$

Work

The common names of the amines are derived from the various alkyl groups bonded to the amine nitrogen atom.

Answer
1. methylamine
2. dimethylamine
3. methyl-*sec*-butylamine
4. diethylamine

Reactions Involving Amines

Preparation of Amines

Amines are prepared by the reduction of amides and nitro compounds.

If two of the R groups are of an amide are hydrogen atoms, the product will be a *primary amine*. If only one R group is a hydrogen atom and the other is an organic group, the product will be a *secondary amine*. If both R groups are organic groups, the product will be a *tertiary amine*.

The following equation represents the reduction of an amide:

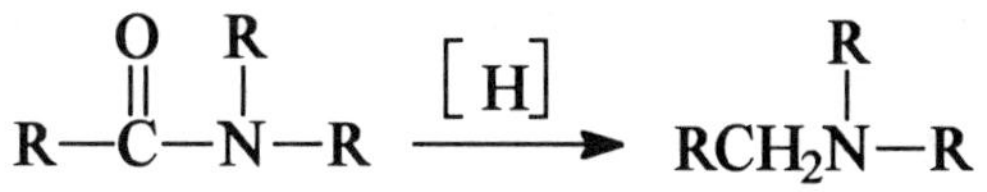

$$\underset{\text{Amide}}{R-\overset{\overset{\displaystyle O}{\|}}{C}-\overset{\overset{\displaystyle R}{|}}{N}-R} \xrightarrow{[H]} \underset{\text{Amine}}{RCH_2\overset{\overset{\displaystyle R}{|}}{N}-R}$$

(R may represent a hydrogen atom
or an organic group.)

Nitro compounds are aromatic compounds. The product of reduction of a nitro compound will be an aromatic amine.

Basicity

Amines have a nonbonding pair of electrons that can be shared with an electron-deficient group to form a new bond. For instance, an amine may react with a proton (H^+), producing a new N–H bond. The original unshared pair of electrons on the nitrogen atom has been shared with the electron-deficient proton. The product is an **alkylammonium ion**.

Neutralization

Since amines are moderately strong bases, they react with most acids to form alkylammonium salts. The generalized reaction is represented here:

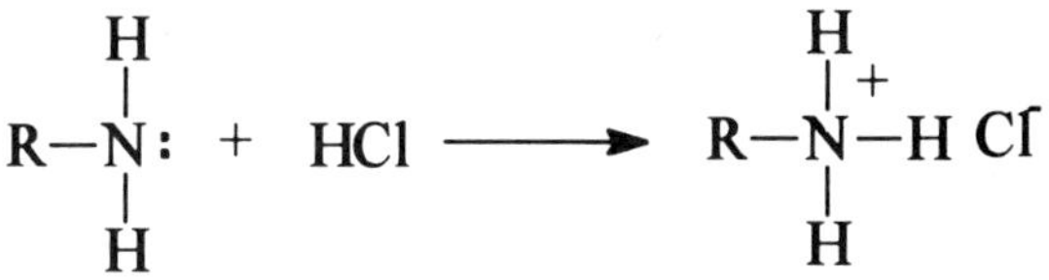

$$R-\overset{\overset{\displaystyle H}{|}}{\underset{\underset{\displaystyle H}{|}}{N}}: \ + \ HCl \longrightarrow R-\overset{\overset{\displaystyle H}{|}}{\underset{\underset{\displaystyle H}{|}}{N}}\overset{+}{-}H \ Cl^-$$

The product, an *alkylammonium salt*, is named by replacing the term *amine* with the term *ammonium* followed by the name of the anion. The salts are ionic species and hence are quite soluble in water.

Example 8
Complete the following reactions and name the products.

1. $CH_3CH_2NH_2 + HCl \ \rightarrow \ ?$
2. $(CH_3)_2NH + HI \ \rightarrow \ ?$
3. $CH_3NH_2 + H_2O \ \rightleftharpoons \ ?$

Work

Amines are weak to moderately strong bases and will react with most acids to form salts.

Answer

1. $CH_3CH_2NH_3^+Cl^-$ ethylammonium chloride
2. $(CH_3)_2NH_2^+I^-$ dimethylammonium iodide
3. $CH_3NH_3^+OH^-$ methylammonium hydroxide

Because amines act as bases (proton acceptors) and amine salts act as acids (proton donors), they are useful as *buffers*. Many naturally occurring amines serve as biological buffers. For instance, the protein hemoglobin helps maintain the acid/base (pH) balance of the blood.

Several important drugs are amines. They are generally administered as alkylammonium salts because the salts are more soluble in water and body fluids.

Quaternary Ammonium Salts

Quaternary ammonium salts are ammonium salts having four organic groups bonded to the nitrogen. They have the general structure R_4N^+A.

Quaternary ammonium salts with very long carbon chains are often used as disinfectants and antiseptics because they have detergent activity.

18.2 Heterocyclic Amines

Heterocyclic amines are cyclic amines in which one of the atoms of the backbone of the ring is nitrogen. They are found in many important cellular macromolecules, including DNA, RNA, and coenzymes. Many drugs, including cocaine, nicotine, quinine, morphine, heroin, and LSD (lysergic acid diethylamide) are heterocyclic amines, as is the artificial sweetener saccharin. Several examples are provided in Figure 18.3 in the text.

18.3 Amides

Amides are an important class of nitrogen-containing organic compounds with the functional group shown here:

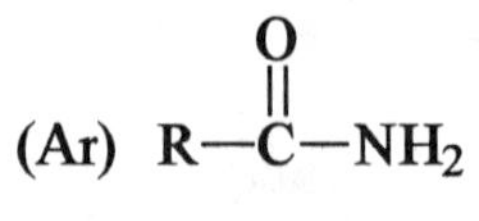

$$\text{(Ar)} \quad R-\overset{\overset{\displaystyle O}{\|}}{C}-NH_2$$

As careful inspection of this structure shows, amides are carboxylic acid derivatives.

The **amide bond** is the central feature in the structure of proteins. The amide bond between two amino acids is called a **peptide bond**. The structure and function of proteins is the subject of the next chapter.

Structure and Physical Properties

Most amides are solids at room temperature, with boiling points even higher than the corresponding carboxylic acid. The simpler amides are quite soluble in water. Both of these properties reflect the strong intermolecular hydrogen bonding. The water solubility decreases as the molecular weight increases.

Nomenclature

Amides are named by removing the *-ic acid* ending of the common name or the *-oic acid* ending of the I.U.P.A.C. name of the carboxylic acid and replacing it with *-amide*. If there are substituents on the nitrogen, they are placed as prefixes and are indicated by *N-*, followed by the name of the substituent. There are no spaces between the prefix and the amide name.

Example 9

Determine the I.U.P.A.C. names of the following amides:

$$1. \quad CH_3—\overset{\overset{\displaystyle O}{\|}}{C}—NH_2$$

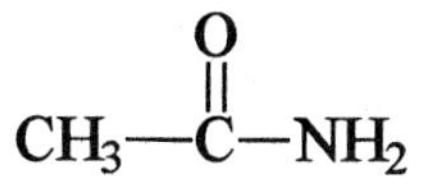

$$2. \quad CH_3CH_2—\overset{\overset{\displaystyle O}{\|}}{C}—\overset{\overset{\displaystyle H}{|}}{N}—CH_3$$

$$3. \quad CH_3CH_2CH_2CH_2—\overset{\overset{\displaystyle O}{\|}}{C}—\overset{\overset{\displaystyle H}{|}}{N}—CH_2CH_3$$

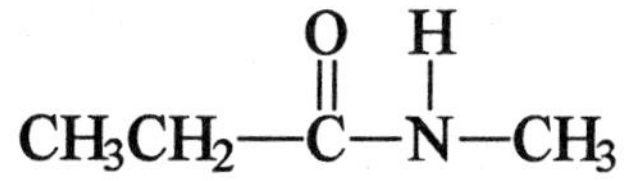

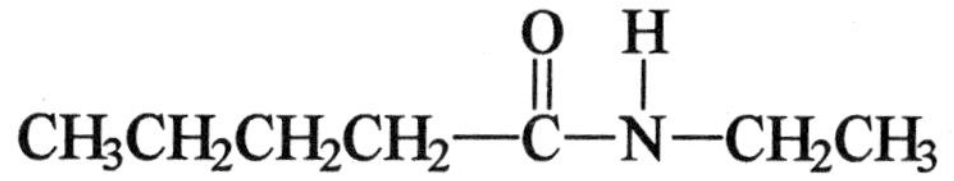

$$4. \quad CH_3CH_2CONH_2$$

Work

1. The common and I.U.P.A.C. names of the amides are derived from the common and I.U.P.A.C. names of the corresponding carboxylic acids.
2. Amides are named by removing the *-ic acid* ending of the common name or the *-oic acid* ending of the I.U.P.A.C. name of the carboxylic acid and replacing it with *-amide*.
3. Substituents on the nitrogen atom are placed as prefixes and are indicated by *N-*, followed by the name of the substituent.

Answer
1. ethanamide
2. *N*-methylpropanamide
3. *N*-ethylpentanamide
4. propanamide

Reactions Involving Amides

Preparation of Amides

Amides can be prepared from carboxylic acid derivatives. The product of the reaction between an acid chloride and either ammonia or an amine produces an amide and either ammonium chloride or an alkylammonium chloride. Two molar equivalents of ammonia or amine are required for this reaction, which is termed an acyl group transfer because the **acyl group** of the acid chloride is transferred from the Cl atom of the acid chloride to the N atom of the amine.

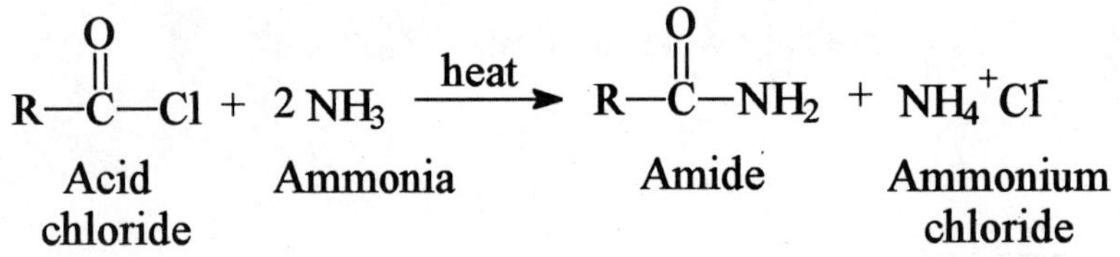

$$
\underset{\substack{\text{Acid} \\ \text{chloride}}}{R-\overset{\overset{\textstyle O}{\|}}{C}-Cl} + \underset{\text{Ammonia}}{2\,NH_3} \xrightarrow{\text{heat}} \underset{\text{Amide}}{R-\overset{\overset{\textstyle O}{\|}}{C}-NH_2} + \underset{\substack{\text{Ammonium} \\ \text{chloride}}}{NH_4^+Cl^-}
$$

Amides may also be prepared by the reaction between an acid anhydride and ammonia or an amine.

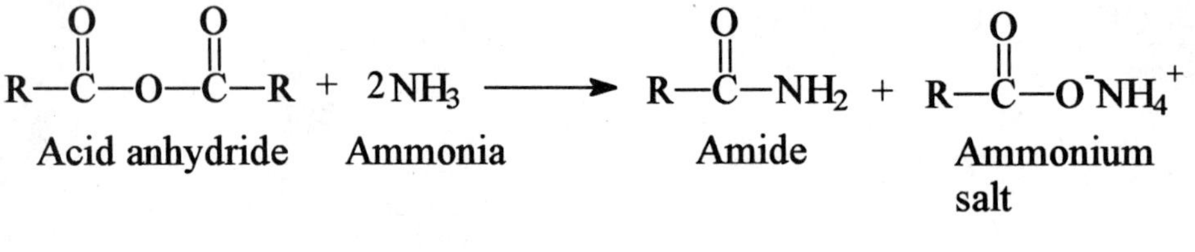

$$
\underset{\text{Acid anhydride}}{R-\overset{\overset{\textstyle O}{\|}}{C}-O-\overset{\overset{\textstyle O}{\|}}{C}-R} + \underset{\text{Ammonia}}{2\,NH_3} \longrightarrow \underset{\text{Amide}}{R-\overset{\overset{\textstyle O}{\|}}{C}-NH_2} + \underset{\substack{\text{Ammonium} \\ \text{salt}}}{R-\overset{\overset{\textstyle O}{\|}}{C}-O^-NH_4^+}
$$

When subjected to heat, the ammonium salt loses a water molecule and produces a second amide molecule.

Hydrolysis of Amides

Hydrolysis of an amide results in breaking the amide bond to produce the carboxylic acid and ammonia or an amine. This reaction requires heat and the presence of a strong acid or base.

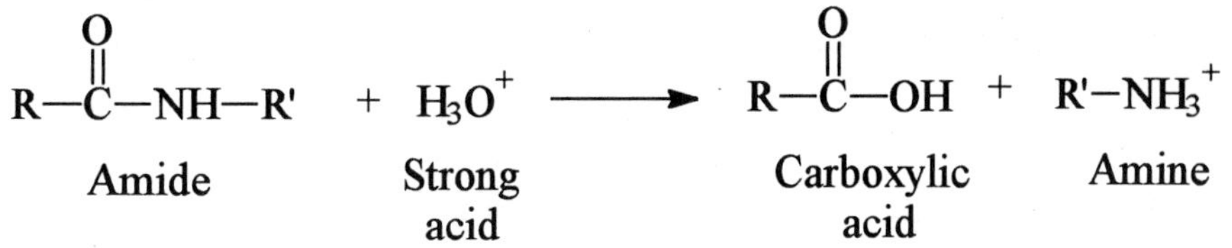

$$R-\overset{\overset{\textstyle O}{\|}}{C}-NH-R' \quad + \quad H_3O^+ \quad \longrightarrow \quad R-\overset{\overset{\textstyle O}{\|}}{C}-OH \quad + \quad R'-NH_3^+$$

| Amide | Strong acid | | Carboxylic acid | Amine |

Example 10

Complete the following reactions:

1. 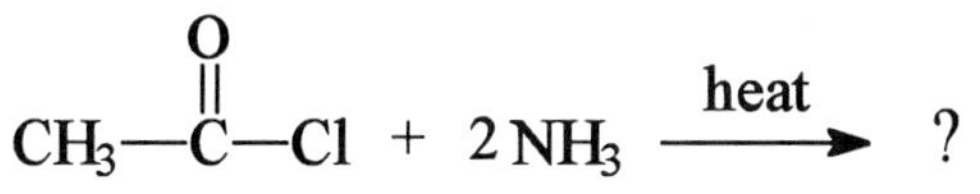 $CH_3-\overset{\overset{\textstyle O}{\|}}{C}-Cl \; + \; 2\,NH_3 \; \xrightarrow{\text{heat}} \; ?$

2. $CH_3CH_2-\overset{\overset{\textstyle O}{\|}}{C}-NH-CH_3 \; + \; H_3O^+ \; \longrightarrow \; ?$

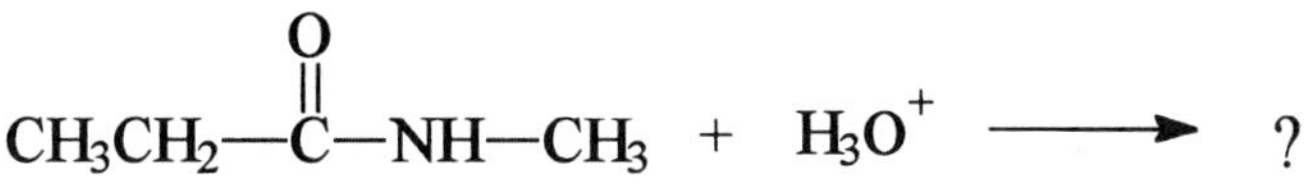

Answer

1. 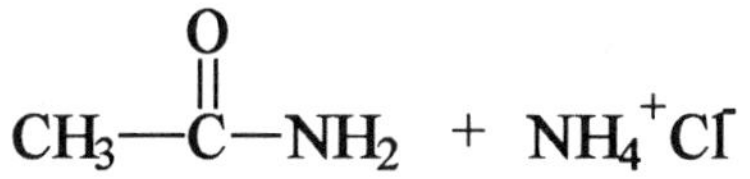 $CH_3-\overset{\overset{\textstyle O}{\|}}{C}-NH_2 \; + \; NH_4^+\,Cl^-$

2. $CH_3CH_2-\overset{\overset{\textstyle O}{\|}}{C}-OH \; + \; CH_3N^+H_3$

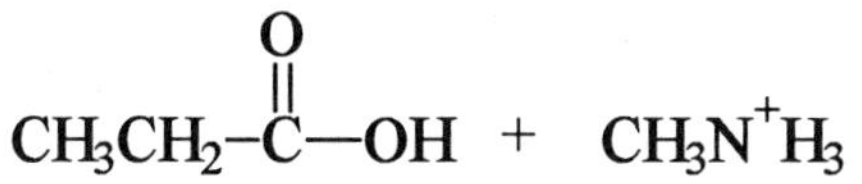

18.4 A Preview of Amino Acids, Proteins, and Protein Synthesis

Proteins are polymers of amino acids. Amino acids have an amino group and a carboxyl group and typically have the following general structure:

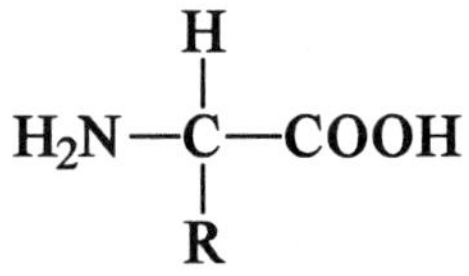

$$H_2N-\overset{\overset{\textstyle H}{|}}{\underset{\underset{\textstyle R}{|}}{C}}-COOH$$

The R group may be a hydrogen atom or an organic group. The primary structure of a protein consists of many amino acids bonded to one another by amide bonds called **peptide**

bonds. Each peptide bond is formed by a reaction joining the amino group of one amino acid and the carboxyl group of another.

Key Terms

acyl group

alkaloid

alkylammonium ion

amide bond

amide group

amide

amine

aminoacyl group

analgesis

anesthetic

heterocyclic amine

peptide bond

primary (1°) amine

quaternary ammonium salt

secondary (2°) amine

tertiary (3°) amine

transfer RNA

Self Test

1. List the following as either a primary, secondary, or tertiary amine or as a quaternary ammonium salt.
    a. $(CH_3)_3N$
    b. $(CH_3)_3N^+CH_2CH_3$ Cl^-
    c. $CH_3CH_2NH_2$
    d. CH_3CH_2-NH-CH_2CH_3
2. What functional groups of amino acids react to form the peptide bond?
3. Provide the *Chemical Abstracts* system names of the following amine compounds:
    a. $CH_3CH_2NH_2$
    b. $(CH_3)_2CHNH_2$
    c. $(CH_3)_3N$
4. What is the name of the amine of benzene?
5. Complete each of the following reactions:
    a. $CH_3CONH_2 + NaOH$
    b. CH_3CH_2-$CONH$-CH_3
6. Complete the following reactions:
    a. $(CH_3)_2NH + HI$?
    b. $(CH_3)_3N + HCl$?
7. Write the general formulas for the following:
    a. primary amine
    b. tertiary amine
8. Determine the I.U.P.A.C. names of the following amides:
    a. acetamide
    b. $CH_3(CH_2)_4$-$CONH$-$CH_2CH_2CH_3$
    c. $CH_3CH_2CH_2$-$CONH$-CH_3

9. What compounds are formed in the reaction between a secondary amine and nitrous acid?
10. The central nervous system refers to what parts of the body?
11. Name two important neuotransmitters.
12. Why are the most drugs that are amines delivered as alkylammonium salts?
13. What disease is caused by a deficiency of dopamine in the human body?
14. Which functional group is found in amines?
15. Amines may be viewed as compounds that are substituted products of which inorganic compound?
16. Amines are classified according to the number of alkyl or aryl groups directly attached to what atom?
17. In what biological molecules are the heterocyclic compounds, pyrimidines and purines, found?
18. What tertiary amine is produced by a reaction between ammonia and methanol?
19. What product is formed in the reaction of an amine with an acid?
20. Name the product that results when methylamine reacts with HCl.
21. The ability of the amines to act as bases and of amine salts to act as acids makes a mixture of these useful as what in solution?
22. Which functional group is composed of one portion from a carboxylic acid and a second portion from an amine?
23. What is the physical state of most amides at room temperature?
24. What two products are released by the hydrolysis of an amide?
25. Why are quaternary ammonium salts effective as disinfectants and antiseptics?
26. What is the term for an ammonium salt that has four organic groups bonded to the nitrogen?
27. The amide bond is the central feature in the structure of what class of biological molecules?
28. What is the term for a substance that can cause cancer?
29. What is the general term for a drug, such a morphine, that acts as a painkiller?

19 *Protein Structure and Function*

Learning Goals

1. Conceptual Goals
· Be familiar with the cellular functions of proteins.
· Describe the primary structure of proteins and the structure of the peptide bond.
· Understand how the primary structure, the amino acid sequence, is used to determine evolutionary relatedness.
· Know the rules for the structure and nomenclature of small peptides.
· Describe the types of secondary protein structure.
· Understand the forces that maintain secondary structure.
· Be familiar with tertiary and quaternary protein structure.
· Understand the R group interactions that maintain the three-dimensional conformation of a protein.
· Know why the correct three-dimensional structure of a protein is essential to its function.
· Understand how extremes of pH and temperature cause denaturation of proteins.

2. Performance Goals
· Draw the general structure of an amino acid.
· Know the classification of side chains (R groups) of the amino acids.
· Define conjugated proteins and prosthetic groups.

3. Health Applications
· Describe the roles of hemoglobin and myoglobin in oxygen transport and storage.
· Recognize the serious consequences that can occur when even a single amino acid in a peptide is altered, as in sickle cell anemia.
· Be familiar with the structures of fibrous proteins, like collagen, and their functions as structural components of the body.
· Describe the role of vitamin C in collagen metabolism.
· Understand the mode of action of the opiate drugs on the brain.
· Appreciate the function of immunoglobulins produced by the immune system to protect us against infectious agents.

Chapter Overview

Proteins are polymers made up of 20 α-amino acids. Most proteins are enormous molecules, composed of many amino acids, and therefore have very high molecular weights. Proteins carry out all the enzymatic reactions of the cell and are essential structural elements of the body.

19.1 Cellular Functions of Proteins

Enzymes are proteins that serve as biological catalysts. **Immunoglobulins** or **antibodies**, are specific protein molecules produced in the immune system in response to foreign **antigens**. **Transport proteins** carry materials from one place to another in the body. They include permeases and **receptors** that move molecules across cell membranes and soluble proteins, such as hemoglobin that transports oxygen from the lungs to the tissues. **Regulatory proteins** control cell function and include hormones such as insulin and glucagon. **Structural proteins** provide mechanical support and outer covering to animals. Proteins are necessary for all forms of movement. Our muscles contract and expand through the interaction of actin and myosin proteins. **Nutrient proteins** are sources of amino acids for embryos and infants.

Example 1

Provide a summary of the cellular functions of proteins.

Answer

1. All the *enzymes* are proteins. Enzymes catalyze almost all the chemical reactions that occur in living cells. Reactions that would take days in the laboratory occur in a matter of seconds or minutes in the presence of enzymes.
2. *Antibodies* are proteins that help stop infections by binding specifically to an antigen and then causing its destruction or removal from the body.
3. *Transport proteins* carry materials from one place to another in the body.
4. *Regulatory proteins* control many aspects of cell function, including metabolism and reproduction.
5. *Structural proteins* provide mechanical support to large animals and provide them with their outer coverings.
6. *Proteins of motion* allow a single-celled organism or higher organism to have different types of motility.
7. *Nutrient proteins* are sources of amino acids for infants and embryos.

19.2 The α-Amino Acids

The general structure of an α-**amino acid** is shown below:

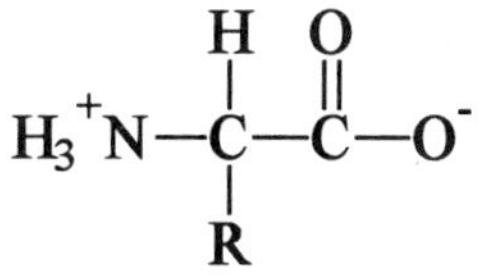

$$H_3{}^+N-\underset{\underset{R}{|}}{\overset{\overset{H}{|}}{C}}-\overset{\overset{O}{\|}}{C}-O^-$$

The α-carbon is attached to a carboxylate group ($CO_2{}^-$), a protonated amino group ($NH_3{}^+$), a hydrogen, and an R group. The carboxylate group and protonated amino groups are necessary for the covalent binding of amino acids to one another to form a protein. The R groups cause the proteins to fold into precise, three-dimensional configurations that determine their ultimate function.

The α-carbon of all the α-amino acids, except glycine, is attached to four different groups and is therefore chiral. The configuration of all the naturally occurring α-amino acids isolated from proteins is L-.

Example 2

Draw and compare the structures of glyceraldehyde and alanine.

Answer

In living organisms we find that glyceraldehyde is in the D-isomer form, but the amino acids exist in the L-isomer form.

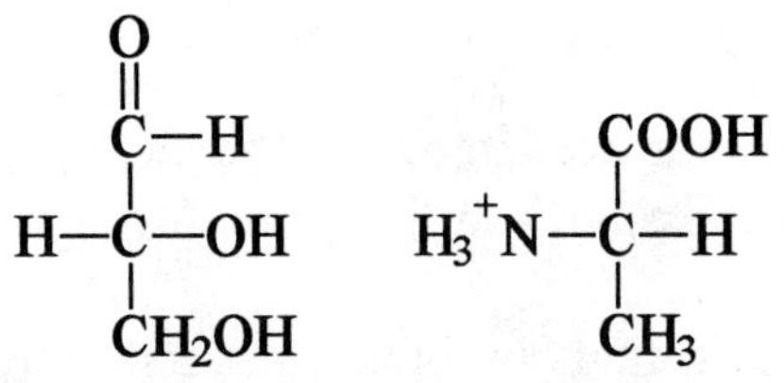

$$
\begin{array}{cc}
\overset{\displaystyle O}{\underset{\displaystyle }{\overset{\displaystyle \|}{C}}}\!\!-\!H & \\
\text{H--C--OH} & \text{H}_3{}^+\text{N--C--H} \\
\text{CH}_2\text{OH} & \text{CH}_3
\end{array}
$$

D-Glyceraldehyde   L-Alanine

The amino acids are grouped according to the polarity of their side chains. The side chains of some amino acids are nonpolar. These amino acids are **hydrophobic** (water-fearing), and they are generally found buried in the interior of proteins. The nonpolar amino acids include alanine, valine, leucine, isoleucine, proline, glycine, cysteine, methionine, phenylalanine, and tryptophan.

Example 3

List the amino acids that contain R groups (side chains) that are hydrophobic. Where are these usually found in a protein molecule?

Answer

Alanine, valine, leucine, isoleucine, proline, glycine, cysteine, methionine, phenylalanine and tryptophan contain hydrophobic R groups. These amino acids are generally found buried in the interior of proteins, where they can associate with one another and remain isolated from interaction with water molecules.

The side chains of the remaining amino acids are polar and therefore are **hydrophilic** (water-loving); they are often found on the surfaces of proteins. These polar amino acids can be subdivided into three classes.

1. Polar, neutral amino acids
2. Negatively charged (acidic) amino acids
3. Positively charged (basic) amino acids

The amino acids are often referred to using an abbreviation that is the first three letters of their names.

Example 4

List and categorize the amino acids that have hydrophilic side chains.

Answer

1. Arginine, aspartic acid, glutamic acid, histidine, lysine, asparagine, glutamine, serine, threonine and tyrosine have side chains that are hydrophilic.
2. These polar amino acids are divided into three classes:

   a. Polar, neutral amino acids: These have side chains that have a high affinity for water, but they are not ionic. Serine, threonine, tyrosine, asparagine, and glutamine fall into this category.

   b. Acidic amino acids are those that have R groups that can ionize to form the negatively charged carboxylate ion ($-COO^-$ ion). Aspartic acid (also called aspartate) and glutamic acid (also called glutamate) are in this category.

   c. Basic amino acids are those with R groups that can ionize to form the positively charged amine ion ($-NH_3^+$). These amino acids act like bases in water, since the side chains react with water, picking up a proton and releasing a hydroxide ion. Lysine, arginine, and histidine are in this category.

19.3 The Peptide Bond

Proteins are polymers of L-α-amino acids joined by **peptide bonds**. This is the covalent bond formed between the α-carboxylate group of one amino acid and the α-amino group of another amino acid. The amino acid with a free α-N^+H_3 group is known as the amino terminal, or simply the N-terminal amino acid, and the amino acid with a free a-CO_2^- group is known as the carboxyl, or C-terminal amino acid. The **N-terminal amino acid** is always drawn on the left, and the **C-terminal amino acid** on the right when depicting a series of covalently linked amino acids.

Example 5
Draw the tripeptide made from alanine + glycine + serine.

Answer

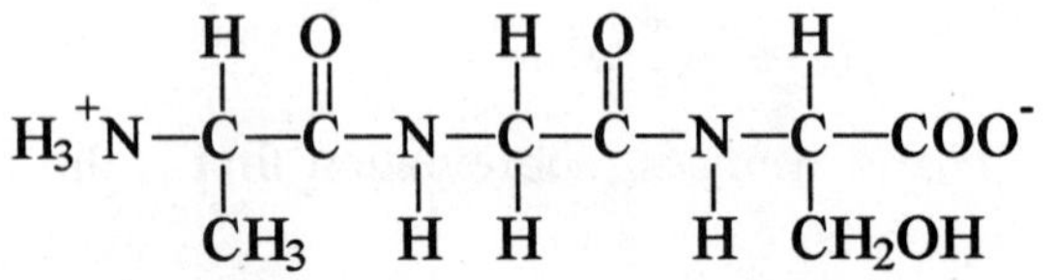

$$H_3{}^+N-\overset{\displaystyle \overset{H}{|}}{\underset{\displaystyle \underset{CH_3}{|}}{C}}-\overset{\displaystyle O}{\overset{\|}{C}}-\overset{\displaystyle \overset{H}{|}}{\underset{\displaystyle \underset{H}{|}}{N}}-\overset{\displaystyle \overset{H}{|}}{\underset{\displaystyle \underset{H}{|}}{C}}-\overset{\displaystyle O}{\overset{\|}{C}}-\overset{\displaystyle \overset{H}{|}}{\underset{\displaystyle \underset{H}{|}}{N}}-\overset{\displaystyle \overset{H}{|}}{\underset{\displaystyle \underset{CH_2OH}{|}}{C}}-COO^-$$

The names of peptides are derived from the C-terminal amino acid, which receives its entire name. For all other amino acid, the ending *-ine*, is changed to *-yl*. Thus the dipeptide tryptophanyl-leucine has leucine as the C-terminal amino acid and tryptophan as the N-terminal amino acid. Example 19.1 in the text describes the procedure for drawing the structure of a peptide chain.

Example 6
Show the hydrolysis reaction for alanyl-glycyl-cysteine.

Answer

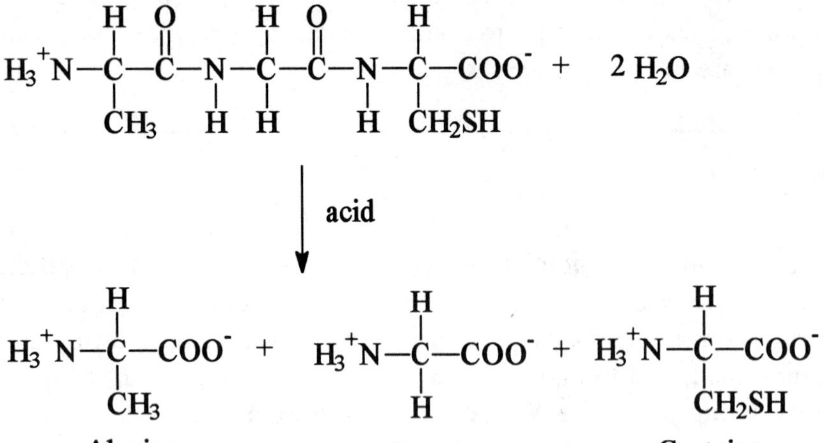

$$H_3{}^+N-\overset{\overset{H}{|}}{\underset{\underset{CH_3}{|}}{C}}-\overset{O}{\overset{\|}{C}}-\overset{\overset{H}{|}}{\underset{\underset{H}{|}}{N}}-\overset{\overset{H}{|}}{\underset{\underset{H}{|}}{C}}-\overset{O}{\overset{\|}{C}}-\overset{\overset{H}{|}}{\underset{\underset{H}{|}}{N}}-\overset{\overset{H}{|}}{\underset{\underset{CH_2SH}{|}}{C}}-COO^- \quad + \quad 2\,H_2O$$

↓ acid

$$H_3{}^+N-\overset{\overset{H}{|}}{\underset{\underset{CH_3}{|}}{C}}-COO^- \quad + \quad H_3{}^+N-\overset{\overset{H}{|}}{\underset{\underset{H}{|}}{C}}-COO^- \quad + \quad H_3{}^+N-\overset{\overset{H}{|}}{\underset{\underset{CH_2SH}{|}}{C}}-COO^-$$

Alanine   Glycine   Cysteine

The lone pair of electrons of the nitrogen atom of the peptide bond interacts with the carbon and oxygen of the carbonyl group. Thus, a resonance structure is formed which gives the peptide bond a partially double bond character. As a result, the peptide bond is planar, and the two adjacent α-carbons lie *trans* to it. The hydrogen of the amide nitrogen is also *trans* to the oxygen of the carbonyl group.

19.4 Primary Structure of Proteins

The **primary protein structure** is the linear sequence of amino acids. This is dictated by the genetic information in the DNA. Over time, mutations occur in the genes. These result in changes in the primary sequence of a protein. The greater the number of amino acid differences between proteins from two different organisms, the greater the evolutionary distance between them. Examination of the amino acid sequences of proteins such as cytochrome c allows scientists to construct evolutionary trees.

19.5 Secondary Structure of Proteins

Secondary protein structure results from folding of the chain of covalently linked amino acids into regularly repeating structures. The folding pattern is maintained by numerous hydrogen bonds.

α-Helix

The most common type of secondary structure is a right-handed helical conformation known as the α-**helix**. There are 3.6 amino acids in each turn of the helix. Each carbonyl oxygen in the helix is hydrogen bonded to an amide hydrogen four amino acids away in the chain, producing an array of hydrogen bonds that are parallel to the long axis of the helix.

The α-**keratins** are α-helical proteins. These **fibrous proteins** form the covering (hair, wool, and fur) of most land animals. The individual α-helices of the keratins coil together in a bundle, producing a three-stranded protofibril that is part of an array known as a microfibril. These molecular pigtails possess great mechanical strength. For example, **myosin** is one of the major proteins of muscle.

β-Pleated Sheet

The second common secondary structure in proteins resembles the pleated folds of drapery, and is known as β-**pleated sheet**. In this secondary structure the polypeptide chain is nearly completely extended, with all the carbonyl oxygens and amide hydrogens involved in hydrogen bonds. The polypeptide chains in a β-pleated sheet may be *parallel* or *antiparallel*. In the parallel structure the N-termini are aligned head to head, and in the antiparallel structure, the N-terminus of one chain is aligned with the C-terminus of a second chain (head to tail). Silk fibroin is an example of a protein whose structure is an antiparallel β-pleated sheet.

19.6 Tertiary Structure of Proteins

Most cellular proteins are **globular**. This globular, three-dimensional structure is called the **tertiary protein structure**. The peptide chain with its regions of secondary structure further folds on itself to achieve the tertiary structure.

The forces that maintain the tertiary structure of a protein include the following:

1. Van der Waals attractions between the R groups of nonpolar amino acids.
2. Hydrogen bonds between the polar R groups of the polar amino acids.
3. Ionic bonds between the R groups of oppositely charged amino acids.
4. Covalent bonds between the thiol-containing amino acids. Two cysteines can be oxidized to a dimeric amino acid, cystine, that can be a cross-link between different proteins or hold two segments within a protein together.

The tertiary structure of the protein determines its biological function; therefore the weak interactions that hold the protein in its correct three-dimensional shape are extremely important.

19.7 Quaternary Structure of Proteins

The active form of some proteins is an aggregate of two or more smaller globular proteins. This is the **quaternary protein structure**. The attractions that hold two or more peptides together are the same as those that maintain tertiary structure: hydrogen bonds, ionic bridges, disulfide bridges, and van der Waals forces. Some proteins must be bound to a nonprotein **prosthetic group** in order to be functional. Such proteins are called **conjugated proteins**. Hemoglobin is an example of a protein with quaternary structure and is also a conjugated protein, bound to heme groups. **Glycoproteins** are conjugated proteins with covalently bonded sugar groups.

Example 7

Summarize the four types of protein structure and their relationship to one another.

Answer

1. The primary structure of a protein results from the amide structure between various amino acids. The primary structure includes the amino acids that are present and the order in which they appear.

$$H_3^+N-\underset{\underset{R}{|}}{\overset{\overset{H}{|}}{C}}-\overset{\overset{O}{\|}}{C}-N-\underset{\underset{H}{|}}{\overset{\overset{H}{|}}{C}}-\overset{\overset{O}{\|}}{C}-N-\underset{\underset{R''}{|}}{\overset{\overset{H}{|}}{C}}-COO^-$$

amino acid 1   amino acid 2   amino acid 3

2. The secondary structure involves the hydrogen bonding that can occur between different amide linkages. The α-helix and the β-pleated sheet structures are the most common kinds.
3. The tertiary structure involves the gross overall folding of the entire protein molecule. Both noncovalent interactions between the side chains (R groups) and covalent −S−S− bridges play a role in determining the overall tertiary structure.
4. The quaternary structure involves the spatial arrangements of two or more peptide chains with respect to one another. The quaternary structure is maintained by R group interactions.

19.8 Myoglobin and Hemoglobin

Myoglobin and Oxygen Storage

Myoglobin is the oxygen storage protein of skeletal muscle. It is bound to a **heme** group that serves as the site for oxygen binding. Myoglobin has a greater affinity for oxygen than hemoglobin does, and therefore it serves as an efficient molecule to receive and store oxygen in the muscle.

Hemoglobin and Oxygen Transport

Hemoglobin (Hb) is the blood protein responsible for oxygen transport from the lungs to other tissues. It is composed of four peptide subunits, two α subunits and two β subunits, each of which contains a heme group.

Example 8
Explain the general structure and function of hemoglobin.

Answer
Hemoglobin is the oxygen transport protein of the blood. It is composed of four separate peptide subunits. There are two identical α chains and two identical β chains. In addition, each subunit of hemoglobin contains a heme group. A hemoglobin molecule therefore has the ability to bind and carry four molecules of oxygen.

Oxygenation of hemoglobin in the lungs is favored by the high partial pressure of oxygen (pO_2) in the air we breath and the low pO_2 in the blood. Thus, oxygen diffuses from the region of high pO_2 to the region of low pO_2. When the blood reaches actively metabolizing tissues this situation is reversed and oxygen diffuses into tissues from the blood.

Oxygen Transport from Mother to Fetus

A fetus receives oxygen from its mother by simple diffusion across the placenta. The fetus makes fetal hemoglobin (Hb F), which has a greater affinity for oxygen than the hemoglobin of the mother, Hb A. Thus the fetus is assured of a constant and adequate supply of oxygen.

Sickle Cell Anemia

Sickle cell anemia is a human genetic disease caused by a mutation in the gene that encodes the β-subunit of hemoglobin . This mutation results in the synthesis of sickle cell hemoglobin (Hb S). There is only a single amino acid difference between normal and sickle cell hemoglobin. But that single change causes deoxyhemoglobin to polymerize, causing the red blood cells to sickle. This results in damage to many organs and death at an early age.

In individuals with sickle cell anemia, both copies of the β-globin gene produce the mutant protein. Individuals with *sickle cell trait* have one mutant and one normal β-globin gene. Typically they do not suffer the symptoms of the disease, however, they may pass the mutant gene on to their children.

Example 9
Explain what causes sickle cell anemia.

Answer
Sickle cell anemia is a human genetic disease that first appeared in tropical west and central Africa. It afflicts about 0.4% of Black Americans. These individuals produce a mutant hemoglobin known as sickle cell hemoglobin (Hb S). Sickle cell anemia receives its name from the sickled appearance of the red blood cells that form in this condition.

Sickle cell hemoglobin differs from normal hemoglobin by a single amino acid that has been replaced in each of the β-chains. A valine has replaced the amino acid glutamic acid in each of the two chains. This substitution provides a basis for the binding of hemoglobin S molecules to one another. When oxyhemoglobin S molecules release their oxygen, they bind to one another as long fibers. These fibers radically alter the shape of the red blood cell, resulting in the sickling effect. The sickled cells are unable to pass through the small capillaries and cause severe medical problems.

19.9 Collagen, An Important Protein in Humans

Collagen, a family of related proteins, makes up about a third of the proteins in humans. It is a structural protein that provides mechanical strength to bone, tendon, and skin. Collagen fibers are found in bone where they provide a scaffold on which **hydroxyapatite** (a calcium phosphate polymer) crystals are arranged. Skin and blood vessels contains loosely woven collagen fibers that can expand in all directions.

The Structure of Collagen

Collagen, composed of left-handed triple-stranded fibers, is unusual because nearly 25% of the amino acids are hydroxylated. These are *4-hydroxyproline* (Hyp) and *5-hydroxylysine*

(Hyl). The triple-stranded helical fiber of collagen is known as **tropocollagen**.

The Role of Vitamin C in Collagen Metabolism

Vitamin C is essential for the hydroxylation of prolines and lysines in the collagen protein. Vitamin C deficiency causes **scurvy**, a disease of collagen metabolism. The symptoms of scurvy include skin lesions, fragile blood vessels, and bleeding gums.

Example 10

Explain the role of vitamin C in collagen formation.

Answer

When the collagen protein is being formed, the amino acids proline and lysine are incorporated into the protein chain. These two amino acids account for nearly one-fourth of the amino acid composition of collagen. They are later modified by two enzymes to form 4-hydroxyproline and 5-hydroxylysine. The two enzymes require vitamin C in order to carry out these important hydroxylation reactions.

19.10 An Overview of Protein Structure and Function

Primary structure is the linear sequence of amino acids covalently joined by peptide bonds. **Secondary structure** is the result of folding of the peptide chain into an α-helix or a β-pleated sheet. It is maintained by hydrogen bonding. **Tertiary structure** is the further folding of a peptide to produce a globular structure. It is maintained by a variety of noncovalent interactions between the R groups of amino acids, including hydrogen bonding, ionic bonding, and van der Waals forces. **Quaternary structure** is the association of two or more peptides to form the functional protein. It is maintained by the same noncovalent R group interactions that are responsible for tertiary structure.

19.11 The Effect of Temperature Proteins

When a protein solution is exposed to temperatures that are too high, the bonds within the protein molecules begin to vibrate violently. The weak interactions that hold the three-dimensional structure of the protein are disrupted, and the protein molecules are **denatured**. These molecules then **coagulate** as they clump together and precipitate out of solution.

19.12 The Effect of pH on Proteins

If the pH of a solution of proteins becomes too high or too low, the characteristic electric charge on the surface of the protein is neutralized. When the net charge on the protein is zero, it is said to be **isoelectric**, and it can no longer interact with the surrounding water molecules. This causes the proteins to aggregate with one another, and coagulation occurs.

When the pH rises too high, proteins become polyanions. When the pH is too low, proteins become polycations. In either case, proteins are denatured due to charge repulsion.

If the pH of the blood becomes too acidic or too basic, the proteins dissolved in the blood and the proteins of the cells in the blood can be denatured. Thus it is critical that the body have mechanisms to maintain the pH of the blood within a narrow range.

19.13 Dietary Protein and Protein Digestion

Proteins are the third major type of energy source in the diet. They are hydrolyzed to amino acids that may be oxidized to provide energy or used directly in protein synthesis. Amino acids are the precursors of a large, diverse group of nitrogen compounds which includes some hormones and the heme group.

Example 11
What are some common dietary sources of proteins?

Answer
Meat is an excellent source of protein. Also, dried beans, peas, and nuts are good sources.

Essential amino acids are those that cannot be synthesized by the body and are required in the diet. **Nonessential amino acids** are those amino acids which can be synthesized by the body and need not be included in the diet.

Proteins are also classified as **complete** or **incomplete**. Animal protein is complete, providing all of the essential and nonessential amino acids in approximately the correct amounts for biosynthesis. Vegetable protein is generally incomplete, lacking a sufficient amount of one or more essential amino acids.

Example 12
How can vegetable proteins be mixed to provide all of the essential amino acids?

Answer
A complete protein provides all of the essential amino acids needed by the human body. Proteins derived from animal sources are generally complete. In contrast, proteins derived from vegetable sources are generally incomplete, because they lack a sufficient amount of one or more amino acids. No one vegetable source is a good source of all the essential amino acids, but by mixing many vegetable sources, an adequate amount of each essential amino acid may be obtained.

Protein digestion begins in the stomach. The enzyme pepsin begins the degradation of dietary protein to amino acids. Further digestion occurs in the small intestine. These hydrolysis reactions are carried out by enzymes such as trypsin and chymotrypsin.

Key Terms

α-amino acid	isoelectric
antibody	α-keratin
antigen	myoglobin
C-terminal amino acid	myosin
coagulation	N-terminal amino acid
complete protein	nonessential amino acid
conjugated protein	peptide bond
denaturation	pH optimum
essential amino acid	β-pleated sheet
enzyme	primary protein structure
fibrous protein	prosthetic group
globular protein	protein
glycoprotein	quaternary protein structure
α-helix	regulatory protein
heme group	secondary protein structure
hemoglobin	sickle cell anemia
hydrophilic amino acid	structural protein
hydrophobic amino acid	tertiary protein structure
hydroxyapatite	transport protein
incomplete protein	vitamin C

Self Test

1. What elements found in dietary proteins are not found in fats and carbohydrates?
2. Name the simplest amino acid.
3. Which amino acid contains a $-CH_2-$ benzene ring structure as its R group?
4. What are enzymes?
5. Where and when are antibodies produced?
6. What two proteins are responsible for the transport and storage of oxygen in higher organisms?
7. What protein is found in hair and fingernails?
8. What is produced by the controlled hydrolysis of proteins?
9. Draw the structure of serine as it appears in a living cell.
10. Which amino acids have negatively charged side chains?
11. Which amino acids have positively charged side chains?
12. Give the three-letter abbreviations used for the following amino acids:

a. alanine   d. histidine
b. cysteine   e. lysine
c. glycine

13. What is the term for the linkage between two amino acids in a peptide?

14. Draw the structure of alanyl-valyl-glycyl-cysteine.

15. What level of protein structure is defined by the sequence of amino acids in the protein?

16. Which level of protein structure is caused by hydrogen bonding between the different peptide bonds?

17. What level of protein structure is defined by the binding of two or more peptides to produce an active functional protein?

18. What two components make up a conjugated protein?

19. A polypeptide or protein has equal numbers of positive and negative charges. What is the term used to describe this situation.

20. What kind of secondary structure characterizes the fibrous proteins of muscles?

21. Amino acids in water at pH = 7 exist as dipolar ions. What are these dipolar ions called?

22. What is the term used to describe amino acids that must be acquired from the diet?

23. What dietary source provides complete protein?

24. What enzyme begins the digestion of protein in the stomach?

25. What is the term for a protein that serves as a source of amino acids for infants or embryos?

26. Which amino acid is a secondary amine?

27. What type of secondary structure characterizes silk fibroin?

28. What type of covalent bond is involved in maintaining the tertiary structure of a protein?

29. What is meant by the term hydrophobic?

30. What happens to the overall charge on a protein when the pH drops too low?

20 *Enzymes*

Learning Goals

1. Conceptual Goals
· Describe the effect enzymes have on the activation energy of a reaction
· Understand the effect of substrate concentration on enzyme-catalyzed reactions
· Discuss the role of the active site and the importance of enzyme specificity
· Describe the difference between the lock-and key-Model and the induced fit model of enzyme-substrate complex formation
· Differentiate between the catalytic and binding groups of the active site
· Describe the four types of enzymes based on specificity
· Understand the transition state and the mechanisms by which it facilitates the chemical reaction
· Discuss the roles of cofactors and coenzymes in enzyme activity
· Understand the mechanisms used by cells to regulate enzyme activity
· Discuss the mechanisms by which certain chemicals inhibit enzyme activity
· Recognize how pH and temperature affect the rate of an enzyme-catalyzed reaction
· Discuss the role of the enzyme chymotrypsin
· Describe the roles of serine proteases

2. Performance Goals
· Recognize the correlation between an enzyme's common name and its function
· Know how to classify enzymes on the basis of their function

3. Health Applications
· Describe the process of blood coagulation and the role of vitamin K in the formation of blood clots
· Explain the role of acetylcholinesterase in nerve transmission
· Provide examples of medical uses of enzymes
· Appreciate the value of enzymes as diagnostic tools, for instance as indicators of heart and liver disease

Chapter Overview

Enzymes are proteins that function as catalysts in cell processes. They are essential for the thousands of metabolic reactions that allow life to exist. Enzymes are remarkably specific, usually recognizing and binding to only a single type of **substrate**, or reactant, and facilitating its conversion to a **product**.

Example 1
Describe some characteristics of enzymes.

233

Answer

1. The life of the cell depends upon the simultaneous occurrence of hundreds of chemical reactions, which must take place rapidly under mild conditions. The fragile cell must carry out its chemical reactions near pH 7, at body temperature (37°C), and in the absence of strong acids or bases. Enzymes allow these critical reactions to occur under the mild conditions required for life.

2. The enzymes are protein molecules that facilitate a biological chemical reaction by lowering the energy of activation and increasing the rate of the reaction.

3. An enzyme is very specific. It generally recognizes only one, or occasionally a few, molecules (substrates) upon which it will work its magic.

4. Enzyme-catalyzed reactions often occur from 1 to 100 million times faster than the uncatalyzed reactions.

20.1 Nomenclature and Classification

Nomenclature of Enzymes

An enzyme's name often tells us the substrate of the reaction and the nature of the reaction. For instance, the enzyme sucrase hydrolyzes the disaccharide sucrose and the enzyme lactate dehydrogenase removes hydrogen atoms from lactate ions. Some enzymes have historical names that do not reveal the nature of the substrate or of the reaction. The substrate and reaction catalyzed by such enzymes as catalase and trypsin must simply be memorized.

Example 2

Explain how the common names of enzymes are derived and give four examples.

Answer

1. The common name of an enzyme is derived from the name of the substrate with which the enzyme interacts, and/or the type of reaction that it catalyzes. Most enzyme names end in -*ase*.

2. The following are specific examples of common names for enzymes:
   a. Urea is the substrate acted on by the enzyme urease.
   b. Lactose is the substrate of lactase.
   c. Dehydrogenase is an enzyme that removes hydrogen atoms from a substrate.
   d. Decarboxylase is an enzyme that removes a carboxyl group from a substrate.

Classification of Enzymes

Enzymes may be classified according to the type of reaction that they catalyze. These six classes are as follows:

1. **Oxidoreductases**
2. **Transferases**
3. **Hydrolases**
4. **Lyases**
5. **Isomerases**
6. **Ligases**

Example 3
Explain completely the functions of oxidoreductases.

Answer
These enzymes catalyze oxidation-reduction reactions. They are responsible for the removal of protons and electrons from from a substrate to cause its oxidation, or the addition of protons and electrons to a substrate to cause its reduction. *Lactate dehydrogenase* is a good example. This enzyme transiently binds the coenzyme NAD^+ which accepts $2H^+ + 2e^-$ from the substrate lactate. The product is pyruvate, as shown below:

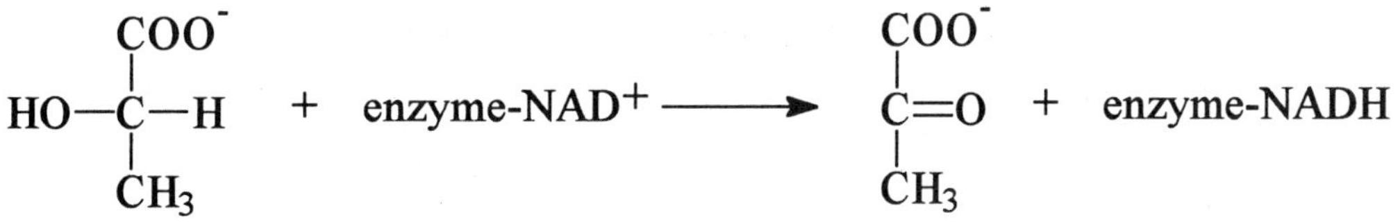

$$\underset{\substack{\text{lactate} \\ \text{(reduced} \\ \text{form)}}}{HO-\underset{\underset{CH_3}{|}}{\overset{\overset{COO^-}{|}}{C}}-H} \;+\; \underset{\substack{\text{lactate} \\ \text{dehydrogenase} \\ \text{(oxidized form)}}}{\text{enzyme-NAD}^+} \longrightarrow \underset{\substack{\text{pyruvate} \\ \text{(oxidized} \\ \text{form)}}}{\underset{\underset{CH_3}{|}}{\overset{\overset{COO^-}{|}}{C}}{=}O} \;+\; \underset{\substack{\text{lactate} \\ \text{dehydrogenase} \\ \text{(reduced form)}}}{\text{enzyme-NADH}}$$

Example 4
List the six types of enzymes classified according to the type of reaction that they catalyze. Briefly describe the function of each class.

Answer
1. *Oxidoreductases*: these enzymes catalyze electron transfers from one molecule to another.
2. *Transferases:* these enzymes catalyze the transfer of functional groups from one molecule to another.
3. *Isomerases:* these enzymes catalyze the rearrangement of functional groups within a molecule to convert the substrate into a different isomeric form.
4. *Hydrolases:* these enzymes catalyze hydrolysis reactions.
5. *Lyases:* these enzymes catalyze the addition of a group to a double bond or the removal of a group to form a double bond.
6. *Ligases:* these enzymes catalyze the condensation or joining of two molecules.

Example 5
To which class does each of the following enzymes belong?
1. pyruvate kinase
2. lipase (hydrolysis of triglycerides)
3. pyruvate carboxylase
4. triose isomerase
5. lactate dehydrogenase

Answer
1. transferase
2. hydrolase
3. lyase
4. isomerase
5. oxidoreductase

20.2 The Effect of Enzymes on the Activation Energy of a Reaction

Every chemical reaction is characterized by an equilibrium constant.   Enzymes do not alter that equilibrium constant. They do, however, change the path of the reaction, providing a lower energy route for the conversion of the substrate into the product. Thus, enzymes increase the rate of a chemical reaction by lowering the activation energy and therefore increasing the rate at which the reaction reaches equilibrium.

20.3 Effect of Substrate Concentration on Enzyme Catalyzed Reactions

For an enzyme-catalyzed reaction, the rate of the reaction increases as the substrate concentration increases. However, at a certain substrate concentration, the reaction rate reaches a maximum. At its maximum rate, the active sites of all the enzyme molecules are occupied by substrate molecules. The reaction rate is then limited by the speed with which the substrate is converted to product and product is released.

20.4 The Enzyme-Substrate Complex

The first step in an enzyme-catalyzed reaction is the binding of the substrate by the enzyme to form the **enzyme-substrate complex**. The groove or pocket in the enzyme that binds to the substrate is the **active site**. The active site of an enzyme is small compared to the overall size of the enzyme. The substrate is held within the active site by weak, noncovalent interactions. The conformation (shape and charge distribution) of the active site is complementary to the conformation of the substrate. **Thus, the conformation of the active site determines the specificity of the enzyme**. Only those substrates that fit into the active site can bind the enzyme.

Example 6

Describe the formation of the enzyme-substrate complex.

Answer

The first step in an enzyme-catalyzed reaction involves the enzyme binding to the substrate to form the enzyme-substrate complex. The portion of the enzyme that is in contact with the substrate is called the active site.

Example 7

Distinguish between the catalytic and binding groups of an enzyme active site.

Answer

1. The R groups of the amino acids that are involved in catalysis are known as the *catalytic groups*.
2. The specific groups of the active site that hold the substrate inthe correct position are called the *binding groups*.

20.5 Specificity of the Enzyme-Substrate Complex

Originally it was thought that the substrate simply snapped into place in the active site, like a piece of a jigsaw puzzle. This view was called the **lock-and-key model.** Our current model describes enzymes ase flexible molecules and an active site that is not a rigid pocket into which the substrate fits precisely. Actually the active site is thought to be a flexible pocket that approximates the shape of the substrate. When the substrate binds, the active site molds itself around the substrate. This is called the **induced fit model.**

Enzymes show a high degree of specificity. For instance, the enzyme urease catalyzes the decomposition of urea to carbon dioxide and ammonia, but it will not accept the related molecule methylurea. Enzyme specificities differ and often enzymes are classified into the following four groups based on this property:

1. **Absolute specificity**
2. **Group specificity**
3. **Linkage specificity**
4. **Stereochemical specificity.**

Example 8

Describe the terminology used to classify enzymes according to their degree of specificity.

Answer

1. *Absolute specificity*—an enzyme that catalyzes the reaction of only one substrate into product is absolutely specific.
2. *Group specificity*—an enzyme that catalyzes processes involving similar molecules containing the same functional group is group specific.
3. *Linkage specificity*—in this type of specificity an enzyme always catalyzes the formation or breakage of only one kind of bond.
4. *Stereochemical specificity*—in this type of specificity an enzyme is capable of catalyzing a reaction involving only one enantiomer.

20.6 The Transition State and Product Formation.

The overall process of an enzymatic reaction can be summarized by the following set of reversible reactions:

$$E + S \overset{\text{Step I}}{\rightleftharpoons} ES \overset{\text{Step II}}{\rightleftharpoons} ES^* \overset{\text{Step III}}{\rightleftharpoons} EP \overset{\text{Step IV}}{\rightleftharpoons} E + P$$

| Enzyme + Substrate | Enzyme-Substrate Complex | Transition State | Enzyme-Product Complex | Enzyme + Product |

In the **transition state** the shape of the substrate is altered, due to its interaction with the enzyme, into an intermediate form having features of both the substrate and the final product. This favors the conversion of the substrate into the product, which is subsequently released. There are several ways in which an enzyme could cause a reaction to proceed more quickly. In some cases, the enzyme exerts "pressure" on a bond, thereby facilitating bond breakage. An enzyme may simply bringing reactants into close proximity and in the proper orientation for reaction to occur. Alternatively, the active site of an enzyme may modify the pH of the microenvironment surrounding the substrate by serving as a donor or acceptor of H^+.

Example 9

Summarize the three ways in which an enzyme may lower the activation energy of a reaction.

Answer

1. The enzyme might weaken a bond in a substrate and therefore facilitate bond breakage.
2. An enzyme may facilitate a reaction by bringing two reactant molecules close together and in the proper orientation, so that a reaction easily occurs.
3. The active site of an enzyme may so modify the pH in the microenvironment of the substrate that a reaction will occur quickly.

20.7 Cofactors and Coenzymes

Some enzymes require an additional nonprotein **prosthetic group** in order to function. The protein portion of such an enzyme is called the **apoenzyme** and the nonprotein prosthetic group is called the **cofactor**. Usually cofactors are metal ions that bind to the enzyme and help maintain the correct shape of the active site.

Other enzymes require the transient binding of a **coenzyme**. Coenzymes are organic molecules that generally serve as carriers of electrons or chemical groups. They take part in a reaction by either donating or accepting chemical groups. Most coenzymes contain modified vitamins as part of their structure.

Nicotinamide adenine dinucleotide (NAD^+) is a coenzyme that accepts hydride ions (a hydrogen atom with two electrons) from the substrate that is oxidized. The portion of NAD^+ derived from the vitamin niacin is reduced to produce NADH.

20.8 Environmental Effects

Effect of pH

As we learned in the previous chapter, if the pH of a solution becomes too acidic or too basic, a protein is denatured. The pH at which an enzyme functions best is called the **pH optimum**. Enzyme function decreases as the pH rises above or falls below the pH optimum and at extremes of pH, they are denatured and cease to function.

Most cellular enzymes function optimally at a pH near 7. However, **pepsin**, a proteolytic enzyme found in the stomach where the pH is very low, has a pH optimum of 2. Another proteolytic enzyme, **trypsin**, functions under the conditions of higher pH found in the intestine and thus has a pH optimum around 8.5.

Effect of Temperature

Enzymes are rapidly denatured if the temperature of the solution rises much above 37°C, but they remain stable at much lower temperatures. For this reason enzymes used for clinical assays are stored in refrigerators or freezers. Since heating enzymes destroys their activity, cells can't survive extremes of temperatures. Thus, heat is an effective means of sterilizing medical instruments and solutions.

Some bacteria and yeast can survive very high temperatures, living in active volcanoes or in hot springs where the temperature is near the boiling point of water. The proteins of these bacteria have a structure that is stable at these extraordinary temperatures.

20.9 Regulation of Enzyme Activity.

Often enzyme activity is regulated by the cell as a means of controlling metabolic processes. One mechanism of regulation, used by bacteria, is to produce the enzyme only when the substrate is present.

Allosteric Enzymes

A more complex level of enzyme regulation involves enzymes that have more than a single binding site. **Allosteric enzymes** are composed of more than a single peptide and have more than one binding site. One subunit binds a substrate in the active site and catalyzes a biochemical reaction. The second subunit has a pocket that binds an **effector molecule**. Effector binding alters the shape of the active site. If the active site is converted to an inactive form, **negative allosterism** occurs. If effector binding converts the active site to an active configuration, **positive allosterism** is the result.

Feedback Inhibition

Feedback inhibition usually involves a biosynthetic pathway of many enzymatic steps. If the product of this multistep pathway builds up to high levels, it inhibits the entire pathway by serving as a negative allosteric effector of an enzyme early in the pathway.

Zymogens

Zymogens are enzymes produced in an inactive form. They are converted to the active form when they reach the site of their activity. This is a protective mechanism. Most of the proteolytic digestive enzymes are produced in an inactive form so that they do not destroy the cells that make them. These enzymes include elastase, trypsin, chymotrypsin, and pepsin.

20.10 Inhibition of Enzyme Activity

Enzyme inhibitors are chemicals that bind to enzymes and either eliminate or drastically reduce their catalytic ability. They are classified on the basis of whether the inhibition is reversible or irreversible, competitive or noncompetitive.

Irreversible Inhibition

Irreversible inhibitors, such as arsenic, bind very tightly, sometimes even covalently, to the enzyme. This binding irreversibly blocks substrate binding or enzyme catalysis.

Reversible, Competitive Inhibitors

Generally these inhibitors are **structural analogs**, molecules that "look like" the structure of the natural substrate for an enzyme because of similarities in shape and charge distribution. Because of this resemblance, the inhibitor can occupy the enzyme active site, but no reaction can occur. Since it is only bound by weak interactions, the inhibitor is easily removed from the active site, providing an opportunity for the substrate to bind. This is competitive inhibition because the degree of inhibition depends on the relative concentrations of substrate and inhibitor and their relative affinities for the active site.

Reversible, Noncompetitive Inhibitors

These inhibitors bind only weakly to R groups of amino acids or perhaps to the metal ion cofactors. Enzyme activity is restored when the inhibitor dissociates from the enzyme-inhibitor complex.

20.11 Proteolytic Enzymes

Proteolytic enzymes are enzymes that catalyze the hydrolysis of the peptide bonds that maintain the primary protein structure. **Chymotrypsin** is a proteolytic enzyme found in the small intestine that cleaves peptide bonds on the carbonyl side of tyrosine, tryptophan and phenylalanine. The specificity of chymotrypsin results from the hydrophobic pocket, a portion of the active site into which the flat aromatic side chains of these amino acids can fit.

Trypsin, chymotrypsin, and **elastase** are proteolytic enzymes produced in the pancreas. They all digest protein in the small intestine, have similar primary and tertiary structures, and share the same mechanism of action. They evolved by divergent evolution of a single gene that was probably duplicated. Each copy of the gene underwent mutation, producing many different proteolytic enzymes with different specificities. Chymotrypsin cleaves peptide bonds on the carbonyl side of aromatic amino acids; trypsin cleaves peptide bonds on the carbonyl side of basic amino acids; and elastase cleaves peptide bonds on the carboxyl side of the amino acids glycine and alanine.

20.12 Proteolytic Enzymes and Blood Coagulation

Blood clotting requires many enzymes acting sequentially. Clotting must occur only when tissues are damaged, because clots formed at the wrong time can cause heart attack or stroke. Thus, the blood clotting enzymes are found in the blood as zymogens, inactive forms that can be activated at the appropriate time. The process begins with the formation of a prothrombin activator in response to either damage to a blood vessel **(intrinsic pathway)** or in response to chemical signals secreted by cells that migrate to the damaged site **(extrinsic pathway)**. The cascade involves a **common pathway** in which blood clotting proteins are sequentially cleaved, thus activating one another. In the final step, the soluble protein **fibrinogen** is cleaved by thrombin to yield the insoluble protein **fibrin**. Polymers of fibrin form a meshwork that becomes attached to blood cells, plasma proteins, and blood vessel walls to make up the clot.

Hemophiliacs have a genetic deficiency of one of the protein factors involved in blood clotting. As a result, they suffer spontaneous bleeding following even minor wounds, blood in the urine, and painful and disfiguring joint hemorrhages.

Vitamin K and Its Role In Blood Clotting

Vitamin K is required for the synthesis of prothrombin, and blood clotting factors VII, IX, and X in the liver. It serves as a coenzyme for the enzyme that catalyzes the γ-carboxylation of ten glutamates near the N-terminus of these clotting factors. The γ-carboxyglutamates bind

Ca^{2+} ions, which is required for proper cleavage in the clotting process.

Vitamin K is a fat-soluble vitamin obtained from green leafy vegetables and liver and is also produced by our intestinal bacteria.

The **coumarin** anticoagulants, "blood thinners", are competitive inhibitors of vitamin K. They reduce the amount of prothrombin and factors VII, IX, and X synthesized by the liver.

20.13 Acetylcholinesterase and Nerve Transmission

Acetylcholine (ACh) is a chemical messenger (**neurotransmitter)** which transmits a message from nerve to muscle cells. It is stored in membrane bound bags, called synaptic vesicles, in the nerve cell ending. ACh is released into the the **nerve synapse** (the space between the nerve and muscle cells) and binds to the ACh receptor in the **postsynaptic membrane** of the muscle cell. This event opens pores in the membrane through which Na$^+$ and K$^+$ ions flow into and out of the cell, respectively, and generates the nerve impulse. This causes the muscle to contract. The enzyme **acetylcholinesterase** destroys the ACh and allows the muscle to relax.

Organic fluorophosphates, such as **diisopropyl fluorophosphate** (DIFP) are poisons that irreversibly inhibit acetylcholinesterase. **Pyridine aldoxime methiodide** is an antidote for DIFP poisoning that displaces the organophosphate group from the active site of the enzyme, alleviating the affects of the poison.

Succinylcholine, a structural analogue of ACh, competes with ACh for receptor binding and can be used as a muscle relaxant in surgical procedures

20.14 Use of Enzymes In Medicine

Enzymes are used as diagnostic tools in medicine and have been commonplace in the food industry for decades.

The serum concentration of some enzymes is used in disease diagnosis. An enzyme assay is a test that is performed to measure the activity of concentration of an enzyme, expressed in international units. The international unit is equal to the amount of enzyme that will catalyze conversion of one μmole (10^{-6} moles) of substrate to product in one minute at standard conditions of temperature and pH.

Liver disease is indicated by elevated levels of one of the isoenzymes of lactate dehydrogenase (LDH,) as well as elevated levels of alanine aminotransferase/serum glutamate-pyruvate transaminase (ALT/SGPT) and aspartate aminotransferase/serum glutamate-oxaloacetate transaminase (AST/SGOT) in blood serum. Elevated blood serum concentrations of amylase and lipase are indications of pancreatitis, or an inflammation of the pancreas.

In the clinical laboratory, enzymes are valuable analytical reagents. For example, enzymes are used in the BUN test (Blood Urea Nitrogen Test). Direct measurement of urea levels is difficult due to the complexity of blood. So the enzyme urease is added which converts each molecule of urea into two molecules of ammonia (the indicator of urea). Ammonia concentration is easily measured.

Key Terms

absolute specificity
acetylcholine
acetylcholinesterase
active site
allosteric enzymes
apoenzyme.
coenzyme
cofactor
common clotting pathway
competitive inhibitor
enzyme
enzyme specificity
enzyme-substrate complex
enzymology
extrinsic pathway of blood clotting
feedback inhibition
fibrin
fibrinogen
group specificity
hemophilia
hydrolase
induced fit model
intrinsic pathway of blood clotting
irreversible enzyme inhibitor
isomerase

ligase
linkage specificity
lock and key model
lyase
negative allosterism
oxidoreductases
pancreatic serine proteases
pH optimum
positive allosterism
product
proteolytic enzymes
prothrombin
reversible, competitive inhibitor
reversible, noncompetitive inhibitor
stereochemical specificity
structural analog
substrate
thrombin
transferase
transition state
vitamin
zymogen

Self Test

1. How does an enzyme speed up a biological chemical reaction?
2. What is the term for the reactant in an enzyme-catalyzed reaction?
3. Write a balanced equation for the reaction catalyzed by catalase.
4. What is the substrate for each of the following enzymes:
    a. succinate dehydrogenase
    b. sucrase
    c. glycogen phosphorylase
5. Write the specific reaction catalyzed by lactate dehydrogenase.
6. What do kinase enzymes do?
7. Which enzyme catalyzes the hydrolysis of triglycerides?
8. Write an equation representing the reaction catalyzed by pyruvate decarboxylase.
9. What effect does an enzyme have on the equilibrium constant of a reaction?
10. What is the first step in an enzyme-catalyzed reaction?

11. List the different classes of enzyme specificity.

12. What are apoenzymes and cofactors?

13. What is the coenzyme of an enzyme that is a dehydrogenase?

14. Why are the water-soluble vitamins important?

15. What is the term for the pH at which an enzyme has the greatest activity?

16. At extreme pH values an enzyme may lose the normal three-dimensional shape of the active site. What is the term used to describe this condition?

17. List two classes of enzyme inhibitors.

18. How do the sulfa drugs work as inhibitors?

19. What is the function of the pancreatic enzymes trypsin, chymotrypsin, and elastase?

20. In enzyme-catalyzed reactions what happens as the concentration of the substrate is increased?

21. At the maximum rate of an enzyme-catalyzed reaction, what may be said about all the enzymes and substrate molecules?

22. What is the term used to describe an enzyme that catalyzes the reaction of only one substrate?

23. Where are the catalytic groups of enzymes located?

24. Most enzymes are rapidly destroyed if the temperature is much higher than what value?

25. What process regulates the activity of allosteric enzymes?

26. In feedback inhibition, which compound causes the first reaction of a pathway to be shut off?

27. What is the term for an enzyme that is first produced in an inactive form?

28. A blood clot requires a sequential series of proteolytic reactions. The process begins with the formation of what enzyme?

21 *Carbohydrate Metabolism*

Learning Goals

1. Conceptual Goals
- · Understand the importance of ATP in cellular energy transfer processes.
- · Understand protein, fat, and carbohydrate digestion.
- · Know what is meant by a catabolic process, and summarize the three major steps in catabolism.
- · List the conditions under which pyruvate is converted to lactate, ethanol, or acetyl coenzyme A.
- · Know the functions of fermentations.
- · Explain the process of gluconeogenesis.
- · Note the major difference between glycolysis and gluconeogenesis.
- · Compare the processes of glycogenesis and glycogenolysis.
- · Summarize how glycogen synthesis and degradation are compatible.

2. Performance Goals
- · Describe the three steps in the degradation of glucose.
- · Describe glycolysis in terms of its two segments.
- · Indicate some practical uses of the lactic acid and alcohol fermentations.
- · Name the enzymes responsible for glycogenesis and glycogenolysis.

3. Health Applications
- · Understand the role of insulin and glucagon in glycogen metabolism.
- · Describe several glycogen storage diseases and their relationship to glycogen metabolism.

Chapter Overview

Cells need a constant supply of energy to maintain essential life processes such as *active transport, biosynthesis*, and *mechanical work*. A supply of energy-rich food molecules, carbohydrates, protein, and fats is required to provide this needed cellular energy. Carbohydrates are the most readily used energy source, and glycolysis is the pathway for the first stages of carbohydrate degradation.

21.1 ATP: The Cellular Energy Currency

Catabolism is the degradation of fuel molecules. The energy released in catabolism is stored as *chemical bond energy*. The molecule used for the storage of chemical energy, and often called the universal energy currency, is **adenosine triphosphate (ATP)**.

ATP is a **nucleotide** composed of the nitrogenous base adenine bonded in N-glycosidic linkage to the sugar ribose. Ribose is bonded to one (AMP), two (ADP), or three (ATP) phosphoryl groups. The molecule is a high-energy compound because the phosphoanhydride bonds holding the terminal phosphoryl groups are *high-energy bonds*. This means that a large amount of energy is released when these bonds are hydrolyzed. Such high-energy bonds are indicated as squiggles (~).

ATP has a higher energy content than the compounds to which it donates energy, but, as must be the case, it contains less energy than the compounds that are involved in forming it. The secret to the success of ATP as the energy currency is that both the reactions that produce ATP and the hydrolysis of ATP to provide energy for cellular work are energetically favored reactions. ATP is an ideal "go-between," shuttling energy from **exothermic reactions** to **endothermic reactions**.

Example 1
What is the function of ATP?

Answer
Adenosine triphosphate (ATP) is a "go-between" molecule that can store or release chemical energy. The secret to the function of ATP as a go-between lies in its chemical structure. The molecule is a high-energy compound because of the phosphoanhydride bonds holding the terminal phosphoryl groups. When these bonds are hydrolyzed, they release a large amount of energy that can be used for cellular work. Then other pathways allow the resynthesis of ATP by providing the energy needed to put ADP + P_i together as ATP.

Example 2
Show how the hydrolysis of ATP is coupled with the synthesis of glucose-6-phosphate.

Answer
In the first step of glycolysis there is a transfer of a phosphoryl group ($-PO_3^{2-}$) from ATP to the C-6 hydroxyl group of a glucose molecule. This reaction is catalyzed by the enzyme hexokinase. This reaction can be "dissected" to reveal the role of ATP as a source of energy. The first thing that happens is the hydrolysis of ATP to ADP and an inorganic phosphate group, abbreviated P_i. This reaction releases 7 kcal/mole of energy:

$$ATP + H_2O \rightarrow ADP + P_i + 7 \text{ kcal/mole}$$

Next, the synthesis of glucose-6-phosphate from glucose and phosphate requires 3.0 kcal/mole:

$$3.0 \text{ kcal/mole} + glucose + P_i \rightarrow glucose\text{-}6\text{-}phosphate + H_2O$$

Thus, by the coupling of a reaction needing energy with the hydrolysis of ATP to release energy, the cell can undergo hundreds of reactions.

21.2 Overview of Catabolic Processes

The first stage of catabolism is the hydrolysis of dietary macromolecules into small subunits. Large polymeric molecules are degraded into their constituent subunits, which are taken into the cells. Polysaccharides are hydrolyzed to monosaccharides. Proteins are hydrolyzed to oligopeptides and amino acids, and fats are hydrolyzed into fatty acids, glycerol, and monoglycerides. All these molecules are then absorbed by the cells lining the intestine. The second stage of catabolism is the conversion of monomers into a form that can enter one of the catabolic pathways and be degraded to yield energy. The final stage of catabolism is the complete oxidation of nutrients and the production of ATP.

21.3 Glycolysis

General Considerations

Glycolysis, the first stage of carbohydrate catabolism, was probably the first successful pathway for energy generation that evolved on earth. It is an anaerobic process, requiring no oxygen, and it is carried out by enzymes in the cytoplasm of the cell. The degradation of glucose by glycolysis yields chemical energy in the form of ATP. Four ATP molecules are produced by **substrate-level phosphorylation**, a reaction in which high-energy phosphoryl groups from one of the substrates in glycolysis are transferred to ADP to form ATP. Chemical energy in the form of reduced **nicotinamide adenine dinucleotide**, NADH, is also produced. Under aerobic conditions, the electrons are donated to an electron transport system for the generation of ATP by **oxidative phosphorylation**; but under anaerobic conditions, NADH is used as a source of electrons in fermentation reactions. The final product of glycolysis is two pyruvate molecules. Under anaerobic conditions, pyruvate is used as a substrate in fermentation reactions.

Example 3
Describe the first stage of carbohydrate metabolism.

Answer
The pathway called glycolysis is the beginning stage for the use of glucose as a source of chemical energy. This pathway is anaerobic, which means that no oxygen is required. The glycolysis pathway releases only about 2% of the potential energy of the glucose molecule. In the pathway, glucose is finally transformed after many steps into two units of pyruvate.

Reactions of Glycolysis

The structures of the intermediates of glycolysis can be found in Figure 21.8 of the text. The following is a brief description of those reactions.

247

Reaction 1

Glucose is phosphorylated by the enzyme *hexokinase*. A phosphoryl group from ATP is transferred to C-6 of glucose, and the product is glucose-6-phosphate.

Reaction 2

The enzyme *phosphoglucoisomerase* isomerizes the glucose-6-phosphate to produce fructose-6-phosphate.

Reaction 3

The enzyme *phosphofructokinase* transfers a phosphoryl group from ATP to the C-1 hydroxyl of fructose-6-phosphate, producing fructose-1, 6-bisphosphate.

Reaction 4

Aldolase splits the fructose-1, 6-bisphosphate into glyceraldehyde-3-phosphate (3-PGAL) and dihydroxyacetone phosphate (DHAP). Only 3-PGAL can be used in glycolysis, and the DHAP is isomerized by the enzyme *triose isomerase*. This produces a second molecule of 3-PGAL.

Reaction 5

Glyceraldehyde-3-phosphate dehydrogenase catalyzes the oxidation of the aldehyde group of 3-PGAL to a carboxylic acid. The coenzyme in this reaction is NAD^+, which is reduced to NADH. Next, an inorganic phosphate group is transferred to the carboxylate group to produce 1,3-bisphosphoglycerate.

Reaction 6

Phosphoglycerokinase catalyzes the transfer of a phosphoryl group from 1,3-bisphosphoglycerate to ADP. This is the first substrate-level phosphorylation of glycolysis to produce ATP.

Reaction 7

Phosphoglyceromutase catalyzes the isomerization of 3-phosphoglycerate to 2-phosphoglycerate.

Reaction 8

Enolase catalyzes the dehydration of 2-phosphoglycerate, producing the energy-rich product phosphoenolpyruvate.

Reaction 9

Pyruvate kinase catalyzes the last substrate-level phosphorylation, in which a phosphoryl group from phosphoenolpyruvate is transferred to ADP to produce ATP. The final product of glycolysis is pyruvate.

Reactions 5 through 9 occur twice per glucose molecule; thus, the final products of glycolysis are 2 NADH, 2 pyruvate, and 4 ATP. However, since 2 ATP/glucose were invested early in glycolysis, the net yield is 2 ATP/glucose molecule.

Example 4
List the steps in glycolysis in which ATP molecules are hydrolyzed to provide energy for specific reactions.

Answer
1. Reaction 1 is a coupled reaction in which the enzyme *hexokinase* catalyzes the transfer of a phosphoryl group from ATP to glucose. The product is glucose-6-phosphate.
2. Reaction 3 is a coupled reaction in which the enzyme *phosphofructokinase* catalyzes the transfer of a phosphoryl group from ATP to fructose-6-phosphate. The product is fructose-1,6-bisphosphate.

Example 5
List the two steps of the anaerobic glycosis pathway that result in the production of ATP molecules.

Answer
1. Reaction 6: This reaction is the first step of the pathway in which energy is harvested in the form of ATP. The phosphoryl group of 1,3-bisphosphoglycerate is transferred to an ADP molecule in the first substrate-level phosphorylation of glycolysis. This reaction is catalyzed by the enzyme *phosphoglycerokinase*.
2. Reaction 9: The final substrate-level phosphorylation in glycolysis involves the transfer of a phosphoryl group from phosphoenolpyruvate to ADP to form ATP. This reaction is catalyzed by the enzyme *pyruvate kinase*.

21.4 Fermentations

In order for glycolysis to continue to degrade glucose and produce ATP, the NADH must be reoxidized, and the pyruvate must be utilized and removed. Under anaerobic conditions, these two requirements are met through fermentation reactions.

Example 6
Explain how the NADH produced by glycolysis in reaction 5 is regenerated into NAD^+.

Answer
1. If the cell is functioning under aerobic conditions, the NADH will be reoxidized, and pyruvate will be completely oxidized by aerobic respiration.
2. Under anaerobic conditions, cells of different types employ a variety of fermentation reactions to accomplish the conversion of NADH to NAD^+.

Lactate Fermentation

Some cells, such as muscle cells and dairy bacteria, utilize lactate fermentation. Under anaerobic conditions, the enzyme *lactate dehydrogenase* reduces pyruvate to lactate. NADH is the reducing agent for this reaction, and thus NAD^+ is regenerated, while the pyruvate is simultaneously used up.

This reaction occurs in muscle cells during strenuous exercise, when the body cannot provide sufficient oxygen for aerobic ATP production. Lactate eventually builds up in the muscle to the extent that glycolysis can no longer proceed, and the muscle cells can no longer function. This point of exhaustion is called the **anaerobic threshold**.

When exercise ceases, the liver takes up the lactate from the blood and converts it back to pyruvate, and the ATP stores are replenished by aerobic energy-harvesting reactions, such as the citric acid cycle, which will be discussed in the next chapter.

Lactate fermentation is used by bacteria that produce yogurt and some cheeses. Lactate is one of the molecules that gives these dairy products their tangy flavor.

Alcohol Fermentation

Under anaerobic conditions, yeast cells ferment sugars produced by fruit and grains to produce ethanol. First, *pyruvate decarboxylase* removes CO_2 from the pyruvate, producing CO_2 and acetaldehyde. Then, *alcohol dehydrogenase* catalyzes the reduction of acetaldehyde to ethanol, reoxidizing NADH in the process. Eventually the stable fermentation end-product, ethanol, builds up to a concentration that kills the yeast cells. It is characteristic of fermentations that the end-product eventually builds up to levels that inhibit the cells carrying out the reaction.

Example 7
Summarize the lactate and ethanol fermentation pathways.

Answer
1. *Lactate fermentation*: This pathway occurs when you have exercised beyond the capacity of your lungs and circulatory system to deliver enough oxygen to the working muscles. The aerobic energy-generating pathways will no longer be able to supply enough ATP, but the muscles still demand energy. Under anaerobic conditions, pyruvate can be reduced by the NADH-lactate dehydrogenase coenzyme system to form lactate plus NAD^+, which is required for continued anaerobic functioning of the glycolysis pathway.
2. *Alcohol fermentation*: Under anaerobic conditions in yeast (usually not humans!) the pyruvate formed by glycolysis of glucose is converted into ethanol and CO_2 in two enzyme-catalyzed steps. First pyruvate is converted into acetaldehyde as CO_2 is released. Then the acetaldehyde is reduced by NADH. This produces ethanol and NAD^+ that allows anaerobic glycolysis to continue.

21.5 The Pentose Phosphate Pathway

The **pentose phosphate pathway** is an alternative pathway for glucose degradation that is particularly abundant in the liver and adipose tissue. It can be divided into three stages, each of which provides a unique biosynthetic requirement for the cell.

In the reactions of the *oxidative stage*, glucose-6-phosphate is oxidized and decarboxylated to produce ribulose-5-phosphate, NADPH, and CO_2. NADPH is the reducing agent required for many biosynthetic pathways. The second stage involves isomerization reactions that convert ribulose-5-phosphate into other pentose phosphates. Among these is ribose-5-phosphate which is required for nucleotide biosynthesis. In the final stage, a complex series of reactions occurs that involve carbon-carbon bond breakage and formation. The products are two molecules of fructose-6-phosphate and one molecule of glyceraldehyde-3-phosphate. Among the intermediates of the third stage is erythrose-4-phosphate, a precursor in the biosynthesis of the aromatic amino acids.

21.6 Gluconeogenesis: The Synthesis of Glucose

During starvation and following extended exercise, the body must make glucose. **Gluconeogenesis** is the process by which glucose is produced from noncarbohydrate precursors. Lactate, all amino acids, except leucine and lysine, and glycerol from fats all serve as precursors for glucose biosynthesis. Gluconeogenesis, shown in Figure 21.11 in the text, appears to be the reverse of glycolysis. Certainly most of the intermediates of the two pathways are identical. However, steps 1, 3, and 9 of glycolysis are irreversible and must be bypassed by other enzymes. Glucose-6-phosphate is dephosphorylated by the enzyme *glucose-6-phosphatase*, found in the liver but not in muscle. Similarly, the phosphorylation of fructose-6-phosphate by phosphofructokinase is irreversible. *Fructose bisphosphatase* catalyzes the removal of the phosphoryl group from fructose-1,6-bisphosphate. Finally, step 9 of glycolysis is bypassed by two enzymes. *Pyruvate carboxylase* carboxylates pyruvate by the addition of atmospheric CO_2, producing oxaloacetate. Then *phosphoenolpyruvate carboxykinase* removes the CO_2 and adds a phosphoryl group, producing phosphoenolpyruvate. The phosphoryl group donor is **guanosine triphosphate**, GTP.

In the Cori Cycle, lactate produced by working muscle is transported to the liver by the bloodstream. In the liver the lactate is converted to pyruvate which may be used to produce glucose by gluconeogenesis. The glucose produced in the liver may be degraded for energy or stored as glycogen.

21.7 Glycogen Synthesis and Degradation

The liver helps regulate the blood glucose level. One of the ways this is accomplished is through the uptake and storage of excess glucose as glycogen (**glycogenesis**). Alternatively, liver cells may degrade glycogen (**glycogenolysis**) and release glucose into the blood. **Glycogen** is a highly branched glucose polymer. The primary chain is linked by α (1 → 4) glycosidic bonds. The branches are linked to the primary chain by α (1 → 6) glycosidic bonds. Glycogen is stored in the cytoplasm of liver and muscle cells as **glycogen granules**. These granules are complexes of glycogen and the enzymes that carry out glycogenesis and glycogenolysis.

Glycogenolysis: Glycogen Degradation

Glycogenolysis is controlled by two hormones, **glucagon** and *epinephrine*. Glucagon is released from the pancreas in response to low blood glucose, and epinephrine is released from the adrenal glands in response to a threat or a stress. Both hormones regulate the activity of *glycogen phosphorylase* and *glycogen synthase*. Glycogen phosphorylase, involved in glycogenolysis, is activated; glycogen synthetase, involved in glycogenesis, is inactivated. The steps in glycogenolysis are summarized below:

> **Step 1:** The enzyme *glycogen phosphorylase* catalyzes phosphorolysis of a glucose unit at one end of glycogen, producing glucose-1-phosphate.
>
> **Step 2:** The enzyme α *(1 $\rightarrow$ 6) glycosidase* hydrolyzes the α (1 $\rightarrow$ 6) glycosidic bond at a branch point, thereby removing the branches. Branches are further degraded by glycogen phosphorylase to produce glucose-1-phosphate.
>
> **Step 3:** *Phosphoglucomutase* converts glucose-1-phosphate to glucose-6-phosphate, which can be degraded by glycolysis or dephosphorylated for transport into the bloodstream.

Example 8

Explain why glycogenolysis is important.

Answer

Glucose is the sole source of energy for mammalian red blood cells and the major source of energy for the brain. Neither of these can store glucose; thus, a constant supply must be available as blood glucose. This is provided by dietary glucose and by the production of glucose either by gluconeogenesis or by glycogenolysis, the degradation of glycogen molecules. Glycogen is a long, branched-chain polymer of glucose that is stored in the liver and skeletal muscles.

The total amount of glucose in the average 150-pound adult male is about 20 grams, but the brain alone consumes 5 to 6 grams of glucose per hour. Breakdown of glycogen in the liver mobilizes the glucose when hormonal signals register a need for increased levels of blood glucose.

Glycogenesis: Glycogen Synthesis

The hormone **insulin**, produced by the pancreas in response to high blood glucose levels, stimulates **glycogenesis**. It accelerates the uptake of glucose by most cells of the body. In the liver, insulin promotes glycogenesis by inhibiting glycogen phosphorylase, thus inhibiting glycogenolysis and stimulating glycogen synthase and glucokinase, two enzymes involved in glycogen synthesis. The following is a summary of the steps of glycogenesis:

Step 1: *Glucokinase* phosphorylates glucose, using ATP as a phosphoryl group donor and forming glucose-6-phosphate.

Step 2: *Phosphoglucomutase* isomerizes glucose-6-phosphate to glucose-1-phosphate.

Step 3: *Pyrophosphorylase* catalyzes bond formation between the C-1 phosphoryl group of glucose and the a phosphoryl group of **uridine triphosphate** (UTP) to produce UDP-glucose. A pyrophosphate group (PP_i) is released in the reaction.

Step 4: *Glycogen synthase* breaks the phosphoester linkage of UDP-glucose and forms an α $(1{\rightarrow}4)$ glycosidic bond between the glucose and the growing glycogen chain, releasing UDP.

Step 5: Finally, the branches are added. The *branching enzyme* removes sections of the linear α $(1 \rightarrow 4)$-linked glycogen and reattaches them elsewhere in the chain by α $(1 \rightarrow 6)$ glycosidic linkages.

Compatibility of Glycogenesis and Glycogenolysis

Glycogenesis and glycogenolysis are regulated by hormonal controls. When blood glucose levels are too high (**hyperglycemia**), insulin stimulates glucose uptake. It also stimulates glucokinase and glycogen synthetase activity and inhibits the first enzyme in glycogen degradation, glycogen phosphorylase. The net effect is the removal of glucose from the blood and its conversion into glycogen in the liver.

Glucagon is produced in response to low blood glucose levels (**hypoglycemia**). Glycogenolysis is stimulated because the activity of glycogen phosphorylase, which catalyzes the first stage of glycogenolysis, is accelerated.

21.8 Conversion of Pyruvate to Acetyl CoA

Under aerobic conditions, cells use oxygen and completely oxidize glucose to CO_2 in a metabolic pathway called the *citric acid cycle*. **Acetyl CoA** is the molecule that carries two-carbon fragments (acetyl groups) produced from pyruvate into the citric acid cycle.

The coenzyme A portion of acetyl CoA is derived from ATP and the vitamin pantothenic acid. The acetyl groups are linked to the thiol group of coenzyme A by a high-energy thioester bond. Acetyl CoA is an "activated" form of the acetyl group.

The reaction that converts pyruvate to acetyl CoA is carried out by the *pyruvate dehydrogenase complex.* Pyruvate is decarboxylated and oxidized. The acetyl group is linked to coenzyme A by a thioester bond.

Acetyl CoA plays a central role in cellular metabolism. It is the product of the degradation of glucose, fatty acids, and some amino acids. It is used in the citric acid cycle to produce large amounts of ATP energy. It is also used for *anabolic* or biosynthetic reactions to produce cholesterol and fatty acids. Thus, acetyl CoA is the intermediate through which all energy sources (fats, proteins, and carbohydrates) are interconvertible.

Key Terms

acetyl CoA	glycolysis
adenosine triphosphate (ATP)	guanosine triphosphate (GTP)
anaerobic threshold	hyperglycemia
catabolism	hypoglycemia
coenzyme A	insulin
Cori Cycle	nicotinamide adenine dinucleotide (NAD$^+$)
fermentation	nucleotide
glucagon	oxidative phosphorylation
gluconeogenesis	pentose phosphate pathway
glycogen	pyruvate dehydrogenase complex
glycogen granule	substrate-level phosphorylation
glycogenesis	uridine triphosphate(UTP)
glycogenolysis	

Self Test

1. Which pathway is thought to be the first successful energy-generating pathway available to early organisms?

2. What molecule in a cell couples exothermic and endothermic reactions?

3. How many kcal/mole of energy are released by the hydrolysis of ATP?

4. Describe glycolysis in terms of its oxygen requirements.

5. What is the net yield of ATP produced by anaerobic glycolysis?

6. Which coenzyme must be reoxidized so that glycolysis can continue?

7. What is the final fermentation end-product in yeast cells?

8. What is the final fermentation end product produced in muscle cells under anaerobic conditions?

9. Of the three classes of food molecules, which one is the most readily used by the human body?

10. Which pathway of carbohydrate degradation is thought to be the most ancient, having existed for at least 3.5 billion years?

11. What subunits are released by digestion of proteins?

12. Which aspect of metabolism involves the degradation of absorbed food molecules to produce energy for the body?

13. Of the several high-energy compounds produced in cells, which one is the principal energy storage compound?

14. What type of chemical reaction is involved in the process of digestion?

15. What happens in the first step of glycolysis that causes glucose to be trapped within the cell?

16. Phosphofructokinase converts what substrate into fructose-1,6-bisphosphate?

17. Where in the cell does glycolysis occur?

18. During glycolysis 1,3-bisphosphoglycerate is converted into 3-phosphoglycerate.

Simultaneously, an ATP is produced. What is the general term for this kind of reaction?

19. What compound is hydrolyzed to produce adenosine-5'-diphosphate, P_i, and energy?

20. Provide the missing product in the following reaction:

$$ATP + water + glucose \rightarrow ADP + 4 \text{ kcal/mole} + \text{?}$$

21. What is the name of the enzyme found in saliva that begins the hydrolysis of starch?

22. What enzyme hydrolyzes milk sugar?

23. What enzyme in the stomach starts the breakdown of proteins?

24. Where does the digestion of triglycerides begin?

25. What is the first portion of the small intestines called?

26. Hexokinase is used to convert what substrate into glucose-6-phosphate?
    a. glucose
    b. fructose
    c. glycogen
    d. pyruvate
    e. lactate

27. The enzyme triose isomerase converts dihydroxyacetone phosphate into what compound?
    a. pyruvate
    b. fructose-1,6-bisphosphate
    c. glucose-1,6-bisphosphate
    d. lactate
    e. glyceraldehyde-3-phosphate

28. Following strenuous exercise, the liver takes up excess lactate from the blood. Into what compound is this lactate converted?
    a. oxaloacetate
    b. glucose
    c. pyruvate
    d. acetyl CoA
    e. glycogen

29. Gluconeogenesis means the production of glycogen from a noncarbohydrate source. Answer T or F.

30. The liver is the only tissue that contains glucose-6-phosphatase. Answer T or F.

Aerobic Respiration and Energy Production

Learning Goals

1. Conceptual Goals
· Recognize that ATP production in the cell is principally due to oxidative metabolic reactions.
· Know that these aerobic energy-harvesting reactions occur in the mitochondria.
· Be aware of the importance of the citric acid cycle in catabolism and anabolism.
· Describe oxidative phosphorylation.
· Know the importance of the urea cycle, and describe its essential steps.
· Summarize the role of the citric acid cycle in catabolism and anabolism.

2. Performance Goals
· Name the major regions of the mitochondria and describe their functions.
· Summarize the major reactions of the citric acid cycle.
· Describe oxidative phosphorylation and its relationship to the electron transport system.
· Describe the fate of amino acids in the citric acid cycle.

3. Health Applications
· Understand energy metabolism in terms of exercise physiology.
· Appreciate the difference between adipose tissue (white fat) and brown fat.
· Know the cause and effect of hyperammonemia.

Chapter Overview

The energy-harvesting reactions that produce the greatest energy yield are aerobic reactions that occur in the **mitochondria**. Mitochondria are the organelles that serve as the cellular "powerplants." An organelle is a membrane-enclosed compartment within the cytoplasm that has a specialized function.

In the mitochondria, the enzymes of the citric acid cycle strip electrons from substrates in the pathway. These are passed through the **electron transport system**, and the energy of the electrons is used to produce ATP. This process is called **oxidative phosphorylation**.

Example 1
What is the function of the mitochondria?

Answer
The oxidative reactions of metabolism are responsible for most cellular ATP production. These reactions occur in metabolic pathways located in the mitochondria. This membrane-enclosed compartment within the cytoplasm is the power plant of the cell, producing most of the ATP for cellular processes. The mitochondria are responsible for the

final oxidation of the acetyl group of acetyl CoA from glycolysis, fatty acid degradation, and amino acid catabolism. The final result is the generation of many ATP molecules.

22.1 Mitochondria

Structure and Function

Mitochondria are bounded by an **outer mitochondrial membrane** and an **inner mitochondrial membrane**. The region between the two membranes is known as the **intermembrane space**, and the region enclosed by the inner membrane is known as the **matrix space**.

The outer membrane is freely permeable to substances of molecular weight less than 10,000 amu. Thus, metabolites to be oxidized via the citric acid cycle easily enter the intermembrane space through channel proteins.

The inner membrane is a highly folded, continuous structure. The individual folds are known as **cristae**. The inner mitochondrial membrane is virtually impermeable to most substances and contains three types of proteins. *Transport proteins* allow the transport of metabolites across the inner mitochondrial membrane into the matrix. *Electron transport system proteins* are involved in electron transfers, for which O_2 serves as the terminal electron acceptor. The third protein is a very large multiprotein complex known as **ATP synthase**, which is responsible for phosphorylation of ADP.

The enzymes of the citric acid cycle and those for the oxidation of fatty acids and amino acids are located in the matrix space.

Example 2
Describe the structure of the mitochondria.

Answer
Mitochondria are football-shaped organelles that are roughly the size of bacteria. This organelle has both an outer and an inner membrane. The region between the two membranes is called the intermembrane space, and the region enclosed by the inner membrane is known as the matrix space. The outer membrane is freely permeable to substances of less than 10,000 amu. This is because of the presence of a large number of transport proteins that form large pores in the membrane.

The inner membrane is a continuous structure that is highly folded. These folds of the inner membrane are called cristae. This membrane is virtually impermeable to most substances. The inner membrane contains transport proteins to allow the transport of metabolites across the membrane. Protein complexes for electron transport, ATP synthesis, the citric acid cycle, and fatty acid oxidation are also located in the inner membrane or matrix space.

Origin of the Mitochondria

Mitochondria are roughly the size of a bacterium, and are thought to have evolved from free-living bacteria that were captured by eukaryotic cells. Some of the evidence for this is that they have their own DNA and protein-synthesizing system. In addition, they are self-replicating, growing and dividing to produce new mitochondria.

22.2 An Overview of Aerobic Respiration

Aerobic respiration is the oxygen-requiring degradation of food molecules and production of ATP. Acetyl groups derived from the breakdown of sugars, amino acids, or lipids are completely oxidized to CO_2 in the reactions of the citric acid cycle. The electrons harvested in these oxidation reactions are used to reduce 3 NAD^+ and 1 FAD, producing 3 NADH and 1 $FADH_2$.

These electrons are then passed through an electron transport system that simultaneously pumps protons (H^+) into the H^+ reservoir in the mitochondrial intermembrane space. The energy of the H^+ reservoir is used by ATP synthase to produce ATP. The entire process is called **oxidative phosphorylation** because the energy of electrons from the oxidation of substrates is used to phosphorylate ADP and produce ATP.

22.3 The Citric Acid Cycle (The Krebs Cycle)

Reactions of the Citric Acid Cycle

The following is a summary of the reactions of the **citric acid cycle**.
1. The acetyl group of acetyl CoA is transferred to oxaloacetate by the enzyme *citrate synthase*, forming citrate.
2. *Aconitase* catalyzes the dehydration of citrate, producing *cis*-aconitate.
3. *Aconitase* then adds a water molecule to the *cis*-aconitate, converting it to isocitrate. The result of these two steps is the isomerization of citrate to isocitrate.
4. *Isocitrate dehydrogenase* catalyzes the first oxidative reaction of the citric acid cycle. This is a complex reaction in which three things happen. First, the hydroxyl group of isocitrate is oxidized to a ketone; then carbon dioxide is released; finally NAD^+ is reduced to NADH. The product of this oxidative decarboxylation reaction is α-ketoglutarate.
5. The *α-ketoglutarate dehydrogenase complex*, an enzyme complex very similar to the pyruvate dehydrogenase complex, mediates the next reaction. Once again, three chemical events occur. First, α-ketoglutarate loses a carboxylate group as CO_2^-; next, NAD^+ is reduced to NADH; and finally, coenzyme A combines with the product to form succinyl CoA. The bond between succinate and coenzyme A is a high-energy thioester linkage.
6. Succinyl CoA is converted to succinate by the enzyme *succinyl CoA synthase*, which removes the CoA group and uses the energy of the thioester bond to add an inorganic phosphate group to GDP, producing GTP. *Dinucleotide diphosphokinase*

then shifts the phosphoryl group from GTP to ADP, producing ATP.

7. *Succinate dehydrogenase* then catalyzes the oxidation of succinate to fumarate. The oxidizing agent, flavin adenine dinucleotide, is reduced in this step.

8. *Fumarase* catalyzes the addition of H_2O to the double bond of fumarate, producing malate.

9. Finally, *malate dehydrogenase* reduces NAD^+ to NADH and oxidizes malate to oxaloacetate. Since the citric acid cycle began with the addition of an acetyl group to oxaloacetate, we have come full circle.

Summary of the Energy Yield

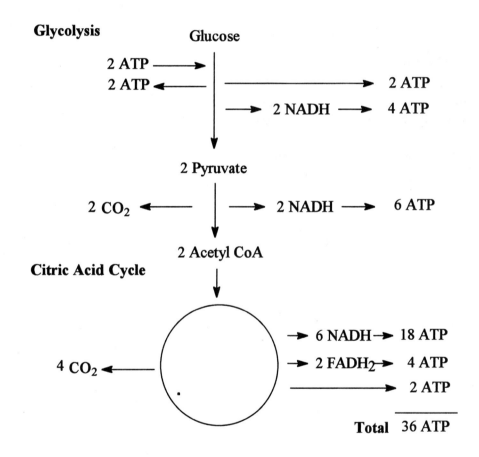

Example 3

Write the specific reactions in the citric acid cycle in which the coenzymes NAD^+ and FAD act as oxidizing agents.

Answer

1. The first oxidative step, Step 4, is catalyzed by isocitrate dehydrogenase:

   isocitrate + NAD^+ $\rightarrow$ α-ketoglutarate + CO_2 + NADH

2. The next step involves the oxidation of α-ketoglutarate into succinate:

   α-ketoglutarate + NAD^+ $\rightarrow$ succinate + CO_2 + NADH

3. In Step 7, succinate is oxidized to fumarate:

   succinate + FAD $\rightarrow$ fumarate + $FADH_2$

4. Then in the final step of the citric acid cycle, malate is oxidized to produce oxaloacetate.

   malate + NAD^+ $\rightarrow$ oxaloacetate + NADH

22.4 Oxidative Phosphorylation

Oxidative phosphorylation is a series of reactions that couples the oxidation of NADH and $FADH_2$ to the phosphorylation of ADP to produce ATP.

Electron Transport Systems and the Hydrogen Ion Gradient

Embedded within the mitochondrial inner membrane are a series of electron carriers called the **electron transport system**. Prominent among these electron carriers are the cytochromes, which carry a heme group. These molecules are arranged within the membrane so that they pass electrons from one to the next. Some of these electron carriers are able to carry hydrogen atoms, while others can only carry electrons. Those that do not pass protons deposit them within the intermembrane space.

NADH carries electrons to the first carrier of the electron transport system, *NADH dehydrogenase*. There it is oxidized to NAD^+, donating a pair of hydrogen atoms, and returns to the site of the citric acid cycle to be reduced again. The pair of electrons is passed to the next electron carrier, but the protons are pumped into the intermembrane compartment. The electrons are passed sequentially through the electron transport system, and at two additional points, protons from the matrix are pumped into the intermembrane compartment. In aerobic organisms, the **terminal electron acceptor** is molecular oxygen, O_2, and the product is water. $FADH_2$ donates its electrons to a carrier of later in the electron transport system, so only four protons are pumped into the intermembrane space.

As the electron transport system continues to function, a high concentration of hydrogen ions accumulates in the intermembrane space, creating a concentration gradient across the inner membrane. Thus, the system generates a high concentration of protons in the intermembrane space; a lower concentration of protons is found in the matrix.

ATP Synthase and the Production of ATP

The last component needed for oxidative phosphorylation is a multiprotein complex called

ATP synthase, or F_0F_1 complex. The F_0 portion provides a channel in the membrane through which protons may pass. The F_1 portion phosphorylates ADP to produce ATP, using the energy of the proton gradient.

ATP synthase harvests the energy of this gradient and uses it to produce ATP. As the protons pass into the matrix through F_0, some of their energy is used by the enzymatic portion of ATP synthase (F_1) to catalyze the phosphorylation of ADP to ATP.

Example 4
What are the two different ways that ATP can be synthesized by the cell? Where are these pathways located, and how efficient are they?

Answer
1. The two ways that ATP can be synthesized by the cell are substrate-level phosphorylation and oxidative phosphorylation.
2. Substrate-level phosphorylation occurs in the cytoplasm of the cell and produces only a few ATP molecules per glucose molecule.
3. Oxidative phosphorylation occurs in the mitochondria and produces a large number of ATP molecules per glucose molecule.

Example 5
Explain how the electron transport system converts NADH back into NAD^+, so that it may be used again by the citric acid cycle.

Answer
The NADH carries a hydride anion, which was originally from glucose or other food molecules, to the first carrier of the electron transport system. There it is oxidized to NAD^+ and returns to be used again in the citric acid cycle.

Example 6
Write an equation to represent the last step of the electron transport system.

Answer
$$2H^+ + 2e^- + 1/2\ O_2 \quad \rightarrow \quad H_2O$$

22.5 Control of the Citric Acid Cycle

The citric acid cycle is regulated so that cellular energy demand determines the rate of cellular energy production. This regulation occurs because some of the enzymes of the citric acid cycle are allosteric enzymes. Binding of an effector, such as ATP, ADP, or NADH, alters the shape of the active site. Effector binding may turn the enzyme on (positive allosterism) or it may inhibit the enzyme (negative allosterism). The pyruvate dehydrogenase complex is inhibited by high concentrations of ATP, acetyl CoA, and NADH. Citrate synthase is an allosteric enzyme inhibited by ATP. Isocitrate dehydrogenase is an allosteric enzyme that is stimulated by ADP binding and is also inhibited by high levels of NADH and ATP. The α-ketoglutarate dehydrogenase complex is inhibited by high levels NADH, succinyl CoA and ATP.

22.6 The Degradation of Amino Acids

Amino acids obtained from the degradation of dietary protein may also be oxidized as energy sources. Amino acid degradation occurs mainly in the liver and takes place in two stages: (1) the removal of the α-amino group and (2) the degradation of the remaining carbon skeleton. In land mammals, the amino group is excreted in the urine as urea. The carbon skeletons of amino acids can be converted into a variety of compounds.

Example 7
When is body protein used for energy?

Answer
Carbohydrates are not the only source of energy. Dietary protein is digested to amino acids, which may also be used as an energy source. Most of the amino acids used for energy come from dietary protein. Only under starvation conditions, when stored glycogen and lipids have been greatly depleted, does the body begin to use its own muscle protein for energy.

Removal of α-Amino Groups: Transamination

An aminotransferase, or **transaminase**, catalyzes the transfer of the α-amino group from an α-amino acid to an α-keto acid. This general reaction is summarized below.

$$
\underset{\substack{\text{Donor}\\\text{amino}\\\text{acid}}}{H-\overset{\overset{+}{NH_3}}{\underset{R}{C}}-COO^-} \; + \; \underset{\substack{\text{Acceptor}\\\text{keto}\\\text{acid}}}{\overset{\overset{O}{\|}}{\underset{R'}{C}}-COO^-} \longrightarrow \underset{\substack{\text{Carbon}\\\text{skeleton}\\\text{of amino acid}}}{\overset{\overset{O}{\|}}{\underset{R}{C}}-COO^-} \; + \; \underset{\substack{\text{New amino}\\\text{acid}}}{H-\overset{\overset{+}{NH_3}}{\underset{R'}{C}}-COO^-}
$$

The α-amino groups of many amino acids are transferred to α-ketoglutarate to produce the amino acid glutamate. This glutamate family of aminotransferases is especially important because the α-keto acid corresponding to glutamate is the citric acid cycle intermediate, α-ketoglutarate. The glutamate aminotransferases thus provide a direct link between amino acid degradation and the citric acid cycle. Alanine aminotransferase and aspartate aminotransferase are important members of this family.

All aminotransferases require the prosthetic group pyridoxal phosphate, a coenzyme derived from vitamin B_6 (pyridoxine). During transamination, the α-amino group is first transferred to pyridoxal phosphate and then from pyridoxal phosphate to the α-keto acid.

Example 8

Explain how the *transamination* of amino acids provides a direct linkage between amino acid degradation and the citric acid cycle.

Answer

Aminotransferases catalyze the transfer of the α-amino group from an α-amino acid to an α-keto acid. The α-amino groups of a great many of the amino acids are transferred to α-ketoglutarate to produce the amino acid glutamate. The glutamate family of transaminases is especially important because the α-keto acid corresponding to glutamate is the citric acid cycle intermediate, α-ketoglutarate. Thus, the glutamate transaminases, as well as others, provide a direct link to the citric acid cycle.

Removal of α-Amino Groups: Oxidative Deamination

Ammonium ion is now liberated from the glutamate by oxidative deamination, catalyzed by *glutamate dehydrogenase*:

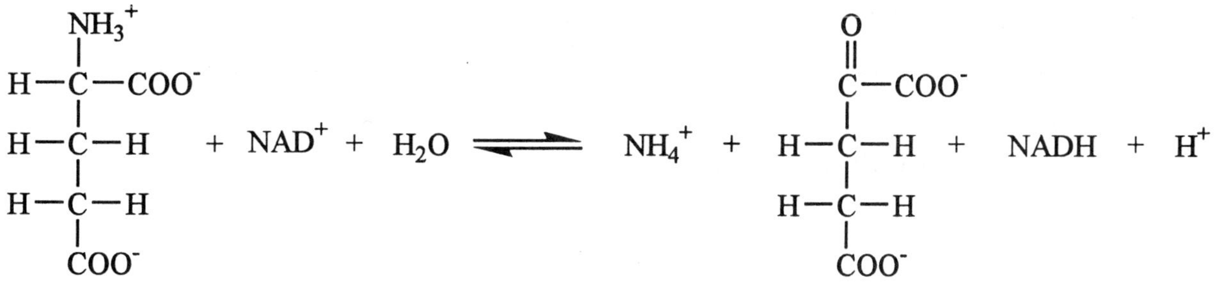

Glutamate α-Ketoglutarate

The Fate of Amino Acid Carbon Skeletons

The carbon skeletons produced by these and other deamination reactions enter the energy-harvesting pathways at many steps, as seen in Figure 22.5 in the text.

Oxidative deamination of glutamate and deamination of other amino acids produce considerable quantities of ammonium ion. This must be incorporated into a biological molecule and removed from the body so that it does not reach toxic levels.

22.7 The Urea Cycle

It is critically important to the survival of the organism to be able to excrete ammonium ions, regardless of the energy required. In man and most terrestrial vertebrates, the means of ammonium ion removal is the **urea cycle**, which occurs in the liver. This cycle keeps excess ammonium ion out of the blood and allows the excretion of the excess in the form of urea.

Reactions of the Urea Cycle

The five reactions of the urea cycle, summarized below, are shown in detail in Figure 22.7 of the text.

Step 1. CO_2 and NH_4^+ react to form carbamoyl phosphate. This reaction requires ATP and H_2O.

Step 2. Carbamoyl phosphate condenses with the amino acid ornithine to produce the amino acid citrulline.

Step 3. Citrulline now condenses with aspartate to produce argininosuccinate. This reaction requires energy released by the hydrolysis of ATP.

Step 4. Argininosuccinate is cleaved to produce the amino acid arginine and the citric acid cycle intermediate fumarate.

Step 5. Finally, arginine is hydrolyzed, producing urea, which will be excreted, and ornithine, the original reactant in the cycle. Note that one of the amino groups of urea is the ammonium ion, and the second is derived from the amino acid aspartate.

A deficiency of urea cycle enzymes causes an elevation of the concentration of NH_4^+, a condition known as **hyperammonemia**. If there is a complete deficiency of one of the enzymes of the urea cycle, the result is death in early infancy. A partial deficiency results in less severe symptoms and can be treated with a low-protein diet.

Example 9

Give a summary of the reactions of the urea cycle.

Answer

1. *Step 1*: The first step involves the reaction of the waste product CO_2 with ammonium ions from the amino acids to form carbamoyl phosphate.

$$CO_2 + NH_4^+ + 2\,ATP \longrightarrow H_2N-\overset{\displaystyle O}{\overset{\displaystyle \|}{C}}-O-\overset{\displaystyle O}{\underset{\displaystyle \underset{O^-}{|}}{\overset{\displaystyle \|}{P}}}-O^- + 2ADP + P_i + 3\,H^+$$

2. *Step 2*: The carbamoyl phosphate then condenses with the amino acid ornithine to produce citrulline (another amino acid).
3. *Step 3*: Citrulline condenses with aspartate to produce argininosuccinate.
4. *Step 4*: The argininosuccinate is cleaved to produce arginine and fumarate.
5. *Step 5*: Finally, arginine is hydrolyzed to generate urea and ornithine. The ornithine can then return to be used again in Step 2.

Example 10
Describe the genetically transmitted diseases that result from the deficiency of an enzyme in the urea cycle.

Answer
A deficiency of a urea cycle enzyme causes an elevation of the concentration of ammonium ion, a condition known as *hyperammonemia*. If there is a complete deficiency of one of these enzymes, the result is death in early infancy. If, however, there is a partial deficiency of one of the cycle's enzymes, the result can be retardation, convulsions, and vomiting. If caught early by health workers, a diet low in protein may lead to less severe clinical symptoms.

22.8 Overview of Anabolism: The Citric Acid Cycle as a Source of Biosynthetic Intermediates

The citric acid cycle plays a key role in biosynthetic reactions, as well as ATP production. Glycolysis and the citric acid cycle provide precursors for the biosynthesis of amino acids, lipids, and the nitrogenous bases found in DNA. They also generate precursors for heme, the prosthetic group required for hemoglobin, myoglobin, cytochrome c, and catalase. Metabolic pathways that function in both **anabolism** and **catabolism** are called **amphibolic pathways**. Cellular energy demand sometimes exceeds the supply of citric acid intermediates. Mammalian cells can produce more oxaloacetate by the carboxylation of pyruvate, a reaction that is also important in gluconeogenesis. This reaction is called an **anaplerotic reaction**, meaning "to fill up or replenish."

Key Terms

aerobic respiration
aminotransferase
amphibolic pathways
anabolism
anaplerotic reactions
ATP synthase
catabolism
citric acid cycle
cristae
electron transport system
F_0F_1 complex
hyperammonemia
inner mitrochondrial membrane

intermembrane space
matrix space
mitochondria
outer mitochondrial membrane
oxidative deamination
oxidative phosphorylation
pyridoxal phosphate
terminal electron acceptor
transaminase
transamination
urea cycle

Self Test

1. The oxidative reactions of metabolism provide most cellular energy. In what specific part of the cell do these reactions occur?

2. The number of mitochondria in a eukaryotic cell varies widely. What need is reflected by this variation?

3. To what compound is the acetyl group transferred in the first step of the citric acid cycle?

4. Malate dehydrogenase reduces NAD^+ to NADH and oxidizes malate into what product in the final step of the citric acid cycle?

5. How many molecules of CO_2 are produced by the complete oxidation of an acetyl group by the citric acid cycle?

6. How many molecules of NADH are formed by the complete oxidation of an acetyl group by the citric acid cycle?

7. How many molecules of $FADH_2$ are produced by the complete oxidation of an acetyl group by the citric acid cycle?

8. In glycolysis, how many NADH are generated per glucose molecule?

9. In what organelle do the reactions of the Krebs cycle, fatty acid oxidation, and oxidative phosphorylation occur?

10. What is the term for the series of electron and proton carriers embedded within the mitochondrial inner membrane?

11. Which of the electron carriers in the electron transport system contain a heme unit?

12. When NADH dehydrogenase picks oxidizes NADH in the mitochondrial matrix, it can pass the electrons to the next carrier. What ion cannot be transferred?

13. Spanning the inner mitochondrial membrane is a protein complex (F_0) that serves a channel protein. What specific particle passes through this channel?

14. Protruding into the mitochondrial matrix is a spherical protein complex (F_1). What is the enzymatic activity of F_1?

15. NADH carries electrons, originally from glucose, to the first carrier of the electron

transport system. What compound is formed by the oxidation of NADH at this site?

16. After NADH is oxidized to NAD^+ in the first step of the electron transport system, what happens to the NAD^+?

17. In the last step of the electron transport system, the electrons have too little energy to accomplish any more work. In aerobic organisms what is the terminal electron acceptor?

18. What proteins of the electron transport system are poisoned by cyanide?

19. How many ATP are produced for each NADH generated in the mitochondria?

20. How many ATP are produced for each $FADH_2$ generated in the mitochondria?

21. What is the source of most of the amino acids that are used by the human body?

22. Where in the human body does the degradation of amino acids primarily occur?

23. The α-amino groups of a great many amino acids are transferred to α-ketoglutarate. What amino acid is produced in this reaction?

24. The glutamate aminotransferases provide a direct link between amino acid degradation and which other process?

25. One of the most important aminotransferases is aspartate aminotransferase, which catalyzes the transfer of the α-amino group of aspartate to α-ketoglutarate. What are the two new compounds produced in this reaction?

26. Where in the human body does the conversion of ammonium ions to urea occur?

27. Provide the missing product in the following reaction:

$$CO_2 + NH_4^+ + 2ATP + H_2O \quad \rightarrow \quad 2ADP + P_i + 3H^+ + ?$$

28. What is the term used to describe metabolic pathways that function in both anabolism and catabolism?

29. Where in the mitochondrion do the reactions of the citric acid cycle occur?

30. What is the first step of amino acid catabolism?

23 *Fatty Acid Metabolism*

Learning Goals

1. Conceptual Goals
· Recognize the pivotal role of acetyl CoA in fatty acid and lipid metabolism.
· Summarize the role of bile salts and lipases in the digestion of lipids.
· Understand the importance of fatty acid metabolism in ATP production in the cell.
· Understand the role of "ketone body" production in β-oxidation.
· Describe the major differences between β-oxidation and fatty acid biosynthesis.

2. Performance Goals
· Describe the degradation of fatty acids, β-oxidation.

3. Health Applications
· Describe the regulation of lipid and carbohydrate metabolism in relation to
  The liver
  Adipose tissue
  Muscle tissue
  The brain
· Summarize the antagonistic effects of glucagon and insulin.
· Relate insulin and insulin production to the disease diabetes mellitus.

Chapter Overview

Acetyl CoA is the key molecule of lipid metabolism. Fatty acids are degraded to acetyl CoA, which is oxidized by the citric acid cycle, and acetyl CoA is the precursor for the biosynthesis of fatty acids and cholesterol.

23.1 Lipid Metabolism in Animals

Digestion and Absorption of Dietary Fats

Fats are hydrophobic molecules and must be extensively processed before they can be digested and absorbed. The enzymes that hydrolyze lipids are called **lipases**. The most effective lipid digestion occurs in the small intestine, where **bile** causes formation of small lipid micelles. **Micelles** are aggregations of molecules having a polar region facing the aqueous exterior and an internal nonpolar region that dissolves lipid. Bile is produced in the liver and stored in the gall bladder. The presence of lipid in the small intestine stimulates secretion of bile into the duodenum.

    Triglycerides are the major lipids in the micelles. A protein called **colipase** binds to the surface of the micelles and facilitates hydrolysis of the triglycerides by pancreatic lipases. The fatty acids and monoglycerides produced by hydrolysis are absorbed by cells of the intestinal epithelium.

Example 1
Explain the digestion and absorption of dietary fats.

Answer
Most dietary fats arrive in the duodenum in the form of fat globules. The globules stimulate the secretion of bile from the gall bladder. The bile salts emulsify the fat globules into tiny droplets. The protein colipase binds to the surface of the lipid droplets and helps pancreatic lipases adhere to the surface and hydrolyze the triglycerides into monoglycerides plus free fatty acids. These products are absorbed through the membranes of the intestinal epithelial cells.

In intestinal cells the monoglycerides and fatty acids are reassembled into triglycerides and combined with protein, producing **chylomicrons**. These chylomicrons eventually enter the blood, and the lipids are once again hydrolyzed to products that can be absorbed by cells of the body.

Example 2
Explain what happens to lipid digestion products after absorption.

Answer
The monoglycerides and fatty acids are reassembled into triglycerides, which are combined with protein to produce the class of plasma lipoproteins called *chylomicrons*. The chylomicrons are secreted into small lymphatic vessels and eventually arrive in the bloodstream. In the bloodstream the triglycerides are once again hydrolyzed to produce glycerol and free fatty acids, which can then be absorbed by the cells and used either for energy or lipid storage.

Lipid Storage
Fatty acids are stored as triglycerides within **adipocytes** (fat cells) that compose **adipose tissue**. When the body demands energy, these triglycerides can be hydrolyzed, and the fatty acids are oxidized to generate ATP.

Example 3
Explain lipid storage in the human body.

Answer
Fatty acids are stored in the form of triglycerides. Most of the body's triglyceride molecules are stored as fat droplets in the cytoplasm of the cells of adipose tissue. A fat cell, or

adipocyte, contains a large fat droplet, which accounts for nearly the entire volume of the cell.

23.2 Fatty Acid Degradation

Overview of Fatty Acid Degradation

Fatty acids are degraded by a pathway called β-**oxidation**, which consists of a set of five reactions. Each trip through the β-oxidation pathway releases acetyl CoA and returns a fatty acyl CoA molecule that contains two fewer carbons. One $FADH_2$ (2 ATP) and one NADH (3 ATP) are produced for each cycle of β-oxidation. Acetyl CoA can enter the citric acid cycle, resulting in production of 12 more ATP.

Example 4
Give an overview of fatty acid degradation.

Answer
The β-oxidation cycle for fatty acid degradation occurs in the mitochondrial matrix. The fatty acids are acted upon by an enzyme system known as the β-oxidation cycle until the fatty acids are all converted into acetyl coenzyme A units.

The Reactions of β-Oxidation

Special transport mechanisms bring fatty acid molecules into the mitochondrial matrix, where β-oxidation occurs. The reactions of β-oxidation are summarized below.

> 1. *Activation:* A fatty acyl CoA molecule is formed between coenzyme A and the fatty acid. This reaction requires ATP, which is cleaved to AMP and pyrophosphate. The bond between the fatty acid and coenzyme A is a high-energy thioester bond.
> 2. *Oxidation:* A pair of hydrogen atoms are removed from the fatty acid, and FAD is reduced to $FADH_2$.
> 3. *Hydration:* The double bond produced in step 2 undergoes a hydration reaction, and the β-carbon is hydroxylated.
> 4. *Oxidation:* The hydroxyl group of the β-carbon is now dehydrogenated, and NAD^+ is reduced to form NADH.
> 5. *Thiolyosis:* A molecule of coenzyme A attacks the β-carbon, releasing acetyl CoA and a fatty acyl CoA that is two carbons shorter than the original fatty acid.

The fatty acyl CoA is further oxidized by cycling through steps 2 through 5 until the fatty acid carbon chain is completely degraded to acetyl CoA. The acetyl CoA is completely oxidized in the citric acid cycle. The balance sheet for ATP production in the complete oxidation of a C_{16}-fatty acid, palmitic acid, is found in Figure 23.8 of the text.

Example 5

Why does the β-oxidation of a fatty acid like palmitic acid produce more energy than the oxidation of an equivalent amount of glucose?

Answer

The complete oxidation of the C_{16}-fatty acid palmitic acid produces 129 ATP molecules. This is 3.5 times more energy than results from the complete oxidation of the same amount of glucose.

23.3 Ketone Bodies

If there is not a sufficient supply of citric acid cycle intermediates to allow the complete oxidation of the acetyl CoA produced in β-oxidation, acetyl CoA is converted to the so-called **ketone bodies**, such as β-hydroxybutyrate, acetoacetate, and acetone.

Example 6

Explain why ketone bodies are produced in the human body.

Answer

The β-oxidation of fatty acids produces a steady supply of acetyl CoA molecules. If glycolysis and the β-oxidation cycle are functioning at the same rate, there will be a steady supply of pyruvate, which can be converted into oxaloacetate. However, if the supply of oxaloacetate is too low to allow all of the acetyl CoA to enter the citric acid cycle, the acetyl CoA molecules are converted into the so-called ketone bodies. The ketone bodies are β-hydroxybutyrate, acetoacetate, and acetone.

Ketosis

Ketosis is the abnormal elevation of blood ketone body concentration. This may occur due to starvation, a diet that is extremely low in carbohydrates, or uncontrolled **diabetes mellitus**. In diabetes this can lead to **ketoacidosis**, because the ketone acids are relatively strong acids and dissociate to release H^+. This causes the pH of the blood to become acidic.

Ketogenesis

The first step in the pathway for ketone body synthesis is the fusion of two molecules of acetyl CoA, catalyzed by the enzyme thiolase. The product of this reaction is acetoacetyl CoA. Acetoacetyl CoA reacts with a third acetyl CoA to yield β-hydroxy-β-methylglutaryl CoA (HMG-CoA). HMG-CoA is cleaved to yield acetoacetate and acetyl CoA. While some acetoacetate spontaneously loses carbon dioxide, producing acetone, most of it undergoes

271

NADH-dependent reduction to produce β-hydroxybutyrate.

Acetoacetate and β-hydroxybutyrate, produced in the liver, are circulated to other tissues through the blood. There they may be reconverted to acetyl CoA and used to produce ATP.

Example 7
Explain ketogenesis.

Answer
The pathway for the production of ketone bodies begins with the reversal of the last step of the β-oxidation cycle. When oxaloacetate levels are low, the enzyme that mediates the last reaction of β-oxidation, thiolase, now mediates the fusion of two acetyl CoA molecules to produce acetoacetyl CoA. The acetoacetyl CoA reacts with another molecule of acetyl CoA to produce HMG-CoA. The HMG-CoA is cleaved to yield *acetoacetate* and acetyl CoA. Some of the acetoacetate spontaneously loses carbon dioxide to produce *acetone*. Some of the acetoacetate is reduced to form β-*hydroxybutyrate*.

23.4 Fatty Acid Synthesis

All organisms possess the ability to synthesize fatty acids. Fatty acid synthesis appears to be simply the reverse of β-oxidation. In fact, the fatty acid chain is constructed by the sequential addition of two carbon acetyl groups, as seen in Figure 23.11 of the text. However, there are several differences between the two processes, including intracellular location, acyl group carriers, enzymes involved, and the electron carriers used. Fatty acid biosynthesis occurs in the cytoplasm. The acyl group carrier is **acyl carrier protein (ACP)**. The portion of ACP that binds the fatty acid is the **phosphopanthetheine** group also found in coenzyme A. A multienzyme complex called *fatty acid synthase* carries out fatty acid biosynthesis and NADPH is used as the reducing agent.

Example 8
Compare fatty acid synthesis with fatty acid degradation.

Answer
On first examination of fatty acid synthesis, it appears that it is simply the reverse of β-oxidation. Although fatty acid synthesis and breakdown are similar, there are several major differences:

1. The enzymes responsible for fatty acid biosynthesis are located in the cytoplasm of vertebrates, whereas those responsible for the degradation of fatty acids are found in the mitochondria.

2. The activated intermediates of fatty acid synthesis are bound to a thiol group of acyl carrier protein. Thus, the thioester intermediates of fatty acid synthesis are not

derivatives of coenzyme A.

3. The seven steps of fatty acid biosynthesis are carried out by the multienzyme complex known as the fatty acid synthase. The enzymes responsible for fatty acid degradation are not associated with one another as a complex.

4. NADH and $FADH_2$ are produced by fatty acid oxidation, whereas NADPH is the reducing agent for fatty acid biosynthesis.

23.5 The Regulation of Lipid and Carbohydrate Metabolism

Fatty acid and carbohydrate metabolism occur at different levels in different organs. The regulation of these two processes is of great physiological importance.

The Liver

The liver plays a central role in the regulation of blood glucose concentration, providing a steady supply of glucose for muscle and brain. It also plays a central role in lipid metabolism. When excess fuel is available, the liver synthesizes fatty acids, which are transported to adipose tissue and stored. During starvation, the liver converts fatty acids to ketone bodies that are exported to other organs. Yet the liver obtains most of its metabolic energy from the carbon skeletons of amino acids.

Example 9
Explain the central role of the liver in lipid metabolism.

Answer
When excess fuel is available, the liver synthesizes fatty acids. These are esterified to produce triglycerides that are transported to the adipose tissues by very low-density lipoprotein complexes. This transport is very active when more calories are eaten than are burned. During fasting or starvation conditions, however, the liver converts fatty acids to acetoacetate and other ketone bodies. These ketone bodies are transported to other organs for energy usage.

Adipose Tissue

Adipose tissue is the major storage depot of fatty acids. Triglycerides produced in the liver are transported through the blood in very low-density lipoprotein complexes. The triglycerides are hydrolyzed, and the fatty acids are absorbed by adipose tissue. In adipocytes, triglycerides are resynthesized from the absorbed fatty acids and glycerol-3-phosphate, which is produced from intermediates of glycolysis.

Muscle Tissue

The energy demand of resting muscle is generally supplied by the β-oxidation of fatty acids. The heart muscle actually prefers ketone bodies over glucose. Working muscle obtains energy by degradation of glycogen.

The Brain

Normally, the brain uses glucose as its sole source of metabolic energy. However, under starvation conditions, the brain converts to the use of ketone bodies produced by the liver. This is possible because ketone bodies can cross the *blood-brain barrier*.

Example 10
Which compounds can be used by the brain for production of ATP?

Answer
Under normal conditions the brain uses glucose as its sole source of metabolic energy. If conditions of starvation occur, the glycogen stores drop sharply, and blood glucose concentrations drop. The ketone bodies, acetoacetate and β-hydroxybutyrate, derived from fatty acid degradation, then provide an alternative source of metabolic energy.

23.6 The Effects of Insulin and Glucagon on Cellular Metabolism

Insulin is a polypeptide hormone secreted by the β-cells of the islets of Langerhans in the pancreas in response to an increase in the blood sugar level. Insulin lowers the concentration of blood glucose by stimulating storage of glucose as glycogen or triglycerides. In general, insulin activates biosynthetic processes and inhibits catabolism.

Insulin acts only on those *target cells* that possess a specific insulin receptor protein in their membranes. The major target cells for insulin are liver, adipose, and muscle cells. Insulin lowers blood glucose levels by altering cellular carbohydrate, protein, and lipid metabolism.

A second hormone, **glucagon**, is secreted by the α-cells of the islets of Langerhans in response to decreased blood glucose levels. The effects of glucagon, generally the opposite of the effects of insulin.

The effects of insulin and glucagon on a variety of metabolic reactions are summarized in the following table:

Actions	Insulin	Glucagon
Cellular glucose transport	Increased	No effect
Glycogen synthesis	Increased	Decreased
Glycogenolysis in liver	Decreased	Increased
Gluconeogenesis	Decreased	Increased
Amino acid uptake and protein synthesis	Increased	No effect
Inhibition of amino acid release and protein degradation	Decreased	No effect
Lipogenesis	Increased	No effect
Lipolysis	Decreased	Increased
Ketogenesis	Decreased	Increased

Key Terms

acyl carrier protein (ACP)
adipocyte
adipose tissue
bile
chylomicron
colipase
diabetes mellitus
fatty acid synthase
glucagon
insulin

ketoacidosis
ketone bodies
ketosis
lipase
micelle
β-oxidation
phosphopantetheine
triglyceride

Self Test

1. The metabolism of fatty acids and lipids revolves around which specific compound?
2. Which compound is the starting material for the synthesis of half of the amino acids?
3. In what form do we find dietary fat when it arrives in the duodenum?
4. What mixture is composed of micelles of lecithin, cholesterol, protein, bile salts, inorganic ions, and bile pigments?
5. What is the term for an aggregate of molecules having polar and nonpolar regions?
6. Name the two major bile salts. (Refer to the textbook)
7. Which protein molecule binds to the surface of lipid droplets and helps pancreatic lipases to adhere to the surface and hydrolyze the ester linkages?
8. Monoglycerides and fatty acids are reassembled into triglycerides and combined with protein in the cells of the small intestine. What class of lipoproteins is formed in this process?
9. What tissue stores most of the body's triglyceride molecules?
10. How many NADH and FADH$_2$ are produced in each round of β-oxidation?

11. What is another term for a fat cell?

12. The complete oxidation of a molecule of palmitic acid, using the β-oxidation cycle and the other oxidation pathways, produces how many molecules of ATP?

13. What is ketosis?

14. List three conditions that can produce ketosis.

15. In patients with uncontrolled diabetes the very high concentration of ketone acids in the blood leads to what condition?

16. Where in the body are acetoacetate and β-hydroxybutyrate produced?

17. Which muscle derives more of its energy from the metabolic use of ketone bodies than it does from glucose?

18. What is the name of the multienzyme complex used for biosynthesis of fatty acids?

19. As a general rule, NADH is produced by catabolic reactions, but what acts as the reducing agent for biosynthetic reactions?

20. Which organ provides a steady supply of glucose for muscles and brain and plays a major role in the regulation of blood glucose concentration?

21. Adipose tissue is unable to make glycerol-3-phosphate. What pathway must it use for its supply of this precursor?

22. What lipoprotein complex transports triglycerides produced in the liver through the blood stream?

23. What is the preferred fuel of liver tissue?

24. What compound is required for the synthesis of triglycerides in adipose tissue?

25. What compounds are oxidized to supply the energy demand of resting muscles?

26. What compound is produced in the first step of β-oxidation of fatty acids?

27. What compound is released in the final step of β-oxidation of fatty acids?

28. List the names of the two ketone bodies that are also acids.

29. Muscle tissues produce lactate under what conditions?

30. What are the major target cells for insulin?

24   *Introduction to Molecular Genetics*

Learning Goals

1. Conceptual Goals
· Understand, in general terms, the role of DNA in the transmission of genetic information.
· Describe the template model of DNA replication.
· Know what is meant by mutation and how mutations can cause cancer.
· Explain how visible radiation can be used by bacteria to repair ultraviolet radiation-induced damage.
· Recognize the consequences of thymine dimers and other mutations.
· Summarize the central dogma, the path of information flow in organisms.
· Describe the essential elements of the genetic code, and develop a "feel" for its elegance.
· Understand, in general terms, the fundamental processes involved in protein synthesis.
· Describe the tools used in the study of DNA and in genetic engineering.

2. Performance Goals
· Describe the general composition of DNA.
· Know the role of nucleotides in DNA structure.
· Recognize the major properties of the double helix, the basic structure of DNA.
· Describe the structure of RNA and how it differs from DNA.
· List three classes of RNA molecules.
· Describe the biosynthesis of messenger RNA.
· Explain the difference between continuous and discontinuous genes.

3. Health Applications
· Describe how ultraviolet light functions as a germicide, a mutagen, and a carcinogen.
· Understand the way nucleoside analogues, like AZT, are used as antiviral chemotherapeutic agents.
· Appreciate the Ames Test, a test for carcinogenic compounds that does not require the use of animals.
· Recognize the potential of pharmaceuticals produced by recombinant DNA technology for use in medicine.
· Understand the basis for DNA fingerprinting.

Chapter Overview

Deoxyribonucleic acid (DNA) carries all the genetic information in the cell. The primary structure of all proteins is dictated by the sequence of nucleotides in the DNA. **Ribonucleic acid (RNA)** molecules are the intermediates that carry out the translation of the genetic code into the structure of a protein. Any change in the nucleotide sequence of the DNA is a **mutation**. Mutations may be spontaneous or may result from the action of a mutagen. Recombinant DNA technology has allowed a better understanding of human genetic diseases and has produced protein products used for the treatment of some of these diseases.

24.1 The Structure of the Nucleotide

Chemical Composition of DNA and RNA

The chemical composition of DNA and RNA is summarized in the following table.

	DNA	**RNA**
Sugar	2'Deoxyribose	Ribose
Purine Nitrogenous Bases	Adenine (A) Guanine (G)	Adenine (A) Guanine (G)
Pyrimidine Nitrogenous Bases	Cytosine (C) Thymine (T)	Cytosine (C) Uracil (U)

In addition, both DNA and RNA contain phosphoryl groups.

Example 1
Explain the chemical composition of DNA and RNA.

Answer
The treatment of DNA with a strongly acidic solution releases deoxyribose, phosphoric acid, and the following four heterocyclic bases: adenine, guanine, cytosine, and thymine. The same treatment of RNA releases ribose, phosphoric acid, and the following four heterocyclic bases: adenine, guanine, cytosine, and uracil.

Nucleotide Structure

Nucleotides are made up of a sugar, a nitrogenous base and at least one phosphoryl group. Because this large structure contains two cyclic molecules, the sugar and the base, the ring atoms of the sugar are designated with a prime to distinguish them from atoms in the nitrogenous base. The covalent bond between the sugar and the phosphoryl group is a phosphoester bond. The bond between the base and the sugar is called a β-N-glycosidic linkage.

278

Example 2
Explain how nucleosides and nucleotides are formed.

Answer
Nucleosides are produced by the combination of a pentose (ribose or deoxyribose) with a purine or a pyrimidine base.

A nucleotide is produced if a hydroxyl group of the pentose portion of a nucleoside is converted into a phosphate ester.

To name a nucleotide remove the *-ine* ending and replace it with either *-osine*, for purines, or *-idine*, for pyrimidines. (Uracil is the only exception to this rule; the *-acil* ending is replaced with *-idine*, producing the name uridine.) Nucleotides with the sugar ribose are **ribonucleotides**, and those having the sugar 2'-deoxyribose are **deoxyribonucleotides**. Finally, a prefix is added to indicate the number of phosphoryl groups that are attached.

Names and abbreviations of the ribonucleotides and deoxyribonucleotides containing guanine are given below:

Nucleotide	Abbreviation
deoxyguanosine monophosphate	dGMP
deoxyguanosine diphosphate	dGDP
deoxyguanosine triphosphate	dGTP
guanosine monophosphate	GMP
guanosine diphosphate	GDP
guanosine triphosphate	GTP

Example 3
Name the following nucleotides:

1. adenine + deoxyribose + phosphate
2. guanine + ribose + phosphate
3. adenosine-phosphate-phosphate-phosphate

Answer
1. deoxyadenosine-5'-monophosphate [dAMP]
2. guanosine-5'-monophosphate [GMP]
3. adenosine-5'-triphosphate [ATP]

24.2 The Structure of DNA and RNA

DNA Structure: The Double Helix

A strand of DNA is a polymer of 2'-deoxyribonucleotide units bonded to one another by 3'-5' phosphodiester bonds. The backbone of the polymer is called the sugar-phosphate backbone because it is composed of alternating deoxyribose and phosphoryl groups in phosphodiester linkage. A nitrogenous base is bonded to each sugar by an *N*-glycosidic linkage.

Example 4
Summarize the primary structure of DNA.

Answer
The primary structure of DNA is the linear sequence of its 2'-deoxyribonucleotides. The nucleotides of DNA are linked by 3'-5' phosphodiester bonds. Each single strand of DNA is characterized by a backbone of alternating deoxyribose and phosphoryl groups in phosphodiester linkages. Each deoxyribose is also linked to one of the nitrogenous bases by an *N*-glycosidic bond.

DNA is a **double helix** of two strands of DNA held together by hydrogen bonds. The sugar-phosphate backbone spirals around the outside of the helix, and the nitrogenous bases extend into the center at right angles to the axis of the helix. Adenine forms hydrogen bonds with thymine, and cytosine forms hydrogen bonds with guanine. The hydrogen-bonded bases are called **base pairs**. It is the hydrogen bonds between the base pairs that hold the two strands of the double helix together. The two strands of DNA are complementary, because the sequence of bases on one strand automatically determines the sequence of bases on the opposite strand. There are 10 base pairs in each turn of the helix. The two strands of the double helix are **antiparallel** to one another; they proceed in opposite directions.

Example 5
Summarize the double helix structure of DNA.

Answer
DNA consists of two strands wound around each other. Each strand has a helical conformation, and the resultant structure is called a double helix. The major properties of the DNA double helix are as follows:
  1. The sugar phosphate backbone winds around the outside of the bases like the handrails of a spiral staircase.
  2 Each purine base is hydrogen bonded to a pyrimidine base in the interior of the double helix. Adenine is always paired with thymine, and guanine is always paired with cytosine. Each base pair lies at nearly right angles to the long axis of the helix,

like the stairs of the spiral staircase. Due to the base pairing of A to T and G to C, the two strands are complementary to one another. Therefore, the sequence of bases on one strand automatically determines the sequence of the other strand.

3. The two strands of the DNA double helix are antiparallel. One strand advances in the 5' → 3' direction, and the second strand advances in the 3'→ 5' direction.

4. The double helix of DNA completes one full turn every 10 nucleotides.

RNA Structure

RNA molecules are single stranded. The sugar-phosphate backbone of RNA consists of ribonucleotides linked by 3'-5' phosphodiester bonds. The sugar in RNA is ribose rather than 2'-deoxyribose, and uracil (U) replaces thymine.

24.3 DNA Replication

DNA must be replicated before cell division so that each daughter cell inherits a copy of each gene. The first step in replication is the separation of the strands of DNA. Proteins do this by breaking the hydrogen bonds between the base pairs. Then the enzyme **DNA polymerase III** "reads" each parental strand, also called the template, and catalyzes the polymerization of a complementary daughter strand. Thus, each daughter DNA molecule consists of one parental strand and one newly synthesized daughter strand. This mode of DNA replication is called **semiconservative replication**.

Bacterial chromosomes are circular DNA molecules. Replication begins at a **replication origin** and proceeds bidirectionally around the circular chromosome. The site at which the new DNA is being synthesized is called the **replication fork**.

Example 6
Summarize DNA replication.

Answer
DNA replication occurs each time a cell divides. In this way all of the genetic information is passed from one generation to the next. As a result of specific base pairing, the sequence of bases along the pentose-phosphate backbone of each strand of DNA automatically specifies the sequence of bases in the complementary strand.

The first event that occurs in DNA replication is the slow separation of the two strands as two new strands are formed, using the parental strands as templates.

24.4 Information Flow in Biological Systems

The genetic information in the DNA must be expressed to produce the proteins that actually carry out the work of the cell. Gene expression involves two steps. First, DNA is transcribed to produce a variety of RNA molecules. This process is called transcription. Then the RNA molecules participate in translation, a process in which proteins are produced. This unidirectional expression of the genetic information is called the central dogma of molecular biology and can be summarized as follows:

$$DNA \quad \rightarrow \quad RNA \quad \rightarrow \quad PROTEIN$$

Example 7
Explain the following terms: *central dogma*, *transcription*, and *translation*.

Answer
The *central dogma* of molecular biology states that in cells the flow of genetic information contained in DNA is a one-way street that leads from DNA to RNA to protein synthesis. The process by which a single strand of DNA serves as a template for the synthesis of an RNA molecule is called *transcription*. In this process part of the information of DNA is copied into a strand of RNA. The process by which the message is converted into protein is called *translation*.

Classes of RNA Molecules

There are three classes of RNA found in the cell, and they are classified by their function in gene expression.

Messenger RNA (mRNA) carries the genetic information for a protein from DNA to the ribosomes. It is a complementary RNA copy of a gene on the DNA.

Ribosomal RNA (rRNA) is a structural and functional component of the ribosomes, which are "platforms" on which protein synthesis occurs.

Transfer RNA (tRNA) translates the genetic code of the mRNA into the primary sequence of amino acids in the protein. In addition to the primary structure, tRNA molecules have a cloverleaf-shaped secondary structure resulting from base-pair hydrogen bonding (A–U and G–C), and a tertiary structure. The sequence CCA is found at the 3'-end of the tRNA. The 3'-adenosine of this sequence can be covalently attached to an amino acid. Three nucleotides at the base of the cloverleaf structure form the **anticodon**. This triplet of bases forms hydrogen bonds to a **codon** (complementary sequence of bases) in a messenger RNA (mRNA) molecule on the surface of a ribosome during protein synthesis. This interaction assures that the correct amino acid is brought to the site of protein synthesis at the appropriate location in the growing peptide chain.

Example 8
Describe the different classes of RNA molecules.

Answer
An RNA molecule is classified by its cellular location and by its function.

1. *Messenger RNA (mRNA)* carries the genetic information for a protein from DNA to the ribosomes. A messenger RNA molecule is a copy of the information encoded in a gene: like the other classes of RNA molecules, it is synthesized by the process of transcription. In eukaryotic cells messenger RNA molecules are synthesized in the nucleus and transported to the cytoplasm, where they are translated.

2. *Ribosomal RNA (rRNA)* is a structural and functional component of the ribosomes. The ribosomes serve as platforms on which protein synthesis occurs. Ribosomal RNA is a structural component of the ribosomes, but also participates in the synthesis of the protein from the code carried by the mRNA.

3. *Transfer RNA (tRNA)* is responsible for translating the genetic code of the mRNA into the primary sequence of amino acids in protein. The tRNA molecules carry the needed amino acids to the ribosome for use in protein synthesis.

Transcription

Transcription is catalyzed by the enzyme **RNA polymerase**. RNA polymerase binds to a specific nucleotide sequence, the **promoter**, at the beginning of a gene. It then separates the two strands of DNA so that it can "read" the sequence of the DNA. Chain elongation begins as the RNA polymerase reads the DNA template strand and catalyzes the polymerization of a complementary RNA copy. The final stage of transcription is termination. The RNA polymerase finds a termination sequence at the end of the gene and releases the newly formed RNA molecule.

Post-Transcriptional Processing of RNA

In eukaryotes, transcription produces a **primary transcript** the undergoes extensive **post-transcriptional modification** before it is exported from the nucleus for translation. The first modification is the addition of a **cap structure** to the 5' end of the RNA. This facilitates efficient translation. The second modification is the addition of a **poly(A) tail** to the 3' end of the RNA. This protects the mRNA from enzymatic degradation.

The third modification is **RNA splicing**. Bacterial genes are continuous, and all the nucleotide sequences of the gene are found in the mRNA. Eukaryotic genes are discontinuous; there are extra DNA sequences within the genes. The initial mRNA, or primary transcript, carries both the protein-coding sequences and these extra sequences, which are termed intervening sequences or introns. The **introns** are removed by the process of RNA splicing. Splicing is carried out with the assistance of ribonucleoprotein particles called **spliceosomes**. They aid in the recognition of specific sequences at the splice boundaries and stabilize the splicing complex. The splicing reactions involves two transesterification reactions.

24.5 The Genetic Code

The **genetic code** is a triplet code. Each code word (codon) consists of three nucleotides. The three-letter genetic code contains 64 codons, but there are only 20 amino acids. Thus, there are 44 more codons than are required. Three of the codons (UAA, UAG, and UGA) are translation termination signals, leaving 41 additional codons. The genetic code is said to be highly **degenerate**. Methionine and tryptophan are the only amino acids encoded by only a single codon. All others are encoded by at least two codons, and serine and leucine each have six codons. As a result of this, the genetic code is quite resistant to mutation. For those amino acids that have multiple codons, the first two bases define the amino acid, and the third position is variable. A **point mutation** (the change of a single base) in the third position, therefore, often has no effect upon the amino acid that is incorporated into a protein.

24.6   Protein Synthesis

Protein synthesis, or **translation**, occurs on ribosomes. Ribosomes are complexes of ribosomal RNA and proteins consisting of two subunits, a small and a large ribosomal subunit.

Small Ribosomal Subunit: 1 rRNA + 33 proteins
Large Ribosomal Subunit; 3 rRNA + 49 proteins

Many ribosomes simultaneously translate each mRNA. These structures are called polyribosomes or **polysomes**.

The Role of Transfer RNA

The molecule that decodes the information on the mRNA molecule into the primary structure of a protein is transfer RNA (tRNA). In order to do this, the tRNA must be covalently linked to one specific amino acid. The enzyme that binds the specific tRNA to its specific amino acid is an **aminoacyl tRNA synthetase**. The product is called an **aminoacyl tRNA**.
    The tRNA recognizes the appropriate codon on the mRNA by codon-anticodon hydrogen bonding.

The Process of Translation

Initiation

Initiation factors mediate the formation of a translation initiation complex composed of an mRNA molecule, the small and large ribosomal subunits, and the initiator tRNA. The initiator tRNA specifically recognizes the **initiation codon** AUG on the mRNA. The ribosome has two sites for binding tRNA molecules. The first site is called the **peptidyl tRNA binding site (P-site)**. It holds the initiator tRNA in the initiation complex and then carries the tRNA bound to the growing peptide during the remainder of protein synthesis. The second site, called the **aminoacyl tRNA binding site (A-site),** holds the aminoacyl tRNA carrying the next amino acid to be added to the peptide chain.

Chain Elongation

This occurs in three steps that are repeated until protein synthesis is complete.

1. Binding of an aminoacyl tRNA molecule to the empty A-site.
2. Peptidyl transferase catalyzes the formation of the peptide bond, and the peptide chain is shifted to the tRNA that occupies the A-site.
3. The uncharged tRNA molecule is discharged, and the ribosome changes positions (**translocation**), so that the next codon on the mRNA occupies the A-site, shifting the new peptidyl tRNA from the A-site to the P-site.

Termination

When a "stop," or **termination codon** (UAA, UAG, or UGA) is encountered, translation is terminated. A **release factor** finds the empty A-site and causes release of the newly formed peptide chain and the ribosomal subunits.

The newly synthesized protein folds into its characteristic three-dimensional shape, and if it has quaternary structure, it may associate with other protein subunits. Some proteins are further modified following protein synthesis by the addition of carbohydrate or lipid molecules.

24.7 Mutations, Ultraviolet Light, and DNA Repair

The Nature of Mutations

Any change in the nucleotide sequence of a DNA molecule is called a **mutation**. Mutations can arise from errors during DNA replication or may be the result of chemicals, called **mutagens**, that damage the DNA.

Mutations are classified by the kind of change that occurs in the DNA. The substitution of a single nucleotide for another is called a **point mutation**. Loss of one or more nucleotides is a **deletion mutation**. Addition of one or more nucleotides to a DNA sequence is an **insertion mutation**.

The Results of Mutations

Some mutations are **silent**; that is, they cause no change in the organism. Often, however, the result of a mutation has a negative effect on the health of the organism. The effect of a mutation depends on how it alters the genetic code for a protein. There are approximately 4000 human genetic diseases that are known to be caused by mutations, including sickle-cell anemia, hemophilia, cystic fibrosis, and color-blindness.

Mutagens and Carcinogens

Often mutagens are also **carcinogens**, cancer causing chemicals. Most cancers result from one or more mutations in a single normal cell. These mutations result in the loss of normal growth control, causing the abnormal cell to proliferate. If that growth is not controlled or destroyed, it will result in the death of the individual.

Ultraviolet Light Damage and DNA Repair

Ultraviolet (UV) light causes damage to DNA by inducing the formation of **thymine dimers**. Such dimers consist of adjacent pairs of thymine molecules covalently bonded to one another. This interferes with normal hydrogen bonding between these thymine molecules and the complementary adenine bases and this region of DNA cannot be replicated or transcribed. As a result, the cell dies.

Bacteria have four different mechanisms to repair ultraviolet light damage. Mutations occur when the repair system makes an error and causes a change in the nucleotide sequence of the DNA.

Ultraviolet lights (germicidal lamps) are used in hospitals to kill bacteria in the air and on environmental surfaces, such as in a vacant operating room. In addition, excessive exposure to UV light (as during sun tanning) has been correlated with an increased incidence of skin cancer.

Consequences of Defects in DNA Repair

The human repair system for thymine dimers requires at least five enzymes. A mutation in the gene for one of these enzymes results in the genetic skin disorder *xeroderma pigmentosum*. Individuals with this disorder are extremely sensitive to the UV rays of sunlight and usually develop multiple skin cancers before the age of 20.

24.8 Recombinant DNA

Tools Used in the Study of DNA

Many of the techniques and tools used to study DNA and to clone genes were developed or discovered during basic studies on bacterial DNA replication and gene expression. These include many enzymes that catalyze reactions of DNA molecules, gel electrophoresis, cloning vectors, and hybridization techniques.

Restriction Enzymes

Restriction enzymes are bacterial enzymes that "cut" the sugar-phosphate backbone of DNA molecules at specific nucleotide sequences. Often these enzymes leave short single-stranded stretches at the ends of the DNA molecules. These are called "sticky ends" because they can reassociate with one another by hydrogen bonding. This is a property of the DNA fragments generated by restriction enzymes that is very important to gene cloning.

These enzymes are used to digest large DNA molecules into smaller fragments of specific size. DNA from an individual will produce a reproducible set of DNA fragments when digested with a particular restriction enzyme. This is essential for the study or cloning of DNA from any source.

Agarose Gel Electrophoresis

One means of studying the DNA fragments produced by restriction enzyme digestion is agarose gel electrophoresis. The digested DNA sample is placed in a sample well in the gel and an electric current is applied. The DNA fragment will move through the gel away from the negative electrode (cathode) and toward the positive electrode (anode). Smaller DNA fragments move more rapidly than larger ones and as a result the DNA fragments end up distributed throughout the gel according to their size.

Hybridization

Hybridization is a technique used to identify the presence of a gene on a particular DNA fragment. This technique is based on the fact that complementary DNA or RNA sequences will hydrogen bond, or hybridize, to one another. It doesn't matter whether the single-stranded sequences come from the same organism or from two different kinds of organisms. If the nucleotide sequences are complementary they will hybridize with one another.

One hybridization technique, called *Southern Blotting*, involves hybridization of DNA fragments from an agarose gel. This technique allows the detection of specific nucleotide sequences on a DNA fragment of known size.

DNA Cloning Vectors

A **cloning vector** is a piece of DNA having its own replication origin, so that it can be replicated inside a host cell. Often the bacterium *Escherichia coli* serves as the host cell in which the vector carrying the cloned DNA is replicated in abundance.

There are two major kinds of cloning vectors: *phage vectors* and *plasmid vectors*. Phage vectors are specially modified bacterial viruses and plasmids are extra pieces of circular DNA found in most kinds of bacteria. Both can carry a cloned DNA fragment into a bacterial cell and carry out replication that allows the amplification of the vector carrying the cloned DNA fragment.

Genetic Engineering

To clone a gene, vector DNA and target DNA (the source of the gene to be cloned) are digested with the same restriction enzyme. They are then mixed together under conditions that allow the sticky ends of the target and vector DNA to hybridize with one another. DNA ligase is added to covalently join the ends of the DNA molecules.

Finally, these recombinant DNA molecules are introduced into a bacterial cell. This can be done by transformation, a process by which bacterial cells are specially treated to favor the entry of DNA into the cells. Antibiotic selection and hybridization can be used to detect those clones carrying the target gene.

Genetic engineers have had to overcome many obstacles to clone eukaryotic genes of special interest medically. For instance, if the researcher desires to produce the protein product of a eukaryotic gene, he or she cannot simply clone cellular DNA because there are introns within eukaryotic genes. Molecular biologists found that a DNA copy of a eukaryotic mRNA could be made using the enzyme reverse transcriptase of a class of viruses called retroviruses. This DNA copy of the mRNA carries all the protein-coding sequences of a gene, but none of the intron sequences. Thus bacteria are able to transcribe and translate the cloned DNA and produce valuable products for use in medicine and other applications.

Great progress has been made by applying genetic engineering to genetic disease, but most of these products represent only a treatment, not a cure. The ultimate dream for the future is to be able to remove "bad" genes from the human population entirely. It will be many years, or even decades, before that dream will become a reality.

Key Terms

aminoacyl tRNA

aminoacyl tRNA binding
  site of ribosome (A-site)

aminoacyl tRNA synthetase

anticodon

antiparallel strands

base pairs

cap structure

carcinogen

central dogma

cloning vector

codon

complementary strands

degenerate code

deletion mutation

deoxyribonucleic acid (DNA)

deoxyribonucleotide

DNA polymerase III

double helix

exon

hybridization

initiation factors

insertion mutation

intron

messenger RNA (mRNA)

mutagen

mutation

nucleotide

peptidyl tRNA binding
  site of ribosome (P-site)

point mutation

poly(A) tail

polysome

post-transcriptional modification

primary transcript

promoter

purine

pyrimidine

release factors

replication fork

replication origin

restriction enzyme

ribonucleic acid (RNA)

ribonucleotide

ribosomal RNA (rRNA)

ribosome

RNA polymerase

RNA splicing

semiconservative DNA replication

silent mutation

termination codon

thymine dimers

transcription

transfer RNA (tRNA)

translation

translocation

Self Test

1. Which molecule in living things is the carrier of genetic information?
2. In which type of nucleic acid is thymine usually found?
3. Give the correct abbreviations for the following compounds:
    a. adenine + deoxyribose + triphosphate unit
    b. cytosine + ribose + triphosphate unit
    c. thymine + deoxyribose + monophosphate unit
4. State the central dogma of molecular biology.
5. What two men elucidated the structure of DNA in 1953?
6. In naming the ring atoms of a pentose, what designation is used to distinguish them from atoms in the base?
7. What type of bonding holds the two strands of DNA together as a double helix?
8. Each base pair in the DNA molecule lies at what angle to the long axis of the helix?
9. Which base in one strand of DNA is always hydrogen-bonded to adenine in the other strand?
10. One strand of the DNA molecule has a 5' → 3' orientation. What is the orientation of the opposite strand?
11. As a result of specific base pairing in the DNA molecule, what is the relationship of the sequence of bases on one strand to the sequence of bases on the other strand?
12. What is the name of the DNA damage that is caused by ultraviolet light?
13. What is the first event to occur during replication of DNA?
14. DNA replication in *E. coli* begins at a unique sequence on the circular chromosome. What is the term for this site?
15. What is the term for the site at which the new deoxyribonucleotides are added to the growing daughter strand?
16. What are the two major functions of the enzyme DNA polymerase III?
17. What disease is caused by the accumulation of mutations that cause uncontrolled cell growth and division?
18. List the classes of different RNA molecules.
19. Which form of RNA carries the needed amino acids to the ribosomes for use in protein synthesis?
20. Name the enzyme that catalyzes the transcription of DNA.
21. What sequence on DNA indicates the site at which the RNA polymerase should begin transcription?
22. In eukaryotic cells the initial mRNA (primary transcript) carries the sequences for a protein but also contains noncoding sequences. What is the term for these noncoding sequences?
23. What is the term for the coding sequences that remain after mRNA splicing?
24. The three-letter genetic code contains 64 words. What is the term for these genetic words?
25. Who first proposed that different triplet codons may serve as code words for the same amino acid?
26. Which two amino acids have only a single codon?

27. When the abnormal gene for the formation of the beta chain in hemoglobin is translated into protein, a valine is substituted for which amino acid?

28. What genetic skin disorder is caused by a mutation in the repair endonuclease gene or in other genes in the repair pathway?

29. What are the terms used to describe the two subunits of the ribosome?

30. What is the name of the complexes of many ribosomes along a single mRNA?

31 What is the term used to describe a mutation that results in no change in the organism?

32. What is the term for an enzyme that recognizes a specific DNA nucleotide sequence and "cuts" the sugar-phosphate backbone at that site?

33. List two types of cloning vectors.

34. What is the technique that relies on specific complementary base pairing to identify a particular gene on a DNA fragment?

35. List five medically important proteins that are currently produced by genetic engineering.

A

Answers to Chapter Self Tests

Chapter 1

1. 1.02 g/mL
2. 1.48×10^{-1}
3. 3.01×10^{-6}
4. 2.24×10^3cm/s
5. 3.95×10^{-22}g
6. 2.99×10^{10}cm/s
7. $-20.3°C$
8. 4.21×10^3g
9. $-40.0°F$
10. 0.545
11. energy
12. data
13. measurement
14. metric system
15. France
16. conversion factors
17. uncertainty
18. Kelvin
19. 310.0 K
20. 3.7×10^{-3}

Chapter 2

1. b, c, d, f, and g
2. b, c, d, e, and f
3. 12 protons and 12 electrons
4. 16 protons and 16 electrons
5. 17 protons and 18 electrons
6. (1) Lithium Atom (Li): 3 protons; 3 electrons
   (2) Lithium Ion (LI^{+1}): 3 protons; 2 electrons
   (3) Li le$^-$ + Li^{+1}
7. gas, liquid, and solid
8. gas
9. gas
10. (1) color
    (2) odor
    (3) taste
    (4) compressibility
    (5) melting point
    (6) boiling point
11. molecule
12. heterogeneous
13. atom

14. nucleus
15. electron
16. atomic number
17. neutron
18. isotopes
19. positive, or cation, or +1

20.
$$^1_1H, \quad ^2_1H, \quad ^3_1H$$

Chapter 3

1. periodic table
2. valence electrons
3. 6
4. helium
5. noble gas
6. a more stable energy state
7. isoelectric ion
8. Mendeleev and Lothar Meyer
9. atomic number
10. representative elements
11. transition elements
12. $N = 1, 2, 3$, etc.
13. two
14. orbital
15. s
16. opposite
17. electron configuration
18. $1s^2, 2s^2, 2p^2$
19. noble gases
20. Group 1A
21. electronegativity

Chapter 4

1. ionic bond
2. covalent bond
3. positive and negative ions
4. 2
5. left
6. right
7. metal or cation
8. sodium oxide
9. lithium sulfide
10. aluminum bromide
11. ferrous ion
12. iron (III) ion
13. higher positive charge
14. ferrous sulfate
15. copper (II) oxide

16. Na_2SO_4
17. 4
18. 8
19. Cl
20. difference in electronegativities
21. H_2O
22. ICl

23. a. 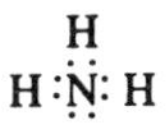
$$H:\overset{..}{\underset{..}{N}}:H$$
with H on top

b. 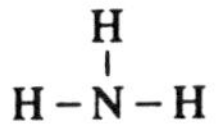
$$H-\overset{\overset{\displaystyle H}{|}}{N}-H$$

24. a.  $H:\overset{..}{\underset{..}{O}}:H$ b. 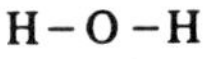 $H-O-H$

25. a. 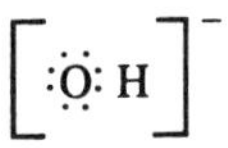 $\left[:\overset{..}{\underset{..}{O}}:H\right]^{-}$ b. 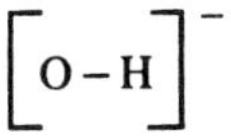 $\left[O-H\right]^{-}$

26. a. 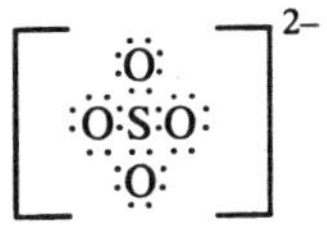 $\left[\overset{\displaystyle :\overset{..}{O}:}{:\overset{..}{O}:\overset{..}{S}:\overset{..}{O}:}_{:\overset{..}{O}:}\right]^{2-}$ b. 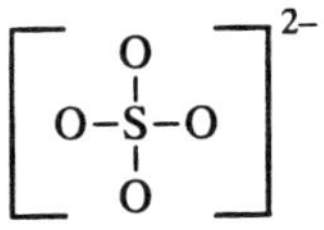 $\left[\overset{\overset{\displaystyle O}{|}}{O-S-O}_{\underset{\displaystyle O}{|}}\right]^{2-}$

27. a. 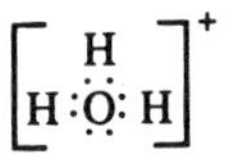 $\left[\overset{\displaystyle H}{H:\overset{..}{O}:H}\right]^{+}$ b. 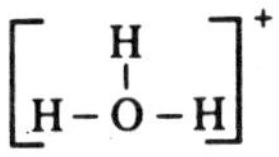 $\left[H-\overset{\overset{\displaystyle H}{|}}{O}-H\right]^{+}$

28. valence electrons
29. covalent bonding
30. $Na^{+} + Cl^{-}$
31. $Ca^{2+} + S^{2-}$
32. electron transfer
33. 2
34. 4

Chapter 5

1. 1
2. 2.0 grams
3. 6.02×10^{23} atoms or Avogadro's number
4. 4.81 grams
5. N_2
6. covalent compound
7. amu
8. 18.02 amu
9. molecular weight
10. reactants
11. 2
12. 0.5
13. coefficients
14. 2
15. 2
16. 1
17. 80.09 grams
18. 169.9 grams
19. 7.149 grams
20. 3.01 moles
21. Avogadro's number

22. O_2

Chapter 6

1. evaporation
2. ideal
3. 759.0 mm of mercury pressure
4. 5.00 atmospheres
5. 2.09 liters
6. 1.96×10^{-2} moles of helium are present
7.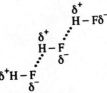
8. nitrogen + oxygen
9. Dalton's
10. surface tension
11. surfactants
12. vapor pressure
13. melting point
14. very high
15. 20.0 liters
16. condensation
17. amorphous
18. barometer
19. calorimeter
20. covalent

Chapter 7

1. 1.10 % (W/V)
2. 800.0 mL
3. 1.25 liters
4. 5.52×10^{-2} moles
5. 0.466 gram
6. 0.19 molar HCl solution
7. 0.185 molar KNO_3 solution
8. 52.1 mL
9. concentration
10. 13.5 grams
11. molarity
12. osmosis
13. hypertonic
14. isotonic
15. crenation
16. electrolyte
17. hypertonic solution
18. hypotonic solution
19. molarity

20. osmosis
21. osmotic pressure
22. saturated solution
23. solute
24. supersaturated solution

Chapter 8

1. 1.40×10^3 calories
2. a.
$$K_{eq} = \frac{[N_2O_4]}{[NO_2]^2}$$

   b.
$$K_{eq} = \frac{[PCl_3][Cl_2]}{[PCl_5]}$$

   c.
$$K_{eq} = \frac{[N_2][H_2O]^2}{[NO]^2[H_2]^2}$$

3. high
4. non-spontaneous
5.. gas
6. positive
7. calorimeter
8. fuel value
9. activated complex
10. equilibrium
11. equilibrium
12. LeChatelier
13. activation energy
14. catalyst
15. endothermic
16. entropy

Chapter 9

1. b
2. 0.0592 molar HCl solution
3. 5.00×10^{-14} moles/liter of H_3O^+ are present
4. pH = 3.0
5. combination
6. combination
7. decomposition
8. decomposition
9. single-replacement
10. double-replacement
11. double-replacement
12. neutralization
13. Brönsted-Lowry
14. dissociation

15. b
16. b

Chapter 10

1. a. $^{0}_{-1}e$ b. $^{1}_{0}n$ c. $^{210}_{83}Bi$

2. 3.125 mg remain
3. 9
4. (1) alpha particles, (2) beta particles, (3) gamma rays
5. gamma ray
6. alpha particle
7. radioactive
8. half-life, or $t_{1/2}$
9. fission
10. cancer or malignant cells
11. thyroid gland
12. background
13. inversely
14. nuclear imaging device
15. film badge
16. gamma ray
17. natural radioactivity
18. C-14 or carbon-14
19. alpha particle
20. radon

Chapter 11

1. b, c, and d
2. tetrahedral
3. 120°, or planar
4. 180°
5. functional group
6. hydroxyl
7. carbonyl group
8. aromatic
9. a
10. c
11. a
12. b
13. a
14. c
15. e
16. inorganic chemistry
17. urea
18. hydrogen
19. carbon-to-carbon double bond
20. C_9H_{20}

21. a. 2-methylbutane

   b. 2,2-dimethylpentane

   c. 2,2,3-trimethylpentane

22. 1-bromo-4-chloro-2-methylpentane

23. 2-bromo-2-methylpropane

24. 2-bromo-4,4-dimethylhexane

25. $C_3H_8 + 5O_2 \rightarrow 3CO_2 + 4H_2O$

26. Two products are formed: 1-chloropropane and 2-chloropropane

27. none or little

28. lower

29. parent compound

30. trichloromethane

31. cycloalkane or cycloalkanes

32. ring

33. combustion

34. 2

35. light, or UV, or high temperature

36. substitution

Chapter 12

1. a. alkane: C_4H_{10} could be butane

   b. alkene: C_4H_8 could be 1-butene or 2-butene

   c. alkyne: C_4H_6 could be 1-butyne or 2-butyne

2. 3-ethyl-2-pentene

3. 4,5-dibromo-2-hexyne

4.

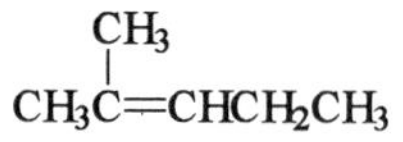

$$\underset{\displaystyle CH_3\overset{\textstyle |}{C}=CHCH_2CH_3}{\overset{\textstyle CH_3}{}}$$

5.

$$\underset{CH_3CH_2}{\overset{CH_3}{}}C=C\underset{H}{\overset{CH_3}{}}$$

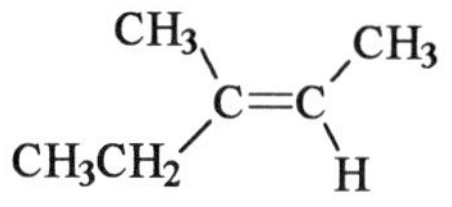

6. *trans*-2-bromo-2-butene

7.

$$CH_3CH_2CH=CH_2 + H_2 \xrightarrow[\text{pressure}]{\text{Ni, heat}} CH_3CH_2CH_2CH_3$$

   1-Butene   Butane

8.

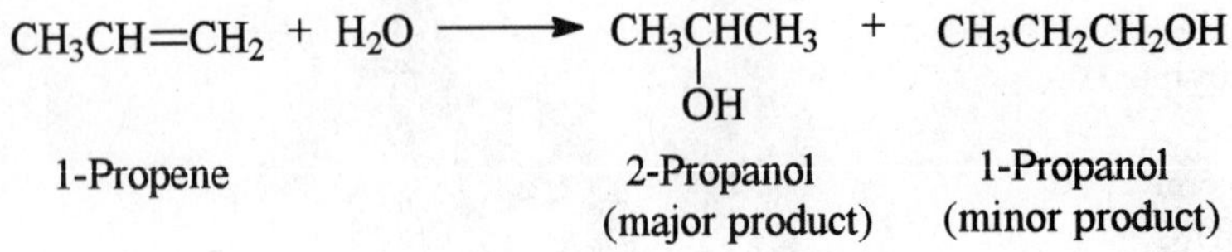

$$CH_3CH{=}CH_2 + H_2O \longrightarrow CH_3\underset{\underset{\displaystyle OH}{|}}{C}HCH_3 \ + \ CH_3CH_2CH_2OH$$

1-Propene   2-Propanol   1-Propanol
                           (major product)   (minor product)

9.

$$CH_3CH_2CH{=}\underset{\underset{\displaystyle CH_3}{|}}{C}CH_3 + HCl \xrightarrow{CCl_4} CH_3CH_2CH_2\underset{\underset{\displaystyle Cl}{|}}{\overset{\overset{\displaystyle CH_3}{|}}{C}}CH_3$$

2-Methyl-2-pentene   2-Chloro-2-methylpentane
                                     (major product)

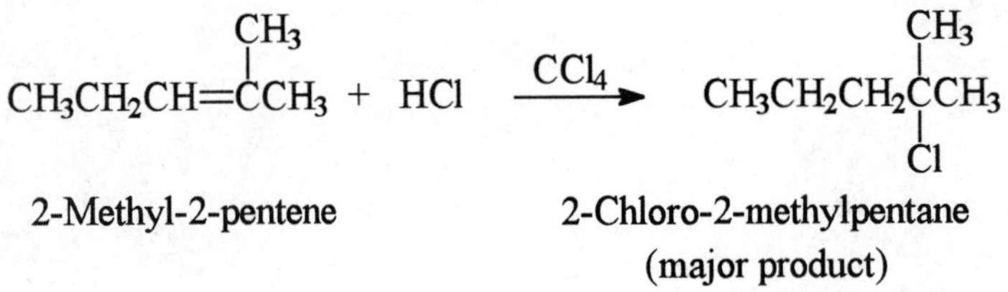

10.   a. benzene
         b. phenol or hydroxybenzene
         c. 1,2-dichlorobenzene or ortho-dichlorobenzene
         d. 1,2-dichloro-3-bromobenzene
         e. 1,4-dibromobenzene or para-dibromobenzene

11.   acetylene

12.   trans-2-butene

13.   trans-1,2-dibromopropene

14.   hydrogenation

15.   palladium and nickel

16. CH_3CH_2OH

17.   benzene

18.   chloroform

19.   carbon-to-carbon double bond

20.   addition

21.   all carbon atoms have four single bonds

22.   at least one carbon-to-carbon triple bond

23.

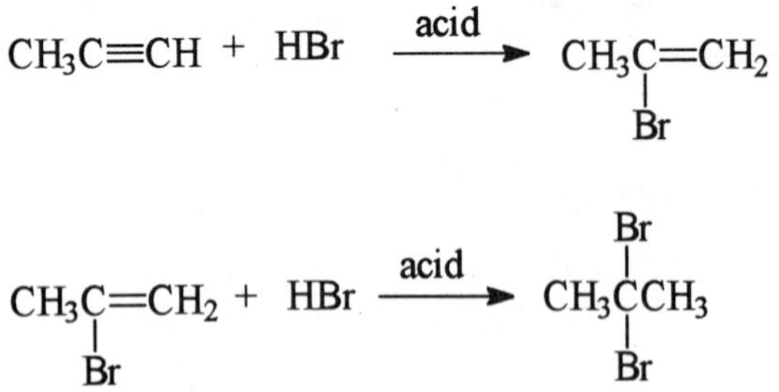

$$CH_3C{\equiv}CH + HBr \xrightarrow{acid} CH_3\underset{\underset{\displaystyle Br}{|}}{C}{=}CH_2$$

$$CH_3\underset{\underset{\displaystyle Br}{|}}{C}{=}CH_2 + HBr \xrightarrow{acid} CH_3\underset{\underset{\displaystyle Br}{|}}{\overset{\overset{\displaystyle Br}{|}}{C}}CH_3$$

24.

$$CH_3CH_2C \equiv CH + H_2 \xrightarrow[\text{pressure}]{\text{Ni, heat}} CH_3CH_2CH = CH_2$$

$$CH_3CH_2CH = CH_2 + H_2 \xrightarrow[\text{pressure}]{\text{Ni, heat}} CH_3CH_2CH_2CH_3$$

25. $CH_3CH_2CH_2CH_3$
26. Vegetable oil

Chapter 13

1. a. 1-nonanol   c. 1-propanol
   b. methanol   d. dimethyl ether
2. a
3. 3-bromo-4-methyl-2-pentanol
4. 5-chloro-2-pentanol
5. 1,2,3-propanetriol
6. a. propyl alcohol
   b. isopropyl alcohol
   c. *tert*-butyl alcohol
7. a. CH_3OH
   b. CH_3CH_2OH

8.

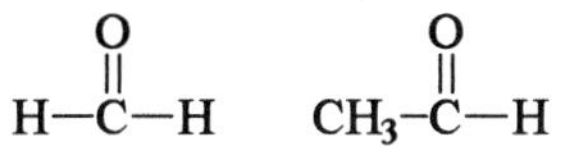

$$\underset{\text{formaldehyde}}{H-\overset{\overset{\textstyle O}{\|}}{C}-H} \quad \underset{\text{acetaldehyde}}{CH_3-\overset{\overset{\textstyle O}{\|}}{C}-H}$$

9. a. ethene + H_2O
   b. propene + H_2O
   c. 1-butene + H_2O + 2-butene + H_2O
      (minor product)   (major product)
10. a. formaldehyde, then to formic acid
    b. acetaldehyde, then to acetic acid
    c. acetone
    d. no reaction

11.

First:

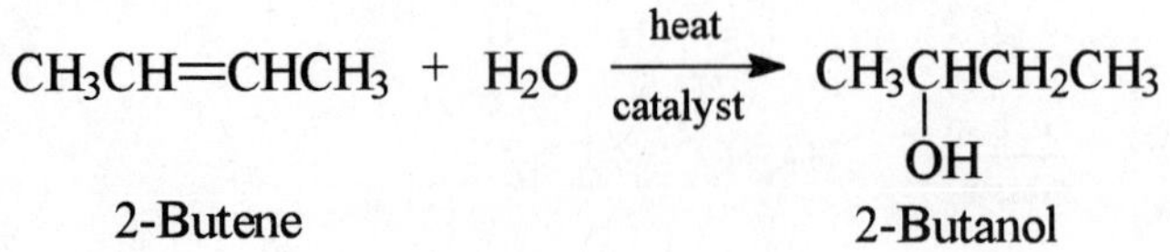

$$CH_3CH{=}CHCH_3 + H_2O \xrightarrow[\text{catalyst}]{\text{heat}} CH_3\underset{\underset{\text{OH}}{|}}{C}HCH_2CH_3$$

2-Butene   2-Butanol

Then:

$$CH_3\underset{\underset{\text{OH}}{|}}{C}HCH_2CH_3 \xrightarrow{[O]} CH_3\overset{\overset{\text{O}}{\|}}{C}CH_2CH_3$$

2-Butanol   2-Butanone

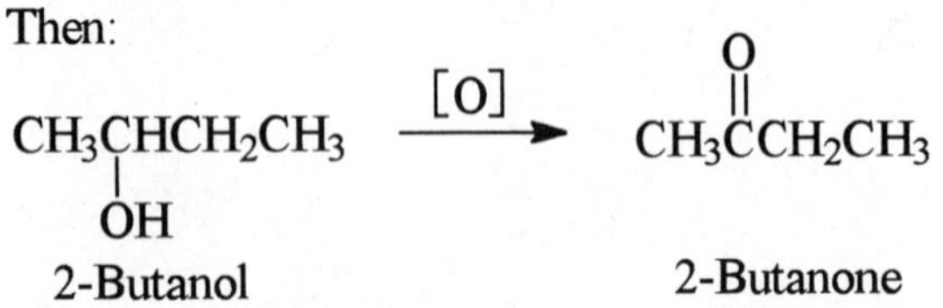

12. a. dimethyl ether
    b. diethyl ether
    c. ethyl methyl ether (or methyl ethyl ether)
13. d. acetic acid
14. a. methane
15. alcohol
16. thiol
17. They are polar and can form hydrogen bonds between alcohol molecules
18. ethanol
19. denatured alcohol
20. glycerol
21. 1
22. tertiary
23. oxidation
24. ethanal (acetaldehyde)
25. thiol or thiols
26. b. ethers
27. 0.1%
28. b. ethers

Chapter 14

1. a. 1-propanol
    b. propanal
2.

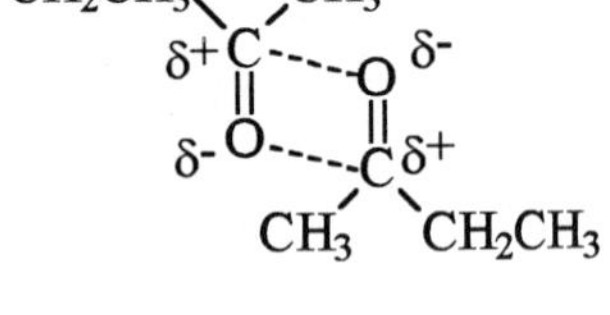

300

3.   a. ethanal   d. 3-pentanone

       b. propanone   e. 4-bromohexanal

       c. propanal

4.   The numbers are not needed.

5.

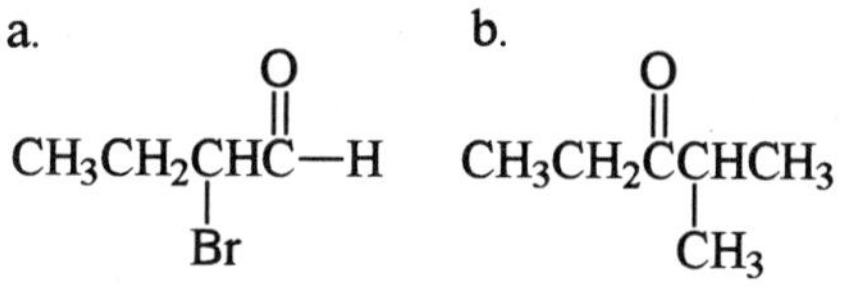

    a.

$$CH_3CH_2\underset{\underset{Br}{|}}{C}H\overset{\overset{O}{\|}}{C}-H$$

    b.

$$CH_3CH_2\overset{\overset{O}{\|}}{C}\underset{\underset{CH_3}{|}}{C}HCH_3$$

6.

$$CH_3OH \longrightarrow \overset{\overset{O}{\|}}{H}CH \longrightarrow HCOOH$$

   Methanol   Methanal   Methanoic acid

$$CH_3CH_2OH \longrightarrow CH_3\overset{\overset{O}{\|}}{C}H \longrightarrow CH_3COOH$$

   Ethanol   Ethanal   Ethanoic acid

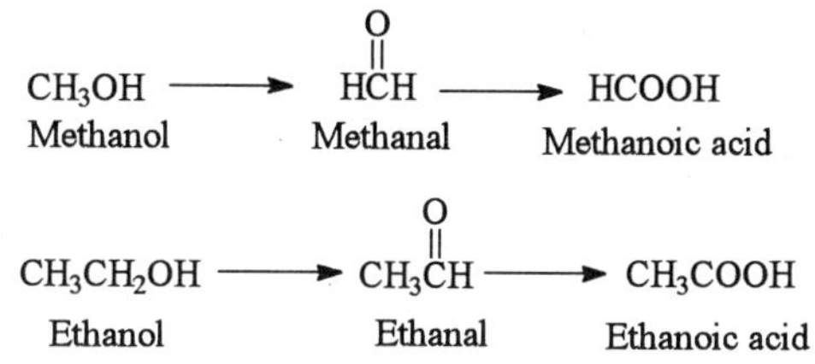

7.   a. HCOOH   d. CH_3CH_2COOH

       b. CH_3COOH e. no reaction

       c. no reaction

8.   a. ethanal (or ethanoic acid)

       b. propanal (or propanoic acid)

       c. propanone

       d. no reaction

9.   a. $CH_3COOH + Ag^0$

       b. No reaction: CH_3COCH_3 is a ketone.

       c. No reaction: $CH_3CH_2COCH_3$ is a ketone.

10. a. oxidation   c. reduction

      b. oxidation   d. oxidation

11.

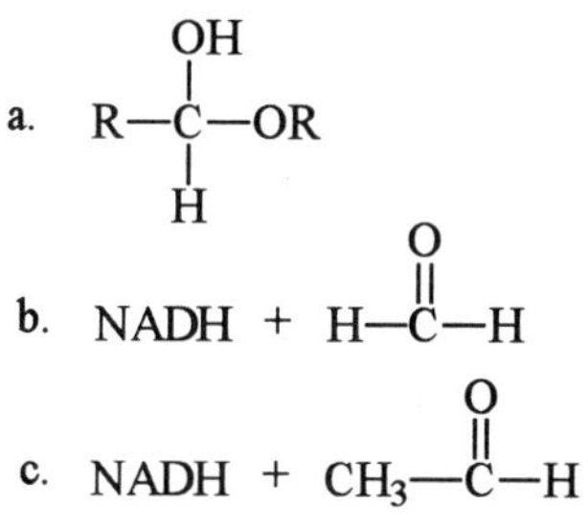

    a. $R-\underset{\underset{H}{|}}{\overset{\overset{OH}{|}}{C}}-OR$

    b. $NADH + H-\overset{\overset{O}{\|}}{C}-H$

    c. $NADH + CH_3-\overset{\overset{O}{\|}}{C}-H$

12.   aldehydes and ketones

13.   ketone

14. acetaldehyde
15. formalin
16. carboxylic acids
17. acetic acid
18. benzoic acid
19. Tollens' Test and Benedict's Test
20. acetal
21. vitamin A
22. the carbonyl group is polar
23. aldehydes
24. phenyl group
25. aldehydes
26. ketone

Chapter 15

1. The monosaccharides are a, b, c and f. Sucrose is a disaccharide, and starch is a polysaccharide.
2. a. aldohexose   d. aldopentose

     b. ketohexose   e. aldopentose

     c. aldohexose   f. aldotriose
3. enantiomers
4. cellulose
5. D-glucose
6. D-glyceraldehyde
7. α-amylase and β-amylase
8. glycosidic bond
9. An aldose has an aldehyde structure and a ketose contains a ketone structure.
10. A triose contains three carbon atoms and a pentose contains five carbon atoms.
11. a. D-glyceraldehyde   c. L-ribose

     b. D-ribose   d. D-deoxyribose
12. grape sugar, dextrose, or blood sugar
13. D-fructose
14. glycogen
15. a. glucose + H_2O → no reaction

     b. lactose + H_2O → glucose + galactose

     c. sucrose + H_2O → glucose + fructose

     d. starch + H_2O → many glucose molecules

     e. cellulose + H_2O → many glucose molecules
16. sucrose, a nonreducing sugar
17. carbohydrate(s)
18. humans cannot make the enzyme cellulase
19. lactose intolerance
20. D-glucose
21. alpha
22. hemiketal
23. ribose or D-ribose
24. $C_5H_{10}O_5$
25. galactosemia
26. reducing sugars
27. disaccharides
28. liver and muscle tissues

29. Benedict's or Tollens'
30. Benedict's or Tollens'

Chapter 16

1. a. ROH d. RCHO
   b. ROR e. RCOOH
   c. RCOOR   f. RCOR

2.

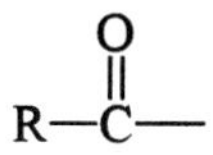

$$R-\overset{\overset{\displaystyle O}{\|}}{C}-$$

3. a. methanoic acid   c. ethanoic acid
   b. hexanoic acid   d. 3-methylbutanoic acid
4. a. CH$_3$COOH
   b. HCOOH
   c.

—COOH

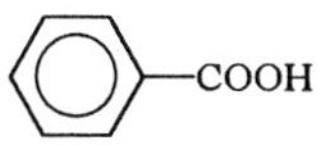

   d. C$_{11}$H$_{23}$COOH
5. a. butyric acid
   b. propionic acid
   c. α-methylpropionic acid
6. a. ethyl formate
   b. ethyl acetate
   c. pentyl acetate
7. a. methyl ethanoate
   b. ethyl propanoate
8. a. acetic acid
9. a. CH$_3$COO$^-$ + H$_3$O$^+$
   b. CH$_3$COO$^-$Na$^+$ + H$_2$O
   c.

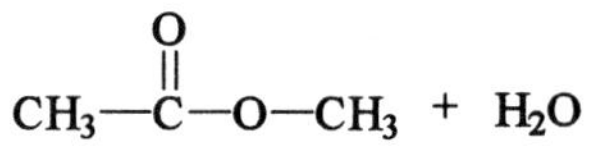

$$CH_3-\overset{\overset{\displaystyle O}{\|}}{C}-O-CH_3 \; + \; H_2O$$

10. alcohol + carboxylic acid
11. carboxylate ion + hydronium ion
12. a. sodium butyrate + water
    b. calcium acetate + water
13. soaps
14. hydrophobic end
15. methanol + propanoic acid
16. ATP
17. saponification

303

18. carboxyl group
19. formic acid
20. thioester
21. methyl ethanoate
22. Oxidation reaction. $KMnO_4$ is an oxidizing agent
23. carboxylic acid
24. Coenzyme A
25. water
26. hydrophilic
27. micelles
28. CH_3OH
29. phosphate ester or phosphoester
30. phosphoric anhydride bond or phosphoanhydride bond

Chapter 17

1. fatty acids
2. They are nonpolar molecules.
3. triglycerides (triacylglycerols)
4. They are long-chain monocarboxylic acids with an even number of carbon atoms.
5. glycerol + stearic acid + lauric acid + palmitic acid
6. long-chain alcohol + fatty acid
7. 1. chylomicrons
    2. very low-density lipoproteins (VLDL)
    3. low-density lipoproteins (LDL)
    4. high-density lipoproteins (HDL)
8. glycerol + 3 fatty acids
9. HDL
10. cholesterol
11. hydrophobic (nonpolar) fatty acid tail
12. cholesterol
13. the degree of saturation and carbon chain length
14. $C_{11}H_{23}COOH + NaOH \rightarrow C_{11}H_{23}COO^-Na^+ + H_2O$
15. the four fused rings known as the steroid nucleus
16. egg yolks, dairy products, liver, and other animal meats
17. progesterone
18. lipids
19. solids
20. ester + water
21. Ca^{2+} or Mg^{2+} or Fe^{3+}
22. saturated solid fats
23. prostaglandins, leukotrienes, and thromboxanes
24. dilation
25. nonionic and nonpolar
26. polar and nonpolar
27. lecithin
28. high-density lipoproteins
29. cortisone
30. birth control

Chapter 18

1. a. tertiary amine   c. primary amine
   b. quaternary ammonium salt   d. secondary amine
2. carboxyl groups and amino groups
3. a. ethanamine
   b. 2-propanamine
   c. N,N-dimethylmethanamine
4. aniline
5. a. $CH_3COO^-Na^+ + NH_3$
   b. $CH_3CH_2COOH + CH_3N^+H_3$
6. a. $(CH_3)_2NH_2^+I^-$
   b. $(CH_3)_3NH^+Cl^-$
7. a. RNH_2
   b. R_3N
8. a. ethanamide
   b. N-propylhexanamide
   c. N-methylbutanamide
9. nitrosoamines
10. brain, spinal cord and all nerves that radiate from the spinal cord
11. serotonin and acetylcholine
12. because the alkylammonium salts are more soluble in water and body fluids
13. Parkinson's Disease
14. amino group
15. ammonia
16. nitrogen
17. nucleic acids or DNA + RNA
18. trimethylamine
19. alkylammonium salt
20. methylammonium chloride
21. buffers
22. amide
23. solid
24. an amine + water
25. because they are detergents
26. quaternary ammonium salt
27. proteins
28. carcinogen
29. analgesic

Chapter 19

1. nitrogen and sulfur
2. glycine
3. phenylalanine
4. They are proteins that act as biological catalysts.
5. Antibodies, also called immunoglobulins, are specific protein molecules produced by special cells of the immune system in response to foreign antigens.
6. hemoglobin and myoglobin
7. keratin
8. a mixture of the amino acids

9.

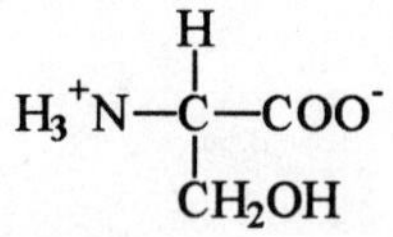

$$H_3{}^+N-\underset{\underset{CH_2OH}{|}}{\overset{\overset{H}{|}}{C}}-COO^-$$

10. aspartic acid (or aspartate) and glutamic acid (or glutamate)
11. lysine, arginine, and histidine
12. a. ala   d. his
    b. cys   e. lys
    c. gly
13. peptide bond or amide bond
14.

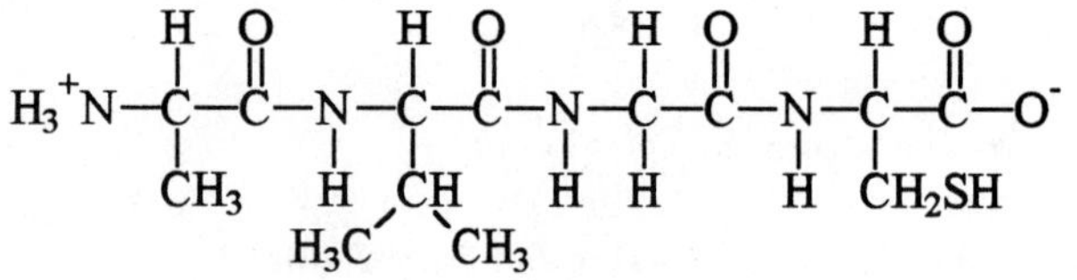

$$H_3{}^+N-\underset{CH_3}{\overset{H}{C}}-\overset{O}{C}-\underset{H}{N}-\underset{\underset{H_3C\ \ CH_3}{CH}}{\overset{H}{C}}-\overset{O}{C}-\underset{H}{N}-\underset{H}{\overset{H}{C}}-\overset{O}{C}-\underset{H}{N}-\underset{CH_2SH}{\overset{H}{C}}-\overset{O}{C}-O^-$$

15. primary structure
16. secondary structure
17. quaternary structure
18. protein molecule + additional nonprotein part
19. isoelectric
20. α-helix or helices
21. zwitterions
22. essential amino acids
23. meats
24. pepsin
25. nutrient protein
26. proline
27. β-pleated sheet
28. disulfide bond
29. A hydrophobic molecule is nonpolar and thus is not water-soluble.
30. When the pH drops too low, a protein becomes a polycation.

Chapter 20

1. The enzyme lowers the activation energy of the reaction.
2. substrate
3. $2H_2O_2\ (l) \rightarrow 2H_2O\ (l) + O_2\ (g)$
4. a. succinate
   b. sucrose
   c. glycogen

5.

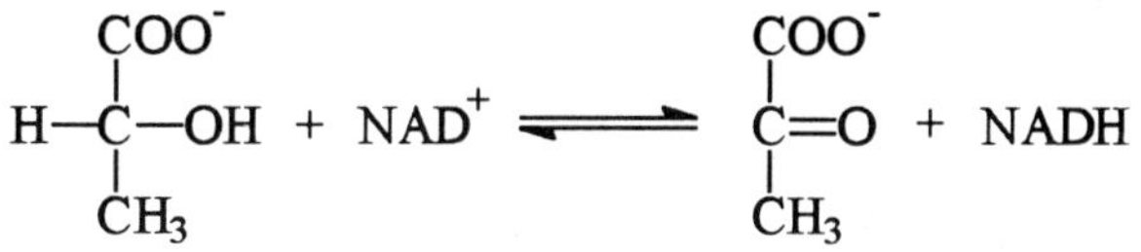

$$\begin{matrix} \text{COO}^- \\ | \\ \text{H}-\text{C}-\text{OH} \\ | \\ \text{CH}_3 \end{matrix} + \text{NAD}^+ \rightleftharpoons \begin{matrix} \text{COO}^- \\ | \\ \text{C}=\text{O} \\ | \\ \text{CH}_3 \end{matrix} + \text{NADH}$$

Lactate   Pyruvate

6. They catalyze the transfer of phosphate groups.
7. lipase
8.

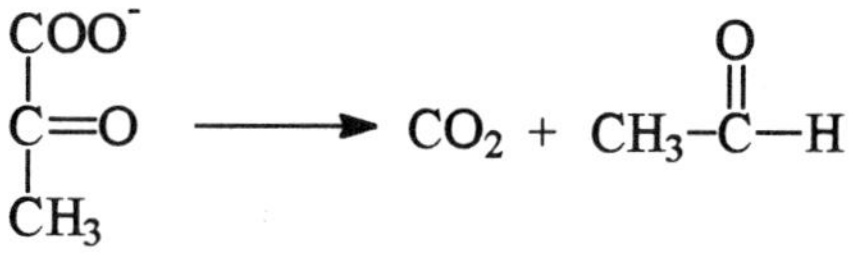

$$\begin{matrix} \text{COO}^- \\ | \\ \text{C}=\text{O} \\ | \\ \text{CH}_3 \end{matrix} \longrightarrow \text{CO}_2 + \text{CH}_3-\overset{\displaystyle \text{O}}{\overset{\displaystyle \|}{\text{C}}}-\text{H}$$

Pyruvate   Acetaldehyde

9. no effect
10. formation of enzyme-substrate complex
11. 1. absolute specificity 3. linkage specificity
    2. group specificity   4. stereochemical specificity
12. The protein portion of a conjugated enzyme is called its apoenzyme. The nonprotein part is known as the cofactor.
13. Usually it is nicotinamide adenine dinucleotide, as either the oxidized form (NAD^+) or its reduced form ($NADH + H^+ = NADH$)
14. Most are used in some form as coenzymes.
15. pH optimum
16. denaturation
17. 1. irreversible inhibitors
    2. competitive inhibitors
18. The sulfa drugs are competitive inhibitors of a bacterial enzyme required for the synthesis of the required vitamin folic acid.
19. They are proteases, digestive enzymes that hydrolyze peptide bonds in proteins.
20. The rate of reaction increases until, at a certain substrate concentration, the rate reaches a maximum.
21. The active sites of all the enzyme molecules in solution are occupied by substrate molecules.
22. absolute specificity
23. in the enzyme active site
24. the optimum temperature (37°C for most enzymes found in the human body)
25. effector binding
26. generally, the end-product of the pathway
27. zymogen
28. a prothrombin activator

Chapter 21

1. glycolysis
2. ATP
3. 7 kcal/mole
4. anaerobic
5. two
6. NADH (NADH + H$^+$)
7. ethanol
8. lactate
9. carbohydrates
10. glycolysis
11. amino acids
12. catabolism
13. adenosine triphosphate, or ATP
14. hydrolysis
15. phosphorylation
16. fructose-6-phosphate
17. cytoplasm
18. substrate-level phosphorylation
19. adenosine triphosphate or ATP
20. glucose-6-phosphate
21. amylase
22. lactase
23. pepsin
24. small intestine or duodenum
25. duodenum
26. a. glucose
27. e. glyceraldehyde-3-phosphate
28. c. pyruvate
29. F
30. T

Chapter 22

1. mitochondria
2. energy need
3. oxaloacetate
4. oxaloacetate
5. two
6. three
7. one
8. two
9. mitochondria
10. electron transport system
11. cytochromes
12. hydrogen ions or protons
13. protons
14. F$_1$ catalyzes the phosphorylation of ADP to produce ATP.
15. NAD$^+$
16. It returns to be used again in the Krebs cycle.

17. oxygen, or O_2
18. cytochromes
19. three
20. two
21. the diet
22. liver
23. glutamate, or glutamic acid
24. citric acid cycle, or Krebs cycle
25. oxalocetate and glutamate
26. liver
27. carbamoyl phosphate
28. amphibolic
29. matrix
30. deamination

Chapter 23

1. acetyl coenzyme A
2. acetyl coenzyme A
3. fat globules
4. bile
5. micelle
6. cholic acid and chenodeoxycholic acid
7. colipase
8. chylomicrons
9. adipose tissue
10. 1 $FADH_2$ and 1 NADH
11. adipocyte
12. 129 ATP molecules
13. The abnormal rise in concentration of blood ketone bodies.
14. Starvation, a diet that is extremely low in carbohydrates, and the disease diabetes mellitus.
15. ketoacidosis
16. liver
17. heart muscle
18. fatty acid synthase
19. NADPH
20. liver
21. Glycerol-3-phosphate is synthesized from glyceraldehyde-3-phosphate produced during glycolysis.
22. VLDL
23. amino acid carbon skeletons
24. glycerol-3-phosphate
25. β-oxidation of fatty acids
26. acyl CoA
27. acetyl CoA
28. β-hydroxybutyrate and acetoacetate
29. anaerobic
30. liver, adipose, and muscle cells

Chapter 24

1. DNA

2. DNA
3. a. dATP
   b. CTP
   c. dTMP
4. DNA → RNA → protein
5. Watson and Crick
6. prime
7. hydrogen bonding between A and T and G and C
8. right angle
9. thymine
10. 3' → 5' direction
11. One strand specifies the sequence of bases on the other.
12. thymine dimers
13. The double helix slowly unwinds.
14. replication origin
15. replication fork
16. (1) It reads the parental strand and produces the daughter strand. (2) It proofreads the newly synthesized daughter strand to ensure that no errors have been made.
17. cancer
18. (1) Messenger RNA—carries the genetic information for protein from DNA to the ribosomes; (2) Ribosomal RNA—a structural and functional component of the ribosomes; (3) Transfer RNA—responsible for translating the genetic code of the mRNA into the primary structure of a protein.
19. transfer RNA
20. RNA polymerase
21. promoter
22. intervening sequences, or introns
23. exons
24. codons
25. Crick
26. methionine and tryptophan
27. glutamic acid
28. xeroderma pigmentosum
29. a small and a large ribosomal subunit
30. polysomes, or polyribosomes
31. silent mutation
32. restriction enzyme
33. phage vectors and plasmid vectors
34. hybridization
35. insulin, human growth hormone, blood clotting factors VIII and IX, interferon, hepatitis B virus vaccine

B

Chapter 1
Chemistry: Methods and Measurement
Solutions to the Odd-Numbered Problems

****

In-Chapter Questions and Problems

****

1.1   a. $1 \text{ L} \times \dfrac{10^3 \text{ mL}}{\text{L}} = 1 \times 10^3 \text{ mL}$

        b. $1 \text{ L} \times \dfrac{10^6 \text{ uL}}{1 \text{ L}} = 1 \times 10^6 \text{ uL}$

        c. $1 \text{ L} \times \dfrac{10^{-3} \text{ kL}}{1 \text{ L}} = 1 \times 10^{-3} \text{ kL}$

        d. $1 \text{ L} \times \dfrac{10^2 \text{ cL}}{1 \text{ L}} = 1 \times 10^2 \text{ cL}$

        e. $1 \text{ L} \times \dfrac{10^{-1} \text{ daL}}{1 \text{ L}} = 1 \times 10^{-1} \text{ daL}$

1.3   a. $0.50 \text{ in} \times \dfrac{2.54 \text{ cm}}{1 \text{ in}} \times \dfrac{1 \text{ m}}{10^2 \text{ cm}} = 1.3 \times 10^{-2} \text{ m}$

        b. $0.75 \text{ qt} \times \dfrac{0.946 \text{ L}}{1 \text{ qt}} = 0.71 \text{ L}$

        c. $56.8 \text{ g} \times \dfrac{1 \text{ lb}}{454 \text{ g}} \times \dfrac{16 \text{ oz}}{1 \text{ lb}} = 2.00 \text{ oz}$

1.5   a.   Three (all non-zero digits)

b. Three (all non-zero digits)

c. Four (zeros between non-zero digits are significant)

d. Two (trailing zero significant due to decimal)

e. Three (leading zeros are not significant)

1.7 a. 2.4×10^{-3}

b. 1.80×10^{-2}

c. 2.24×10^{2}

1.9 a. 8.09 (3 significant figures)

b. 5.9 (2 significant figures)

c. 20.19 (4 significant figures)

1.11 a. 51 (2 significant figures)

b. 8.0×10^{1} (2 significant figures)

c. 1.6×10^{2} (2 significant figures)

1.13 a. 61.404 rounds to 61.4

b. 6.7174 rounds to 6.17

c. 0.066494 rounds to 0.0665 (or, 6.65×10^{-2})

1.15 a. $^{\circ}C = \dfrac{^{\circ}F - 32}{1.8} = \dfrac{32 - 32}{1.8} = 0 \ ^{\circ}C$

b. $K = \ ^{\circ}C + 273 = 0 + 273 = 273 \ K$

1.17 $\text{g alcohol} = 30.0 \ \text{mL alcohol} \times \dfrac{0.789 \ \text{g alcohol}}{1 \ \text{mL alcohol}}$

g alcohol = 23.7 g alcohol

End-of-Chapter Questions and Problems

1.19   a.   Chemistry is the study of matter and the changes that matter undergoes.

      b.   Matter is the material component of the universe.

      c.   Energy is the ability to do work.

      d.   A hypothesis is an "educated guess" at the explanation of observed behavior of our surroundings.

      e.   A theory is a hypothesis developed to explain behavior of matter that has been verified by using the scientific method.

      f.   A law is a statement of observed behavior for which no exceptions have been found.

1.21   a.   gram (or kilogram)
      b.   liter
      c.   meter

1.23   Weight is the force exerted on a body by gravity; mass is a quantity of matter. Mass is an independent quantity while weight is dependent on gravity which may differ from location to location.

1.25   Density is mass per volume. Specific gravity is the ratio of the density of a substance to the density of water at 4°C.

1.27   The scientific method is an organized way of doing science. It uses carefully planned experimentation to study our surroundings.

1.29   a. $2.0 \text{ lb} \times \dfrac{16 \text{ oz}}{1 \text{ lb}} = 32 \text{ oz}$

      b. $2.0 \text{ lb} \times \dfrac{1 \text{ t}}{2000 \text{ lb}} = 1.0 \times 10^{-3} \text{ t}$

      c. $2.0 \text{ lb} \times \dfrac{454 \text{ g}}{1 \text{ lb}} = 9.1 \times 10^{2} \text{ g}$

d. $2.0 \text{ lb} \times \dfrac{454 \text{ g}}{1 \text{ lb}} \times \dfrac{10^3 \text{ mg}}{1 \text{ g}} = 9.1 \times 10^5 \text{ mg}$

e. $2.0 \text{ lb} \times \dfrac{454 \text{ g}}{1 \text{ lb}} \times \dfrac{1 \text{ da}}{10^1 \text{ g}} = 9.1 \times 10^1 \text{ da}$

1.31   a. $3.0 \text{ g} \times \dfrac{1 \text{ lb}}{454 \text{ g}} = 6.6 \times 10^{-3} \text{ lb}$

b. $3.0 \text{ g} \times \dfrac{1 \text{ lb}}{454 \text{ g}} \times \dfrac{16 \text{ oz}}{1 \text{ lb}} = 1.1 \times 10^{-1} \text{ oz}$

c. $3.0 \text{ g} \times \dfrac{1 \text{ kg}}{10^3 \text{ g}} = 3.0 \times 10^{-3} \text{ kg}$

d. $3.0 \text{ g} \times \dfrac{1 \text{ cg}}{10^{-2} \text{ g}} = 3.0 \times 10^2 \text{ cg}$

e. $3.0 \text{ g} \times \dfrac{1 \text{ mg}}{10^{-3} \text{ g}} = 3.0 \times 10^3 \text{ mg}$

1.33   a. $^\circ C = \dfrac{^\circ F - 32}{1.8} = \dfrac{50 - 32}{1.8} = 10^\circ C$

b. $K = {}^\circ C + 273 = 10 + 273 = 283 K$

1.35   a. $K = {}^\circ C + 273 = 20 + 273 = 293 K$

b. If $^\circ C = \dfrac{^\circ F - 32}{1.8}$ then, $1.8 {}^\circ C = {}^\circ F - 32$

and $^\circ F = (1.8 \, {}^\circ C) + 32$ therefore, $^\circ F = [(1.8)(20)] + 32 = 68 \, {}^\circ F$

1.37 $9 \text{ pt} \times \dfrac{1 \text{ qt}}{2 \text{ pt}} \times \dfrac{0.946 \text{ L}}{1 \text{ qt}} = \approx 4 \text{ L (1 significant figure)}$

1.39   If $^\circ C = \dfrac{^\circ F - 32}{1.8}$ then $^\circ F = 1.8 \, {}^\circ C + 32$

$^\circ F = 1.8 \,(38.5) + 32 = 101 \, {}^\circ F$

314

1.41   a. 3
        b. 3
        c. 3
        d. 4
        e. 4
        f. 3

1.43   a. 3.87×10^{-3}

        b. 5.20×10^{-2}

        c. 2.62×10^{-3}

        d. 2.43×10^{-1}

        e. 2.40×10^{2}

        f. 2.41×10^{0}

1.45   a. $(23)(657) = 1.5 \times 10^{4}$

        b. $0.00521 + 0.236 = 2.41 \times 10^{-1}$

        c. $\dfrac{18.3}{3.0576} = 5.99$

        d. $1157.23 - 17.812 = 1139.42$

        e. $\dfrac{(1.987)(298)}{0.0821} = 7.21 \times 10^{3}$

1.47   a. 1.23×10^{1}

        b. 5.69×10^{-2}

        c. -1.527×10^{3}

        d. 7.89×10^{-7}

        e. 9.2×10^{7}

        f. 5.280×10^{-3}

        g. 1.279×10^{0}

h. -5.3177×10^2

1.49   a.   3,240

        b.   0.000150

        c.   0.4579

        d.   -683,000

        e.   -0.0821

        f.   299,790,000

        g.   1.50

        h.   602,000,000,000,000,000,000,000

1.51 $d = \dfrac{m}{V} = \dfrac{3.00 \times 10^2\,g}{50.0\ mL} = 6.00\ g/mL$

1.53 $1.50 \times 10^2\ \cancel{mL} \times \dfrac{7.20\ g}{1\,\cancel{mL}} = 1.08 \times 10^3\ g$

1.55 $d = \dfrac{m}{V} = \dfrac{98\,g}{1.00 \times 10^2\,cm^3} = 9.8 \times 10^{-1}\ g/cm^3$, therefore, teak

1.57   Specific gravity of alcohol $= \dfrac{density\ of\ alcohol}{density\ of\ water\ at\ 4^{oC}}$

        Specific gravity of alcohol $= \dfrac{0.789\ g/mL}{1.00\ g/mL} = 0.789$

1.59 $d_{lead} = \dfrac{5.0 \times 10^1\ g}{6.36\ cm^3} = 7.9\ g/cm^3$

        $d_{uranium} = \dfrac{75\ g}{3.97\ cm^3} = 19\ g/cm^3$

        $d_{platinum} = \dfrac{2140\ g}{1.00 \times 10^2\ cm^3} = 21.4\ g/cm^3$

Lead has the lowest density and platinum has the greatest density.

Chapter 2
The Structure of the Atom
Solutions to the Odd-Numbered Problems

****
In-Chapter Questions and Problems
****

2.1   a.   physical property
       b.   chemical property
       c.   physical property
       d.   physical property
       e.   physical property

2.3   a.   pure substance
       b.   heterogeneous mixture
       c.   homogeneous mixture
       d.   pure substance

2.5   a.   16 protons and 16 electrons (atomic number = 16)
            32 - 16 = 16 neutrons (mass number - atomic number)
       b.   11 protons and 11 electrons (atomic number = 11)
            23 - 11 = 12 neutrons (mass number - atomic number)

2.7   Step 1. Convert each percentage to a decimal fraction.

$$90.51\% \ ^{20}_{10}\text{Ne} \ \text{x} \ \frac{1}{100\%} = 0.9051 \ ^{20}_{10}\text{Ne}$$

$$0.27\% \ ^{21}_{10}\text{Ne} \ \text{x} \ \frac{1}{100\%} = 0.0027 \ ^{21}_{10}\text{Ne}$$

$$9.22\% \ ^{22}_{10}\text{Ne} \ \text{x} \ \frac{1}{100\%} = 0.0922 \ ^{22}_{10}\text{Ne}$$

Step 2.

| contributions to atomic mass by $^{20}_{10}\text{Ne}$ | = | (fraction of all Ne atoms that are $^{20}_{10}\text{Ne}$) | x | (mass of a $^{20}_{10}\text{Ne}$ atom) |

$$= \ 0.9051 \ \text{x} \ 19.99 \ \text{amu}$$
$$= \ 18.091 \ \text{amu}$$

contributions to = (fraction of all Ne atoms x (mass of a $^{21}_{10}Ne$
atomic mass by $^{21}_{10}Ne$ that are $^{21}_{10}Ne$)   atom)

           = 0.0027 x 20.99 amu
           = 0.057 amu

contributions to = (fraction of all Ne atoms x (mass of a $^{22}_{10}Ne$
atomic mass by $^{22}_{10}Ne$ that are $^{22}_{10}Ne$)   atom)

           = 0.9022 x 21.99 amu
           = 2.03 amu

Step 3. The weighted average is:

atomic mass of = (contribution of $^{20}_{10}Ne$) + (contribution of $^{21}_{10}Ne$)
naturally occurring
neon   + (contribution of $^{22}_{10}Ne$)

           = 18.091 + 0.057 amu + 2.03 amu
           = 20.18 amu

2.9   DeBroglie considered electrons to have both wave and particle properties.

****
End-of-Chapter Questions and Problems
****

2.11   A physical property is a characteristic of a substance that can be observed without the substance undergoing a change in chemical composition.

2.13   a. chemical reaction
           b. physical change
           c. physical change
           d. chemical reaction
           e. chemical reaction

2.15   Chemical properties of matter include flammability and toxicity.

2.17   A pure substance has constant composition with only a single substance whereas a mixture is composed of two or more substances.

2.19   Mixtures are composed of two or more substances. A homogeneous mixture has uniform composition while a heterogeneous mixture has non-uniform composition.

2.21   A gas is made up of particles that are widely separated. A gas will expand to fill any container and it has no definite shape or volume.

2.23   An intensive property is a characteristic of a substance that is independent of the quantity of the substance. An extensive depends on the quality of the substance.

2.25   An element is a pure substance that cannot be changed into a simpler form of matter by any chemical reaction. An atom is the smallest unit of an element that retains the properties of that element.

2.27   a.   Atomic number = 8, therefore 8 protons and 8 electrons. Mass number - atomic number = 16 - 8 = 8, therefore 8 neutrons.
        b.   Atomic number = 15, therefore 15 protons and 15 electrons. Mass number - atomic number = 31 - 15 = 16, therefore 16 neutrons.

2.29   Isotopes are atoms of the same element that differ in mass due to the fact that they contain different numbers of neutrons.

2.31

	Particle	Mass	Charge
a.	electron	≈ 0 amu	-1
b.	proton	≈ 1 amu	$+1$
c.	neutron	≈ 1 amu	0

2.33   a.   An ion is a charged atom or group of atoms formed by the loss or gain of electrons.
        b.   A loss of electrons by a neutral species results in a cation.
        c.   A gain of electrons by a neutral species results in an anion.

2.35

	Atomic Symbol	# Protons	# Neutrons	# Electrons	Charge
a.	$^{23}_{11}Na$	11	12	11	0
b.	$^{32}_{16}S^{2-}$	16	16	18	2-
c.	$^{16}_{8}O$	8	8	8	0
d.	$^{24}_{12}Mg^{2+}$	12	12	10	2+
e.	$^{39}_{19}K^{1+}$	19	20	18	1+

2.37   The major postulates of Dalton's atomic theory include the following:

♦   All matter consists of tiny particles called atoms.
♦   Atoms cannot be created, divided, destroyed, or converted to any other type of atom.
♦   All atoms of a particular element have identical properties.
♦   Atoms of different elements have different properties.
♦   Atoms combine in simple whole-number ratios.
♦   Chemical change involves joining, separating, or rearranging atoms.

2.39   a.   James Chadwick demonstrated the existence of the neutron in 1932. He accomplished this with a series of experiments using small nuclei as projectiles to study the nucleus.
        b.   Louis DeBroglie theorized that electrons had wave-like as well as particle-like properties; this theory is known as *wave-particle duality*.
        c.   Hans Geiger provided the basic experimental evidence for the existence of the nucleus. A small, dense, positively charged region within the atom was indicated by his alpha-particle scattering experiment.
        d.   The Bohr theory describes electron arrangement in atoms. Bohr proposed an atomic model that depicted the atom as a nucleus surrounded by fixed energy levels that can be occupied by electrons. He believed that each level was defined by a circular orbit located at some specified distance from the nucleus. Electron promotion from a lower to higher energy level results from absorption of energy that produces an excited state atom. The process of relaxation allows the atom to return to the ground state (the electron falls from a higher to lower energy level) and energy is released.

2.41   Our understanding of the nucleus is based on the gold foil experiment performed by Geiger and interpreted by Rutherford. In this experiment, Geiger bombarded a piece of gold foil with alpha particles, and observed that some alpha particles passed straight through the foil, others were deflected and some simply bounced back. This led Rutherford to propose that the atom consisted of a small, dense nucleus (alpha particles bounced back), surrounded by a cloud of electrons (some alpha particles were deflected). The size of the nucleus is small when compared to the volume of the atom (alpha particles were able to pass through the foil).

2.43   The most important points of Bohr's theory are:

♦   Electrons are found in orbits at discrete distances from the nucleus.
♦   The orbits are quantitized - they are of discrete energies.
♦   Electrons can only be found in these orbits, never in between (they are able to jump instantaneously from orbit to orbit).
♦   Electrons can undergo transitions - if an electron absorbs energy, it will jump to a higher orbit; when the electron falls back to a lower orbit, it will release energy.

2.45 Crookes used the cathode ray tube. He observed particles emitted by the cathode and traveling toward the anode. This ray was deflected by an electric field. Thomson measured the curvature of the ray influenced by the electric field. This measurement provided the mass to charge ratio of the negative particle. Thomson also gave the particle the name, electron.

2.47 A cathode ray is the negatively charged particle formed in a cathode ray tube. It was characterized as an electron, with a mass of zero and a charge of −1.

2.49 a. An isotope of an element differs in mass because the atom has a different number of neutrons.
   b. The atomic number gives the number of protons in the nucleus.
   c. The atomic mass of an atom is due to the number of protons and neutrons in the nucleus.
   d. A charged atom is called a(n) ion.
   e. Electrons surround the nucleus and have a negative charge.

2.51
radiowave	↑	
microwave		
infrared		
visible		Increasing
ultraviolet		Wavelength
x-ray		
gamma ray		

2.53 The DeBroglie hypothesis stated that the electron has both particle-like and wave-like properties. This concept became the basis for developments that ultimately led to our modern atomic theory.

2.55 Bohr's atomic model was the first to successfully account for electronic properties of atoms, specifically, the interaction of atoms and light (spectroscopy). Although significantly modified, it still remains a useful model to predict bonding in simple systems.

2.57 An orbit is a fixed, defined region of space in which an electron can be found. An orbital is a three-dimensional region of space defined by the probability of finding an electron.

Chapter 3
Elements, Atoms, and the Periodic Table
Solutions to the Odd-Numbered Problems

****

In-Chapter Questions and Problems

****

3.1   a. Zr (zirconium)
        b. 2299
        c. Cr (chromium)
        d. Bi (bismuth)

3.3   a. helium, atomic number = 2, mass = 4.00 amu
        b. fluorine, atomic number = 9, mass = 19.00 amu
        c. manganese, atomic number = 25, mass = 54.94 amu

3.5   a. Total electrons = 11, valence electrons = 1
        b. Total electrons = 12, valence electrons = 2
        c. Total electrons = 16, valence electrons = 6
        d. Total electrons = 17, valence electrons = 7
        e. Total electrons = 18, valence electrons = 8

3.7   a. Sulfur: $1s^2, 2s^2, 2p^6, 3s^2, 3p^4$
        b. Calcium: $1s^2, 2s^2, 2p^6, 3s^2, 3p^6, 4s^2$

3.9   a. Ca^{2+} and Ar are isoelectronic
        b. Sr^{2+} and Kr are isoelectronic
        c. S^{2-} and Ar are isoelectronic
        d. Mg^{2+} and Ne are isoelectronic
        e. P^{-3} and Ar are isoelectronic

3.11   a. (smallest) F, N, Be (largest)
        b. (lowest) Be, N, F (highest)
        c. (lowest) Be, N, F, (highest)

****

End-of-Chapter Questions and Problems

****

3.13   a. <u>periodic law</u> - elemental properties are periodic as a function of their atomic number.
        b. <u>period</u> - a horizontal row across the periodic table.
        c. <u>group</u> - a vertical column on the periodic table.
        d. <u>ion</u> - a charged unit resulting from the gain or loss of electrons from a neutral atom or

group of atoms.

3.15   a. true
        b. true

3.17   a. The metals are: Na, Ni, Al
        b. The representative metals are: Na, Al
        c. The elements that tend to form positive ions are: Na, Ni, Al
        d. The element that is inert is: Ar

3.19   a. sodium
        b. potassium
        c. magnesium

3.21   Group IA is known collectively as the alkali metals and consists of lithium, sodium, potassium, rubidium, cesium, and francium.

3.23   Group VIIA is known collectively as the halogens and consists of fluorine, chlorine, bromine, iodine, and astatine.

3.25   The early periodic table contained many fewer elements. We now know of over one hundred elements. The early periodic table was arranged by atomic weight. The modern table is arranged by atomic number.

3.27   A metalloid is an element along the "staircase" boundary between metals and nonmetals; metalloids exhibit both metallic and nonmetallic properties.

3.29   a. one
        b. one
        c. three
        d. seven
        e. zero (or eight)
        f. zero (or two)

3.31   All of the elements in group IA have one valence electron located in an s orbital. All have an outermost electron configuration of ns^1.

3.33   The electron capacity of a shell is $2n^2$, where n is the number of the shell or principal energy level.

3.35   A principal energy level is designated n = 1, 2, 3, and so forth. It is similar to Bohr's orbits in concept. A sublevel is a part of a principal energy level and is designated s, p, d, and f.

3.37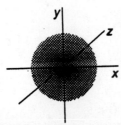

The s orbital represents the probability of finding an electron in a region of space surrounding the nucleus.

3.39 Three p orbitals (p_x, p_y, p_z) can exist in a given principal energy level.

3.41 A 3p orbital is a higher energy orbital than a 2p orbital because it is a part of a higher energy principal energy level.

3.43 $2n^2 = 2(1)^2 = 2\bar{e}$ for n = 1
$2n^2 = 2(2)^2 = 8\bar{e}$ for n = 2
$2n^2 = 2(3)^2 = 18\bar{e}$ for n = 3

3.45
  a. Al (13$\bar{e}$) $1s^2, 2s^2, 2p^6, 3s^2, 3p^1$ (hence, a p orbital)
  b. Na (11$\bar{e}$) $1s^2, 2s^2, 2p^6, 3s^2$ (hence, an s orbital)
  c. Sc (21$\bar{e}$) $1s^2, 2s^2, 2p^6, 3s^2, 3p^6, 4s^2, 3d^1$ (hence, a d orbital)
  d. Ca (20$\bar{e}$) $1s^2, 2s^2, 2p^6, 3s^2, 3p^6, 4s^2$ (hence, an s orbital)
  e. Fe (26$\bar{e}$) $1s^2, 2s^2, 2p^6, 3s^2, 3p^6, 4s^2, 3d^6$ (hence, an d orbital)
  f. Cl (17$\bar{e}$) $1s^2, 2s^2, 2p^6, 3s^2, 3p^5$ (hence, an p orbital)

3.47
  a. Li^+ (Li loses 1$\bar{e}$ to attain outermost octet)
  b. O^{2-} (O gains 2$\bar{e}$ to attain outermost octet)
  c. Ca^{2+} (Ca loses 2$\bar{e}$ to attain outermost octet)
  d. Br^- (Br gains 1$\bar{e}$ to attain outermost octet)
  e. S^{2-} (S gains 2$\bar{e}$ to attain outermost octet)
  f. Al^{3+} (Al loses 3$\bar{e}$ to attain outermost octet)

3.49
  a. O^{2-}, 10$\bar{e}$; Ne, 10$\bar{e}$; Isoelectronic
  b. S^{2-}, 18$\bar{e}$; Cl^-, 18$\bar{e}$; Isoelectronic

3.51 Group IA metals form only a 1+ ion because the loss of one electron produces an electron configuration similar to their nearest noble gas. Group IIA metals form only a 2+ ion because the loss of two electrons produces an electron configuration similar to their nearest noble gas.

3.53
  a. Na^+; it has a noble gas electron configuration
  b. S^{2-}; it has a noble gas electron configuration
  c. Cl^-; it has a noble gas electron configuration

3.55
  a. (Smallest) F, O, N (Largest)

b. (Smallest) Li, K, Cs (Largest)
c. (Smallest) Cl, Br, I (Largest)

3.57   From an inspection of Figure 3.6, Cl.

3.59   a. (Smallest) O, N, F (Largest)
        b. (Smallest) Li, K, Cs (Largest)
        c. (Smallest) Cl, Br, I (Largest)

3.61   A positive ion is always smaller than its parent atom because the positive charge of the nucleus is shared among fewer electrons in the ion. As a result, each electron is pulled closer to the nucleus and the volume of the ion decreases.

3.63   The fluoride ion has a completed octet of electrons and an electron configuration resembling its nearest noble gas.

Chapter 4
Structure and Properties of Ionic and Covalent Compounds
Solutions to the Odd-Numbered Problems

****

In-Chapter Questions and Problems

****

4.1   a. $LiBr$ (one Li^+ and one Br^+)

            b. $CaBr_2$ (one Ca^{2+} and two Br^-)

            c. Ca_3N_2 (three Ca^{2+} and two N^{3-})

4.3   a.   potassium cyanide
            b.   magnesium sulfide
            c.   lithium acetate

4.5   a. $CaCO_3$
            b. $NaHCO_3$

4.7   a.   diboron trioxide
            b.   nitrogen oxide
            c.   iodine chloride
            d.   phosphorus trichloride

4.9   a. P_2O_5
            b. SiO_2

4.11   a.

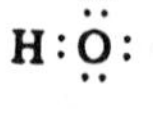

$$
\begin{array}{c}
\text{H} \\
\text{H} : \overset{\textstyle ..}{\underset{\textstyle ..}{\text{O}}} :
\end{array}
$$

            b.

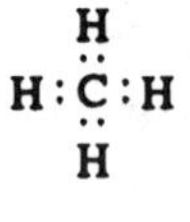

$$
\begin{array}{c}
\text{H} \\
\text{H} : \overset{\textstyle ..}{\underset{\textstyle ..}{\text{C}}} : \text{H} \\
\text{H}
\end{array}
$$

4.13   a.

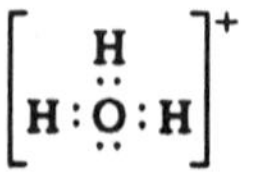

$$
\left[\begin{array}{c}
\text{H} \\
\text{H} : \overset{\textstyle ..}{\underset{\textstyle ..}{\text{O}}} : \text{H}
\end{array} \right]^{+}
$$

            b.

$$
\left[: \overset{\textstyle ..}{\underset{\textstyle ..}{\text{O}}} : \text{H} \right]^{-}
$$

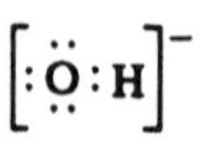

4.15   a.   The bonded nuclei are closer together when a double bond exists, in comparison to a single bond.

b.   The bond strength increases as the bond order increases. Therefore, a double bond is stronger than a single bond.

4.17   The Lewis Structures of the two resonance forms of SeO_2 are:

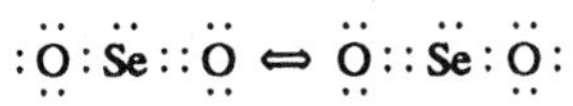

:Ö:Se::Ö ⇌ Ö::Se:Ö:

4.19   a.

H:P̈:H   P̈
   H   H—|—H
            H

Three groups and one lone pair of electrons surround the phosphorus atom; the structure is trigonal pyramidal (similar to the structure of ammonia).

b.
       H   H
H:S̈i:H   Si
       H   H—|—H
                   H

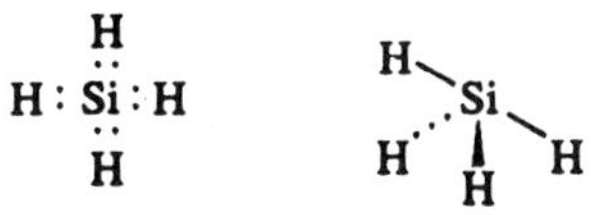

Four groups surround the silicon atom; the structure is tetrahedral (similar to the structure of methane).

4.21   a.   O - S   Oxygen is more electronegative than sulfur; the bond is polar. The electrons are pulled toward the oxygen atom.

b.   C ≡ N   Nitrogen is more electronegative than carbon; the bond is polar. The electrons are pulled toward the nitrogen atom.

c.   Cl - Cl   There is no electronegativity difference between two identical atoms; the bond is nonpolar.

d.   I - Cl   Chlorine is more electronegative than iodine; the bond is polar. The electrons are pulled toward the chlorine atom.

4.23   a.   Cl   Cl
            \   /
             B
             |
             Cl

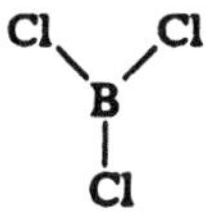

Three groups surround the central atom forming 120° bond angles. Due to the symmetrical arrangement of the three B-Cl bonds, their polarities cancel and the molecule is nonpolar.

b.

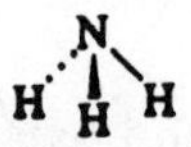

Three groups and a lone pair of electrons surround the central atom. Due to the effect of the lone pair, the molecule is polar.

c.

H—Cl

The H-Cl bond is polar due to the electronegativity difference between hydrogen and chlorine. Since H-Cl is the only bond in the molecule, the molecule is polar.

d.

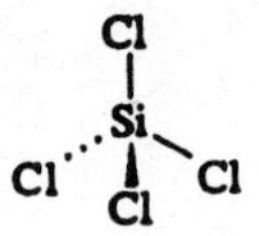

Four groups, all equivalent, surround the central atom. the structure is tetrahedral and the molecule is nonpolar.

4.25   a. H_2O is polar ← higher melting and boiling point.
            C_2H_4 is nonpolar

        b.   CO is polar ← higher melting and boiling point.
            CH_4 is nonpolar

        c. NH_3 is polar ← higher melting and boiling points.
            N_2 is nonpolar

        d. Cl_2 is nonpolar.
            ICl is polar ← higher melting and boiling points.

****

End-of-Chapter Questions and Problems

****

4.27   a.   Ionic
        b.   Covalent
        c.   Covalent
        d.   Covalent

4.29   a. $Li \cdot + \cdot \ddot{Br}: \longrightarrow Li^+ + : \ddot{Br}:^-$

        b. $Mg: + 2 \cdot \ddot{Cl}: \longrightarrow Mg^{2+} + 2 : \ddot{Cl}:^-$

4.31   a.   NCl₃
$$:\overset{..}{\underset{..}{Cl}}:\overset{..}{N}:\overset{..}{\underset{..}{Cl}}:$$
$$:\overset{..}{\underset{..}{Cl}}:$$

N = 5 valence electrons
3 x Cl -> 3 x 7 = 21 valence electrons
Total of 26 valence electrons

This structure satisfies the octet rule for N and Cl.

b.   CH₃OH

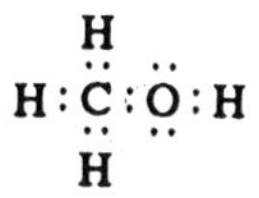

$$\begin{array}{c} H \\ H:\overset{H}{\underset{H}{C}}:\overset{..}{O}:H \end{array}$$

C = 4 valence electrons
O = 6 valence electrons
4 x H -> 4 x 1 = 4 valence electrons
Total of 14 valence electrons

This structure satisfies the octet rule for C and O.

c.   CS₂

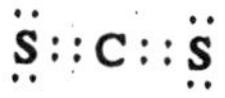

$$\overset{..}{\underset{.}{S}}::C::\overset{..}{\underset{.}{S}}$$

C = 4 valence electrons
2 x S -> 2 x 6 = 12 valence electrons
Total of 16 valence electrons

This structure satisfies the octet rule for C and S.

4.33   a.   Sodium ion
       b.   Copper(I) ion (or cuprous ion)
       c.   Magnesium ion
       d.   Iron(II) ion (or ferrous ion)
       e.   Iron(III) ion (or ferric ion)

4.35   a. K^+
       b. Br^-
       c. Ca^{2+}
       d. Cr^{6+}

4.37   a.   NaCl   [one 1+ cancels one 1-]
       b.   MgBr₂   [one 2+ cancels two 1-]

c.   CuO   [one 2+ cancels one 2-]
d.   Fe$_2$O$_3$ [two 3+ cancels three 2-]
e.   AlCl$_3$ [one 3+ cancels three 1-]

4.39   a.   Magnesium chloride
       b.   Aluminum chloride
       c.   Calcium sulfide
       d.   Sodium oxide
       e.   Iron(III) hydroxide

4.41   a.   Al$_2$O$_3$
       b.   Li$_2$S
       c.   BH$_3$
       d.   Mg$_3$P$_2$

4.43   Ionic solid state compounds exist in regular, repeating, three-dimensional structures; the crystal lattice. the crystal lattice is made up of positive and negative ions. Solid state covalent compounds are made up of molecules which may be arranged in a regular crystalline pattern or in an irregular (amorphous) structure. The melting points of ionic solids are generally much higher than those of covalent solids.

4.45   a.   H·

       b.   He:

       c.   ·Ċ·

       d.   ·N̈·

4.47   a.   Li$^+$

       b.   Mg^{2+}

       c.   :Cl̈:$^-$

       d.   :P̈:$^{3-}$

4.49   Resonance can occur when more than one valid Lewis structure can be written for a molecule. Each individual structure which can be drawn is a resonance form. The true nature of the structure for the molecule is the resonance hybrid, which consists of the "average" of the resonance forms.

4.51
```
      H H
      .. ..
  H:C:C:O:H
      .. ..
      H H
```

330

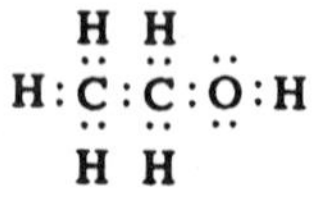

4.53

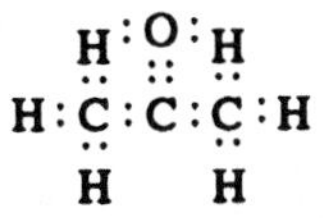

$$H:\overset{\cdot\cdot}{\underset{}{O}}:H$$
$$H:\overset{\cdot\cdot}{\underset{\cdot\cdot}{C}}:\overset{\cdot\cdot}{\underset{}{C}}:\overset{\cdot\cdot}{\underset{}{C}}:H$$
$$\overset{}{\underset{H}{}}\quad\overset{}{\underset{H}{}}$$

4.55

a. C and N   Polar covalent; N is more electronegative than C, resulting in an unequal sharing of electrons in the bond.

b. Si and P   Polar covalent; P is more electronegative than Si, resulting in the unequal sharing of electrons in the bond.

c. Na and Cl   Ionic; the bond is between a metal and a non-metal.

d. Na and O   Ionic; the bond is between a metal and a non-metal.

e. Ca and Br   Ionic; the bond is between a metal and a non-metal.

4.57

a. C and N

$$\left[:C \equiv N: \right]^{-}$$

b. Si and P

$$\left[:Si \equiv P: \right]^{-} \quad \text{(analogous to } CN^{-}\text{)}$$

c), d), and e) are ionic compounds.

4.59   A molecule containing no polar bonds <u>must</u> be nonpolar. A molecule containing polar bonds may or may not itself be polar. It depends upon the number and arrangement of the bonds. For example:

♦   If a molecule contains only one bond, and that bond is polar, the molecule must be polar.

♦   If the molecule contains more than one polar bond, the molecule will be non-polar if the arrangement of the bonds causes their effects to cancel; is not, the molecule will be polar.

♦   If lone pairs of electrons are present, their effect must be considered as well.

♦   If a molecule contains no polar bonds, it cannot be a polar molecule.

4.61   Polar compounds have strong intermolecular attractive forces. Higher temperatures are needed to overcome these forces and convert the solid to a liquid; hence, we predict higher melting points for polar compounds when compared to non-polar compounds.

4.63   a.   A covalent bond is the sharing of two or more electrons between two nonmetals.

        b.   Electron density is a measure of the distribution of electrons in orbitals around atoms (for atomic orbitals) or between atoms (for molecular orbitals).

4.65

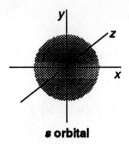

4.67   Molecular orbital theory incorporates subshells; Bohr's theory does not.

Chapter 5
Calculations and the Chemical Equation
Solutions to the Odd-Numbered Problems

****

In-Chapter Questions and Problems

****

5.1　a.　O atoms = $2.50 \text{ mol O} \times \dfrac{6.02 \times 10^{23} \text{ O atoms}}{1 \text{ mol O}}$

O atoms = 1.51×10^{24} oxygen atoms

b.　O atoms = $2.50 \text{ mol O}_2 \times \dfrac{6.02 \times 10^{-23} \text{ O}_2 \text{ molecules}}{1 \text{ mol O}_2} \times \dfrac{2 \text{ O atoms}}{1 \text{ O}_2 \text{ molecule}}$

O atoms = 3.01×10^{24} oxygen atoms

5.3　g He = $3.50 \text{ mol He} \times \dfrac{4.00 \text{ g He}}{1 \text{ mol He}} = 14.0 \text{ g He}$

　　g He = 14.0 g He

5.5　a.　3 atoms of hydrogen x 1.01 amu/atom = 3.03 amu
　　　　1 atom of nitrogen x 14.01 amu/atom = 14.01 amu
　　　　　　　　　　　　　　　　　　　　　17.04 amu

　　　　The mass of a single unit of NH_3 is 17.04 amu/formule unit. Therefore, the
mass of 1 mole of formula units is 17.04 grams or 17.04 g/mol.

　　b.　6 atoms of carbon x 12.01 amu/atom = 72.06 amu
　　　　12 atoms of hydrogen x 1.01 amu/atom = 12.12 amu
　　　　6 atoms of oxygen x 16.00 amu/atoms = 96.00 amu
　　　　　　　　　　　　　　　　　　　　　180.18 amu

　　　　The mass of a single unit of $C_6H_{12}O_6$ is 180.18 amu/formula unit. Therefore,
the mass of 1 mole of formula units is 180.18 grams or 180.18 g/mol.

　　c.　1 atom of cobalt x 58.93 amu/atom　　= 58.93 amu
　　　　2 atoms of chlorine x 35.45 amu/atom = 70.90 amu
　　　　12 atoms of hydrogen x 1.01 amu/atom = 12.12 amu
　　　　6 atoms of oxygen x 16.00 amu/atom　 = 96.00 amu
　　　　　　　　　　　　　　　　　　　　　237.95 amu

The mass of a single unit of $CoCl_2 \bullet 6H_2O$ is 237.95 amu/formula unit. Therefore, the mass of 1 mole of formula units is 237.95 grams or 237.95 g/mol.

5.7　a.　$4Fe(s) + 3O_2(g) \longrightarrow 2Fe_2O_3(s)$

b.　$2C_6H_6(l) + 15O_2(g) \longrightarrow 12CO_2(g) + 6H_2O(g)$

5.9　a.　$5.00 \text{ mol } H_2O \times \dfrac{18.02 \text{ g } H_2O}{1 \text{ mol } H_2O} = 90.1 \text{ g } H_2O$

b.　$25.0 \text{ g LiCl} \times \dfrac{1 \text{ mol LiCl}}{42.39 \text{ g LiCl}} = 0.590 \text{ mol LiCl}$

5.11　a.　$1 \text{ mol } C_2H_5OH \times \dfrac{3 \text{ mol } O_2}{1 \text{ mol } C_2H_5OH} = 3 \text{ mol } O_2$

b.　$1 \text{ mol } C_2H_5OH \times \dfrac{3 \text{ mol } O_2}{1 \text{ mol } C_2H_5OH} \times \dfrac{32 \text{ g } O_2}{1 \text{ mol } O_2} = 96.00 \text{ g } O_2$

5.13　Iron(III) oxide is Fe_2O_3

a.　$4Fe(s) + 3O_2(g) \longrightarrow 2Fe_2O_3(s)$

b.　$5.00 \text{ g } Fe_2O_3 \times \dfrac{1 \text{ mol } Fe_2O_3}{159.7 \text{ g } Fe_2O_3} \times \dfrac{4 \text{ mol Fe}}{2 \text{ mol } Fe_2O_3} \times \dfrac{55.85 \text{ g Fe}}{1 \text{ mol Fe}} = 3.50 \text{ g Fe}$

5.15　a.　Step 1.　Write down information about the reaction:

$Sn(s) + 2HF(aq) \rightarrow SnF_2(s) + H_2(g)$
100.0 g　60.0 g

Step 2.　Convert each mass of reactant to moles:

$100.0 \text{ g Sn} \times \dfrac{1 \text{ mol Sn}}{118.69 \text{ g Sn}} = 0.8425 \text{ mol Sn}$

$60.0 \text{ g HF} \times \dfrac{1 \text{ mol HF}}{20.008 \text{ g HF}} = 3.00 \text{ mol HF}$

Step 3.   Choose one reactant and calculate the number of moles of the other reactant needed to react completely:

$$3.00 \text{ mol HF} \times \frac{1 \text{ mol Sn}}{2 \text{ mol HF}} = 1.50 \text{ mol Sn}$$

Thus 1.50 mol Sn is needed, but only 0.8425 mol Sn is present; Sn is present in insufficient amount and must be the limiting reactant.

Step 4.   Use the limiting reactant to calculate the mass of product:

$$0.8425 \text{ mol Sn} \times \frac{1 \text{ mol SnF}_2}{1 \text{ mol Sn}} \times \frac{156.69 \text{ g SnF}_2}{1 \text{ mol SnF}_2} = 132.0 \text{ g SnF}_2$$

b.   % yield = $\dfrac{\text{actual yield}}{\text{theoretical yield}}$ x 100%

$$= \frac{5.00 \text{ g}}{132.00 \text{ g}} \times 100\% = 3.79\% \text{ yield}$$

****

End-of-Chapter Questions and Problems
****

5.17 $\dfrac{1.66 \times 10^{-24} \text{ g He}}{1 \text{ amu}} \times \dfrac{4.00 \text{ amu}}{1 \text{ atom He}} \times \dfrac{6.02 \times 10^{23} \text{ atom He}}{1 \text{ mol He}} = \dfrac{4.00 \text{ g He}}{\text{mol He}}$

5.19   a. $20.0 \text{ g He} \times \dfrac{1 \text{ mol He}}{4.00 \text{ g He}} = 5.00 \text{ mol He}$

b. $0.040 \text{ kg Na} \times \dfrac{10^3 \text{ g Na}}{1 \text{ kg Na}} \times \dfrac{1 \text{ mol Na}}{22.99 \text{ g Na}} = 1.7 \text{ mol Na}$

c. $3.0 \text{ g Cl} \times \dfrac{1 \text{ mol Cl}}{35.45 \text{ g Cl}} = 8.5 \times 10^{-2} \text{ mol Cl}$

5.21   A molecule is a single unit composed of atoms joined by covalent bonds. An ion-pair is composed of positive and negatively charged ions joined by electrostatic attraction, the ionic bond. The ion pairs, unlike the molecule, do not form single units; the electrostatic charge is directed to other ions in a crystal lattice, as well.

5.23   a.   1 atom Na x $\dfrac{22.99 \text{ amu Na}}{\text{atom Na}}$ = 22.99 amu Na

1 atom Cl x $\dfrac{35.45 \text{ amu Cl}}{\text{atom Cl}}$ = $\underline{35.45 \text{ amu Cl}}$
$\phantom{1 atom Cl x \dfrac{35.45 \text{ amu Cl}}{\text{atom Cl}} = }$ 58.44 amu NaCl

The mass of a single unit of NaCl is 58.44 amu/formula unit. Therefore, the mass of a mole of NaCl formula units is 58.44 g/mol.

b.   2 atoms Na x $\dfrac{22.99 \text{ amu Na}}{\text{atom Na}}$ = 45.98 amu Na

1 atom S x $\dfrac{32.06 \text{ amu S}}{\text{atom S}}$ = 32.06 amu S

4 atoms O x $\dfrac{16.00 \text{ amu O}}{1 \quad \text{atom O}}$ = $\underline{64.00 \text{ amu O}}$
$\phantom{4 atoms O x \dfrac{16.00 \text{ amu O}}{1 atom O} = }$ 142.04 amu Na_2SO_4

The mass of a single unit of Na_2SO_4 is 142.04 amu/formula unit. Therefore, the mass of a mole of Na_2SO_4 formula units is 142.04 g/mol.

c.   3 atoms Fe x $\dfrac{55.85 \text{ amu Fe}}{\text{atom Fe}}$ = 167.55 amu Fe

2 atoms P x $\dfrac{33.97 \text{ amu P}}{\text{atom P}}$ = 61.94 amu P

8 atoms O x $\dfrac{16.00 \text{ amu O}}{\text{atom O}}$ = $\underline{128.00 \text{ amu O}}$
$\phantom{8 atoms O x \dfrac{16.00 \text{ amu O}}{\text{atom O}} = }$ 357.49 amu $Fe_3(PO_4)_2$

The mass of a single unit of $Fe_3(PO_4)_2$ is 357.49 amu/formula unit. Therefore, the mass of a mole of $Fe_3(PO_4)_2$ formula units is 357.49 g/mol.

5.25   a.   The formula weight of NaCl is 58.44 g/mol.

15.0 g NaCl x $\dfrac{1 \text{ mol NaCl}}{58.44 \text{ g NaCl}}$ = 0.257 mol NaCl

b.   The formula weight of Na_2SO_4 is 142.04 g/mol.

$$15.0 \text{ g Na}_2\text{SO}_4 \times \frac{1 \text{ mol Na}_2\text{SO}_4}{142.04 \text{ g Na}_2\text{SO}_4} = 0.106 \text{ mol Na}_2\text{SO}_4$$

5.27   a.   The formula weight of H_2O is 18.02 g/mol.

$$1.000 \text{ mol H}_2\text{O} \times \frac{18.02 \text{ g H}_2\text{O}}{1 \text{ mol H}_2\text{O}} = 18.02 \text{ g H}_2\text{O}$$

          b.   The formula weight of NaCl is 58.44 g/mol.

$$2.000 \text{ mol NaCl} \times \frac{58.44 \text{ g NaCl}}{1 \text{ mol NaCl}} = 116.9 \text{ g NaCl}$$

5.29   a.   The formula weight of He is 4.00 g/mol.

$$10.0 \text{ mol He} \times \frac{4.00 \text{ g He}}{1 \text{ mol He}} = 40.0 \text{ g He}$$

          b.   The formula weight of H_2 = 2.016 g/mol.

$$1.00 \times 10^2 \text{ mol H}_2 \times \frac{2.016 \text{ g H}_2}{1 \text{ mol H}_2} = 2.02 \times 10^2 \text{ g H}$$

5.31   a.   The formula weight of Mg is 24.31 g/mol.

$$0.100 \text{ mol Mg} \times \frac{24.31 \text{ g Mg}}{1 \text{ mol Mg}} = 2.43 \text{ g Mg}$$

          b.   The formula weight of $CaCO_3$ is 100.09 g/mol.

$$0.100 \text{ mol CaCO}_3 \times \frac{100.09 \text{ g CaCO}_3}{1 \text{ mol CaCO}_3} = 10.0 \text{ g CaCO}_3$$

          c.   The formula weight of $C_6H_{12}O_6$ is 180.16 g/mol.

$$0.100 \text{ mol } C_6H_{12}O_6 \times \frac{180.16 \text{ g } C_6H_{12}O_6}{1 \text{ mol } C_6H_{12}O_6} = 18.0 \text{ g } C_6H_{12}O_6$$

d. The formula weight of NaCl is 58.44 g/mol.

$$0.100 \text{ mol NaCl} \times \frac{58.44 \text{ g NaCl}}{1 \text{ mol NaCl}} = 5.84 \text{ g NaCl}$$

5.33 a. The formula weight of KBr is 119.01 g/mol.

$$50.0 \text{ g KBr} \times \frac{1 \text{ mol KBr}}{119.01 \text{ g KBr}} = 0.420 \text{ mol KBr}$$

b. The formula weight of $MgSO_4$ is 120.37 g/mol

$$50.0 \text{ g } MgSO_4 \times \frac{1 \text{ mol } MgSO_4}{120.37 \text{ g } MgSO_4} = 0.415 \text{ mol } MgSO_4$$

c. The formula weight of Br_2 is 159.82 g/mol.

$$50.0 \text{ g } Br_2 \times \frac{1 \text{ mol } Br_2}{159.82 \text{ g } Br_2} = 0.313 \text{ mol } Br_2$$

d. The formula weight of NH_4Cl is 53.49 g/mol.

$$50.0 \text{ g } NH_4Cl \times \frac{1 \text{ mol } NH_4Cl}{53.49 \text{ g } NH_4Cl} = 0.935 \text{ mol } NH_4Cl$$

5.35 The ultimate basis for a correct chemical equation is the law of conservation of mass. No mass may be gained or lost in a chemical reaction; the chemical equation must reflect this fact.

5.37 a. $2C_4H_{10}(g) + 13O_2(g) \longrightarrow 10H_2O(g) + 8CO_2(g)$

b. $Au_2S_3(s) + 3H_2(g) \longrightarrow 2Au(s) + 3H_2S(g)$

c. $Al(OH)_3(s) + 3HCl(aq) \rightarrow AlCl_3(aq) + 3H_2O(l)$

d. $(NH_4)_2Cr_2O_7(s) \rightarrow Cr_2O_3(s) + N_2(g) + 4H_2O(g)$

e. $C_2H_5OH(l) + 3O_2(g) \rightarrow 2CO_2(g) + 3H_2O(g)$

5.39   a. $N_2(g) + 3H_2(g) \rightarrow 2NH_3(g)$

          b. $HCl(aq) + NaOH(aq) \rightarrow NaCl(aq) + H_2O(l)$

5.41   a. $C_6H_{12}O_6(s) + 6O_2(g) \rightarrow 6H_2O(l) + 6CO_2(g)$

          b. $Na_2CO_3(s) \xrightarrow{\Delta} Na_2O(s) + CO_2(g)$

5.43   The formula weight of B_2O_3 is 69.62 g/mol and the formula weight of B_2H_6 is 27.67 g/mol.

$$20.0 \text{ g } B_2H_6 \times \frac{1 \text{ mol } B_2H_6}{27.67 \text{ g } B_2H_6} \times \frac{1 \text{ mol } B_2O_3}{1 \text{ mol } B_2H_6} \times$$

$$\frac{69.62 \text{ g } B_2O_3}{1 \text{ mol } B_2O_3} = 50.3 \text{ g } B_2O_3$$

5.45   The formula weight of $CrCl_3$ is 158.35 g/mol and the formula weight of Cr_2O_3 is 151.99 g/mol.

$$50.0 \text{ g } Cr_2O_3 \times \frac{1 \text{ mol } Cr_2O_3}{151.99 \text{ g } Cr_2O_3} \times \frac{2 \text{ mol } CrCl_3}{1 \text{ mol } Cr_2O_3} \times$$

$$\frac{158.35 \text{ g } CrCl_3}{1 \text{ mol } CrCl_3} = 104 \text{ g } CrCl_3$$

5.47   a. $N_2(g) + 3H_2(g) \rightarrow 2NH_3(g)$

          b.   Three moles of H_2 will react with one mole of N_2, according to the coefficients in the balanced equation.

          c.   One mole of N_2 will produce two moles of the product NH_3, according to the coefficients in the balanced equation.

          d. $140.0 \text{ g } N_2 \times \dfrac{1 \text{ mol } N_2}{28.02 \text{ g } N_2} \times \dfrac{3 \text{ mol } H_2}{1 \text{ mol } N_2} = 1.50 \text{ mol } H_2$

e. $1.50 \text{ mol } H_2 \times \dfrac{2 \text{ mol } NH_3}{3 \text{ mol } H_2} \times \dfrac{17.03 \text{ g } NH_3}{1 \text{ mol } NH_3} = 17.0 \text{ g } NH_3$

5.49   a.   5 C atoms x 12.01 amu/atom C = 60.05 amu
         11 H atoms x   1.008 amu/atom H = 11.088 amu
         1 N atom x 14.01 amu/atom N = 14.01 amu
         2 O atoms x 16.00 amu/atom O = 32.00 amu
         1 S atom x 32.06 amu/atom S = 32.06 amu
                                          149.21 amu

The mass of a single unit of $C_5H_{11}NO_2S$ is 149.21 amu/formula unit. Therefore the mass of a mole of $C_5H_{11}NO_2S$ formula units is 149.21 g a/mol.

b. $1 \text{ mol } C_5H_{11}NO_2S \times \dfrac{2 \text{ mol O atoms}}{1 \text{ mol } C_5H_{11}NO_2S} \times \dfrac{6.02 \times 10^{23} \text{ O atoms}}{1 \text{ mol O atoms}} =$

$$1.20 \times 10^{24} \text{ O atoms}$$

c. $1 \text{ mol } C_5H_{11}NO_2S \times \dfrac{2 \text{ mol O atoms}}{1 \text{ mol } C_5H_{11}NO_2S} \times \dfrac{16.00 \text{ g O}}{1 \text{ mol O atoms}} = 32.00 \text{ g O}$

d. $50.0 \text{ g } C_5H_{11}NO_2S \times \dfrac{1 \text{ mol } C_5H_{11}NO_2S}{149.21 \text{ g } C_5H_{11}NO_2S} \times \dfrac{2 \text{ mol O atoms}}{1 \text{ mol } C_5H_{11}NO_2S} \times$

$$\dfrac{16.00 \text{ g O}}{1 \text{ mol O atoms}} = 10.7 \text{ g O}$$

5.51   The formula weight of HgO is 216.59 g/mol.

$$1.00 \times 10^2 \text{ g HgO} \times \dfrac{1 \text{ mol HgO}}{216.59 \text{ g HgO}} \times \dfrac{1 \text{ mol } O_2}{2 \text{ mol HgO}} \times \dfrac{32.00 \text{ g } O_2}{1 \text{ mol } O_2} = 7.39 \text{ g } O_2$$

5.53   The balanced equation is:

$2C_2H_2(g) + 5O_2(g) \rightarrow 4CO_2(g) + 2H_2O(g)$

The formula weight of C_2H_2 is 26.04 g/mol.

$$20.0 \text{ kg } C_2H_2 \times \frac{10^3 \text{ g } C_2H_2}{1 \text{ kg } C_2H_2} \times \frac{1 \text{ mol } C_2H_2}{26.04 \text{ g } C_2H_2} \times \frac{5 \text{ mol } O_2}{2 \text{ mol } C_2H_2} \times$$

$$\frac{32.00 \text{ g } O_2}{1 \text{ mol } O_2} = 6.14 \times 10^4 \text{ g } O_2$$

5.55   Step 1.   Write down information about the reaction:

$$C_{10}H_{20}(l) + H_2(g) \rightarrow C_{10}H_{22}(s)$$

1.00×10^2 g   1.00 g

Step 2.   Convert each mass of reactant to moles:

$$1.00 \times 10^2 \text{ g } C_{10}H_{20} \times \frac{1 \text{ mol } C_{10}H_{20}}{140.3 \text{ g } C_{10}H_{20}} = 0.713 \text{ mol } C_{10}H_{20}$$

$$1.00 \text{ g } H_2 \times \frac{1 \text{ mol } H_2}{2.016 \text{ g } H_2} = 0.496 \text{ mol } H_2$$

Step 3.   The reaction states that decene and hydrogen react in a 1:1 mole ratio; 0.496 mol H_2 will react with a 0.496 mol H. consequently H_2 is the limiting reactant.

Step 4.   Use the limiting reactant to calculate the mass of product:

$$0.496 \text{ mol } H_2 \times \frac{1 \text{ mol } C_{10}H_{22}}{1 \text{ mol } H_2} \times \frac{142.3 \text{ g } C_{10}H_{22}}{1 \text{ mol } C_{10}H_{22}} = 70.6 \text{ g } C_{10}H_{22}$$

5.57   Step 1.   Write down information about the reaction:

$$N_2O_4(l) + 2N_2H_4(l) \rightarrow 3N_2(g) + 4H_2O(g)$$

1.00 kg   2.00 kg

Step 2.   Convert each mass of reactant to moles:

$$1.00 \text{ kg } N_2O_4 \times \frac{10^3 \text{ g } N_2O_4}{1 \text{ kg } N_2O_4} \times \frac{1 \text{ mol } N_2O_4}{92.02 \text{ g } N_2O_4} = 10.9 \text{ mol } N_2O_4$$

$$2.00 \text{ kg N}_2\text{H}_4 \times \frac{10^3 \text{ g N}_2\text{H}_4}{1 \text{ kg N}_2\text{H}_4} \times \frac{1 \text{ mol N}_2\text{H}_4}{32.05 \text{ g N}_2\text{H}_4} = 62.4 \text{ mol N}_2\text{H}_4$$

Step 3.   The ratio of moles N_2H_4 to N_2O_4 is approximately 6:1, not 2:1, as called for by the balanced equation. The engineer made an error. N_2O_4 is the limiting reactant.

Step 4.   Use the limiting reactant to calculate the mass of product:

$$10.9 \text{ mol N}_2\text{O}_4 \times \frac{2 \text{ mol N}_2\text{H}_4}{1 \text{ mol N}_2\text{O}_4} \times \frac{28.02 \text{ g}}{1 \text{ mol N}_2} = 6.10 \times 10^2 \text{ g N}_2$$

Chapter 6
States of Matter:
Gases, Liquids, and Solids
Solutions to the Odd-Numbered Problems

****

In-Chapter Questions and Problems

****

6.1   a. $725 \text{ mm Hg} \times \dfrac{1 \text{ atm}}{760 \text{ mm Hg}} = 0.954 \text{ atm}$

     b. $29.0 \text{ cm Hg} \times \dfrac{10 \text{ mm Hg}}{1 \text{ cm Hg}} \times \dfrac{1 \text{ atm}}{760 \text{ mm Hg}} = 0.382 \text{ atm}$

     c. $555 \text{ torr} \times \dfrac{1 \text{ atm}}{760 \text{ torr}} = 0.730 \text{ atm}$

6.3   a. $P_i V_i = P_f P_f$

          $P_i = \dfrac{P_f V_f}{V_i}$

          $P_i = \dfrac{(5.0 \text{ atm})(7.5 \text{ L})}{1.0 \text{ L}} = 37.5 \text{ atm} = 38 \text{ atm}$ (2 significant figures)

     b. $P_i V_i = P_f P_f$

          $P_f = \dfrac{P_i V_i}{V_f}$

          $P_f = \dfrac{(5.0 \text{ atm})(1.0 \text{ L})}{0.20 \text{ L}} = 25 \text{ atm}$

6.5   Initial temperature: 25°C + 273 = 298 K

a.   100°C + 273 = 373 K

$$\frac{V_i}{T_i} = \frac{V_f}{T_f}$$

$$V_f = \frac{V_i T_f}{T_i}$$

$$V_f = \frac{(3.00 \text{ L})(373 \text{ K})}{(298 \text{ K})} = 3.76 \text{ L}$$

b. $°C = \frac{5}{9} (°F - 32) = \frac{5}{9} (150 - 32) = 66°C$

K = °C + 273 = 66 + 273 = 339 K

$$\frac{V_i}{T_i} = \frac{V_f}{T_f}$$

$$V_f = \frac{V_i T_f}{T_i}$$

$$V_f = \frac{(3.00 \text{ L})(339 \text{ K})}{(298 \text{ K})} = 3.41 \text{ L}$$

c. $\dfrac{V_i}{T_i} = \dfrac{V_f}{T_f}$

$$V_f = \frac{V_i T_f}{T_I}$$

$$V_f = \frac{(3.00\,L)(273\,K)}{(298\,K)} = 2.75\,L$$

6.7 $\quad \dfrac{P_i V_i}{T_i} = \dfrac{P_f V_f}{T_f}$

$$P_f = \frac{P_i V_i T_f}{V_f T_i}$$

$$P_i = 760\ \text{torr} \times \frac{1\ \text{atm}}{760\ \text{torr}} = 1.00\ \text{atm}$$

$V_i = 2.00\ L$

$T_f = T_i = 25.0°C + 273 = 298\ K$

$V_f = 10.0\ L$

$$P_f = \frac{(1.00\ \text{atm})(2.00\ L)(298\ K)}{(10.0\ L)(298\ K)}$$

$P_f = 0.200\ \text{atm}$

6.9 $\quad PV = nRT$

Solving for volume, V,

$$V = \frac{nRT}{P}$$

$$N = 10.0\ g\ N_2 \times \frac{1\ \text{mol}\ N_2}{28.00\ g\ N_2} = 0.357\ \text{mol}\ N_2$$

$R = 0.0821\ L-atm/K-mol$

$$T = 30°C + 273 = 303 \text{ K}$$

$$P = 750 \text{ torr} \times \frac{1 \text{ atm}}{760 \text{ torr}} = 0.987 \text{ atm}$$

Substituting,

$$V = \frac{(0.357 \text{ mol } N_2)(0.0821 \text{ L} - \text{atm/K} - \text{mol})(303 \text{ K})}{0.987 \text{ atm}}$$

$$V = 9.00 \text{ L}$$

6.11 $PV = nRT$

Solving for number of moles of N_2, n,

$$n = \frac{PV}{RT}$$

$P = 1.00 \text{ atm}$

$V = 5.00 \text{ L}$

$R = 0.0821 \text{ L} - \text{atm/K} - \text{mol}$

$T = 273 \text{ K}$

Substituting,

$$n = \frac{(1.00 \text{ atm})(5.00 \text{ L})}{(0.0821 \text{ L} - \text{atm/K} - \text{mol})(273)} = 0.223 \text{ mol } N_2$$

End-of-Chapter Questions and Problems

6.13   A volume of 5 L (ordinate) corresponds to a pressure of 1 atm (abscissa).

6.15   A volume of 2 L (ordinate) corresponds to a pressure of 2.5 atm (abscissa).

$$PV = k$$

$$(2.5 \text{ atm}) (2 \text{ L}) = 5 \text{ L-atm} = k$$

6.17 $P_i = 1.00$ atm $P_f = ?$ atm

$V_i = 20.9$ L $V_f = 4.00$ L

$$P_i V_i = P_f V_f$$

$$P_f = \frac{P_i V_i}{V_f}$$

$$P_f = \frac{(1.00 \text{ atm})(20.9 \text{ L})}{(4.00 \text{ L})}$$

$$P_f = 5.23 \text{ atm}$$

6.19   Charles's law states that the volume of a gas varies directly with the absolute temperature if pressure and number of moles of gas are constant.

6.21 $V_i = 2.00$ L $V_f = ?$ L
$T_i = 250$ °C $T_f = 500$ °C

$$\frac{V_i}{T_i} = \frac{V_f}{T_f}$$

$$V_f = \frac{V_i T_f}{T_i}$$

$$V_f = \frac{(2.00 \text{ L})(500°C + 273)}{250°C + 273}$$

$$V_f = \frac{(2.00\,L)(773\,K)}{523\,K}$$

$V_f = 2.96\,L$

The change in volume, $\Delta V = V_f - V_i$

$$\Delta V = 2.96\,L - 2.00\,L$$

$$\Delta V = 0.96\,L$$

6.23 $V_i = 1.25\,L$ $V_f = ?\,L$
       $T_i = 20°C$ $T_f = 80°C$

$$\frac{V_i}{T_i} = \frac{V_f}{T_f}$$

$$V_f = \frac{V_i\, T_f}{T_i}$$

$$V_f = \frac{(1.25\,L)(80°C + 273)}{20°C + 273}$$

$$V_f = \frac{(1.25\,L)(353\,K)}{293\,K}$$

$V_f = 1.51\,L$

6.25   Examine each effect separately:

▸ Volume and temperature are <u>directly</u> proportional; increasing T <u>increases</u> V.
▸ Volume and pressure are <u>inversely</u> proportional; decreasing P <u>increases</u> V.

Therefore, both variables work together to <u>increase</u> the volume.

6.27 $\dfrac{P_i V_i}{T_i} = \dfrac{P_f V_f}{T_f}$

$$P_f V_f T_i = P_i V_i T_f$$

$$V_f = \frac{P_i V_i T_f}{P_f T_i}$$

6.29 $P_i = 1.00$ atm $P_f = 125$ atm
        $V_i = 2.25$ L $V_f = ?$ L
        $T_i = 16°C$ $T_f = 20°C$

Using the equation derived in question 6.29,

$$V_f = \frac{P_i V_i T_f}{P_f T_i}$$

and substituting:

$$V_f = \frac{(1.00 \text{ atm})(2.25 \text{ L})(20°C + 273)}{(125 \text{ atm})(16°C + 273)}$$

$$V_f = \frac{(1.00 \text{ atm})(2.25 \text{ L})(293 \text{ K})}{(125 \text{ atm})(289 \text{ K})}$$

$$V_f = 1.82 \times 10^{-2} \text{ L}$$

6.31 $n_i = 1.00 \text{ g He} \times \dfrac{1 \text{ mol He}}{4.00 \text{ g He}} = 0.25$ mol He

        $V_i = 1.00$ L

        $n_f = 6.00 \text{ g He} \times \dfrac{1 \text{ mol He}}{4.00 \text{ g He}} = 1.50$ mol He

        $V_f = ?$ L

$$\frac{V_i}{n_i} = \frac{V_f}{n_f}$$

$$V_f = \frac{V_i n_f}{n_i}$$

$$V_f = \frac{(1.00 \text{ L})(1.50 \text{ mol He})}{(0.25 \text{ mol He})}$$

$$V_f = 6.00 \text{ L}$$

6.33 Avogadro's law states that equal volumes of a gas contain the same number of moles if measured under the same conditions of temperature and pressure.

6.35 $P_i = 750$ torr $P_f = 1.00$ atm
$V_i = 65.0$ mL $V_f = ?$ L
$T_i = 22°C$ $T_f = 273$ K

$$\frac{P_i V_i}{T_i} = \frac{P_f V_f}{T_f}$$

$$V_f = \frac{P_i V_i T_f}{P_f T_i}$$

$$V_f = \frac{\left(750 \text{ Torr} \times \frac{1 \text{ atm}}{760 \text{ torr}}\right)\left(650 \text{ mL} \times \frac{1 \text{ L}}{10^3 \text{ mL}}\right)(273 \text{ K})}{(1.00 \text{ atm})(22°C + 273)}$$

$$V_f = \frac{(0.987 \text{ atm})(6.5 \times 10^{-2} \text{ L})(273 \text{ K})}{(1.00 \text{ atm})(295 \text{ K})}$$

$$V_f = 5.94 \times 10^{-2} \text{ L}$$

6.37 $PV = nRT$

$$V = \frac{nRT}{P}$$

$$V = \frac{(1.00 \text{ mol})(0.0821 \text{ L-atm K}^{-1} - \text{mol}^{-1})(273 \text{ K})}{(1.00 \text{ atm})}$$

$$V = 22.4 \text{ L}$$

6.39 $PV = nRT$

$$V = \frac{nRT}{P}$$

$$V = \frac{(4.00)(0.0821 \text{ L-atm K}^{-1} - \text{mol}^{-1})(27°C + 273)}{\left(8.25 \text{ torr} \times \dfrac{1 \text{ atm}}{760 \text{ torr}}\right)}$$

$$V = \frac{(4.00 \text{ mol})(0.0821 \text{ L-atm K}^{-1} - \text{mol}^{-1})(300 \text{ K})}{(1.09 \times 10^{-2} \text{ atm})}$$

$$V = 9.08 \times 10^3 \text{ L}$$

6.41 $PV = nRT$

$$T = \frac{PV}{nR}$$

$$T = \frac{(1.00 \text{ atm})(2.00 \text{ L})}{\left(1.75 \text{ g O}_2 \times \dfrac{1 \text{ mol O}_2}{32.0 \text{ g O}_2}\right)(0.0821 \text{ L} - \text{atm K}^{-1} - \text{mol}^{-1})}$$

T = 445 K and 445 K - 273 = 172°C

6.43   Gases exhibit more ideal behavior at low pressures. At low pressures, gas particles are more widely separated and therefore the attractive forces between particles are less. The ideal gas model assumes negligible attractive forces between gas particles.

6.45   The kinetic molecular theory states that the average kinetic energy of the gas particles increases as the temperature increases. Kinetic energy is proportional to $(velocity)^2$. therefore, as the temperature increases the gas particle velocity increases and the rate of mixing increases as well.

6.47   Dalton's law states that the total pressure of a mixture of gases is the sum of the partial pressures of the component gases.

6.49 $P_T = P_{N_2} + P_{F_2} + P_{He}$

$P_T = 0.40 \text{ atm} + 0.16 \text{ atm} + 0.18 \text{ atm}$

$P_T = 0.74 \text{ atm}$

6.51   Intermolecular forces in liquids are considerably stronger than intermolecular forces in gases. Particles are, on average, much closer together in liquids and the strength of attraction is inversely proportional to the distance of separation.

6.53   The vapor pressure of a liquid increases as the temperature of the liquid increases.

6.55   Evaporation is the conversion of a liquid to a gas at a temperature lower than the boiling point of the liquid. Condensation is the conversion of a gas to a liquid at a temperature lower than the boiling point of the liquid.

6.57   Viscosity is the resistance to flow caused by intermolecular attractive forces. Complex molecules may become entangled and not slide smoothly across one another.

6.59   Solids are essentially incompressible because the average distance of separation among particles in the solid state is small. there is literally no space for the particles to crowd closer together.

6.61   a. ionic solids - high melting temperature, brittle

        b. covalent solids - high melting temperature, hard

Chapter 7
Reactions and Solutions
Solutions to the Odd-Numbered Problems

****

In-Chapter Questions and Problems

****

7.1   a. DR
      b. SR
      c. DR
      d. D

7.3   a. $KCl(aq) + AgNO_3(aq) \rightarrow KNO_3(aq) + AgCl(s)$

      The solubility rules predict that silver chloride is insoluble; a precipitation reaction occurs.

      b. $CH_3COOK(aq) + AgNO_3(aq) \rightarrow$ no reaction

      The solubility rules predict that both potential products, potassium nitrate (KNO_3) and silver acetate (CH_3COOAg) are soluble; no precipitation reaction occurs.

7.5 $$\% \, (W/V) \; = \; \frac{\text{grams of solute}}{\text{milliliters of solution}} \; x \; 10^2$$

$$\% \, (W/V) \; = \; \frac{10.0 \, g \; NaCl}{0.0600 \, L \; x \; \dfrac{10^3 \, mL}{1 \, L}} \; x \; 10^2$$

$$\% \, (W/V) \; = \; 16.7\% \; NaCl$$

7.7 $$\% \, (W/V) \; = \; \frac{\text{grams of solute}}{\text{milliliters of solution}} \; x \; 10^2$$

$$\% \, (W/V) = \frac{15.0 \text{ g KCl}}{0.200 \text{ L} \times \dfrac{10^3 \text{ mL}}{1 \text{ L}}} \times 10^2$$

$$\% \, (W/V) = 7.50\% \text{ KCl}$$

7.9 $$\% \, (V/V) = \frac{\text{mL solute}}{\text{mL solution}} \times 10^2$$

$$\% \, (V/V) = \frac{5.00 \text{ mL ethanol}}{1.00 \times 10^2 \text{ mL solution}} \times 10^2$$

$$\% \, (V/V) = 5.00\% \text{ ethanol}$$

7.11   a. $$\% \, (W/V) = \frac{\text{grams of solute}}{\text{milliliters of solution}} \times 10^2$$

$$\% \, (W/V) = \frac{20.0 \text{ g } O_2}{78.0 \text{ L} \times \dfrac{10^3 \text{ mL}}{1 \text{ L}}} \times 10^2$$

$$\% \, (W/V) = 2.56 \times 10^{-2} \, \% \text{ oxygen}$$

b.   It is necessary to use the ideal gas equation to determine the volume of the solute, O_2.

$$PV = nRT$$

Solving for volume, V,

$$V = \frac{nRT}{P}$$

For O_2: $n = 20.0 \text{ g } O_2 \times \dfrac{1 \text{ mol } O_2}{32.0 \text{ g } O_2} = 0.625 \text{ mol } O_2$

$R = 0.082 \text{ L}-\text{atm/K}-\text{mol}$

$T = 273 \text{ K (standard temperature)}$

$P = 1.00 \text{ atm (standard pressure)}$

$$V = \frac{(0.625 \text{ mol } O_2)(0.00821 \text{ L}-\text{atm/K}-\text{mol})(273 \text{ K})}{(1.00 \text{ atm})}$$

$V = 14.0 \text{ L } O_2$

Then, $\% \text{ (V/V)} = \dfrac{\text{mL solute}}{\text{mL solution}} \times 10^2$

$$\% \text{ ((V/V)} = \frac{14.0 \text{ L } O_2 \times \dfrac{10^3 \text{ mL } O_2}{1 \text{ L } O_2}}{78.0 \text{ L solution} \times \dfrac{10^3 \text{ mL solution}}{1 \text{ L solution}}} \times 10^2$$

$$\% \text{ (V/V)} = 17.9\%$$

c.

$$\% \text{ (W/W)} = \frac{\text{grams solute}}{\text{grams solution}} \times 10^2$$

$$\% \text{ (W/W)} = \frac{20.0 \text{ g } O_2}{20.0 \text{ g } O_2 + 80.0 \text{ g } N_2} \times 10^2 = 20.0\%$$

7.13

$$M_{HCl} = \frac{mol\ HCl}{L_{solution}}$$

Solving for mol HCl,

$$mol\ HCl = (M_{HCl})(L_{solution})$$

$$mol\ HCl = (0.250M)(5.00 \times 10^2\ mL \times \frac{1\ L}{10^3\ mL})$$

$$mol\ HCl = 0.125\ mol\ HCl$$

7.15

$$(M_1)(V_1) = (M_2)(V_2)$$

$$M_1 = 12.0\ M$$

$$M_2 = 2.0\ M$$

$$V_2 = 1.0 \times 10^2\ mL \times \frac{1\ L}{10^3\ mL} = 1.0 \times 10^{-1}\ L$$

Solving for the initial volume of 12 M HCl, V_1,

$$V_1 = \frac{(M_2)(V_2)}{(M_1)} = \frac{(2.0M)(1.0 \times 10^{-1}L)}{(12M)}$$

$$V_1 = 1.7 \times 10^{-2}\ L\ (or\ 17\ mL)\ of\ 12M\ HCl$$

To prepare the solution dilute 1.7×10^{-2} L of 12 M HCl with sufficient water to produce 1.0×10^2 mL of total solution.

7.17

$$NH_4NO_3 \xrightarrow{H_2O} NH_4^+ + NO_3^-$$

$$\frac{2 \text{ mol particles}}{1 \text{ mol } NH_4NO_3} \times \frac{5.0 \times 10^{-3} \text{ mol } NH_4NO_3}{\text{L solution}} = 1.0 \times 10^{-2} \frac{\text{mol particles}}{\text{L solution}}$$

and $\dfrac{1.0 \times 10^{-2} \text{ mol particles}}{\text{L solution}} = 1.0 \times 10^{-2}$ osm

7.19   In Question 7.17 we found that the solution was 1.0×10^{-2} osm; substituting this value for M (M represents the osmolarity) in the expression:

$$\pi = MRT$$

$$\pi = \frac{1.0 \times 10^{-2} \text{ mol particles}}{\text{L solution}} \times \frac{0.0821 \text{ L } -atm}{\text{K } -mol} \times (25 + 273 \text{ K})$$

$$\pi = 0.24 \text{ atm}$$

****
End-of-Chapter Questions and Problems
****

7.21   a.   Heating an alkaline earth metal carbonate, for example:

$$MgCO_3(s) \triangleq MgO(s) + CO_2(g)$$

b.   The replacement of copper by zinc in copper sulfate,

$$Zn(s) + CuSO_4(aq) \dashrightarrow ZnSO_4(aq) + Cu(s)$$

7.23   Reaction of two soluble substances to form an insoluble product.   For example:

$$2NaOH(aq) + FeCl_2(aq) \dashrightarrow Fe(OH)_2(s) + 2NaCl(aq)$$

7.25   a. $2C_2H_6(g) + 7O_2(g) \dashrightarrow 4CO_2(g) + 6H_2O(g)$

b. $6K_2O(s) + P_4O_{10}(s) \dashrightarrow 4K_3PO_4(s)$

c. $MgBr_2(aq) + H_2SO_4(aq) \dashrightarrow 2HBr(g) + MgSO_4(aq)$

7.27   a. $Ca(s) + F_2(g) \dashrightarrow CaF_2(s)$

b. $2Mg(s) + O_2(g) \dashrightarrow 2MgO(s)$

c. $3H_2(g) + N_2(g) \rightarrow 2NH_3(g)$

7.29 $\%(W/V) = \dfrac{\text{grams of solute}}{\text{milliliters of solution}} \times 10^2$

a.
$$\%(W/V) = \dfrac{20.0 \text{ g NaCl}}{1.00 \text{ L soln} \times \dfrac{10^3 \text{ mL soln}}{1 \text{ L soln.}}} \times 10^2$$

$\%(W/V) = 2.00\% \text{ NaCl}$

b.
$$\%(W/V) = \dfrac{33.0 \text{ g C}_6\text{H}_{12}\text{O}_6}{5.00 \times 10^2 \text{ mL soln}} \times 10^2$$

$\%(W/V) = 6.60\% \text{ C}_6\text{H}_{12}\text{O}_6$

7.31 $\%(V/V) = \dfrac{\text{mL solute}}{\text{mL solution}} \times 10^2$

a.
$$\%(V/V) = \dfrac{50.0 \text{ mL ethanol}}{1.00 \text{ L soln} \times \dfrac{10^3 \text{ mL soln}}{1 \text{ L soln}}} \times 10^2$$

$\%(V/V) = 5.00\% \text{ ethanol}$

b.
$$\%(V/V) = \dfrac{50.0 \text{ mL ethanol}}{5.00 \times 10^2 \text{ mL soln}} \times 10^2$$

$\%(V/V) = 10.0\% \text{ ethanol}$

7.33 $\%(W/W) = \dfrac{\text{grams solute}}{\text{grams solution}} \times 10^2$

a.

$$\%(W/W) = \frac{21.0\,g\ NaCl}{1.00\times10^2\,g\ soln}\ x\ 10^2$$

$$\%(W/W) = 21.0\%\ NaCl$$

b.   Use the density of the sodium chloride solution as a conversion factor to calculate the volume of the sodium chloride solution.

$$5.00\times10^2\ mL\ soln\ x\ \frac{1.12\,g\ soln}{1\ mL\ soln} = 5.60\times10^2\,g\ soln$$

$$Then,\ \%(W/W) = \frac{21.0\,g\ NaCl}{5.60\times10^2\,g\ soln}\ x\ 10^2$$

$$\%(W/W) = 3.75\%\ NaCl$$

7.35 $$\%(W/W) = \frac{g\ solute}{g\ soln}\ x\ 10^2$$

Solve for g solute,

$$g\ solute = \frac{\%(W/W)(g\ soln)}{10^2}$$

a. $$g\ NaCl = \frac{[0.900\%((W/W)](2.50\times10^2\ g\ soln)}{10^2}$$

$$g\ NaCl = 2.25\ g\ NaCl$$

b.   Assume that the density of the solution is 1.00 g/mL; then

$$g\ CH_3COONa = \frac{[1.25\%(W/W)](2.50\times10^2\,g\ soln)}{10^2}$$

$$g\ CH_3COONa = 3.13\ g\ CH_3COONa$$

7.37　a.　$M_{NaCl} = \dfrac{\text{mol NaCl}}{\text{L solution}}$

$$M_{NaCl} = \dfrac{20.0 \text{ g NaCl} \times \dfrac{1 \text{ mol NaCl}}{58.44 \text{ g NaCl}}}{1.00 \text{ L solution}}$$

$M_{NaCl} = 0.342 \text{ M NaCl}$

b.　$M_{C_6H_{12}O_6} = \dfrac{\text{mol } C_6H_{12}O_6}{\text{L solution}}$

$$M_{C_6H_{12}O_6} = \dfrac{33.0 \text{ g } C_6H_{12}O_6 \times \dfrac{1 \text{ mol } C_6H_{12}O_6}{180.0 \text{ g } C_6H_{12}O_6}}{5.00 \times 10^2 \text{ mL soln} \times \dfrac{1 \text{ L soln}}{10^3 \text{ mL soln}}}$$

$M_{C_6H_{12}O_6} = 0.367 \text{ M } C_6H_{12}O_6$

7.39　$M = \dfrac{\text{mol solute}}{\text{L solution}}$

Solve for mol solute,

mol solute = (M)(L solution)

a.　$\text{mol NaCl} = (0.100M)\left(2.50 \times 10^2 \text{ mL soln} \times \dfrac{1 \text{ L soln}}{10^3 \text{ mL soln}} \right)$

mol NaCl = 2.50×10^{-2} mol NaCl

and, 2.50×10^{-2} mol NaCl $\times \dfrac{58.44 \text{ g NaCl}}{1 \text{ mol NaCl}} = 1.46 \text{ g NaCl}$

b. $\text{mol } C_6H_{12}O_2 = (0.200M)(2.50 \times 10^2 \text{ mL soln} \times \dfrac{1 \text{ L soln}}{10^3 \text{ mL soln}})$

$\text{mol } C_6H_{12}O_6 = 5.00 \times 10^{-2} \text{ mol } C_6H_{12}O_6$

and, $5.00 \times 10^{-2} \text{ mol } C_6H_{12}O_6 \times \dfrac{180.0 \text{ g } C_6H_{12}O_6}{1 \text{ mol } C_6H_{12}O_6} = 9.00 \text{ g } C_6H_{12}O_6$

7.41 $M_{C_{12}H_{22}O_{11}} = \dfrac{\text{mol } C_{12}H_{22}O_{11}}{\text{L solution}}$

$M_{C_{12}H_{22}O_{11}} = \dfrac{50.0 \text{ g } C_{12}H_{22}O_{11} \times \dfrac{1 \text{ mol } C_{12}H_{22}O_{11}}{342.3 \text{ g } C_{12}H_{22}O_{11}}}{L}$

$M_{C_{12}H_{22}O_{11}} = 0.146M \; C_{12}H_{22}O_{11}$

7.43 $(M_1)(V1) = (M_2)(V_2)$

$M_1 = 1.00 \text{ M}$

$V_1 = ?$

$M_2 = 0.100 \underline{M}$

$V_2 = 0.500 \text{ L}$

Solve for V_1

$V_1 = \dfrac{(M_2)(V_2)}{(M_1)}$

Substitute,

$V_1 = \dfrac{(0.100)(0.500)}{(1.00)}$

$V_1 = 5.00 \times 10^{-2} \text{ L}$

7.45 $(M_1)(V_1) = (M_2)(V_2)$

$M_1 = ?$

$V_1 = 50.0 \text{ mL} \times \dfrac{1 \text{ L}}{10^3 \text{ mL}} = 5.00 \times 10^{-2} \text{ L}$

$M_2 = 2.00 \text{ M}$

$V_2 = 500.0 \text{ mL} \times \dfrac{1 \text{ L}}{10^3 \text{ mL}} = 5.000 \times 10^{-1} \text{ L}$

Solve for M_1,

$M_1 = \dfrac{(M_2)(V_2)}{(V_1)}$

Substitute,

$M_1 = \dfrac{(2.00)(5.000 \times 10^{-1} \text{ L})}{(5.00 \times 10^{-2} \text{ L})} = 20.0 \text{ M}$

7.47   A colligative property is a solution property that depends on the concentration of solute particles rather than the identity of the particles.

7.49   Salt is an ionic substance that dissociates in water to produce positive and negative ions. These ions (or particles) lower the freezing point of water. If the concentration of salt particles is large, the freezing point may be depressed below the surrounding temperature, and the ice would melt.

Chapter 8
Chemical and Physical Change:
Energy, Rate, and Equilibrium
Solutions to the Odd-Numbered Problems

❋❋

In-Chapter Questions and Problems
❋❋❋❋❋❋❋❋❋❋❋❋❋❋❋❋❋❋❋❋❋❋❋❋❋❋❋❋❋❋❋❋❋❋

8.1   a.   Exothermic. The reaction <u>produces</u> energy used to heat our homes.

       b.   Exothermic. ΔH° is <u>negative</u>, meaning that energy is <u>released</u> by the reaction.

       c.   Exothermic. 18.3 kcal of energy is shown as a <u>product</u> of the reaction.

8.3 $Q = m_w \times \Delta T_w \times SH_w$

Solving for ΔT_w

$$\Delta T_w = \frac{Q}{m_w \times SH_w}$$

Substituting,

$$\Delta T_w = \frac{6.5 \times 10^2 \text{ cal}}{50.0\text{g} \times \dfrac{1.00 \text{ cal}}{\text{g} - {}^\circ\text{C}}}$$

$$\Delta T_w = 13 \ {}^\circ\text{C}$$

8.5 $Q = m_w \times \Delta T_w \times SH_w$

$$Q = 1.00 \times 10^3 \text{g} \times 3.0 \ {}^\circ\text{C} \times \frac{1.00 \text{ cal}}{\text{g} - {}^\circ\text{C}}$$

$Q = 3.0 \times 10^3$ Cal in a $1.0 - g$ sample of candy.

and,

$$3.0 \times 10^3 \text{ cal} \times \frac{1 \text{ nutritional Cal}}{10^3 \text{ cal}} = 3.0 \text{ nutritional Calories in a } 1.0 - g \text{ sample of candy}$$

and,

$$\frac{3.0 \text{ nutritional Cal}}{1.0 \text{ g candy}} \times \frac{454 \text{ g candy}}{1 \text{ lb candy}} \times \frac{1 \text{ lb candy}}{16 \text{ oz candy}} \times \frac{2.5 \text{ oz}}{1 \text{ candy bar}} =$$

$$\frac{2.1 \times 10^2 \text{ nutritional Cal}}{\text{candy bar}}$$

8.7   a.   rate $= k [N_2]^n [O_2]^{n'}$

      b.   rate $= k [C_4H_6]^n$

8.9   a. $K_{eq} = \dfrac{[N_2][O_2]^2}{[NO_2]^2}$

      b. $K_{eq} = [H_2]^2 [O_2]$

8.11   a.   A would decrease; the system would shift to remove excess B. B reacts with A to form products C and D.

      b.   A would increase; addition of excess C shifts the equilibrium to the left producing both A and B.

      c.   A would decrease; A would react with B to compensate for the loss of some D. The equilibrium shifts to the right.

      d.   A would remain the same. the presence of a catalyst increases the rate of attainment of equilibrium, but does not affect the equilibrium position.

End-of-Chapter Questions and Problems

8.13   a.   An exothermic reaction is one in which energy is released during chemical change.

        b.   An endothermic reaction is one in which energy is absorbed during chemical change.

        c.   A calorimeter is a device for measuring heat absorbed or released during chemical change.

8.15   Enthalpy is a measure of heat energy.

8.17

$$Q = m_w \times \Delta T_w \times SH_w$$

$$Q = 2.00 \times 10^2 \text{ g H}_2\text{O} \times 6.00\,^\circ\text{C} \times \frac{1.00 \text{ cal}}{\text{g H}_2\text{O} \,^\circ\text{C}}$$

$$Q = 1.20 \times 10^3 \text{ cal}$$

8.19   a.   Entropy increases. Conversion of a solid to a liquid results in an increase in disorder of the substance. Solids retain their shape while liquids will flow and their shape is determined by their container.

        b.   Entropy increases. Conversion of a liquid to a gas results in an increase in disorder of the substance. Gas particles move randomly with very weak interactions between particles, much weaker than those interactions in the liquid state.

8.21   An increase in stability is equated with a decrease in energy (reaching a lower energy state). The energy of products is less than that of the reactants in an exothermic reaction; energy is given off in an exothermic reaction.

8.23   Isopropyl alcohol quickly evaporates (liquid -> gas) after being applied to the skin. Conversion of a liquid to a gas requires heat energy. The heat energy is supplied by the skin. When this heat is lost, the skin temperature drops.

8.25   The activated complex is the arrangement of reactants in an unstable transition state as a chemical reaction proceeds. The activated complex must form in order to convert reactants to products.

8.27   Assume a generalized exothermic reaction:

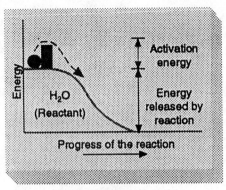

(a) Noncatalyzed reaction

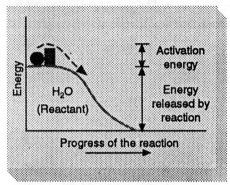

(b) Catalyzed reaction

8.29   Enzymes are biological catalysts. The enzyme lysozyme catalyzes a process that results in the destruction of the cell walls of many harmful bacteria. This helps to prevent disease in organisms.

The breakdown of foods to produce material for construction and repair of body tissue, as well as energy, is catalyzed by a variety of enzymes. For example, amylase begins the hydrolysis of starch in the mouth.

8.31   An increase in concentration of reactants means that there are more molecules in a certain volume. The probability of collision is enhanced because they travel a shorter distance before meeting another molecule. The rate is proportional to the number of collisions per unit time.

8.33   Rate = $k\,[N_2O_4]^n$

8.35   A catalyst speeds up a chemical reaction by facilitating the formation of the activated complex, thus lowering the activation energy, the energy barrier for the reaction.

8.37   A dynamic equilibrium has fixed concentrations of all reactants and products - these concentrations do not change with time. However, the process is dynamic because products and reactants are continuously being formed and consumed. The concentrations do not change because the <u>rates</u> of production and consumption are equal.

8.39 $K_{eq} = \dfrac{[NO_2]^2}{[N_2O_4]}$

8.41   A physical equilibrium describes physical change; examples include the equilibrium between ice and water, or the equilibrium vapor pressure of a liquid.

A chemical equilibrium describes chemical change; examples include the reactions shown in questions 8.33 and 8.34.

8.43   a.   Equilibrium shifts to the left. Increasing the temperature increases the energy, a product of the reaction.

       b.   No change; the number of moles of gaseous products and reactants are equal.

       c.   No change; a catalyst has no effect on the equilibrium position of the reaction.

8.45   a.   a slow reaction is an incomplete reaction -- False; A slow reaction may go to completion, but take a longer period of time.

       b.   the rate of forward and reverse reactions is never the same - - False; The rate of forward and reverse reactions are equal in a dynamic equilibrium situation.

8.47   a. PCl_3 increases. Addition of product shifts the equilibrium to the left, favoring reactants.

       b. PCl_3 decreases. Added Cl_2 reacts with PCl_3 to produce products; the equilibrium shifts to the right.

       c. PCl_3 decreases. Removal of product shifts the equilibrium to the right, favoring the formation of more product.

       d. PCl_3 decreases. Decreasing the temperature removes heat from the system. Heat is a product; therefore the equilibrium shifts to the right.

       e. PCl_3 remains the same. Addition of a catalyst has no effect on the equilibrium position.

Chapter 9
Charge-Transfer Reactions:
Acids and Bases and Oxidation-Reduction
Solutions to the Odd-Numbered Problems

****
In-Chapter Questions and Problems
****

9.1   A 1.0×10^{-3} M HCl solution corresponds to a pH = 3.00 (Example 9.1).

        A solution of hydrochloric acid with a pH = 4.00 corresponds to an HCl concentration of 1.0×10^{-4} M (Example 9.2).

9.3   Referring to the discussion of the decimal - based system, a solution of sodium hydroxide producing $[OH^-] = 1.0 \times 10^{-2}$ M corresponds to a pH of 12.00.

9.5   A pH of 8.50 is a non-integer. The calculation of the $[H_3O^+]$ is most easily accomplished with the aid of a calculator.

$$pH = -\log [H_3O^+]$$

        and, $[H_3O^+] = 10^{-pH}$

        On the calculator:
                Enter 8.50
                Press "change sign" key
                Press 10^X key
                The result is $[H_3O^+] = 3.16 \times 10^{-9}$ M

9.7   The acid-base reaction is:

$$HCl(aq) + NaOH(aq) \rightarrow NaCl(aq) + H_2O(l)$$

$$(M_{acid})(V_{liters\ acid}) = (M_{base})(V_{liters\ base})$$

        Solving for M_{base},

$$M_{base} = \frac{(M_{acid})\ (V_{liters\ acid})}{(V_{liters\ base})}$$

Substituting

$$M_{base} = \frac{(0.2000\,M)(20.00\,mL \times \dfrac{1L}{10^3\,mL})}{(40.00\,mL \times \dfrac{1L}{10^3\,mL})}$$

$M_{base} = 0.1000\ M\ NaOH$

9.9   The equilibrium reaction is:

$$CO_2 + H_2O \rightleftharpoons H_2CO_3 \rightleftharpoons H_3O^+ + HCO_3^-$$

An increase in the partial pressure of CO_2 is a stress on the left side of the equilibrium. The equilibrium will shift to the right in an effort to decrease the concentration of CO_2. This will cause the molar concentration of H_2CO_3 to increase.

9.11   The equilibrium reaction is:

$$CO_2 + H_2O \rightleftharpoons H_2CO_3 \rightleftharpoons H_3O^+ + HCO_3^-$$

In Question 9.9, the equilibrium shifts to the right. Therefore the molar concentration of H_3O^+ should increase.

In Question 9.10, the equilibrium shifts to the left. Therefore the molar concentration of H_3O^+ should decrease.

9.13   Propanoic acid is the acid;
[acid] = 2.00×10^{-1} M

Sodium acetate is the salt;
[salt] = 2.00×10^{-1} M

The equilibrium is

$$C_2H_5COOH(aq) + H_2O(l) \rightleftharpoons H_3O^+(aq) + C_2H_5COO^-(aq)$$
$$\text{acid} \qquad\qquad\qquad\qquad\qquad\quad \text{salt}$$

and the hydronium ion concentration,

$$[H_3O^+] = \frac{[acid]\,Ka}{[salt]}$$

Substituting the values given in the problem

$$[H_3O^+] = \frac{[2.00 \times 10^{-1}]\,1.34 \times 10^{-5}}{[2.00 \times 10^{-1}]}$$

$$[H_3O^+] = 1.34 \times 10^{-5}$$

and since

$$pH = -\log[H_3O^+]$$

$$pH = \log 1.34 \times 10^{-5}$$

$$pH = 4.87$$

****
End-of-Chapter Questions and Problems
****

9.15   a.   An Arrhenius acid is a substance that dissociates, producing hydrogen ions.

       b.   A Brŏnsted-Lowry acid is a substance that behaves as a proton donor.

9.17   The Brŏnsted-Lowry theory provides a broader view of acid-base theory than does the Arrhenius theory. Brŏnsted-Lowry emphasizes the role of the solvent in the dissociation process.

9.19 $[H_3O^+][OH^-] = 1.0 \times 10^{-14}$

Solving for $[H_3O^+]$,

$$[H_3O^+] = \frac{1.0 \times 10^{-14}}{[OH^-]}$$

       a.   Substituting $[OH^-] = 1.0 \times 10^{-7}M$

$$[H_3O^+] = \frac{1.0 \times 10^{-14}}{1.0 \times 10^{-7}} = 1.0 \times 10^{-7}M$$

       b.   Substituting $[OH^-] = 1.0 \times 10^{-3}\,M$

$$[H_3O^+] = \frac{1.0 \times 10^{-14}}{1.0 \times 10^{-3}} = 1.0 \times 10^{-11} \text{ M}$$

9.21   a.   Neutral, $[H_3O^+] = 1.0 \times 10^{-7} \text{ M}$

   b.   Basic, $[H_3O^+]$ is less than $1.0 \times 10^{-7} \text{ M}$

9.23   a.   pH $= -\log [H_3O^+]$

       pH $= -\log [1.0 \times 10^{-7}]$

       pH $= 7.00$

   b.   pH $= -\log [H_3O^+]$

       Since $[H_3O^+][OH^-] = 1.0 \times 10^{-14}$

       and, $[H_3O^+] = \dfrac{1.0 \times 10^{-14}}{[OH^-]}$

       Substituting, $[H_3O^+] = \dfrac{1.0 \times 10^{-14}}{1.0 \times 10^{-9}}$

       $[H_3O^+] = 1.0 \times 10^{-5} \text{M}$

       and, pH $= -\log [1.0 \times 10^{-5}]$

           pH $= 5.00$

9.25   a.   pH $= -\log [H_3O^+]$

       and, $[H_3O^+] = 10^{-pH}$

       On the calculator
           Enter 1.00
           Press "change sign" key
           Press 10^x key
           The result is $[H_3O^+] = 1.0 \times 10^{-1} \text{M}$

       and, $[H_3O^+][OH^-] = 1.0 \times 10^{-14}$

solving for [OH⁻]

$$[OH^-] = \frac{1.0 \times 10^{-14}}{[H_3O^+]} = \frac{1.0 \times 10^{-14}}{1.0 \times 10^{-1}}$$

$$[OH^-] = 1.0 \times 10^{-13} M$$

b. $pH = -\log [H_3O^+]$

and, $[H_3O^+] = 10^{-pH}$

On the calculator:
        Enter 9.00
        Press "change sign" key
        Press 10^x key
        The result is $[H_3O^+] = 1.0 \times 10^{-9} M$

and, $[H_3O^+][OH^-] = 1.0 \times 10^{-14}$

solving for [OH⁻]

$$[OH^-] = 1.0 \times \frac{10^{-14}}{[H_3O^+]} = \frac{1.0 \times 10^{-14}}{1.0 \times 10^{-9}}$$

$$[OH^-] = 1.0 \times 10^{-5} M$$

9.27   a. $pH = -\log [H_3O^+]$

and, $\{H_3O^+] = 10^{pH}$

On the calculator:
        Enter 1.30
        Press "change sign" key
        Press 10^x key
        The result is $[H_3O^+] = 5.0 \times 10^{-2} M$

and, $[H_3O^+][OH^-] = 1.0 \times 10^{-14}$

solving for [OH⁻]

$$[OH^-] = \frac{1.0 \times 10^{-14}}{[H_3O^+]} = \frac{1.0 \times 10^{-14}}{5.0 \times 10^{-2}}$$

$$[OH^-] = 2.0 \times 10^{-13} M$$

b. $pH = -\log [H_3O^+]$

and, $[H_3O^+] = 10^{-pH}$

On the calculator:
    Enter 9.70
    Press "change sign" key
    Press 10^x key
    The result is $[H_3O^+] = 2.0 \times 10^{-10} M$

and, $[H_3O^+][OH^-] = 1.0 \times 10^{-14}$

solving for $[OH^-]$

$$[OH^-] = \frac{1.0 \times 10^{-14}}{[H_3O^+]} = \frac{1.0 \times 10^{-14}}{2.0 \times 10^{-10}}$$

$$[OH^-] = 5.0 \times 10^{-5} M$$

9.29   A neutralization reaction is one in which an acid and a base react to produce water and a salt (a "neutral solution).

9.31   a. $pH = -\log \{H_3O^+]$

and, $[H_3O^+] = 10^{-pH}$

On the calculator:
    Enter 6.00
    Press "change sign" key
    Press 10^x key
    The result is $[H_3O^+] = 1.0 \times 10^{-6} M$

and, $[H_3O^+] [OH^-] = 1.0 \times 10^{-14}$

solving for $[OH^-]$

$$[OH^-] = \frac{1.0 \times 10^{-14}}{[H_3O^+]} = \frac{1.0 \times 10^{-14}}{1.0 \times 10^{-6}}$$

$[OH^-] = 1.0 \times 10^{-8}M$

b. $pH = -\log [H_3O^+]$

and, $[H_3O^+] = 10^{-pH}$

On the calculator:
    Enter 5.20
    Press "change sign" key
    Press 10^x key
    The result is $[H_3O^+] = 6.3 \times 10^{-6}M$

and, $[H_3O^+] [OH^-] = 1.0 \times 10^{-14}$

solving for $[OH^-]$

$$[OH^-] = \frac{1.0 \times 10^{-14}}{[H_3O^+]} = \frac{1.0 \times 10^{-14}}{6.3 \times 10^{-6}}$$

$[OH^-] = 1.6 \times 10^{-9}M$

c. $pH = -\log \{H_3O^+]$

and, $[H_3O^+] = 10^{-pH}$

On the calculator:
    Enter 7.80
    Press "change sign" key
    Press 10^x key
    The result is $[H_3O^+] = 1.6 \times 10^{-8}M$

and, $[H_3O^+] [OH^-] = 1.0 \times 10^{-14}$

solving for $[OH^-]$

$$[OH^-] = \frac{1.0 \times 10^{-14}}{[H_3O^+]} = \frac{1.0 \times 10^{-14}}{1.6 \times 10^{-8}}$$

$$[OH^-] = 6.3 \times 10^{-7} M$$

9.33 One pH unit difference corresponds to a tenfold (10^1) difference in concentration of H_3O^+ or OH^-.

    a. $4 - 2 = 2$ pH units $\therefore$ $(10^1)^2$ or 1×10^2

    b. $11 - 7 = 4$ pH units $\therefore$ $(10^1)^4$ or 1×10^4

    c. $12 - 2 = 10$ pH units $\therefore$ $(10^1)^{10}$ or 1×10^{10}

9.35   a. NH_3 is a weak base.
            NH_4Cl is a salt formed from NH_3.
            Therefore, NH_3 and NH_4Cl can form a buffer solution.

       b. HNO_3 is a strong acid; strong acids are not suitable for buffer preparation; they are completely dissociated.
            Therefore, HNO_3 and KNO_3 cannot form a buffer solution.

9.37   a.   A buffer solution contains components (a weak acid and its salt or a weak base and its salt) that enable the solution to resist large changes in pH when acids or bases are added.

       b.   Acidosis is a medical condition characterized by higher-than-normal levels of CO_2 in the blood and lower-than-normal blood pH.

9.39   a.   Addition of strong acid is equivalent to adding H_3O^+. This is a stress on the right side of the equilibrium and the equilibrium will shift to the left. Consequently the $[CH_3COOH]$ increases.

       b.   Water, in this case, is a solvent and does not appear in the equilibrium expression. Hence, it does not alter the position of the equilibrium.

9.41 $$K_a = \frac{[H_3O^+][salt]}{[acid]}$$

$$[H_3O^+] = \frac{[acid] K_a}{[salt]}$$

$$[H_3O^+] = \frac{[0.200]\,5.80 \times 10^{-7}}{[0.500]}$$

$$[H_3O^+] = 2.32 \times 10^{-7}\ M$$

9.43   a. Oxidation is defined as the loss of electrons, loss of hydrogen atoms, or gain of oxygen atoms.

b. An oxidizing agent removes electrons from another substance. In doing so the oxidizing agent becomes reduced.

9.45   During an oxidation process in an oxidation-reduction reaction the species oxidized <u>loses</u> electrons.

9.47   During an oxidation-reduction reaction the species oxidized is the reducing agent.

9.49 Cl_2 + $2KI$ --> $2KCl + I_2$
             ↑   ↑

        substance reduced   substance oxidized
        oxidizing agent   reducing agent

Chapter 10
Radioactivity and Nuclear Medicine
Solutions to the Odd-Numbered Problems

✴✴✴

In-Chapter Questions and Problems
✴✴✴✴✴✴✴✴✴✴✴✴✴✴✴✴✴✴✴✴✴✴✴✴✴✴✴✴✴✴✴✴✴

10.1   Gamma radiation is very high energy electromagnetic radiation. Other forms of electromagnetic radiation (in descending order of energy) are x-ray, ultraviolet, visible, infra-red, microwave, and radiowave. [See Chapter 2, An Environmental Perspective: Electromagnetic Radiation and its Effects on our Everyday Lives, for more information.

10.3   a. $^{85}_{36}Kr \rightarrow {}^{85}_{37}Rb + {}^{0}_{-1}e$

     b. $^{226}_{88}Ra \rightarrow {}^{4}_{2}He + {}^{222}_{86}Rn$

10.5   The half-life of sodium-24 is 15 hours.
The number (n) of half-lives elapsed is:

$$n = 2.5 \text{ days} \times \frac{24 \text{ hours}}{1 \text{ day}} \times \frac{1 \text{ half-life}}{15 \text{ hours}} = 4 \text{ half-lives}$$

Then,

$$100.0 \text{ ng} \xrightarrow[\text{half-life}]{\text{first}} 50.0 \text{ ng} \xrightarrow[\text{half-life}]{\text{second}} 25.0 \text{ ng} \xrightarrow[\text{half-life}]{\text{third}} 12.5 \text{ ng} \xrightarrow[\text{half-life}]{\text{fourth}} 6.3 \text{ ng}$$

∴ 6.3 ng of sodium-24 remain after 2.5 days.

10.7   The half-life of technetium-99 m is 6 hours.
The number (n) of half-lives elapsed is:

$$n = 12 \text{ hours} \times \frac{1 \text{ half-life}}{6 \text{ hours}} = 2 \text{ half-lives}$$

Then, assume that the original amount is x g.

$$x \text{ g} \xrightarrow[\text{half-life}]{\text{first}} \frac{x}{2} \text{ g} \xrightarrow[\text{half-life}]{\text{second}} \frac{x}{4} \text{ g,}$$

or ¼ of the original amount, x g, remains after 2 half-lives, 12 hours.

10.9   Isotopes with short half-lives release their radiation rapidly. There is much more radiation per unit time observed with short half-life substances; hence, the signal is stronger and the sensitivity of the procedure is enhanced.

10.11   The rem takes into account the relative biological effect of the radiation in addition to the quantity of radiation. This provides a more meaningful estimate of potential radiation damage to human tissue.

****
End-of-Chapter Questions and Problems
****

10.13   a.   Natural radioactivity is the spontaneous decay of a nucleus to produce high-energy particles or rays.
         b.   Background radiation is radiation from natural sources.
         c.   An alpha particle is composed of two protons and two neutrons. It is identical to the nucleus of a helium atom.
         d.   Alpha decay is the release of alpha particles from an unstable nucleus.

10.15   a. $_{2}^{4}\text{He}$

         b. $_{0}^{-1}e$

         c. $_{1}^{1}\text{p}$

         d. $_{92}^{235}\text{U}$

10.17   Alpha and beta particles are matter; gamma radiation is pure energy. Alpha particles are large and relatively slow moving. They are the least energetic and least penetrating. Gamma radiation moves at the speed of light, highly energetic, and most penetrating.

10.19 $_{27}^{60}\text{Co} \rightarrow {_{28}^{60}\text{Ni}} + {_{-1}^{0}\beta} + \delta$

10.21 $_{11}^{23}\text{Na} + {_{1}^{2}\text{H}} \rightarrow {_{11}^{24}\text{Na}} + {_{1}^{1}\text{H}}$

10.23   Natural radioactivity is a spontaneous process; artificial radioactivity is nonspontaneous and results from a nuclear reaction that produces an unstable nucleus.

10.25   •   Nuclei for light atoms tend to be most stable if their neutron/proton ratio is close to 1.
          •   Nuclei with more than 84 protons tend to be unstable.
          •   Isotopes with a "magic number" of protons or neutrons (2, 8, 20, 50, 82, or 126 protons or neutrons) tend to be stable.
          •   Isotopes with even numbers of protons or neutrons tend to be more stable.

10.27   Fission splits nuclei to produce energy.

10.29   Radiocarbon dating is a process used to determine the age of objects. The ratio of the masses of the stable isotope, carbon-12, and unstable isotope, carbon-14, is measured. Using this value and the half-life of carbon-14, the age of the coffin may be calculated.

10.31   The half-life of iodine-131 is 8.1 days.
The number (n) of half-lives elapsed is:

$$n = 24 \text{ days} \times \frac{1 \text{ half-life}}{8.1 \text{ days}} = 3.0 \text{ half-lives (2 } significant \; figures)$$

Then,

$$3.2 \text{ mg} \xrightarrow[\text{half-life}]{\text{first}} 1.6 \text{ mg} \xrightarrow[\text{half-life}]{\text{second}} 0.80 \xrightarrow[\text{half-life}]{\text{third}} 0.40 \text{ mg}$$

∴ 0.40 mg of iodine-131 remains after 24 days.

10.33   The half-life of iron-59 is 45 days.
The number (n) of half-lives elapsed is:

$$n = 135 \text{ days} \times \frac{1 \text{ half-life}}{45 \text{ days}} = 3.0 \text{ half-lives}$$

Then,

$$1.00 \times 10^2 \text{ mg} \xrightarrow[\text{half-life}]{\text{First}} 5.0 \times 10^1 \text{ mg} \xrightarrow[\text{half-life}]{\text{second}} 2.5 \times 10^1 \text{ mg} \xrightarrow[\text{half-life}]{\text{third}} 1.3 \times 10^1 \text{ mg}$$

∴ 13 mg of iron-59 remains after 135 days.

10.35 $^{108}_{47}\text{Ag} + ^4_2\alpha \rightarrow ^{112}_{49}\text{In}$

$^{112}_{49}\text{In}$ is the intermediate isotope of indium.

10.37 a. Technetium-99 m is used to study the heart (cardiac output, size, and shape), kidney (follow-up procedure for kidney transplant), and liver and spleen (size, shape, presence of tumors).

b. Xenon-133 is used to locate regions of reduced ventilation and presence of tumors in the lung.

10.39 a. The fission process involves the breaking down of large, unstable nuclei into smaller, more stable nuclei. This process releases some of the binding energy in the form of heat and/or light.

b. The heat generated during the fusion process could be used to generate steam, which is then used to drive a turbine to create electricity.

10.41 ^3_1H + ^1_1H ----→ ^4_2He + energy

10.43 A "breeder" reactor creates the fuel which can be used by a conventional fission reactor during its fission process.

10.45 a. The level of radiation exposure decreases as the distance from the radioactive source increases.

b. Wearing gloves provides a level of shielding that is very efficient for α and β radiation, but totally ineffective for δ radiation.

10.47 Background radiation, radiation from natural sources, is emitted by the sun as cosmic radiation, and from naturally radioactive isotopes found throughout our environment.

10.49 Relative biological effect is a measure of the damage to biological tissue caused by different forms of radiation.

10.51 a. The curie is the amount of radioactive material needed to produce 3.7×10^{10} atomic disintegrations per second.
b. The roentgen is the amount of radioactive material needed to produce 2×10^{10} ion-pairs when passing through 1 cc of air at 0 °C.

10.53 A film badge detects gamma radiation by darkening photographic film in proportion to the amount of radiation exposure over time. Badges are periodically collected and evaluated for their level of exposure. This mirrors the level of exposure of the personnel wearing the badges.

Chapter 11
An Introduction to Organic Chemistry:
The Saturated Hydrocarbons
Solutions to the Odd-Numbered Problems

****
In-Chapter Questions and Problems
****

11.1   a. 2,3-Dimethylbutane
        b. 2,2-Dimethylpentane
        c. 2,2-Dimethylpropane
        d. 1,2,3-Tribromopropane

11.3   a. The linear isomers of molecular formula C_4H_9Br:

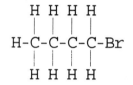

      1-Bromobutane   2-Bromobutane

   b. The linear isomers of molecular formula $C_4H_8Br_2$:

  1,1-Dibromobutane   1,2-Dibromobutane   1,3-Dibromobutane

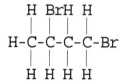

  1,4-Dibromobutane   2,2-Dibromobutane   2,3-Dibromobutane

11.5   a. 1-Bromo-2-ethylcyclobutane
        b. 1,2-Dimethylcyclopropane
        c. Propylcyclohexane

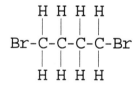

11.7 The following diagram represents the Newman projection of the staggered conformation of propane:

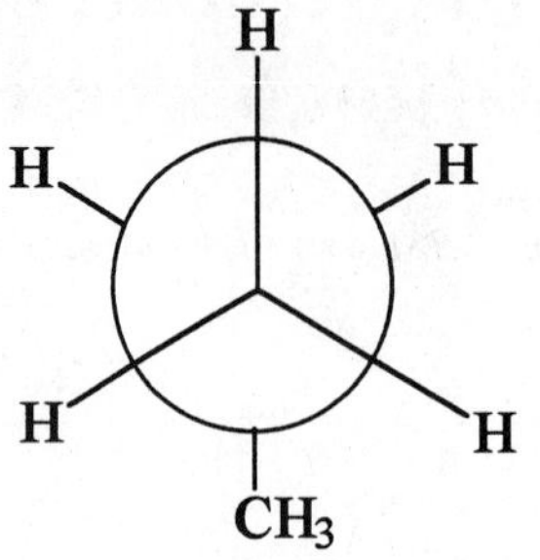

H

H   H

H   H

CH₃

11.9 Three of the six axial hydrogen atoms of cyclohexane lie above the ring. The remaining three hydrogen atoms lie below the ring.

11.11 The following are the three isomers of cyclopropane:

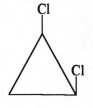

cis-1,2-Dichlorocyclopropane

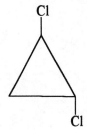

trans-1,2-Dichlorocyclopropane

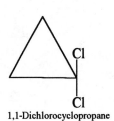

1,1-Dichlorocyclopropane

****

End-of-Chapter Questions and Problems
****

11.13 a. Water-soluble inorganic compounds are good electrolytes because they dissociate into ions in water. The ions conduct an electrical charge.
b. Inorganic compounds exhibit ionic bonding.
c. Organic compounds have lower melting points.
d. Inorganic compounds are more likely to be water-soluble.
e. Organic compounds are flammable.

11.15 a.

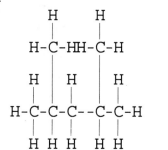

```
        H       H
        |       |
    H-C-HH-C-H
        H   |   H
        |   |   |
    H-C-C-C--C-C-H
        H H H   H H
```

b.
```
    H BrH H
    | | | |
H-C-C-C-C-H
    | | | |
    H H BrH
```

c.
```
                H
                |
            H-C-H
    H H H H |   H H
    | | | | |   | |
H-C-C-C-C-C-C-C-H
    | |   | | | |
    H H   H H H H
        |
    H-C-H
        |
        H
```

d.
```
    H H H H H BrH
    | | | | | | |
H-C-C-C-C-C-C-C-H
    | | | | | | |
    H H H H H H H
```

11.17 Structure b is not possible because there are five bonds to carbon-2. Structure d is not possible because there are five bonds to carbon-3. Structure e is not possible because there are five bonds to carbon-2 and carbon-3. Structure f is not possible because there are five bonds to carbon-3.

11.19 An alcohol - Ethanol:

```
        H H
        | |
    H-C-C-OH
        | |
        H H
```

An aldehyde - Ethanal:

```
        H O
        | ||
    H-C-C-H
        |
        H
```

383

A ketone - Propanone:

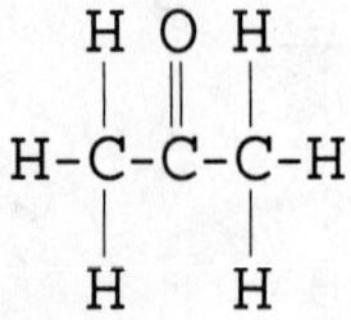

```
      H O H
      | ‖ |
   H-C-C-C-H
      |   |
      H   H
```

A carboxylic acid - Ethanoic acid:

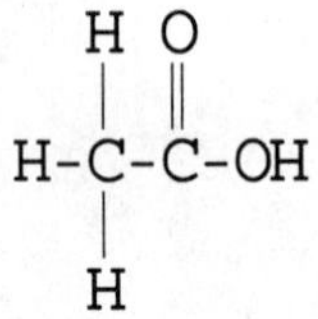

```
      H O
      | ‖
   H-C-C-OH
      |
      H
```

An ester - Methyl ethanoate:

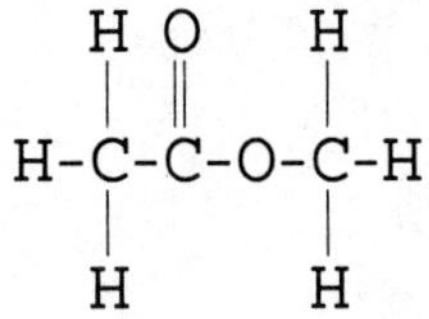

```
      H O   H
      | ‖   |
   H-C-C-O-C-H
      |     |
      H     H
```

An amine - Ethanamine:

```
   H H  H
   | | /
   H-C-C-N
   | | \
   H H  H
```

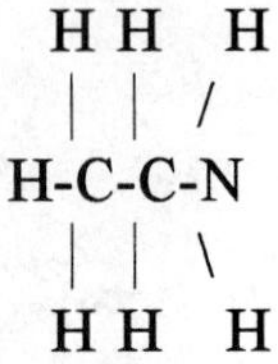

11.21 a. An alkane: C_nH_{2n+2}
   b. An alkyne: C_nH_{2n-2}
   c. An alkene: C_nH_{2n}
   d. A cycloalkane: C_nH_{2n}
   e. A cycloalkene: C_nH_{2n-2}

11.23 Alkanes have only carbon-to-carbon and carbon-to-hydrogen single bonds, as in the molecule ethane:

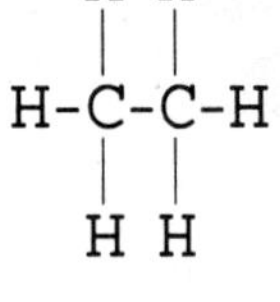

```
      H H
      | |
   H-C-C-H
      | |
      H H
```

Alkenes have at least one carbon-to-carbon double bond, as in the molecule ethene:

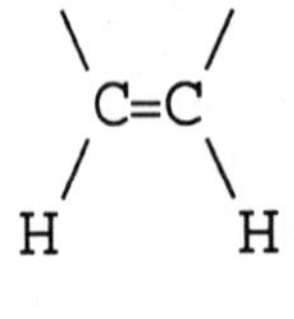

```
   H       H
    \     /
     C=C
    /     \
   H       H
```

Alkynes have at least one carbon-to-carbon triple bond, as in the molecule ethyne:

$$H-C\equiv C-H$$

11.25 a. A carboxylic acid (Propanoic acid):

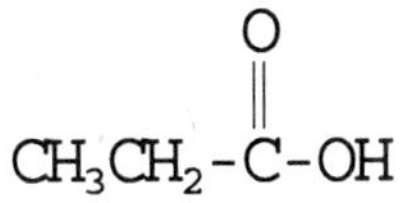

$$CH_3CH_2-\overset{\overset{\displaystyle O}{\|}}{C}-OH$$

b. An amine (Propanamine): $CH_3CH_2CH_2-NH_2$

c. An alcohol (1-Propanol): $CH_3CH_2CH_2-OH$

d. An ester (Ethyl propanoaate):

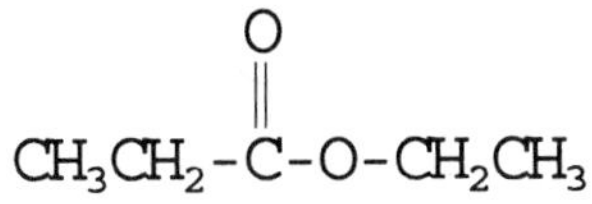

$$CH_3CH_2-\overset{\overset{\displaystyle O}{\|}}{C}-O-CH_2CH_3$$

e. An ether (Diethyl ether): $CH_3CH_2-O-CH_2CH_3$

11.27 a. 2-Bromobutane:

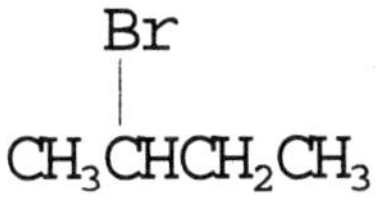

$$\overset{\overset{\displaystyle Br}{|}}{CH_3CHCH_2CH_3}$$

b. 2-Chloro-2-methylpropane:

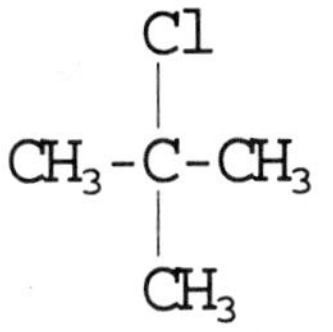

$$CH_3-\underset{\underset{\displaystyle CH_3}{|}}{\overset{\overset{\displaystyle Cl}{|}}{C}}-CH_3$$

c. 2,2-Dimethylhexane:

$$CH_3-\underset{\underset{\displaystyle CH_3}{|}}{\overset{\overset{\displaystyle CH_3}{|}}{C}}-CH_2CH_2CH_2CH_3$$

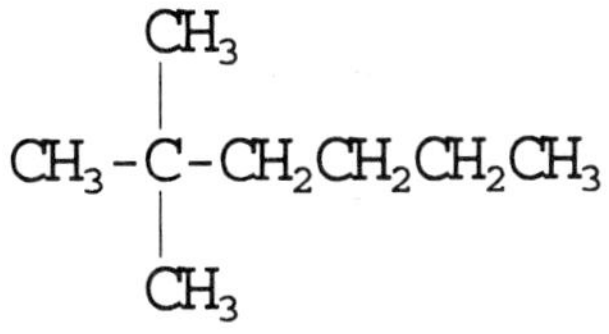

d. Dichlorodiiodomethane:

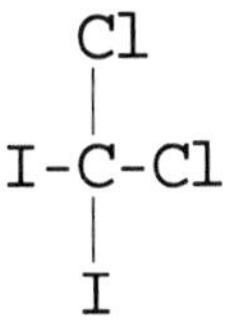

$$I-\underset{\underset{\displaystyle I}{|}}{\overset{\overset{\displaystyle Cl}{|}}{C}}-Cl$$

e. 1,4-Diethylcyclohexane:

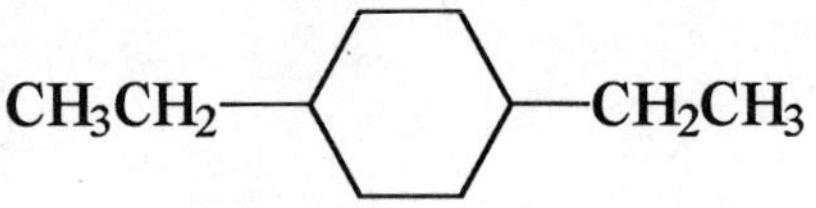

$$CH_3CH_2-\bigcirc-CH_2CH_3$$

f. 2-Iodo-2,4,4-trimethylpentane:

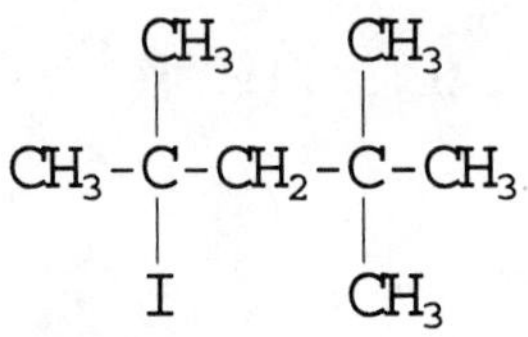

$$
\begin{array}{ccccc}
 & CH_3 & & CH_3 & \\
 & | & & | & \\
CH_3 - & C & -CH_2- & C & -CH_3 \\
 & | & & | & \\
 & I & & CH_3 &
\end{array}
$$

11.29 a. 3-Methylpentane
      b. 1-Bromoheptane
      c. 3-Ethyl-5-methylheptane
      d. 2-Bromo-2-methylpropane
      e. 2,5-Dimethylhexane
      f. 1-Chloro-3-methylbutane
      g. 1,4-Dichloropentane

11.31 a. Identical - both are 2-bromobutane
      b. Identical - both are 3-bromo-5-methylhexane
      c. Identical - both are 2,2-dibromobutane
      d. Isomers of molecular formula $C_6H_{12}Br_2$: 1,3-dibromo-3-methylpentane and 1,4-dibromo-2-ethylbutane

11.33 Structure a is incorrect because there are only three bonds to carbon-3. Structure c is incorrect because there are six bonds to carbon-2.

11.35 First, determine the name of the parent compound, the longest continuous carbon chain in the compound. Number the parent chain to give the lowest number to the carbon bonded to the first substituent encountered. Place the names and numbers of the substituents before the name of the parent compound. Substituents are listed in alphabetical order.

11.37 a. 1,1,3-Trimethylcyclohexane
      b. 1,1-Dimethylcyclopropane
      c. 1,3,5,7-Tetramethylcyclooctane
      d. 1,2,3,4,5-Pentachlorocyclopentane
      e. Ethylcyclopentane

g. 1-Methyl-2-propylcyclobutane

h. 1,2-Dichlorocyclohexane

11.39 The general formula of a cycloalkane is C_nH_{2n}.

11.41 a. Incorrect - 1,2-Dibromocyclobutane
b. Incorrect - 1,2-Diethylcyclobutane
c. Correct
d. Incorrect - 1,2,3-Trichlorocyclohexane

11.43 In the chair conformation the hydrogen atoms and thus the electron pairs of the C-H bonds, are farther from one another. As a result, there is less electron repulsion and the structure is more stable (more energetically favored). In the boat conformation, the electron pairs are more crowded. This causes greater electron repulsion, producing a less stable, less energetically favored conformation.

11.45 Because conformations are freely and rapidly interconverted, they cannot be separated from one another.

11.47 The following is the Newman projection of the staggered conformation of chloroethane:

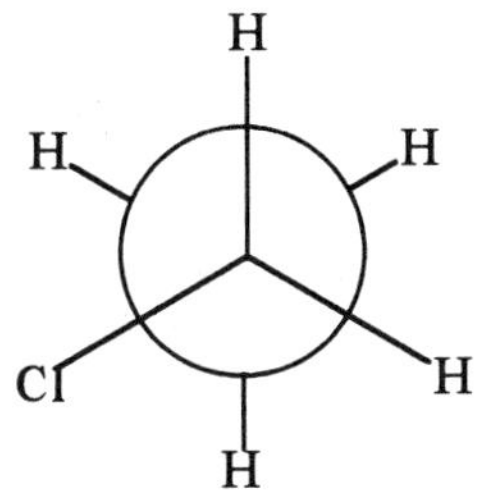

11.49 The staggered conformation is more stable than the eclipsed conformation because the electron pairs of the carbon-hydrogen bonds are farther from one another in the staggered conformation. This results in less electron repulsion and a more stable conformation. In the eclipsed conformation, the electrons are more crowded, resulting in greater electron repulsion and a less stable conformation.

11.51

a. *cis*-1,3-Dibromocyclopentane:

Br
Br

b. *trans*-1,2-Dimethylcyclobutane:

CH3
CH3

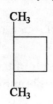

c. *cis*-1,2-Dichlorocyclopropane

Cl
Cl

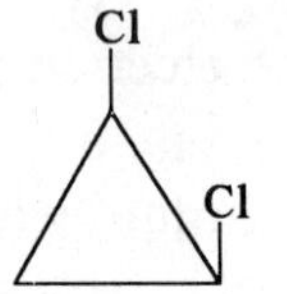

d. trans-1,4-Diethylcyclohexane

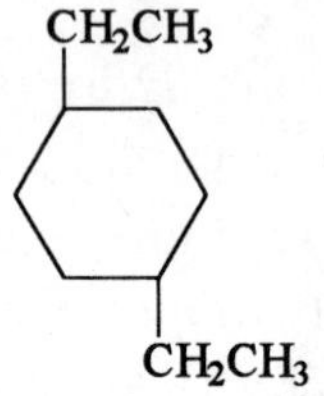

CH2CH3
CH2CH3

11.53 a. *cis*-1,2-Dibromocyclopentane
b. *trans*-1,3-Dibromocyclopentane
c. *cis*-1,2-Dimethylcyclohexane
d. *cis*-1,2-Dimethylcyclopropane

11.55 a. $8\ CO_2 + 10\ H_2O$

b.

CH3
Br-C-CH3
CH3
+
CH3
CH3CHCH2Br
+ 2 HBr

2-Bromo-2-methylpropane 1-Bromo-2-methylpropane

c. Cl_2 + light

11.57 The following molecules are all the constitutional isomers of C_6H_{14}:

CH3CH2CH2CH2CH2CH3
CH3
CH3CHCH2CH2CH3
CH3
CH3CH2CHCH2CH3

Hexane 2-Methylpentane 3-Methylpentane

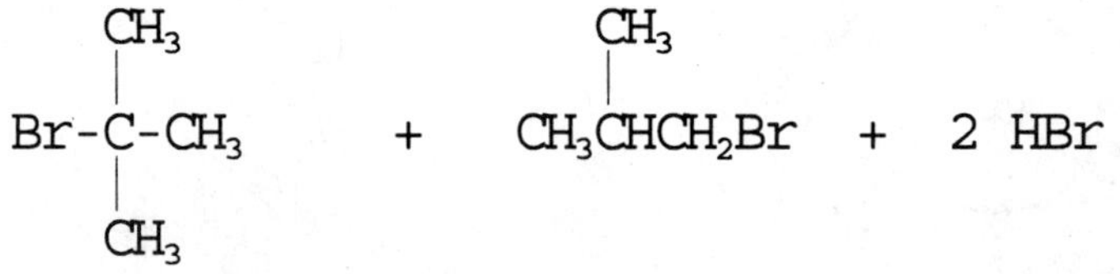

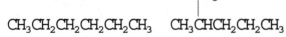

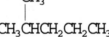

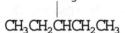

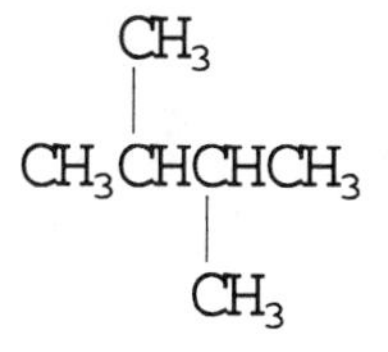

$$CH_3$$
$$|$$
$$CH_3CHCHCH_3$$
$$|$$
$$CH_3$$

$$CH_3$$
$$|$$
$$CH_3CCH_2CH_3$$
$$|$$
$$CH_3$$

2,3-Dimethylbutane 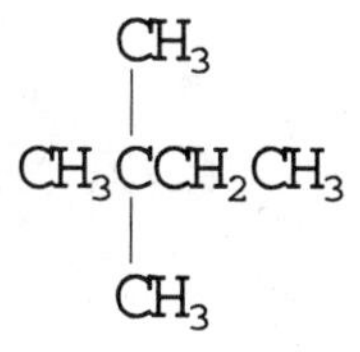   2,2-Dimethylbutane

a. 2,3-Dimethylbutane produces only two monobrominated derivatives: 1-bromo-2,3-dimethylbutane and 2-bromo-2,3-dimethylbutane.

b. Hexane produces three monobrominated products: 1-bromohexane, 2-bromohexane, and 3-bromohexane. 2,2-Dimethylbutane also produces three monobrominated products: 1-bromo-2,2-dimethylbutane, 2-bromo-3,3-dimethylbutane, and 1-bromo-3,3-dimethylbutane.

c. 3-Methylpentane produces four monobrominated products: 1-bromo-3-methylpentane, 2-bromo-3-methylpentane, 3-bromo-3-methylpentane, and 1-bromo-2-ethylbutane.

11.59 The hydrocarbon is cyclooctane, having a molecular formula of C_8H_{16}.

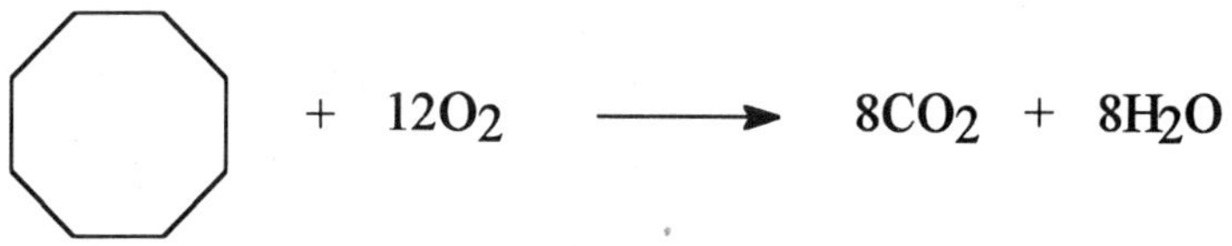

$$\text{(cyclooctane)} + 12O_2 \longrightarrow 8CO_2 + 8H_2O$$

Chapter 12
The Unsaturated Hydrocarbons:
Alkenes, Alkynes, and Aromatics
Solutions to the Odd-Numbered Problems

In-Chapter Questions and Problems
****

12.1   a. 1-Bromo-3-hexyne: c. Dichloroethyne:

$$Br-\overset{\overset{\displaystyle H}{|}}{\underset{\underset{\displaystyle H}{|}}{C}}-\overset{\overset{\displaystyle H}{|}}{\underset{\underset{\displaystyle H}{|}}{C}}-C\equiv C-\overset{\overset{\displaystyle H}{|}}{\underset{\underset{\displaystyle H}{|}}{C}}-\overset{\overset{\displaystyle H}{|}}{\underset{\underset{\displaystyle H}{|}}{C}}-H$$

$$Cl-C\equiv C-Cl$$

b. 2-Butyne: d. 9-Iodo-1-nonyne:

$$H-\overset{\overset{\displaystyle H}{|}}{\underset{\underset{\displaystyle H}{|}}{C}}-C\equiv C-\overset{\overset{\displaystyle H}{|}}{\underset{\underset{\displaystyle H}{|}}{C}}-H$$

$$H-C\equiv C-\overset{\overset{\displaystyle H}{|}}{\underset{\underset{\displaystyle H}{|}}{C}}-\overset{\overset{\displaystyle H}{|}}{\underset{\underset{\displaystyle H}{|}}{C}}-\overset{\overset{\displaystyle H}{|}}{\underset{\underset{\displaystyle H}{|}}{C}}-\overset{\overset{\displaystyle H}{|}}{\underset{\underset{\displaystyle H}{|}}{C}}-\overset{\overset{\displaystyle H}{|}}{\underset{\underset{\displaystyle H}{|}}{C}}-\overset{\overset{\displaystyle H}{|}}{\underset{\underset{\displaystyle H}{|}}{C}}-\overset{\overset{\displaystyle H}{|}}{\underset{\underset{\displaystyle H}{|}}{C}}-I$$

12.3   a. *cis*-3-Octene: b. *trans*-5-Chloro-2-hexene:

$$\underset{\displaystyle H}{\overset{\displaystyle CH_3CH_2}{\diagdown}}C=C\underset{\displaystyle H}{\overset{\displaystyle CH_2CH_2CH_2CH_3}{\diagup}}$$

$$\underset{\displaystyle H}{\overset{\displaystyle CH_3}{\diagdown}}C=C\underset{\displaystyle CH_2CHCH_3}{\overset{\displaystyle H}{\diagup}}\quad\overset{\displaystyle Cl}{\underset{\displaystyle |}{}}$$

c. *trans*-2,3-Dichloro-2-butene:

$$\underset{\displaystyle Cl}{\overset{\displaystyle CH_3}{\diagdown}}C=C\underset{\displaystyle CH_3}{\overset{\displaystyle Cl}{\diagup}}$$

12.5   a. Reactant - *cis*-2-butene;   Only product - butane
       b. Reactant - 1-butene;   Major product - 2-butanol
       c. Reactant - 2-butene;   Only product - 2,3-dichlorobutane
       d. Reactant - 1-pentene;   Major product - 2-bromopentane

12.7

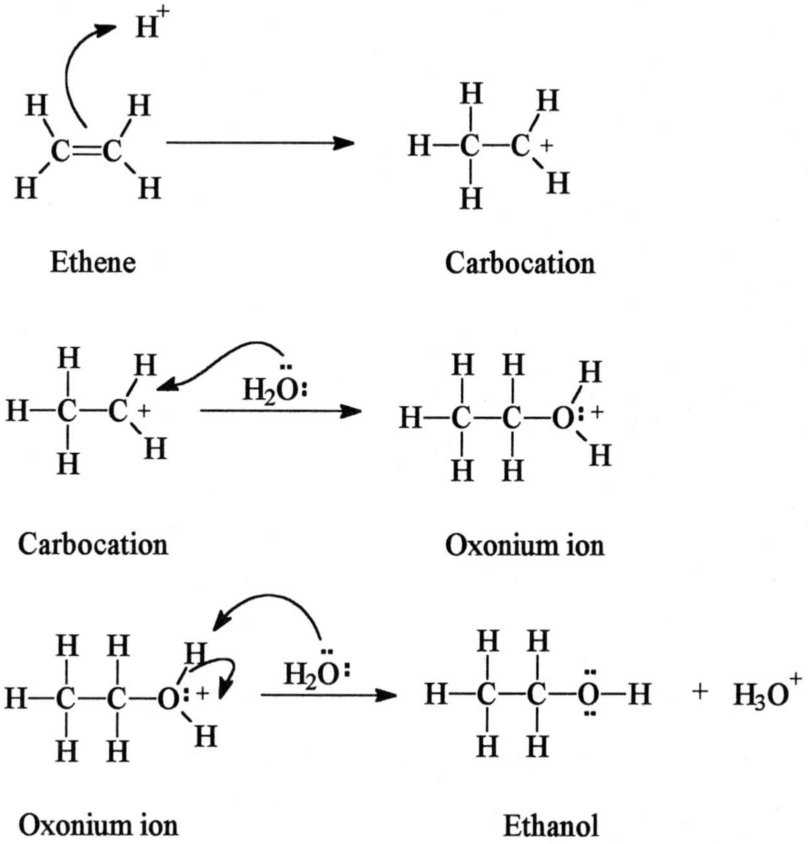

Ethene   Carbocation

Carbocation   Oxonium ion

Oxonium ion   Ethanol

12.9

a. 1,3,5-Trichlorobenzene

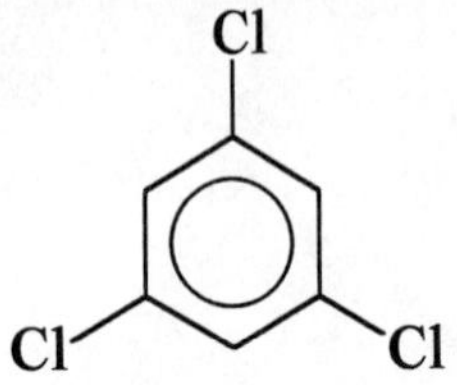

b. *ortho*-Cresol

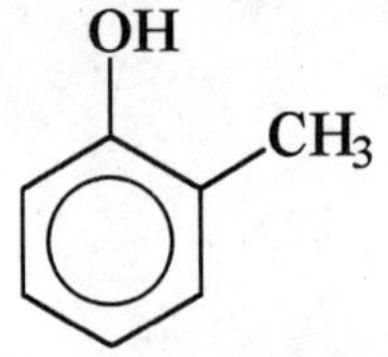

c. 2,5-Dibromophenol

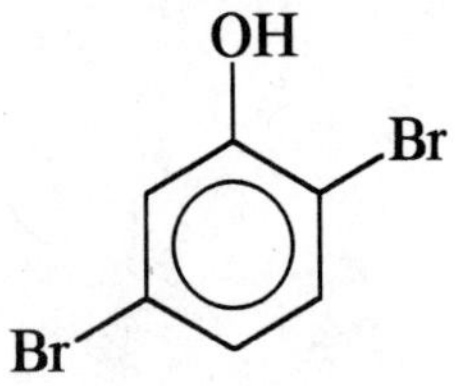

d. *para*-Dinitrobenzene

e. 2-Nitroaniline

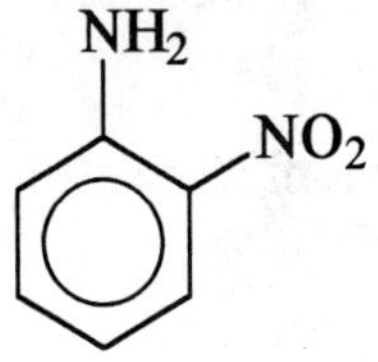

f. *meta*-Nitrotoluene

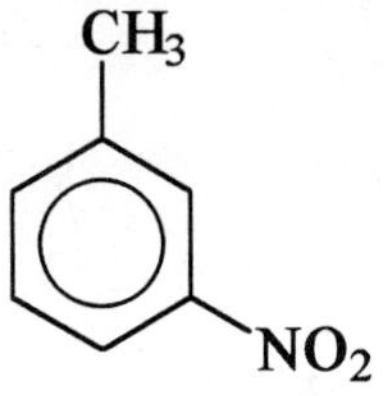

****

End-of-Chapter Questions and Problems

****

12.11 The general formula for an alkane is C_nH_{2n+2}. The general formula for an alkene is C_nH_{2n}. The general formula for an alkyne is C_nH_{2n-2}.

12.13 Ethene is a planar molecule. All the bond angles are 120°.

12.15 Ethyne is a linear molecule. All the bond angles are 180°.

12.17 a. 2-Methyl-2-hexene:

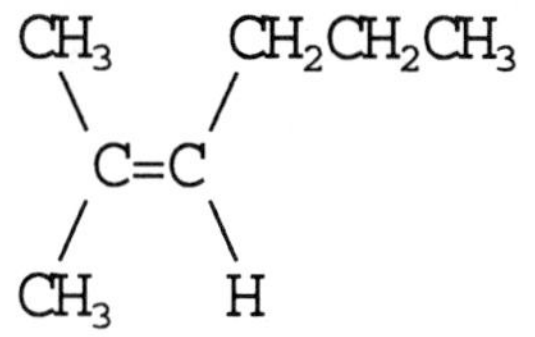

$$CH_3 \quad CH_2CH_2CH_3$$
$$\diagdown \qquad \diagup$$
$$C=C$$
$$\diagup \qquad \diagdown$$
$$CH_3 \qquad H$$

b. *trans*-3-Heptene:

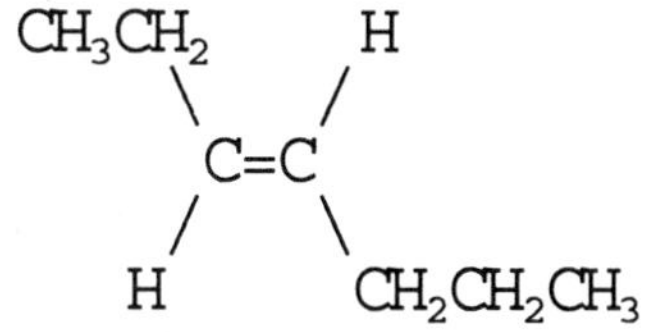

$$CH_3CH_2 \qquad\qquad H$$
$$\diagdown \qquad\qquad \diagup$$
$$C=C$$
$$\diagup \qquad\qquad \diagdown$$
$$H \qquad\qquad CH_2CH_2CH_3$$

c. *cis*-1-Chloro-2-pentene:

$$Cl$$
$$|$$
$$CH_2 \qquad CH_2CH_3$$
$$\diagdown \qquad \diagup$$
$$C=C$$
$$\diagup \qquad \diagdown$$
$$H \qquad\quad H$$

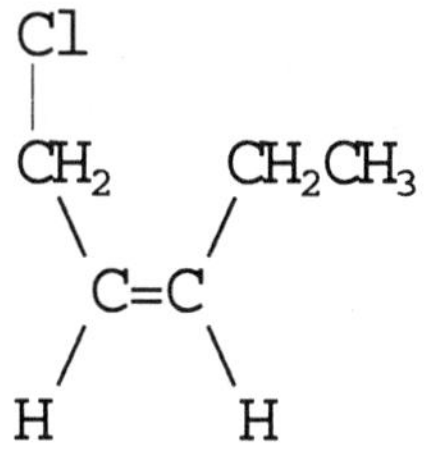

d. *cis*-2-Chloro-2-methyl-3-heptene:

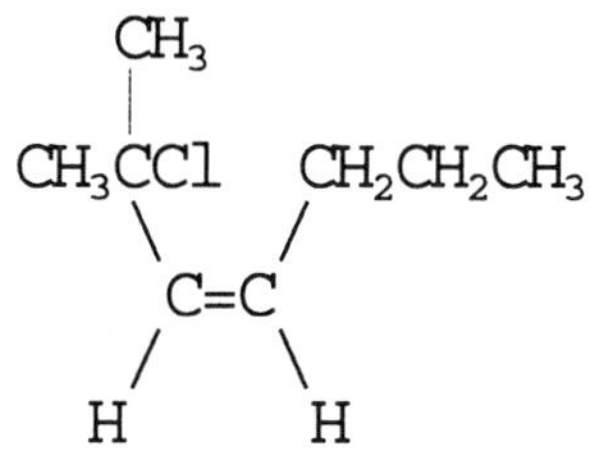

$$CH_3$$
$$|$$
$$CH_3CCl \quad CH_2CH_2CH_3$$
$$\diagdown \qquad\quad \diagup$$
$$C=C$$
$$\diagup \qquad\quad \diagdown$$
$$H \qquad\qquad H$$

e. *trans*-5-Bromo-2,6-dimethyl-3-octene:

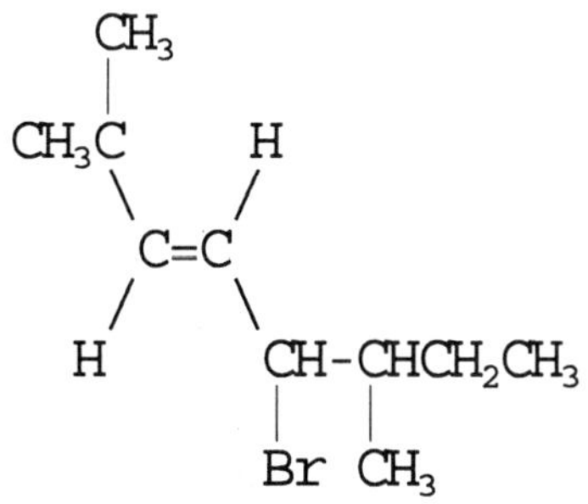

$$CH_3$$
$$|$$
$$CH_3C \qquad H$$
$$\diagdown \qquad\quad \diagup$$
$$C=C$$
$$\diagup \qquad\quad \diagdown$$
$$H \qquad\quad CH-CHCH_2CH_3$$
$$\qquad\qquad | \quad\; |$$
$$\qquad\qquad Br \; CH_3$$

12.19 a. 3-Methyl-1-pentene
       b. 7-Bromo-1-heptene
       c. 5-Bromo-3-heptene

d. 1-*t*-Butyl-4-methylcyclohexene

e. 2,5-Dimethyl-2-hexene

f. 4-Chloro-3-methyl-1-butyne

g. 6-Chloro-1-heptyne

h. 3-Bromo-2-chlorocyclopentene

12.21 a. 2,3-Dibromobutane could not exist as *cis* and *trans* isomers because it is an alkane.

b. 2-Heptene can exist as *cis* and *trans* isomers:

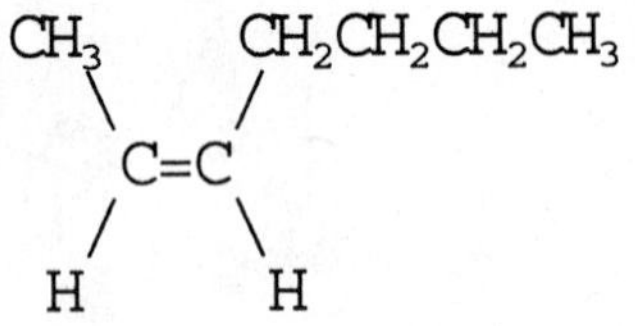

cis-2-Heptene 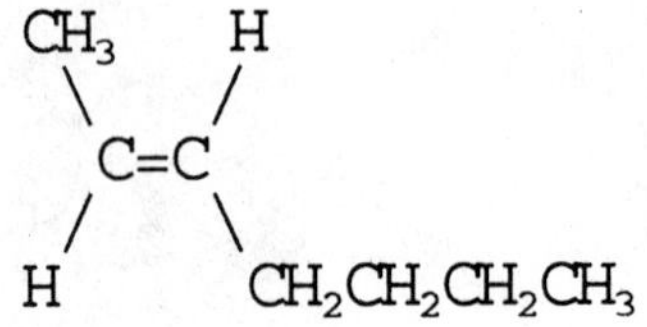   trans-2-Heptene

c. 2,3-Dibromo-2-butene can exist as *cis* and *trans* isomers:

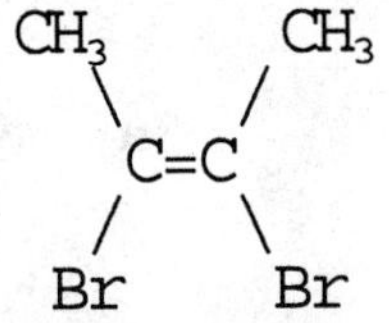

cis-2,3-Dibromo-2-butene 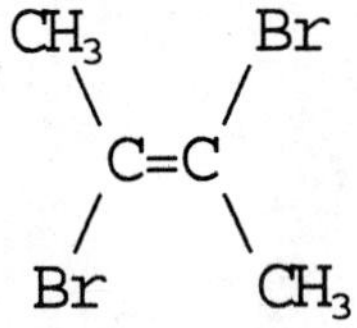   trans-2,3-dibromo-2-butene

d. Propene cannot exist as *cis* and *trans* isomers.

e. 1-Bromo-1-chloro-2-methylpropene cannot exist as *cis* and *trans* isomers.

f. 1,1-Dichloroethene cannot exist as *cis* and *trans* isomers.

g. 1,2-Dibromoethene can exist as *cis* and *trans* isomers.

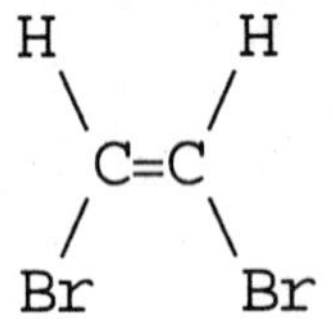

cis-1,2-Dibromoethene 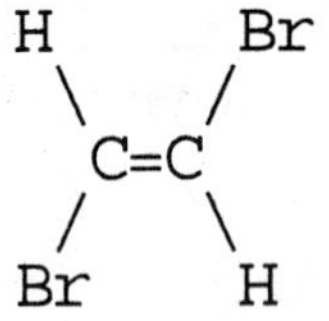   trans-1,2,Dibromoethene

h. 3-Ethyl-2-methyl-2-hexene cannot exist as *cis* and *trans* isomers.

12.23 a. 2-Methyl-4-hexyne is incorrect. The chain should be numbered so that the lowest possible number to a carbon involved in the double bond. The correct name is 4-methyl-2-hexyne.

b. 3-Ethyl-3-hexene is a correct name.

c. 2-Ethyl-7-methyl-4-octyne is incorrect. The parent chain is nine carbons long. The correct name is 2,7-dimethyl-4-nonyne.

d. *trans*-6-Chloro-3-heptene is incorrect. This is the cis isomer, thus the correct name is *cis*-6-chloro-3-heptene.

e. 1-Chloro-4-methyl-2-hexene is incorrect. The complete name of this compound would be trans-1-chloro-4-methyl-2-hexene.

12.25 Addition of bromine (Br_2) to an alkene results in a color change from red to colorless. If equimolar quantities of Br_2 are added to hexene, the reaction mixture will change from red to colorless. This color change will not occur if cyclohexane is used.

12.27 a. H_2

b. H_2O

c. HBr

d. 19 O_2 → 12 CO_2 + 14 H_2O

e. Cl_2

f.

12.29 1-Pentene: $CH_2=CHCH_2CH_2CH_3$

2-Pentene: $CH_3CH=CHCH_2CH_3$

2-Methyl-2-butene:

$$CH_3C=CHCH_3$$
$$|$$
$$CH_3$$

12.31

a. 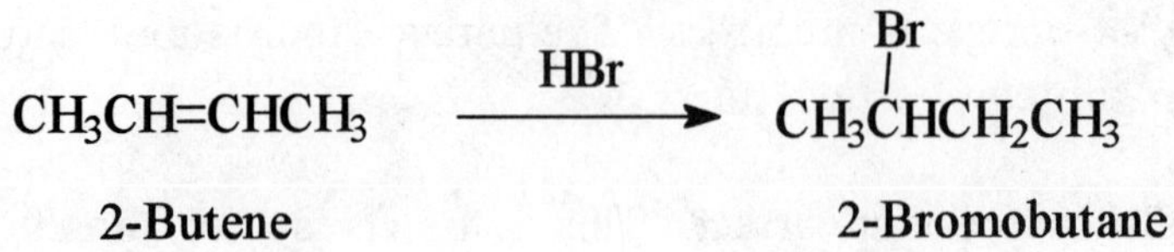 $CH_3CH=CHCH_3$ $\xrightarrow{\text{HBr}}$ $CH_3\overset{\overset{\displaystyle Br}{|}}{C}HCH_2CH_3$

2-Butene   2-Bromobutane

b. 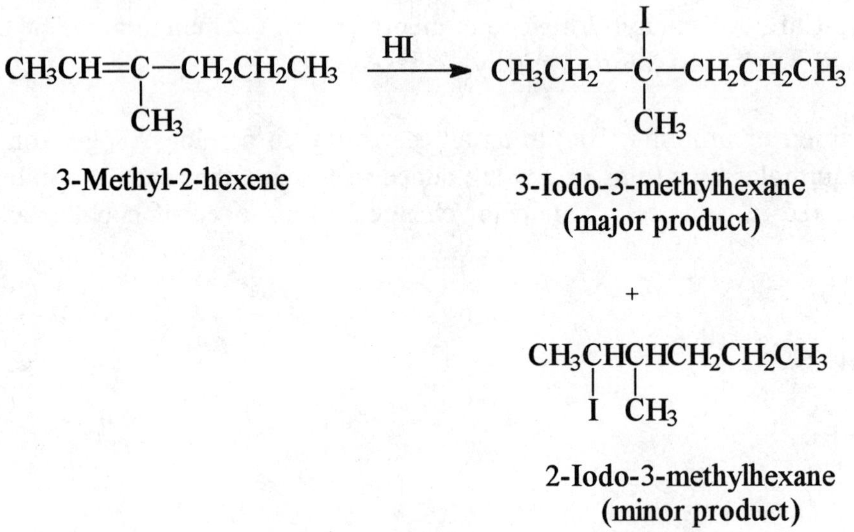 $CH_3CH=\underset{\underset{\displaystyle CH_3}{|}}{C}-CH_2CH_2CH_3$ $\xrightarrow{\text{HI}}$ $CH_3CH_2-\underset{\underset{\displaystyle CH_3}{|}}{\overset{\overset{\displaystyle I}{|}}{C}}-CH_2CH_2CH_3$

3-Methyl-2-hexene   3-Iodo-3-methylhexane
                                (major product)

+

$CH_3\underset{\underset{\displaystyle I}{|}}{C}H\underset{\underset{\displaystyle CH_3}{|}}{C}HCH_2CH_2CH_3$

2-Iodo-3-methylhexane
(minor product)

c.

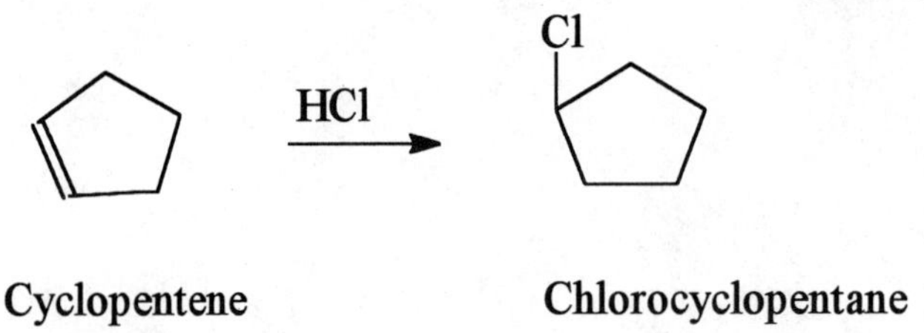

Cyclopentene $\xrightarrow{\text{HCl}}$ Chlorocyclopentane

12.33 A polymer is a macromolecule composed of repeating structural units called monomers.

12.35

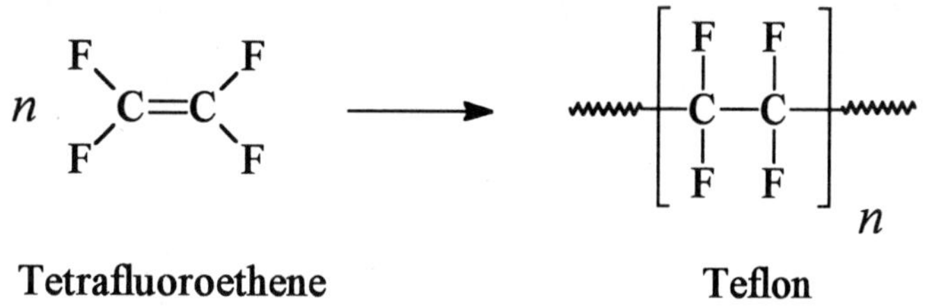

Tetrafluoroethene   Teflon

12.37   A reaction mechanism is a step-by-step explanation of the process by which reactants are converted into products in a chemical reaction. It is useful in understanding the details of the way in which a reaction proceeds.

12.39

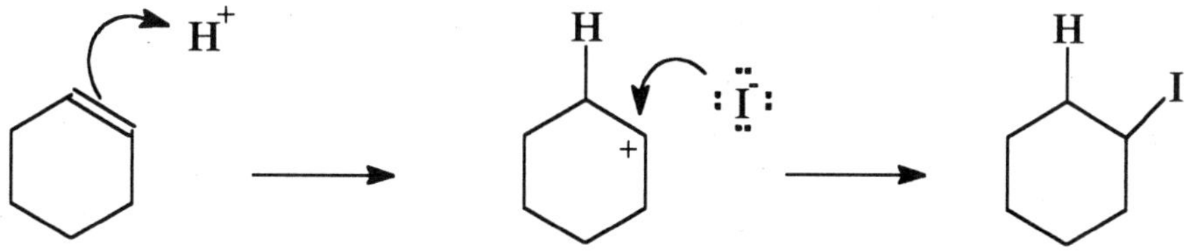

12.41   An electrophile is an electron-poor reactant that seeks electrons from a nucleophile. The carbocation is an electrophile.

12.43

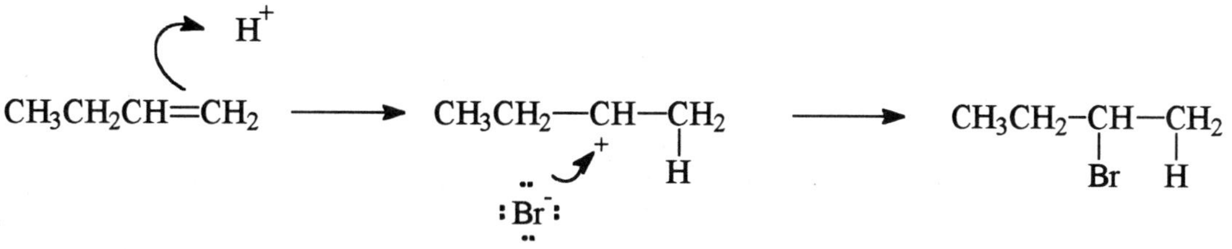

397

12.45 a. 2,4-Dibromotoluene:

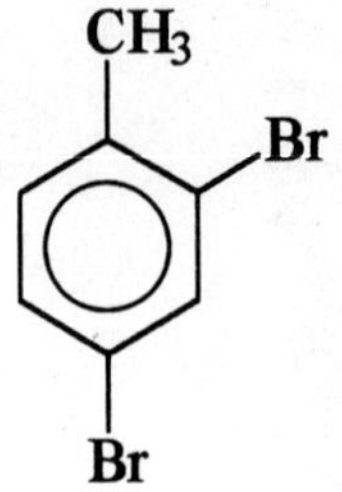

CH$_3$

Br

Br

b. 1,2,4-Triethylbenzene:

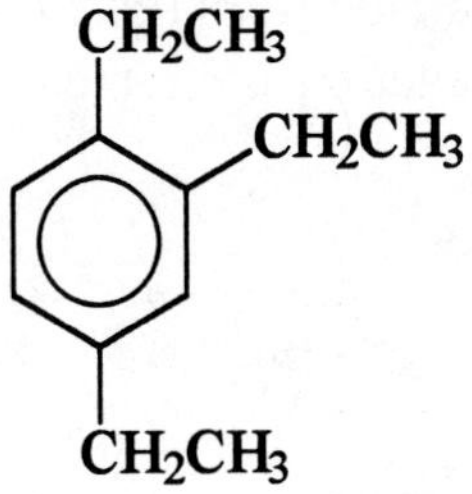

CH$_2$CH$_3$

CH$_2$CH$_3$

CH$_2$CH$_3$

c. Isopropylbenzene:

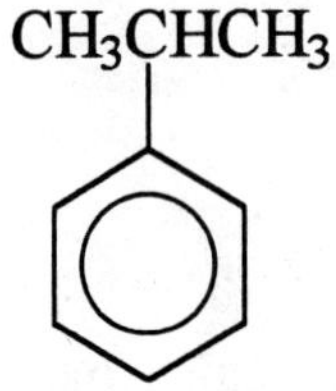

CH$_3$CHCH$_3$

d. 2-Bromo-5-chlorotoluene:

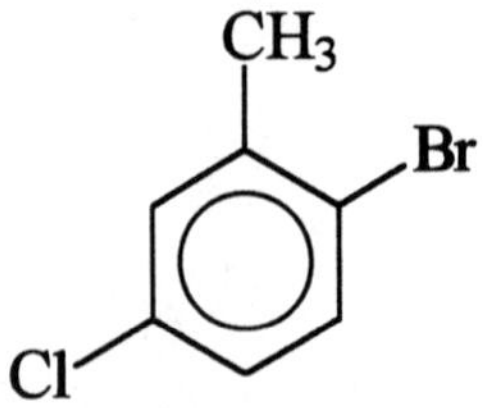

CH$_3$

Br

Cl

12.47

a. *meta*-Cresol:

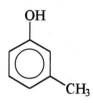

b. Propylbenzene:

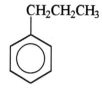

c. 1,3,5-Trinitrobenzene

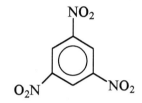

d. *m*-Chlorotoluene

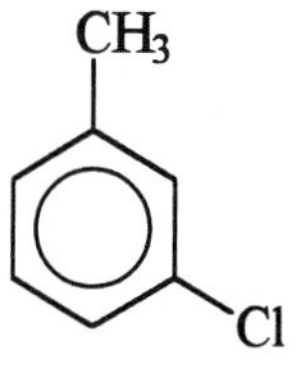

12.49 Kekulé proposed that single and double carbon-carbon bonds alternate around the benzene ring. To explain why benzene does not react like other unsaturated compounds, he proposed that the double and single bonds shift positions rapidly.

12.51 An addition reaction involves addition of a molecule to a double or triple bond in an unsaturated molecule. In a substitution reaction, one chemical group replaces another.

12.53

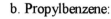

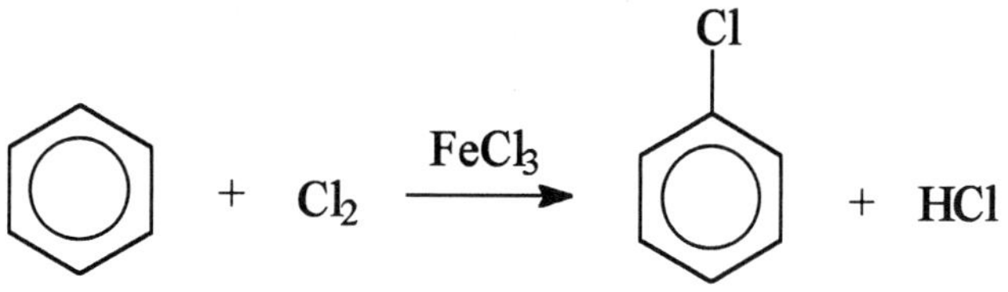

Chapter 13
Alcohols, Phenols, Thiols, and Ethers
Solutions to the Odd-Numbered Problems

****

In-Chapter Questions and Problems

****

13.1   a. 4-Methyl-1-pentanol
        b. 4-Methyl-2-hexanol
        c. 1,2,3-Propanetriol
        d. 4-Chloro-3-methyl-1-hexanol

13.3   a. Primary alcohol
        b. Secondary alcohol
        c. Tertiary alcohol
        d. Aromatic alcohol (phenol)
        e. Secondary alcohol

13.5   a. Ethanol
        b. 2-Propanol is the major product.   1-Propanol is the minor product.
        c. 2-Butanol is the major product.   1-Butanol is the minor product.
        d. 2-Butanol
        e. 2-Methyl-2-propanol is the major product.   2-Methyl-1-propanol is the minor product.

****

End-of-Chapter Questions and Problems

****

13.7 a < d < c < b

13.9   a. CH_3CH_2OH
        b. $CH_3CH_2CH_2CH_2OH$
        c. CH_3CHCH_3
                    |
                   OH

13.11   a. 1-Heptanol
        b. 2-Propanol
        c. 2,2-Dimethylpropanol
        d. 4-Bromo-1-hexanol
        e. 3,3-Dimethyl-2-hexanol
        f. 3-Ethyl-3-heptanol

13.13 a. Cyclopentanol
       b. Cyclooctanol
       c. 3-Methylcyclohexanol

13.15 a. Methyl alcohol
       b. Ethyl alcohol
       c. Ethylene glycol
       d. Propyl alcohol

13.17 Denatured alcohol is 100% ethanol to which benzene or methanol is added. The additive makes the ethanol unfit to drink and prevents illegal use of pure ethanol.

13.19 Fermentation is the anaerobic degradation of sugar that involves no net oxidation. The alcohol fermentation, carried out by yeast, produces ethanol and carbon dioxide.

13.21 When the ethanol concentration in a fermentation reaches 12-13%, the yeast producing the ethanol are killed by it. To produce a liquor of higher alcohol concentration, the product of the original fermentation must be distilled.

13.23 a. Primary alcohol
       b. Secondary alcohol
       c. Tertiary alcohol
       d. Tertiary alcohol
       e. Tertiary alcohol

13.25 a. 2-Pentanol is the major product and 1-pentanol is the minor product.
       b. 2-Pentanol and 3-pentanol
       c. 3-Methyl-2-butanol is the major product and 3-methyl-1-butanol is the minor product.
       d. 3,3-Dimethyl-2-butanol is the major product and 3,3-dimethyl-1-butanol is the minor product.

13.27 a. 2-Butanone
       b. N.R.
       c. Cyclohexanone
       d. N.R.

13.29 a. 3-Pentanone
       b. Propanal (Upon further oxidation, propanoic acid would be formed.)
       c. 4-Methyl-2-pentanone
       d. N.R.
       e. 3-Phenylpropanal (Upon further oxidation, 3-phenylpropanoic acid will be formed.)

13.31

$$CH_3CH_2OH \xrightarrow{\text{liver enzymes}} CH_3\overset{\displaystyle O}{\overset{\|}{C}}-H$$

Ethanol   Ethanal

The product, ethanal, is responsible for the symptoms of a hangover.

13.33 The reaction in which a water molecule is added to 1-butene is a hydration reaction.

$$CH_3CH_2CH{=}CH_2 \;+\; H_2O \xrightarrow{H^+} CH_3CH_2\overset{\displaystyle OH}{\overset{|}{C}}HCH_3$$

        1-Butene   2-Butanol

13.35

$$CH_3CH{=}CH_2 \xrightarrow{H_2O,\ H^+} CH_3\overset{\displaystyle OH}{\overset{|}{C}}{-}CH_3 \xrightarrow{[O]} CH_3{-}\overset{\displaystyle O}{\overset{\|}{C}}{-}CH_3$$

    Propene   2-Propanol   Propanone
 (propylene)   (isopropanol)   (acetone)

13.37

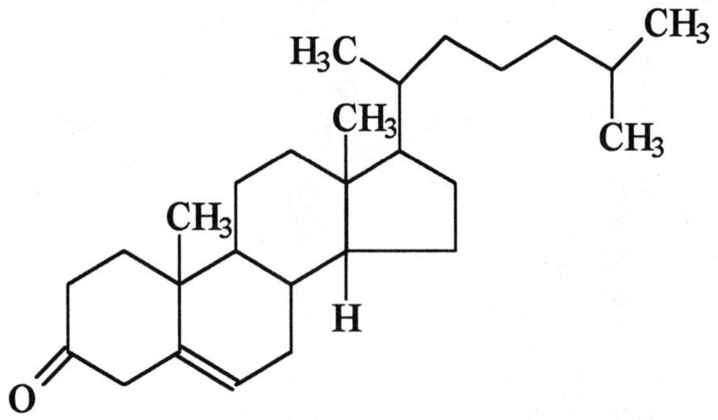

An oxidation product of cholesterol

13.39 When discussing inorganic compounds, oxidation is a loss of electrons, whereas reduction is a gain of electrons.

13.41

$$CH_3CH_2CH_3 \ < \ CH_3CH_2CH_2OH \ < \ CH_3CH_2\overset{\overset{\displaystyle O}{\|}}{C}\!-\!H \ < \ CH_3CH_2\overset{\overset{\displaystyle O}{\|}}{C}\!-\!OH$$

13.43

Picric acid: 2,4,6,-Trinitrotoluene:

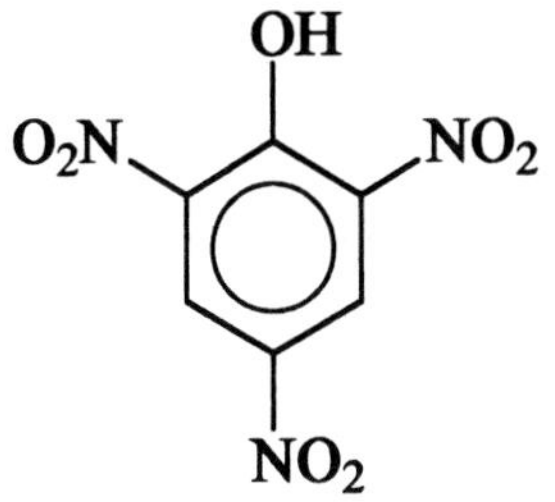

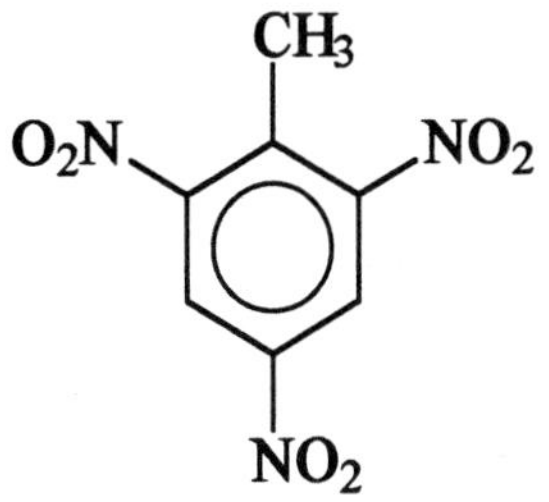

Picric acid is water-soluble because of the polar hydroxyl group that can form hydrogen bonds with water.

13.45 Hexachlorophene, hexylresorcinol, and *o*-phenylphenol are phenol compounds used as antiseptics or disinfectants.

13.47 Alcohols of molecular formula $C_4H_{10}O$

$CH_3CH_2CH_2CH_2OH$

$$CH_3\overset{\overset{\displaystyle OH}{|}}{C}HCH_2CH_3$$

$$CH_3\overset{}{C}HCH_2OH$$
$$\overset{|}{C}H_3$$

$$CH_3-\overset{\overset{\displaystyle OH}{|}}{\underset{\underset{\displaystyle CH_3}{|}}{C}}-CH_3$$

Ethers of molecular formula $C_4H_{10}O$

$CH_3-O-CH_2CH_2CH_3$ $CH_3CH_2-O-CH_2CH_3$

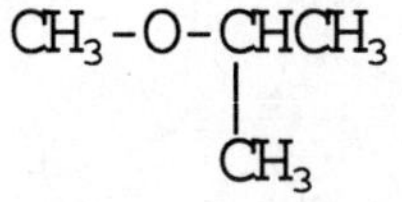

$CH_3-O-\underset{\underset{\displaystyle CH_3}{|}}{C}HCH_3$

13.49   Penthrane: 2,2-Dichloro-1,1-difluoro-1-methoxyethane
        Enthrane: 2-Chloro-1-(difluoromethoxy)-1,1,2-trifluoroethane

13.51

   a. $CH_3CH_2-O-CH_2CH_3$ + H_2O

   b. $CH_3CH_2-O-CH_2CH_3$ + CH_3-O-CH_3 + $CH_3-O-CH_2CH_3$ + H_2O

   c. 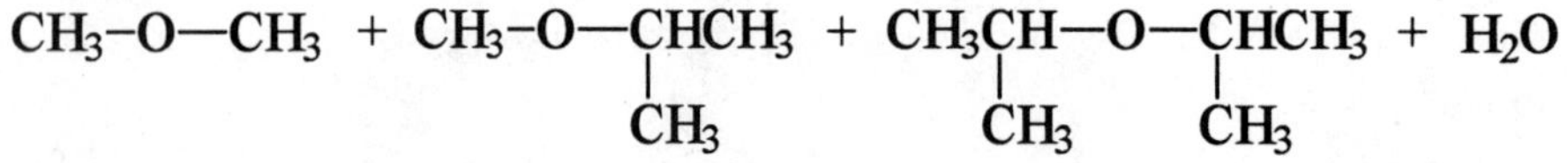 CH_3-O-CH_3 + $CH_3-O-\underset{\underset{\displaystyle CH_3}{|}}{C}HCH_3$ + $CH_3\underset{\underset{\displaystyle CH_3}{|}}{C}H-O-\underset{\underset{\displaystyle CH_3}{|}}{C}HCH_3$ + H_2O

   d.

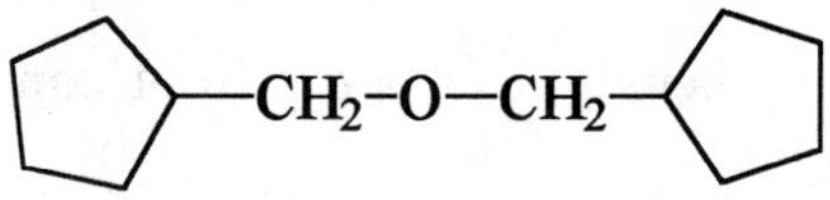

13.53   Cystine:

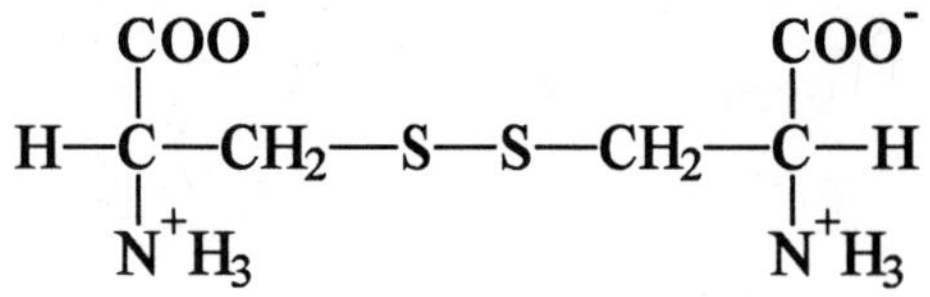

13.55   a. 1-Propanethiol
        b. 2-Butanethiol
        c. 2-Methyl-2-butanethiol
        d. 1,4-Cyclohexanedithiol

Chapter 14
Aldehydes and Ketones
Solutions to the Odd-Numbered Problems

****

In-Chapter Questions and Problems

****

14.1

   a. 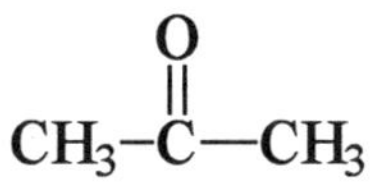 $CH_3-\overset{\overset{\displaystyle O}{\|}}{C}-CH_3$

   b. $CH_3\overset{\overset{\displaystyle OH}{|}}{C}HCH_2CH_2CH_3$

   c. cyclopentane ring with $\overset{\overset{\displaystyle O}{\|}}{C}-H$ substituent

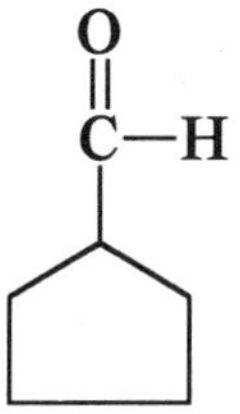

   d. 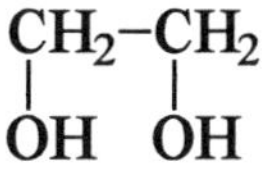 $\underset{\underset{\displaystyle OH}{|}}{CH_2}-\underset{\underset{\displaystyle OH}{|}}{CH_2}$

   e. 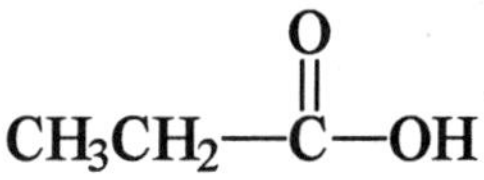 $CH_3CH_2-\overset{\overset{\displaystyle O}{\|}}{C}-OH$

14.3   a. 3-Iodo-2-butanone
       b. 3-Methyl-2-butanone
       c. 2-Fluoro-3-pentanone
       d. 3-Methylheptanal
       e. 2-Methylpropanal

14.5   Ethanal (acetaldehyde) is the aldehyde synthesized from ethanol in the liver.

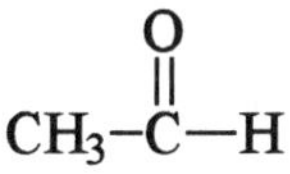

$$CH_3-\overset{\overset{\displaystyle O}{\|}}{C}-H$$

14.7   The following equation represents the oxidation of 1-propanol:

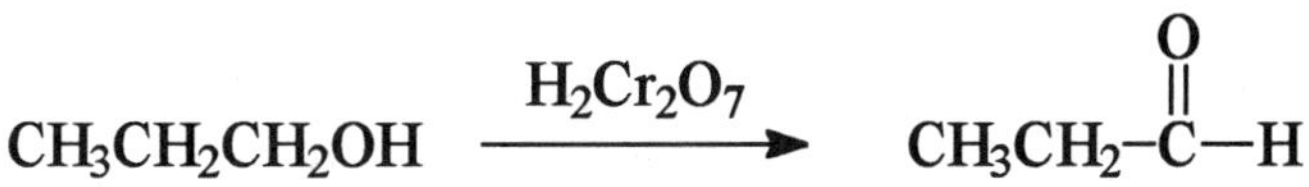

$$CH_3CH_2CH_2OH \xrightarrow{H_2Cr_2O_7} CH_3CH_2-\overset{\overset{\displaystyle O}{\|}}{C}-H$$

      1-Propanol   Propanal

14.9 The following equation represents the reaction between ethanal and Tollens' reagent:

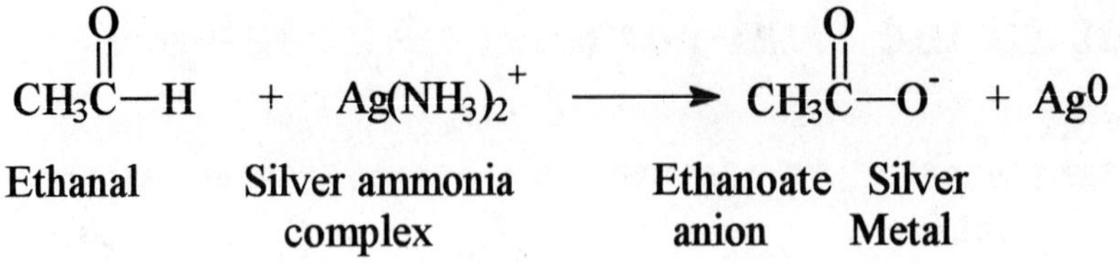

$$\underset{\text{Ethanal}}{CH_3\overset{\overset{\displaystyle O}{\|}}{C}-H} + \underset{\substack{\text{Silver ammonia} \\ \text{complex}}}{Ag(NH_3)_2^+} \longrightarrow \underset{\substack{\text{Ethanoate} \\ \text{anion}}}{CH_3\overset{\overset{\displaystyle O}{\|}}{C}-O^-} + \underset{\substack{\text{Silver} \\ \text{Metal}}}{Ag^0}$$

14.11 The following equation represents the hydrogenation of propanone:

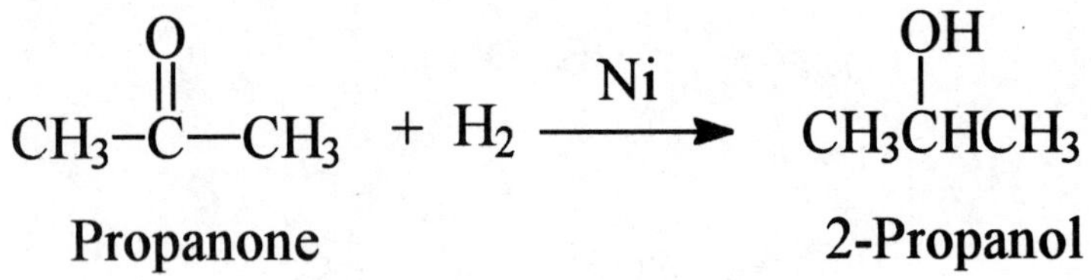

$$\underset{\text{Propanone}}{CH_3-\overset{\overset{\displaystyle O}{\|}}{C}-CH_3} + H_2 \xrightarrow{\text{Ni}} \underset{\text{2-Propanol}}{CH_3\overset{\overset{\displaystyle OH}{|}}{CH}CH_3}$$

14.13 a. Reduction
b. Reduction
c. Reduction
d. Oxidation
e. Reduction

14.15 The following equation represents an aldol condensation of two molecules of propanal:

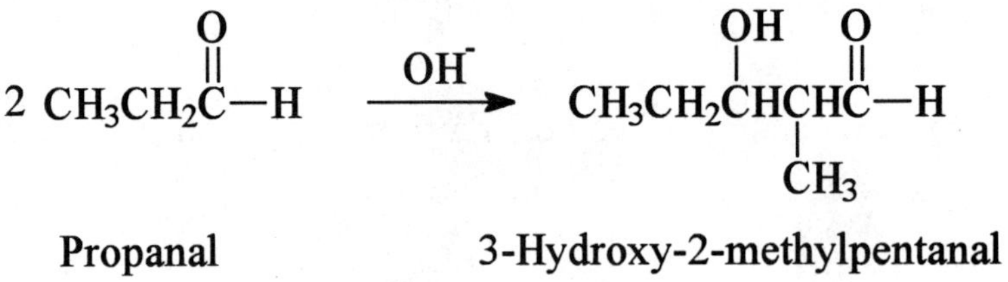

$$2 \underset{\text{Propanal}}{CH_3CH_2\overset{\overset{\displaystyle O}{\|}}{C}-H} \xrightarrow{OH^-} \underset{\text{3-Hydroxy-2-methylpentanal}}{CH_3CH_2\overset{\overset{\displaystyle OH}{|}}{CH}\underset{\underset{\displaystyle CH_3}{|}}{\overset{}{CH}}\overset{\overset{\displaystyle O}{\|}}{C}-H}$$

****
End-of-Chapter Questions and Problems
****

14.17 A good solvent should dissolve a wide range of compounds. Simple ketones are considered to be universal solvents because they have both a polar carbonyl group and nonpolar side chains. As a result, they dissolve organic compounds and are also miscible in water.

14.19

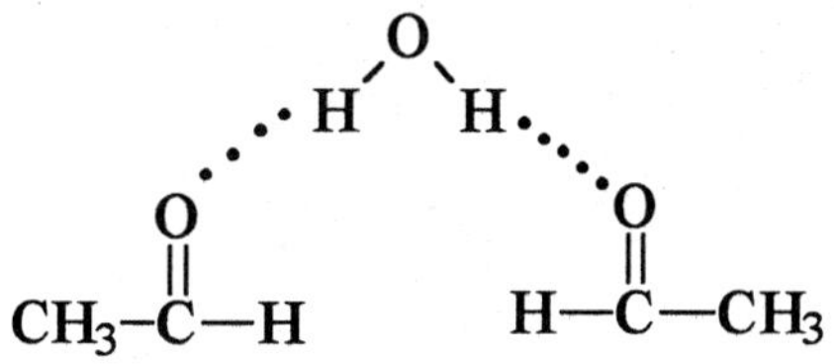

$$\text{O}$$
$$\cdots \text{H} \quad \text{H} \cdots$$
$$\text{O} \qquad \qquad \text{O}$$
$$\| \qquad \qquad \|$$
$$\text{CH}_3-\text{C}-\text{H} \quad \text{H}-\text{C}-\text{CH}_3$$

14.21 Alcohols have higher boiling points than aldehydes or ketones of comparable molecular weights because alcohol molecules can form intermolecular hydrogen bonds with one another. Aldehydes and ketones cannot form intermolecular hydrogen bonds.

14.23 a. Methanal:

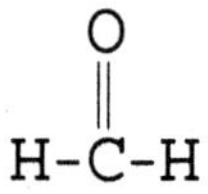

$$\text{O}$$
$$\|$$
$$\text{H-C-H}$$

b. 7,8-Dibromooctanal:

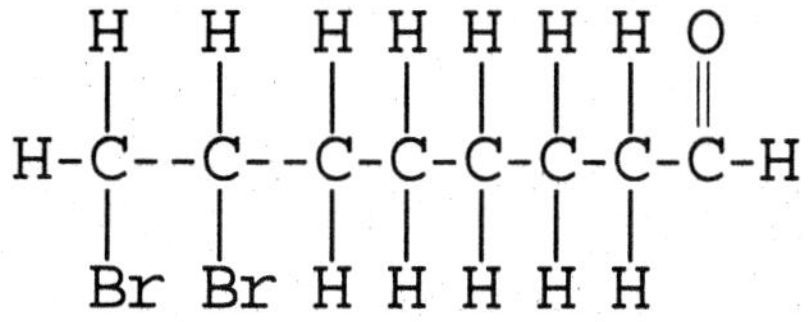

$$\text{H} \quad \text{H} \quad \text{H} \quad \text{H} \quad \text{H} \quad \text{H} \quad \text{H} \quad \text{O}$$
$$| \quad | \quad | \quad | \quad | \quad | \quad | \quad \|$$
$$\text{H-C--C--C-C-C-C-C-C-H}$$
$$| \quad | \quad | \quad | \quad | \quad | \quad |$$
$$\text{Br} \quad \text{Br} \quad \text{H} \quad \text{H} \quad \text{H} \quad \text{H} \quad \text{H}$$

c. Acetone:

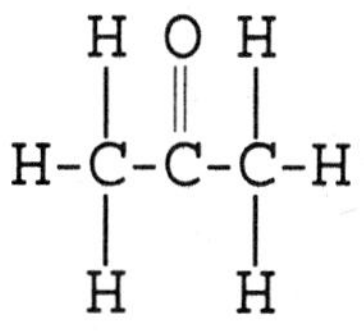

$$\text{H} \quad \text{O} \quad \text{H}$$
$$| \quad \| \quad |$$
$$\text{H-C-C-C-H}$$
$$| \qquad |$$
$$\text{H} \qquad \text{H}$$

d. Hydroxyethanal:

$$\text{H} \quad \text{O}$$
$$| \quad \|$$
$$\text{HO-C-C-H}$$
$$|$$
$$\text{H}$$

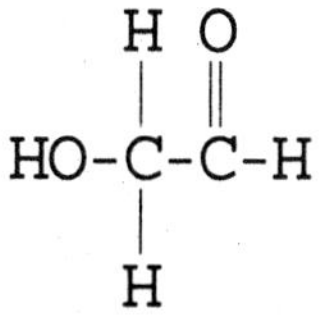

e. 3-Chloro-2-pentanone:

$$\text{H} \quad \text{H} \quad \text{H} \quad \text{O} \quad \text{H}$$
$$| \quad | \quad | \quad \| \quad |$$
$$\text{H-C-C-C-C-C-H}$$
$$| \quad | \quad | \qquad |$$
$$\text{H} \quad \text{H} \quad \text{Cl} \qquad \text{H}$$

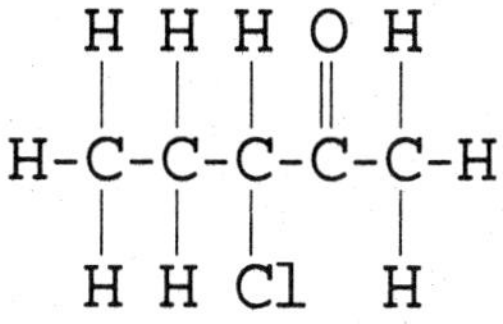

f. Benzaldehyde:

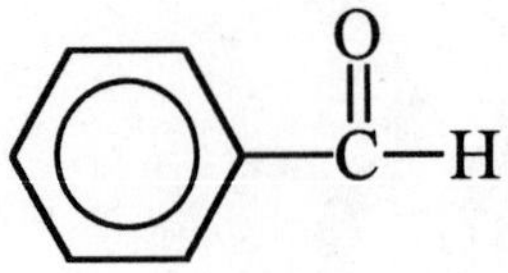

g. 4-Bromo-3-hexanone:

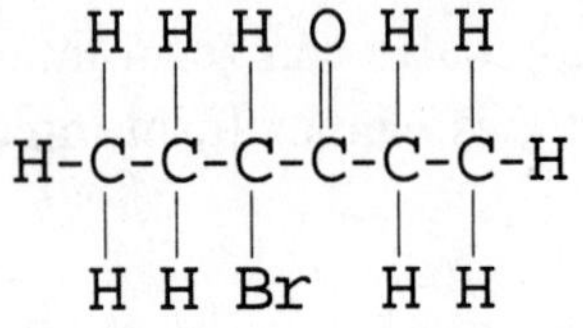

$$
\begin{array}{c}
\text{H}\ \text{H}\ \text{H}\ \ \text{O}\ \text{H}\ \text{H} \\
|\ \ |\ \ |\ \ \ \|\ \ |\ \ | \\
\text{H-C-C-C-C-C-C-H} \\
|\ \ |\ \ |\ \ \ \ \ |\ \ | \\
\text{H}\ \text{H}\ \text{Br}\ \ \text{H}\ \text{H}
\end{array}
$$

14.25 Determine the longest continuous chain containing the carbonyl group. Drop the *-e* ending from the parent alkane and add the suffix *-one*. Number the carbon chain in the direction that gives the carbonyl group the lowest possible number. Name and number the substituents on the chain. Add them as prefixes before the name of the ketone, listing them in alphabetical order.

14.27 a. 3-Bromobutanal
       b. 2-Chloro-2-methyl-4-heptanone
       c. 4,4-Diethyl-2-hexanone
       d. 4,6-Dimethyl-3-heptanone
       e. 3,3-Dimethylcyclopentanone

14.29

a. β-Hydroxybutanal   b. α-Methylpentanal 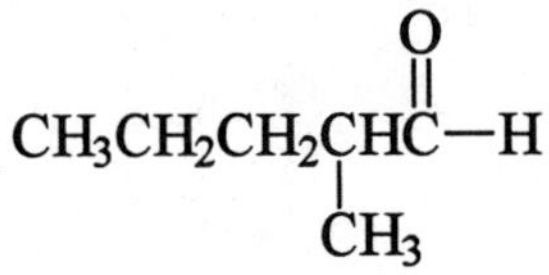   c. γ-Bromohexanal

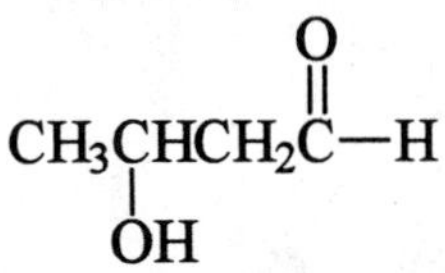

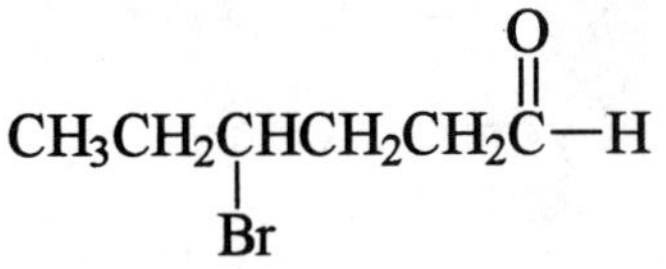

$$\text{CH}_3\text{CHCH}_2\overset{\displaystyle O}{\overset{\|}{\text{C}}}\!\!-\!\text{H}$$
$$\underset{\text{OH}}{}$$

$$\text{CH}_3\text{CH}_2\text{CH}_2\text{CH}\overset{\displaystyle O}{\overset{\|}{\text{C}}}\!\!-\!\text{H}$$
$$\underset{\text{CH}_3}{}$$

$$\text{CH}_3\text{CH}_2\text{CHCH}_2\text{CH}_2\overset{\displaystyle O}{\overset{\|}{\text{C}}}\!\!-\!\text{H}$$
$$\underset{\text{Br}}{}$$

d. β-Iodopentanal 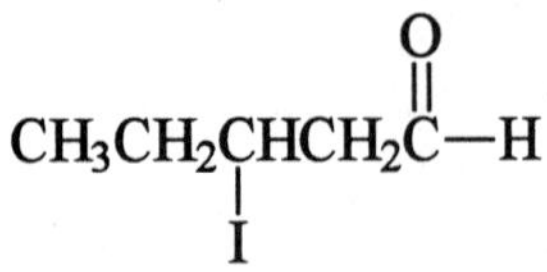   e. α-Hydroxy-β-methylheptanal

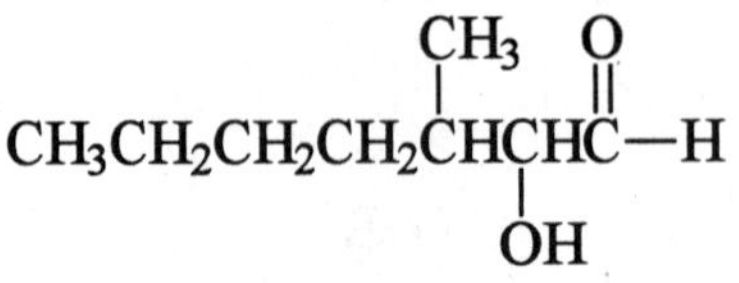

$$\text{CH}_3\text{CH}_2\text{CHCH}_2\overset{\displaystyle O}{\overset{\|}{\text{C}}}\!\!-\!\text{H}$$
$$\underset{\text{I}}{}$$

$$\text{CH}_3\text{CH}_2\text{CH}_2\text{CH}_2\overset{\displaystyle \text{CH}_3}{\underset{}{\text{CH}}}\overset{\displaystyle O}{\overset{\|}{\text{CHC}}}\!\!-\!\text{H}$$
$$\underset{\text{OH}}{}$$

14.31 Acetone is a good solvent because it can dissolve a wide range of compounds. It has both a polar carbonyl group and nonpolar side chains. As a result, it dissolves organic compounds and is also miscible in water.

14.33 The oxidation of ethanol to ethanal occurs in the liver and is catalyzed by the enzyme alcohol dehydrogenase.

14.35 a.

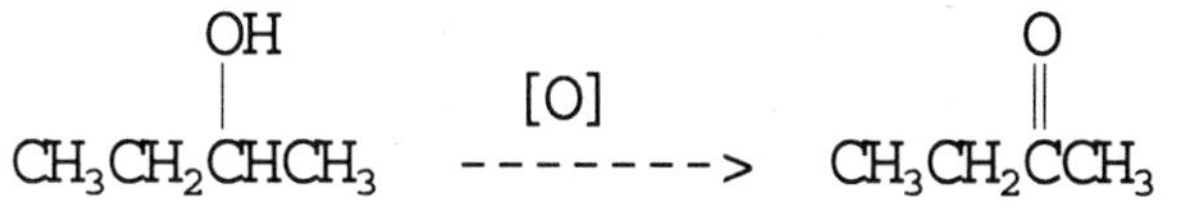

$$
\underset{\text{2-Butanol}}{\underset{|}{\overset{\overset{\text{OH}}{|}}{CH_3CH_2CHCH_3}}} \quad \xrightarrow{[O]} \quad \underset{\text{2-Butanone}}{\overset{\overset{\text{O}}{\parallel}}{CH_3CH_2CCH_3}}
$$

b.

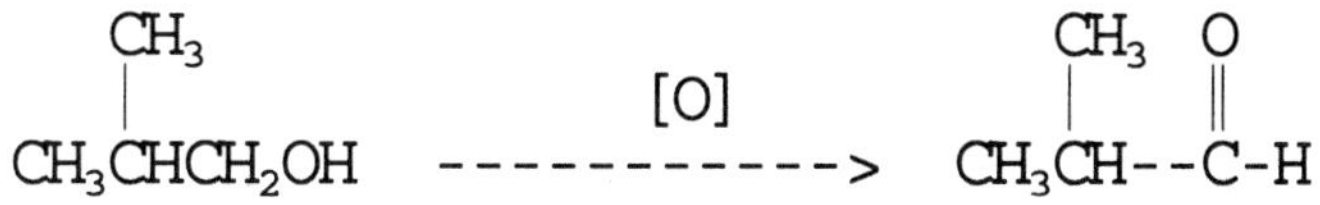

$$
\underset{\text{2-Methyl-1-propanol}}{\overset{\overset{\text{CH}_3}{|}}{CH_3CHCH_2OH}} \quad \xrightarrow{[O]} \quad \underset{\text{2-Methylpropanal}}{\overset{\overset{\text{CH}_3}{|}}{CH_3CH}\text{--}\overset{\overset{\text{O}}{\parallel}}{C}\text{-H}}
$$

Note that 2-methylpropanal can be further oxidized to 2-methylpropanoic acid.

c.

Cyclopentanol   Cyclopentanone

d.

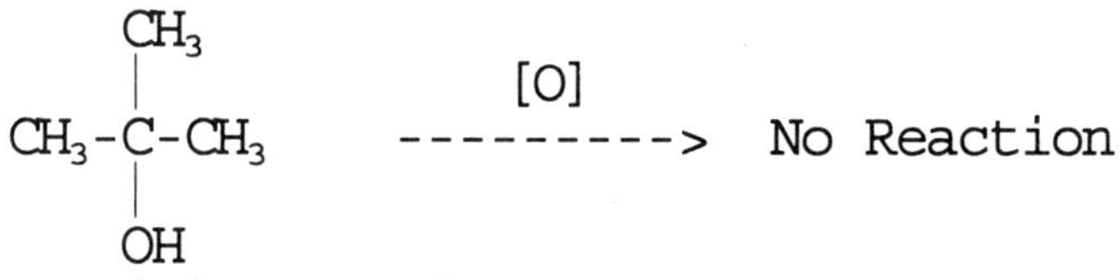

$$
\underset{\underset{\text{2-Methyl-2-propanol}}{\overset{\overset{\text{CH}_3}{|}}{\underset{\overset{|}{\text{OH}}}{CH_3\text{-}C\text{-}CH_3}}}}{} \quad \xrightarrow{[O]} \quad \text{No Reaction}
$$

e.

$$
\underset{\overset{|}{\text{OH}}}{\overset{|}{CH_3CHCH_2CH_2CH_2CH_2CH_2CH_2CH_3}} \quad \xrightarrow{[O]} \quad \underset{\overset{\parallel}{\text{O}}}{CH_3CCH_2CH_2CH_2CH_2CH_2CH_2CH_3}
$$

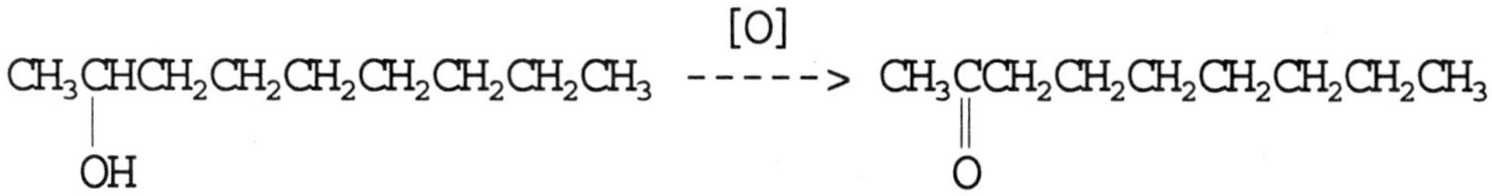

2-Nonanol   2-Nonanone

409

f.

$$CH_2CH_2CH_2CH_2CH_2CH_2CH_2CH_2CH_2CH_3$$
|
$$OH$$

1-Decanol

| [O]
V

$$H-CCH_2CH_2CH_2CH_2CH_2CH_2CH_2CH_2CH_3$$
||
$$O$$

Decanal

Note that decanal can be further oxidized to decanoic acid.

14.37 a. Reduction reaction

$$\overset{O}{\overset{||}{CH_3-C-H}} \quad \text{-------}> \quad CH_3CH_2OH$$

Ethanal   Ethanol

b. Reduction reaction

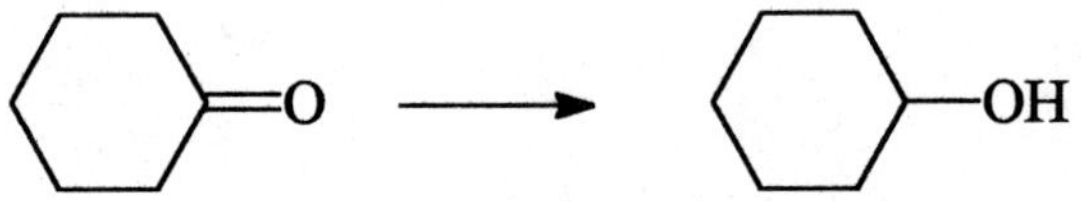

Cyclohexanone   Cyclohexanol

c. Oxidation reaction

$$\overset{OH}{\overset{|}{CH_3CHCH_3}} \quad \text{--------}> \quad \overset{O}{\overset{||}{CH_3-C-CH_3}}$$

2-Propanol   Propanone

14.39 Only (c) 3-methylbutanal and (f) acetaldehyde would give a positive Tollens' test.

14.41 a.

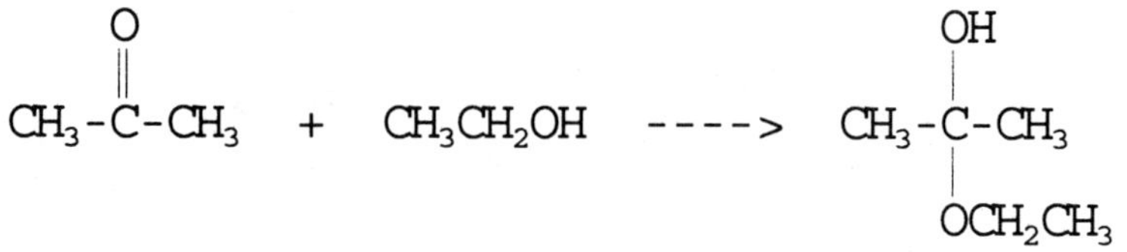

$$CH_3-\overset{\overset{\textstyle O}{\|}}{C}-CH_3 \ + \ CH_3CH_2OH \ ----> \ CH_3-\overset{\overset{\textstyle OH}{|}}{\underset{\underset{\textstyle OCH_2CH_3}{|}}{C}}-CH_3$$

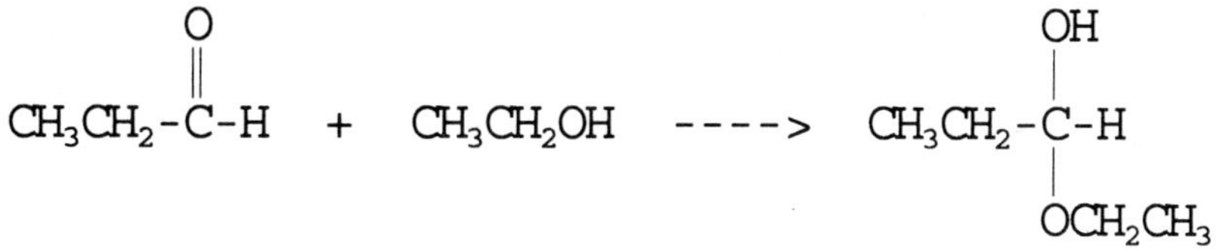

b.

$$CH_3CH_2-\overset{\overset{\textstyle O}{\|}}{C}-H \ + \ CH_3CH_2OH \ ----> \ CH_3CH_2-\overset{\overset{\textstyle OH}{|}}{\underset{\underset{\textstyle OCH_2CH_3}{|}}{C}}-H$$

c.

$$CH_3-\overset{\overset{\textstyle O}{\|}}{C}-H \ + \ CH_3CH_2OH \ ----> \ CH_3-\overset{\overset{\textstyle OH}{|}}{\underset{\underset{\textstyle OCH_2CH_3}{|}}{C}}-H$$

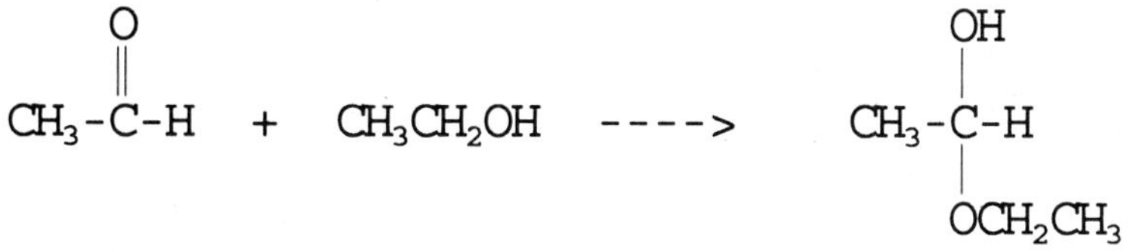

d.

$$CH_3-\overset{\overset{\textstyle O}{\|}}{C}-CH_2CH_2CH_3 \ + \ CH_3CH_2OH \ ---> \ CH_3-\overset{\overset{\textstyle OH}{|}}{\underset{\underset{\textstyle OCH_2CH_3}{|}}{C}}-CH_2CH_2CH_3$$

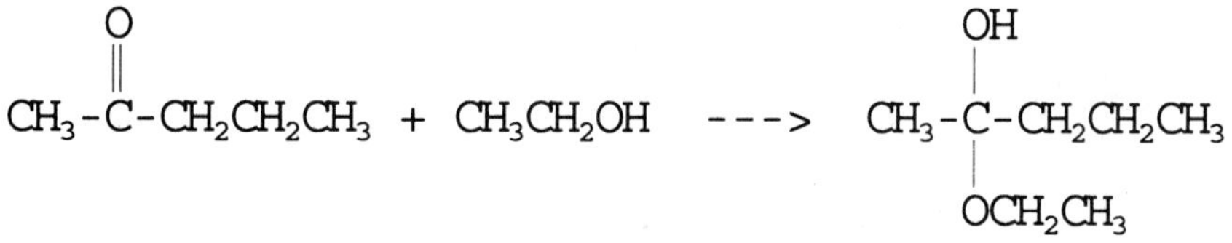

14.43 An acetal is formed when two molecules of alcohol react with an aldehyde. A ketal is formed when two molecules of an alcohol react with a ketone.

14.45 a.

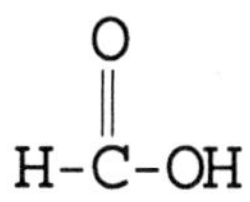

$$H-\overset{\overset{\textstyle O}{\|}}{C}-OH$$

b.

$$CH_3-\overset{\overset{\textstyle O}{\|}}{C}-OH$$

c.

$$CH_3CH_2-\overset{\overset{\textstyle O}{\|}}{C}-OH$$

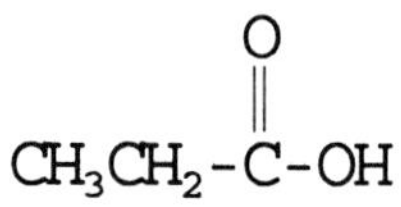

d.

$$CH_3CH_2CH_2-\overset{\overset{\textstyle O}{\|}}{C}-OH$$

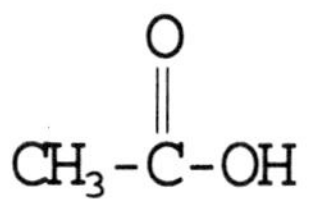

411

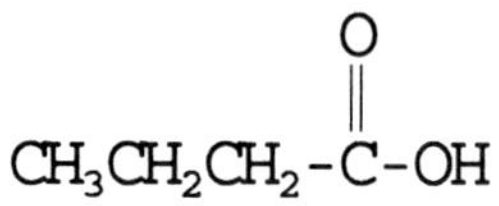

14.47 a. False
       b. True
       c. True
       d. True

14.49

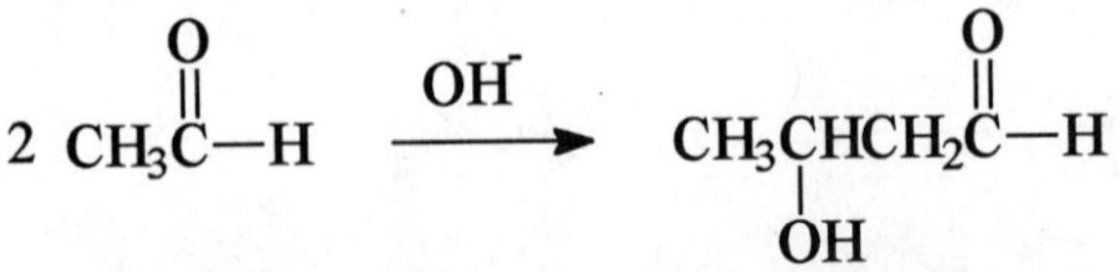

$$2 \; CH_3\overset{\displaystyle O}{\overset{\|}{C}}-H \quad \xrightarrow{\;OH^-\;} \quad CH_3\underset{\displaystyle OH}{CH}CH_2\overset{\displaystyle O}{\overset{\|}{C}}-H$$

Ethanal   3-Hydroxybutanal

14.51

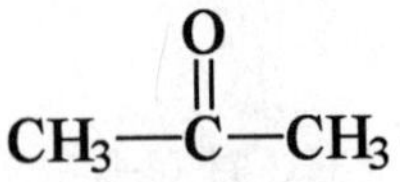

$$CH_3-\overset{\displaystyle O}{\overset{\|}{C}}-CH_3 \qquad \overset{H}{\underset{H}{\Large{>}}}C=C\overset{OH}{\underset{CH_3}{}}$$

Keto form of   Enol form of
Propanone   Propanone

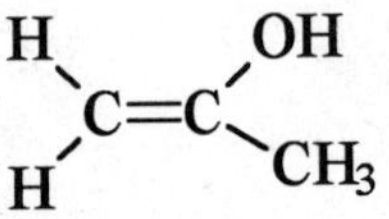

14.53

a. 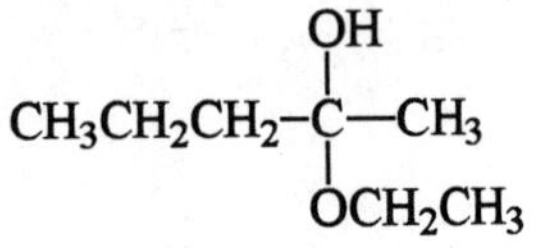 $CH_3CH_2CH_2-\overset{\displaystyle OH}{\underset{\displaystyle OCH_2CH_3}{C}}-CH_3$ b. $\overset{\displaystyle OH}{\underset{\displaystyle OCH_2CH_3}{C}}-CH_3$ c. $\overset{OH}{\underset{OCH_2CH_3}{}}$

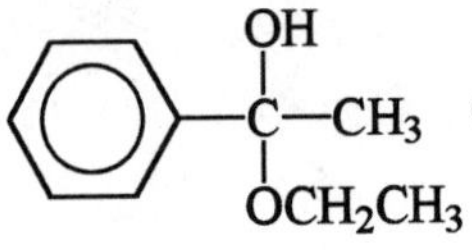

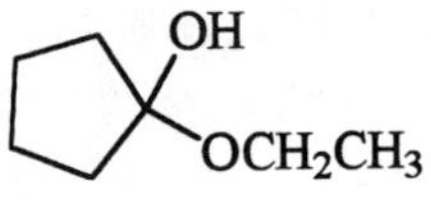

14.55 (1) 2 CH_3CH_2OH
       (2) $KMnO_4/OH^-$
       (3) $CH_3CH=CH_2$

Chapter 15
Carbohydrates
Solutions to the Odd-Numbered Problems

In-Chapter Questions and Problems

15.1 It is currently recommended that 58% of the calories in the diet should be carbohydrates. Of that amount, no more than 10% should be simple sugars.

15.3 An aldose is a sugar with an aldehyde functional group. A ketose is a sugar with a ketone functional group.

15.5 a. ketose
b. aldose
c. ketose
d. aldose
e. ketose
f. aldose

15.7

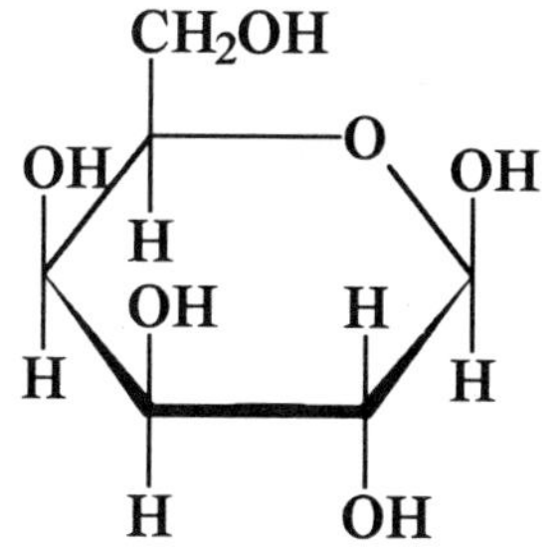

β-D-Galactose 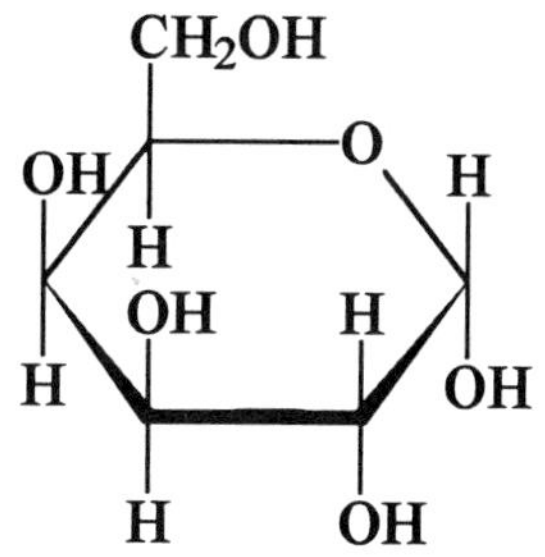   α-D-Galactose

15.9 α-Amylase and β-amylase are digestive enzymes that break down the starch amylose. α-Amylase cleaves glycosidic bonds of the amylose chain at random, producing shorter polysaccharide chains. β-Amylase sequentially cleaves maltose (a disaccharide of glucose) from the reducing end of the polysaccharide chain.

End-of-Chapter Questions and Problems

15.11 A monosaccharide is the simplest sugar and consists of a single saccharide unit. A disaccharide is made up of two monosaccharides joined covalently by a glycosidic bond.

15.13 Mashed potato flakes, rice, and corn starch would contain amylose and amylopectin, both of which are polysaccharides. A candy bar contains sucrose, a disaccharide. Orange juice contains fructose, a monosaccharide. It may also contain sucrose if the label indicates that sugar has been added.

15.15 Four kilocalories of energy are released for each gram of carbohydrate "burned" or oxidized.

15.17

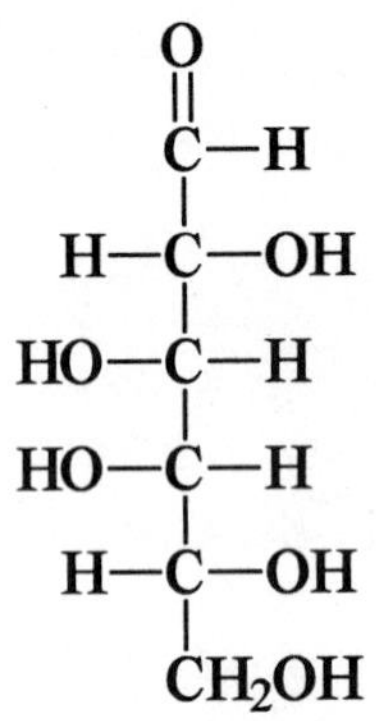

D-Galactose
(An aldohexose)

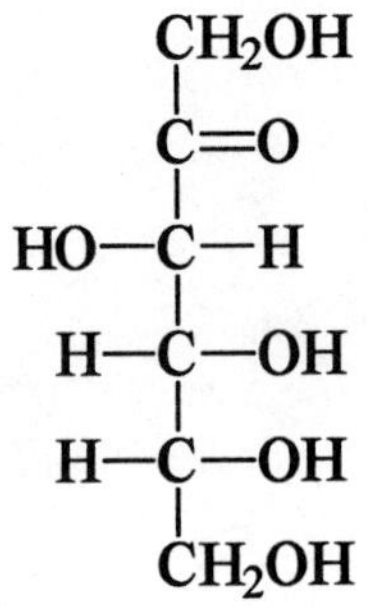

D-Fructose
(A ketohexose)

15.19 a. β-D-Glucose is a hemiacetal.
b. β-D-Fructose is a hemiketal.
c. α-D-Galactose is a hemiacetal.

15.21 The two aldotrioses of molecular formula C_3H_6O:

$$\begin{array}{cc}
\begin{array}{c}
\text{O}\\
\|\\
\text{C}-\text{H}\\
|\\
\text{H}-\text{C}-\text{OH}\\
|\\
\text{CH}_2\text{OH}
\end{array}
&
\begin{array}{c}
\text{O}\\
\|\\
\text{C}-\text{H}\\
|\\
\text{HO}-\text{C}-\text{H}\\
|\\
\text{CH}_2\text{OH}
\end{array}
\end{array}$$

D-Glyceraldehyde   L-Glyceraldehyde

15.23 Dextrose is a common name used for D-glucose.

15.25 D- and L-Glyceraldehyde are a pair of enantiomers, that is, they are nonsuperimposable mirror images of one another. In D-glyceraldehyde the hydroxyl group on the chiral carbon farthest from the aldehyde group (C-2) is on the right of the structure. In L-glyceraldehyde the hydroxyl group on C-2 is on the left of the structure.

15.27 When the carbonyl group at C-1 of D-glucose reacts with the C-5 hydroxyl group, a new chiral carbon is created (C-1). In the α-isomer of the cyclic sugar the C-1 hydroxyl group is below the ring and in the β-isomer the C-1 hydroxyl group is above the ring.

15.29 β-Maltose and α-lactose would give positive Benedict's tests. Glycogen would give only a weak reaction because it is a long polymer and thus there are fewer reducing ends for a given mass of the carbohydrate.

15.31 Enantiomers are stereoisomers that are nonsuperimposable mirror images of one another. For instance:

$$\begin{array}{cc}
\begin{array}{c}
\text{O}\\
\|\\
\text{C}-\text{H}\\
|\\
\text{H}-\text{C}-\text{OH}\\
|\\
\text{CH}_2\text{OH}
\end{array}
&
\begin{array}{c}
\text{O}\\
\|\\
\text{C}-\text{H}\\
|\\
\text{HO}-\text{C}-\text{H}\\
|\\
\text{CH}_2\text{OH}
\end{array}
\end{array}$$

D-Glyceraldehyde   L-Glyceraldehyde

15.33 The linear structure of an aldehyde sugar forms a cyclic structure by formation of an intramolecular hemiacetal. The carbonyl group of the monosaccharide reacts with a hydroxyl group on one of the other carbon atoms. The product is a cyclic intramolecular hemiacetal.

15.35

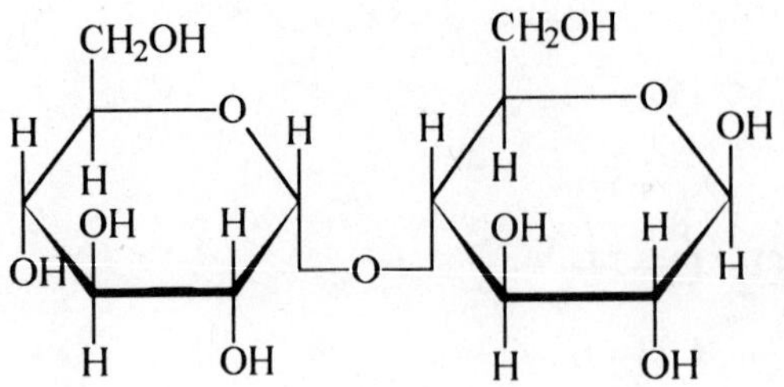

CH$_2$OH

CH$_2$OH

β-Maltose

15.37 Milk is the major source of lactose.

15.39 Since galactose is one of the two monosaccharides making up lactose (milk sugar), eliminating milk and milk products from the diet allows the patient to avoid most of the ill-effects of galactosemia.

15.41 Lactose intolerance is the inability to produce the enzyme lactase that hydrolyzes the milk sugar lactose into its component monosaccharides, glucose and galactose. As a result of the undigested lactose in the intestine, uncomfortable symptoms, including abdominal cramps and diarrhea, occur.

15.43 Both amylose and cellulose are linear polymers of glucose units. However, the glucose units of amylose are joined by $\alpha(1 \rightarrow 4)$ glycosidic bonds and those of cellulose are bonded together by $\beta(1 \rightarrow 4)$ glycosidic bonds.

15.45 The major physiological purpose of glycogen is to serve as a storage molecule for glucose. This represents an energy reservoir for the body. Glycogen synthesis and degradation in the liver are involved in regulation of blood glucose levels.

15.47 α-Amylase and β-amylase are produced in the salivary glands and in the pancreas.

Chapter 16
Carboxylic Acids and Carboxylic Acid Derivatives
Solutions to the Odd-Numbered Problems

****

In-Chapter Questions and Problems

****

16.1   a. 3-Hexanone
       b. 3-Hexanone
       c. Hexane
       d. Dipropyl ether
       e. Hexanal

16.3   a. 2,4-Dimethylpentanoic acid
       b. 2,4-Dichlorobutanoic acid
       c. 3-Methylcyclohexanecarboxylic acid
       d. 2-Ethylcyclopentanecarboxylic acid

16.5   a. *o*-Toluic acid:

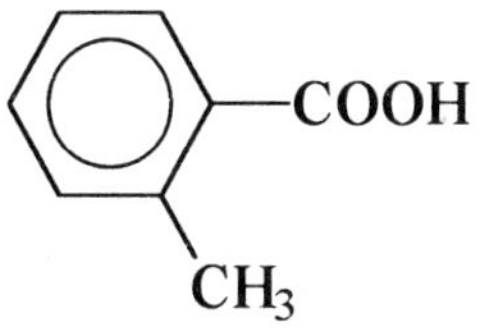

b. 2,4,6-Tribromobenzoic acid:

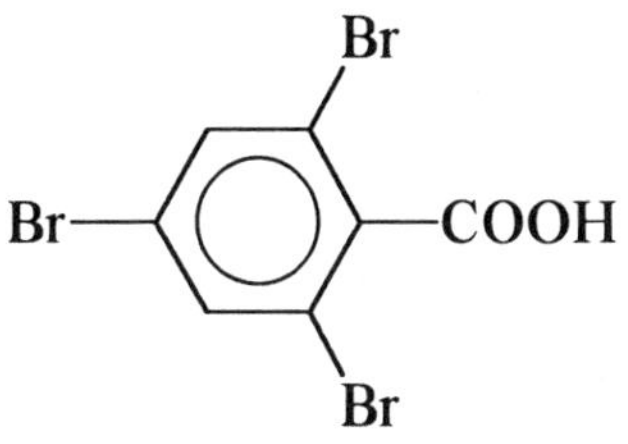

c. 2,2,2-Triphenylethanoic acid:

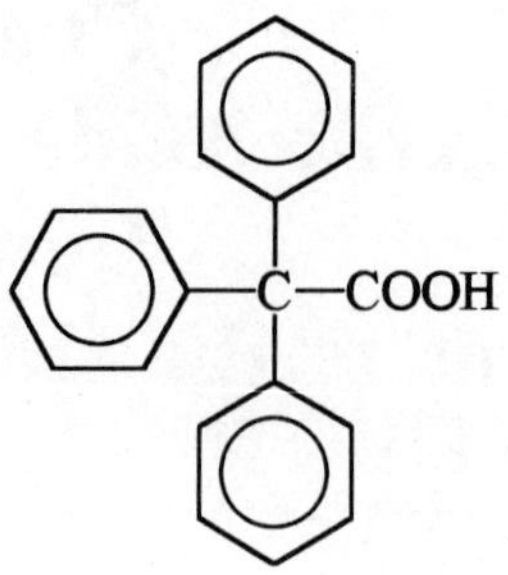

$$\text{(C}_6\text{H}_5)_3\text{C}-\text{COOH}$$

16.7   a.

$$\underset{\text{O}}{\overset{\text{O}}{\text{CH}_3\text{CH}_2-\overset{\|}{\text{C}}-\text{H}}} \quad ----> \quad \text{CH}_3\text{CH}_2-\overset{\|}{\text{C}}-\text{OH}$$

Propanal would be the first oxidation product. However, it would quickly be oxidized further to propanoic acid.

b.

$$\text{HO}-\overset{\overset{\text{O}}{\|}}{\text{C}}-\text{CH}_2\text{CH}_2\text{CH}_2\text{CH}_3$$

c.

$$\text{CH}_3\text{CH}_2-\overset{\overset{\text{O}}{\|}}{\text{C}}-\text{O}^- \ \text{K}^+$$

d. $[\text{CH}_3\text{CH}_2\text{CH}_2\text{COO}]^-_2 \ \text{Ba}^{2+}$

16.9   a. CH_3COOH + $\text{CH}_3\text{CH}_2\text{CH}_2\text{OH}$

Ethanoic acid   1-Propanol

b. $\text{CH}_3\text{CH}_2\text{CH}_2\text{CH}_2\text{CH}_2-\text{COO}^- \ \text{K}^+$ + $\text{CH}_3\text{CH}_2\text{CH}_2\text{CH}_2\text{OH}$

Potassium hexanoate   1-Butanol

c. $\text{CH}_3\text{CH}_2\text{CH}_2\text{CH}_2\text{COO}^- \ \text{Na}^+$ + CH_3OH

Sodium pentanoate   Methanol

d. $\text{CH}_3\text{CH}_2\text{CH}_2\text{CH}_2\text{CH}_2-\text{COOH}$ + $\text{CH}_3\text{CHCH}_2\text{CH}_2\text{CH}_3$
                                                                    |
                                                                   OH
Hexanoic acid   2-Pentanol

End-of-Chapter Questions and Problems

16.11 a. 2-Bromopentanoic acid:

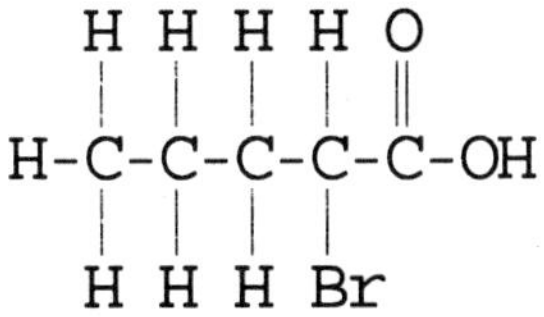

```
    H  H  H  H  O
    |  |  |  |  ||
 H--C--C--C--C--C--OH
    |  |  |  |
    H  H  H  Br
```

b.   2-Bromo-3-methylbutanoic acid:

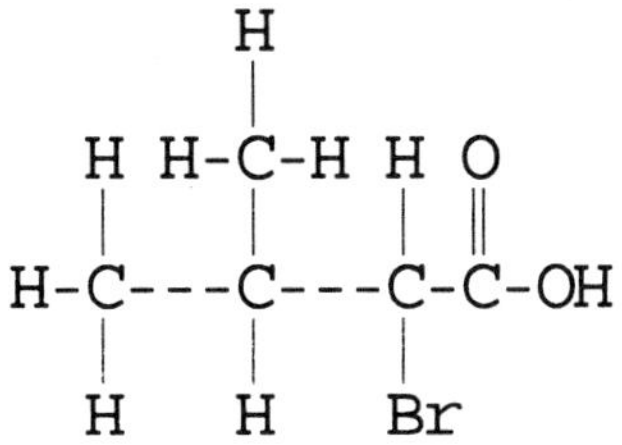

```
          H
          |
    H  H--C--H  H  O
    |     |     |  ||
 H--C-----C-----C--C--OH
    |     |     |
    H     H     Br
```

c.   2-Bromocyclohexanecarboxylic acid:

```
        ┌──────────COOH
        │
        └──Br
```

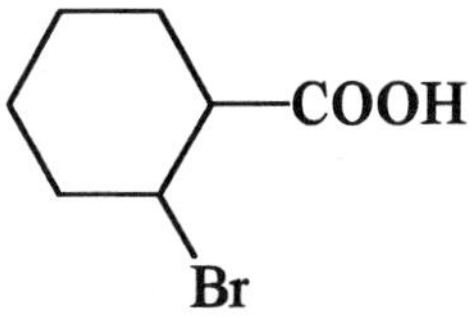

d.   2,6-Dichlorocyclohexanecarboxylic acid:

```
         Cl
          │
          ├──────COOH
          │
         Cl
```

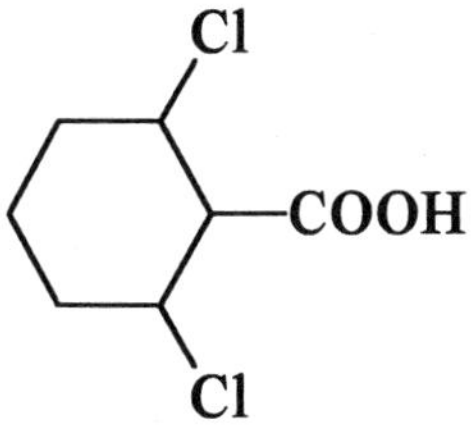

e.   2,4,6-Trimethylstearic acid:

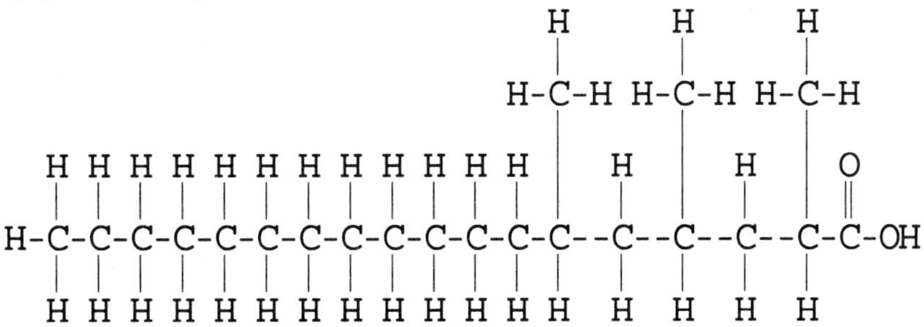

```
                              H       H       H
                              |       |       |
                            H-C-H   H-C-H   H-C-H
    H H H H H H H H H H H H  |   H   |   H   |   O
    | | | | | | | | | | | |  |   |   |   |   |   ||
 H--C-C-C-C-C-C-C-C-C-C-C-C--C---C---C---C---C-C-OH
    | | | | | | | | | | | |  |   |   |   |
    H H H H H H H H H H H H  H   H   H   H
```

f.   Propenoic acid:

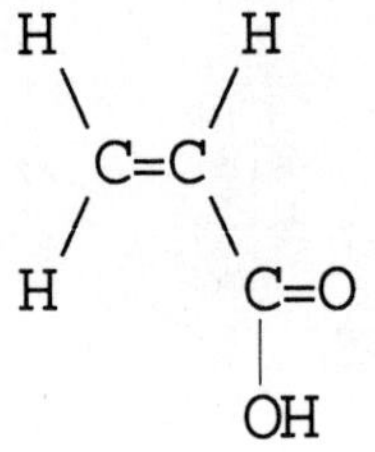

$$H_2C=CH-C(=O)-OH$$

(structure shown: C=C with two H on top carbon's left, H and H, bottom left H, and C=O with OH)

16.13

Butanoic acid:

$$H-\underset{\underset{H}{|}}{\overset{\overset{H}{|}}{C}}-\underset{\underset{H}{|}}{\overset{\overset{H}{|}}{C}}-\underset{\underset{H}{|}}{\overset{\overset{H}{|}}{C}}-\overset{\overset{O}{\|}}{C}-OH$$

2-Methylpropanoic acid:

$$H-\overset{\overset{H-\overset{\overset{H}{|}}{C}-H}{|}}{\underset{\underset{H}{|}}{C}}-\underset{\underset{H}{|}}{C}-\overset{\overset{O}{\|}}{C}-OH$$

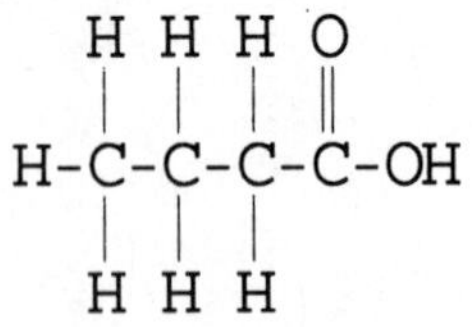

Butanoic acid 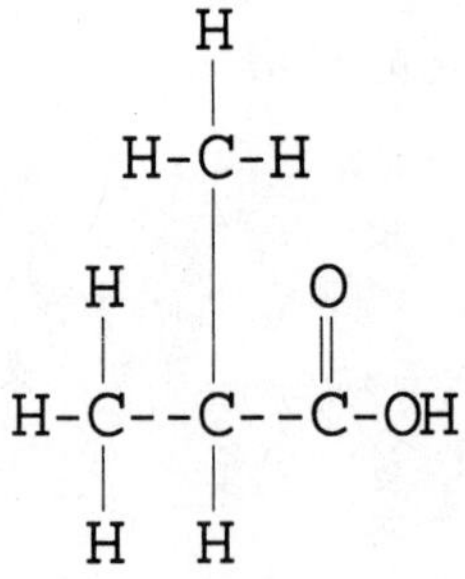   **2-Methylpropanoic acid**

16.15   a. 4,4-Dimethylhexanoic acid:

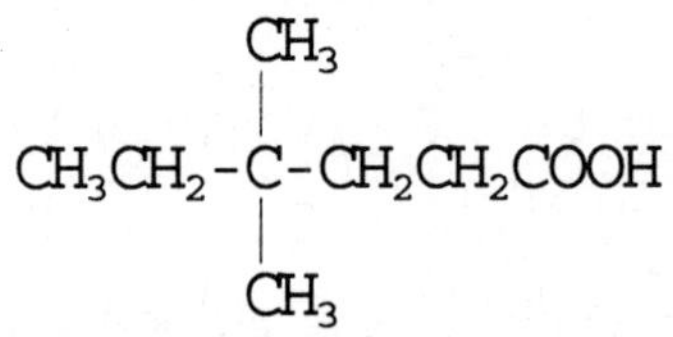

$$CH_3CH_2-\underset{\underset{CH_3}{|}}{\overset{\overset{CH_3}{|}}{C}}-CH_2CH_2COOH$$

b. 3-Bromo-4-methylpentanoic acid:

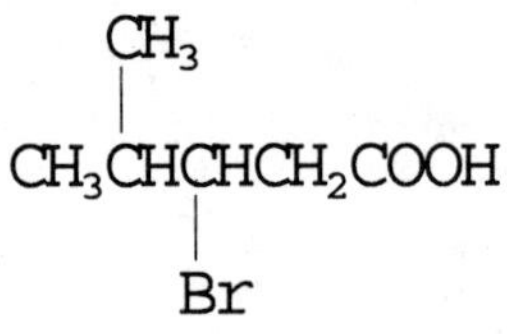

$$CH_3\underset{\underset{Br}{|}}{\overset{\overset{CH_3}{|}}{CHCHCH_2COOH}}$$

c. 2,3-Dinitrobenzoic acid:

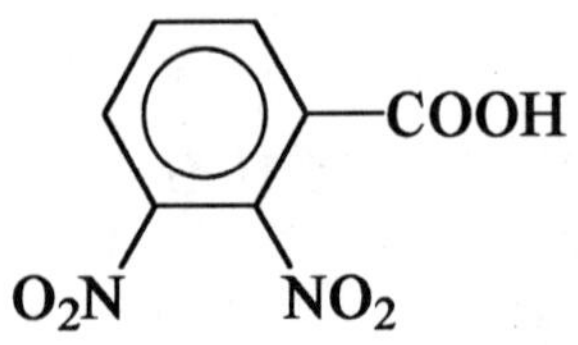

(benzene ring with —COOH, and O_2N and NO_2 substituents)

d. 3-Methylcyclohexanecarboxylic acid:

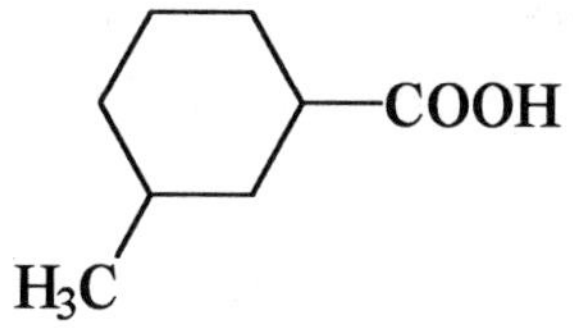

16.17 a. I.U.P.A.C. name: 2-Hydroxypropanoic acid
      Common name: α-Hydroxypropionic acid

   b. I.U.P.A.C. name: 3-Hydroxybutanoic acid
      Common name: β-Hydroxybutyric acid

   c. I.U.P.A.C. name: 4,4-Dimethylpentanoic acid
      Common name: γ,γ-Dimethylvaleric acid

   d. I.U.P.A.C. name: 3,3-Dichloropentanoic acid
      Common name: β,β-Dichlorovaleric acid

16.19 a. Heptanoic acid
      b. 1-Propanol
      c. Pentanoic acid
      d. Butanoic acid

16.21 The smaller carboxylic acids are water-soluble. They have sharp, sour tastes and unpleasant aromas.

16.23 Citric acid is found naturally in citrus fruits. It is added to foods to give them a tart flavor or to act as a food preservative and anti-oxidant. Adipic acid imparts a tart flavor to soft drinks and is a preservative.

16.25 Carboxylic acids are produced commercially by the oxidation of the corresponding alcohol or aldehyde, as seen in the following example:

$$CH_3CH_2OH \xrightarrow{[O]} CH_3-\overset{\overset{\displaystyle O}{\|}}{C}-H \xrightarrow{[O]} CH_3-\overset{\overset{\displaystyle O}{\|}}{C}-OH$$

   Ethanol   Ethanal   Ethanoic acid
   (ethyl alcohol)   (acetaldehyde)   (acetic acid)

16.27 Soaps are made from water, a strong base, and natural fats or oils. The fats and oils are triesters of glycerol. In the presence of the strong base, the ester bonds are hydrolyzed and the salts of the long chain fatty acids are formed. The salts of fatty acids are soaps.

16.29 a. CH_3COOH

b.

$$H_3CH_2CH_2-\overset{\overset{\displaystyle O}{\|}}{C}-O-CH_3 \quad + \quad H_2O$$

c. CH_3OH

16.31 a. The oxidation of 1-pentanol in the presence of an oxidizing agent yields pentanal.
b. Continued oxidation of pentanal yields pentanoic acid.

16.33 a. Methyl benzoate:

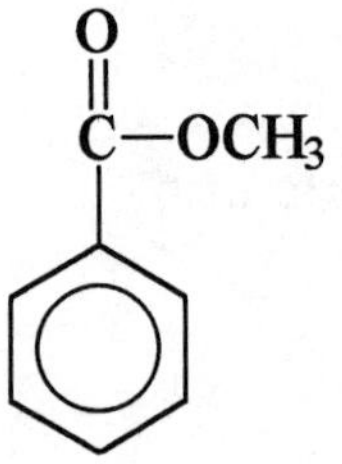

$$\overset{\overset{\displaystyle O}{\|}}{C}-OCH_3$$

b. Butyl decanoate:

$$CH_3CH_2CH_2CH_2CH_2CH_2CH_2CH_2CH_2-\overset{\overset{\displaystyle O}{\|}}{C}-O-CH_2CH_2CH_2CH_3$$

c. Methyl propionate:

$$CH_3CH_2-\overset{\overset{\displaystyle O}{\|}}{C}-O-CH_3$$

d. Ethyl propionate:

$$CH_3CH_2-\overset{\overset{\displaystyle O}{\|}}{C}-O-CH_2CH_3$$

e. Ethyl-*m*-nitrobenzoate:

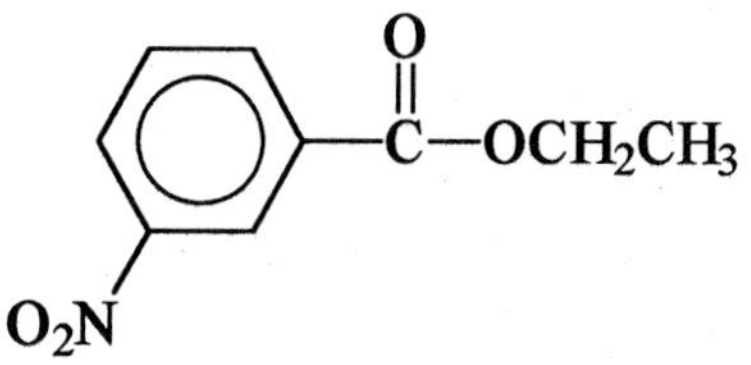

$$\text{(benzene ring)}-\overset{\overset{\displaystyle O}{\parallel}}{C}-OCH_2CH_3$$

with O$_2$N substituent

f. Isopropyl acetate:

$$CH_3-\overset{\overset{\displaystyle O}{\parallel}}{C}-O-\overset{\overset{\displaystyle CH_3}{|}}{C}HCH_3$$

g. Methyl butyrate:

$$CH_3CH_2CH_2-\overset{\overset{\displaystyle O}{\parallel}}{C}-O-CH_3$$

16.35 a.

$$CH_3CH_2CH_2-\overset{\overset{\displaystyle O}{\parallel}}{C}-O-CH_2CH_3$$

b.

$$CH_3CH_2-\overset{\overset{\displaystyle O}{\parallel}}{C}-OH \quad + \quad CH_3CH_2OH$$

c. $CH_3CH_2CH_2OH$

d.

$$CH_3CH_2\overset{\overset{\displaystyle Br}{|}}{C}HCH_2-\overset{\overset{\displaystyle O}{\parallel}}{C}-O^- \quad + \quad CH_3CH_2OH$$

16.37 Saponification is a reaction in which a soap is produced. More generally, it is the hydrolysis of an ester in the presence of a base. The following reaction shows the base-catalyzed hydrolysis of an ester:

$$CH_3(CH_2)_{14}-\overset{\overset{\displaystyle O}{\parallel}}{C}-O-CH_3 \quad + \quad NaOH \quad ----> \quad CH_3(CH_2)_{14}-\overset{\overset{\displaystyle O}{\parallel}}{C}-O^- \; Na^+ \quad + \quad CH_3OH$$

16.39

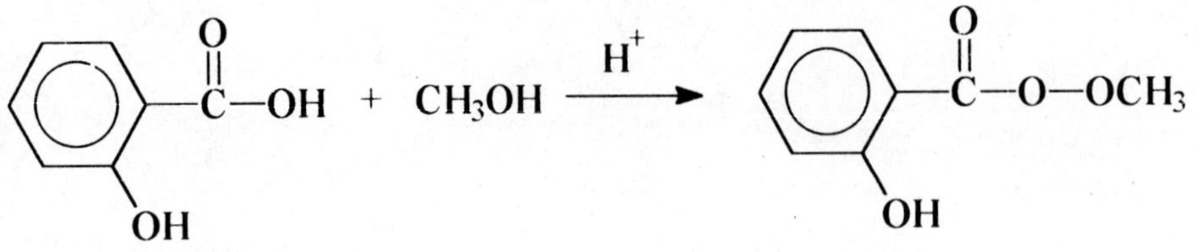

Salicylic acid   Methyl salicylate

16.41 Compound A is

$$CH_3CH_2CH_2CH_2-\overset{\overset{\displaystyle O}{\|}}{C}-O-CH_3$$

Compound B is

$$CH_3CH_2CH_2CH_2-\overset{\overset{\displaystyle O}{\|}}{C}-OH$$

Compound C is CH_3OH

16.43   a. PCl_3, PCl_5, or $SOCl_2$

b .

$$CH_3-\overset{\overset{\displaystyle O}{\|}}{C}-OH$$

c.

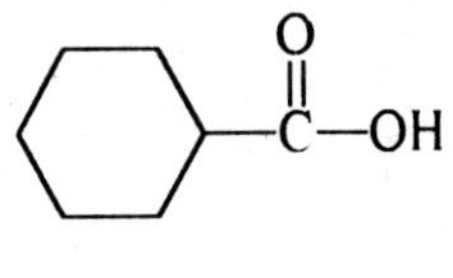

16.45   a.

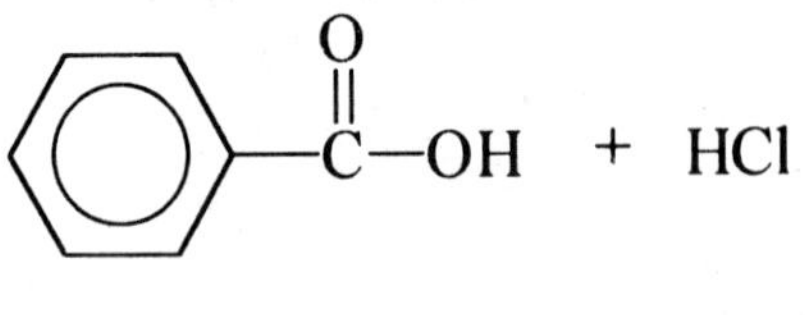

424

b.

$$2 \text{ CH}_3 - \overset{\overset{\displaystyle O}{\|}}{C} - OH$$

16.47 a. Decanoic anhydride:

$$\text{CH}_3(\text{CH}_2)_8 - \overset{\overset{\displaystyle O}{\|}}{C} - O - \overset{\overset{\displaystyle O}{\|}}{C} - (\text{CH}_2)_8\text{CH}_3$$

b. Acetic anhydride:

$$\text{CH}_3 - \overset{\overset{\displaystyle O}{\|}}{C} - O - \overset{\overset{\displaystyle O}{\|}}{C} - \text{CH}_3$$

c. Valeric anhydride:

$$\text{CH}_3(\text{CH}_2)_3 - \overset{\overset{\displaystyle O}{\|}}{C} - O - \overset{\overset{\displaystyle O}{\|}}{C} - (\text{CH}_2)_3\text{CH}_3$$

d. Benzoyl chloride:

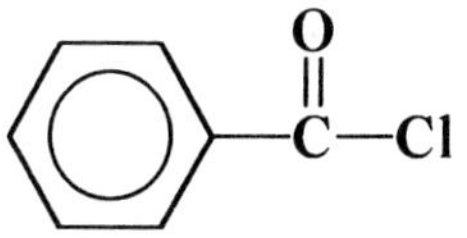

16.49 Acid chlorides are noxious, irritating chemicals. They are slightly polar and have boiling points similar to comparable aldehydes or ketones. They cannot be dissolved in water because they react violently with it.

16.51 Monoester:

$$\text{HO} - \overset{\overset{\displaystyle O}{\|}}{\underset{\underset{\displaystyle OH}{|}}{P}} - O - \text{CH}_2\text{CH}_3$$

b. Diester:

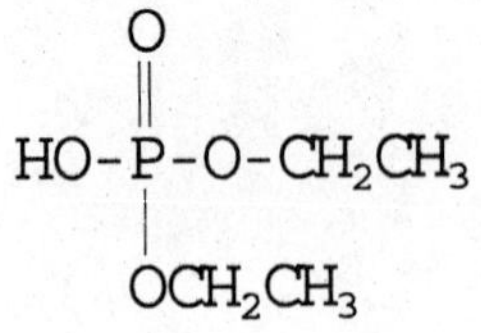

$$HO-\overset{\overset{\displaystyle O}{\|}}{\underset{\underset{\displaystyle OCH_2CH_3}{|}}{P}}-O-CH_2CH_3$$

c. Triester:

$$CH_3CH_2-O-\overset{\overset{\displaystyle O}{\|}}{\underset{\underset{\displaystyle OCH_2CH_3}{|}}{P}}-O-CH_2CH_3$$

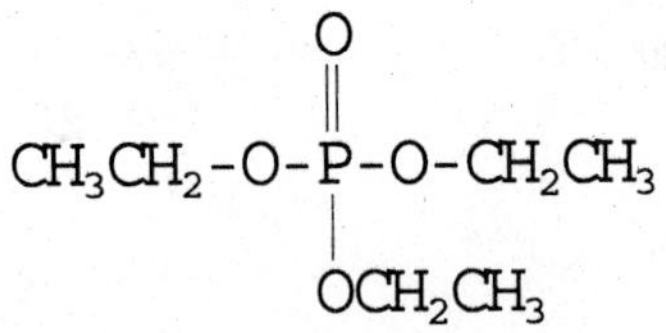

16.53 ATP is the molecule used to store the energy released in metabolic reactions. The energy is stored in the phosphoanhydride bonds between two phosphoryl groups. The energy is released when the bond is hydrolyzed. A portion of the energy can be transferred to another molecule if the phosphoryl group is transferred from ATP to the other molecule.

16.55

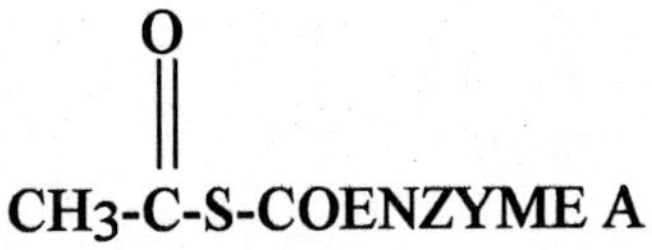

$$CH_3-\overset{\overset{\displaystyle O}{\|}}{C}-S-COENZYME\ A$$

16.57 The structure of nitroglycerine:

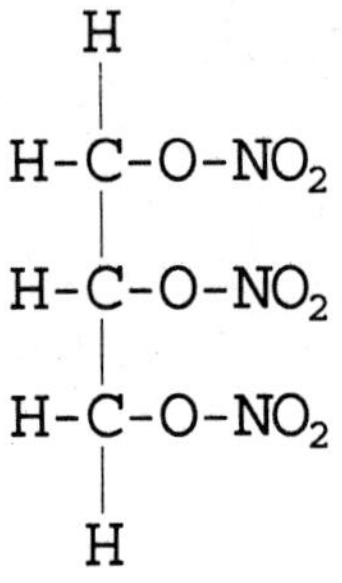

$$\begin{array}{c} H \\ | \\ H-C-O-NO_2 \\ | \\ H-C-O-NO_2 \\ | \\ H-C-O-NO_2 \\ | \\ H \end{array}$$

426

Chapter 17
Lipids and Their Functions in Biochemical Systems
Solutions for the Odd-Numbered Problems

****
In-Chapter Questions and Problems
****

17.1   a. Oleic acid:

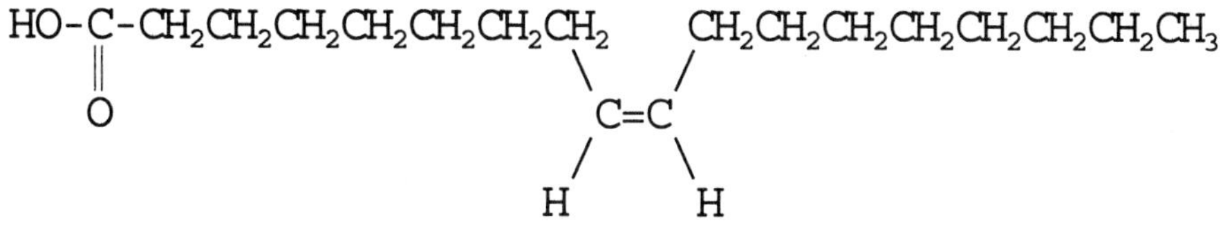

HO-C-CH₂CH₂CH₂CH₂CH₂CH₂CH₂   CH₂CH₂CH₂CH₂CH₂CH₂CH₂CH₃
      ‖   \   /
      O   C=C
                                /   \
                              H   H

   b. Lauric acid:

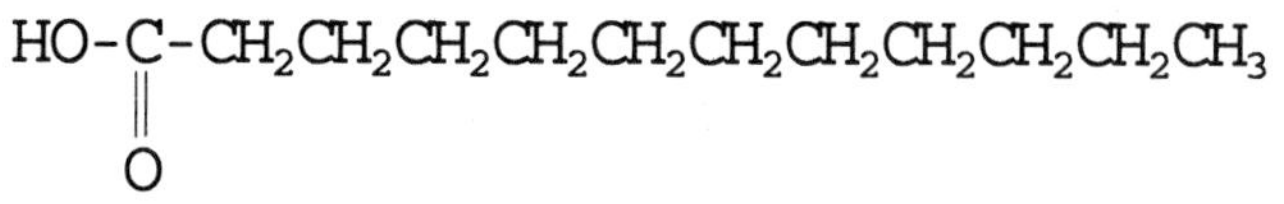

HO-C-CH₂CH₂CH₂CH₂CH₂CH₂CH₂CH₂CH₂CH₂CH₃
   ‖
   O

   c. Linoleic acid:

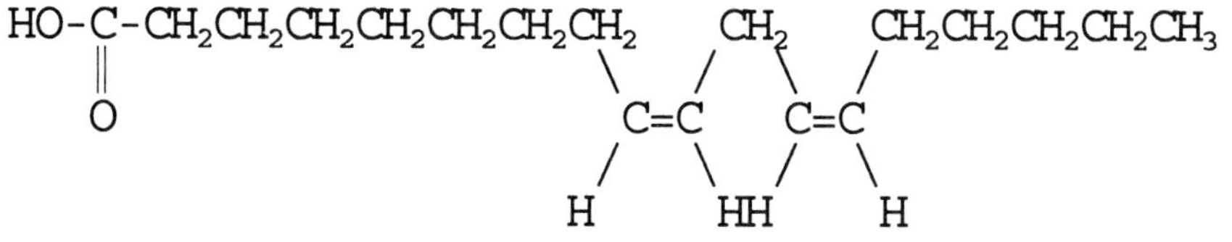

HO-C-CH₂CH₂CH₂CH₂CH₂CH₂CH₂   CH₂   CH₂CH₂CH₂CH₂CH₃
   ‖   \   / \   /
   O   C=C   C=C
                          /   \   /   \
                        H   HH   H

   d. Stearic acid:

HO-C-CH₂CH₂CH₂CH₂CH₂CH₂CH₂CH₂CH₂CH₂CH₂CH₂CH₂CH₂CH₂CH₂CH₃
   ‖
   O

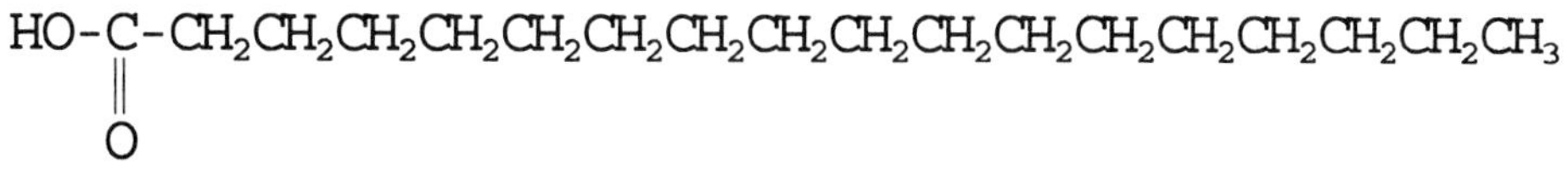

17.3   a. Esterification of lauric acid and ethanol:

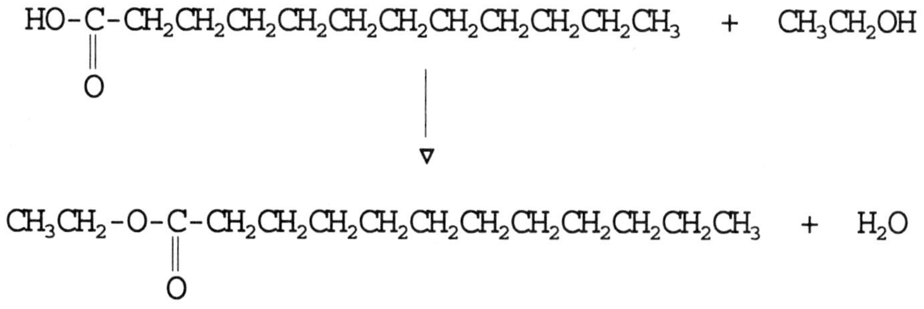

HO-C-CH₂CH₂CH₂CH₂CH₂CH₂CH₂CH₂CH₂CH₂CH₃   +   CH₃CH₂OH
   ‖
   O
                                    |
                                    ▽

CH₃CH₂-O-C-CH₂CH₂CH₂CH₂CH₂CH₂CH₂CH₂CH₂CH₂CH₃   +   H₂O
         ‖
         O

427

b. Reaction of oleic acid with NaOH:

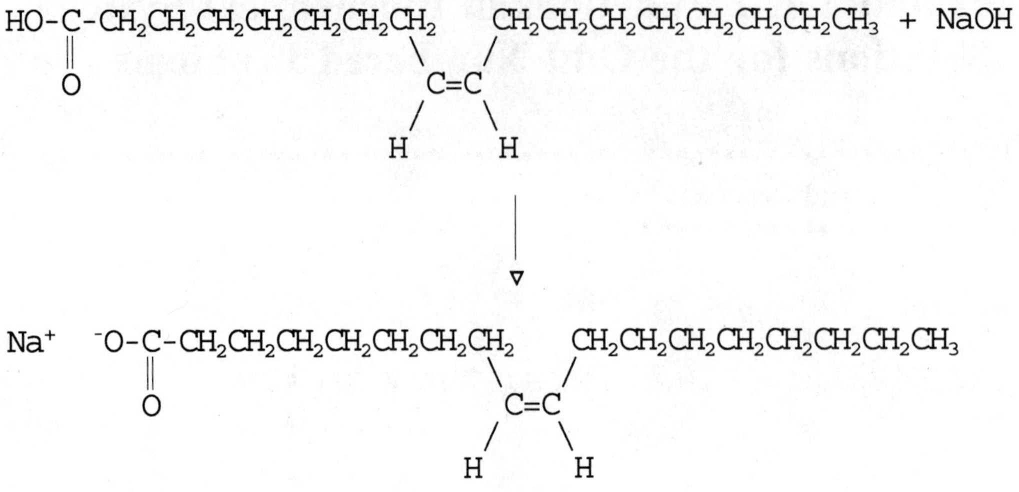

$$HO-\overset{\overset{\displaystyle }{\|}}{\underset{O}{C}}-CH_2CH_2CH_2CH_2CH_2CH_2CH_2 \quad CH_2CH_2CH_2CH_2CH_2CH_2CH_2CH_3 \;+\; NaOH$$

$$\underset{\underset{H}{/}}{C}=\underset{\underset{H}{\backslash}}{C}$$

↓

$$Na^+ \quad {}^-O-\overset{\overset{\displaystyle }{\|}}{\underset{O}{C}}-CH_2CH_2CH_2CH_2CH_2CH_2CH_2 \quad CH_2CH_2CH_2CH_2CH_2CH_2CH_2CH_3$$

$$\underset{\underset{H}{/}}{C}=\underset{\underset{H}{\backslash}}{C}$$

c. Hydrogenation of arachidonic acid:

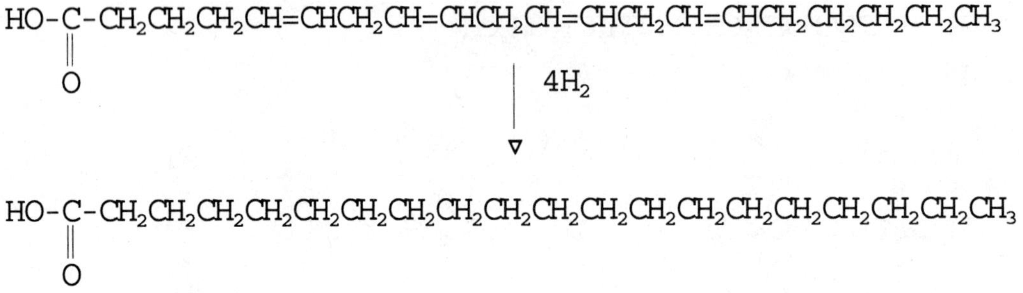

$$HO-\overset{\overset{\displaystyle }{\|}}{\underset{O}{C}}-CH_2CH_2CH_2CH=CHCH_2CH=CHCH_2CH=CHCH_2CH=CHCH_2CH_2CH_2CH_2CH_3$$

$4H_2$

↓

$$HO-\overset{\overset{\displaystyle }{\|}}{\underset{O}{C}}-CH_2CH_2CH_2CH_2CH_2CH_2CH_2CH_2CH_2CH_2CH_2CH_2CH_2CH_2CH_2CH_2CH_2CH_2CH_3$$

17.5 a. Mono, di, and triglycerides of oleic acid:

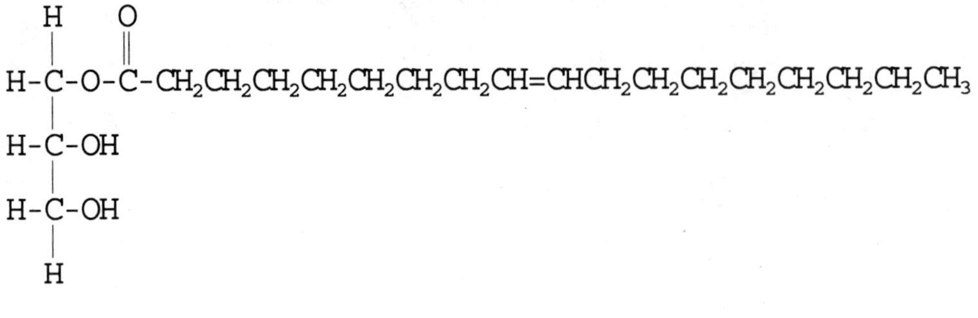

$$\begin{array}{l}
\underset{H}{\overset{H\qquad O}{\underset{|}{\overset{|\qquad\|}{}}}} \\
H-\overset{|}{\underset{|}{C}}-O-\overset{\|}{C}-CH_2CH_2CH_2CH_2CH_2CH_2CH_2CH=CHCH_2CH_2CH_2CH_2CH_2CH_2CH_2CH_3 \\
H-\overset{|}{\underset{|}{C}}-OH \\
H-\overset{|}{\underset{|}{C}}-OH \\
\;\;\;\;\;\overset{|}{H}
\end{array}$$

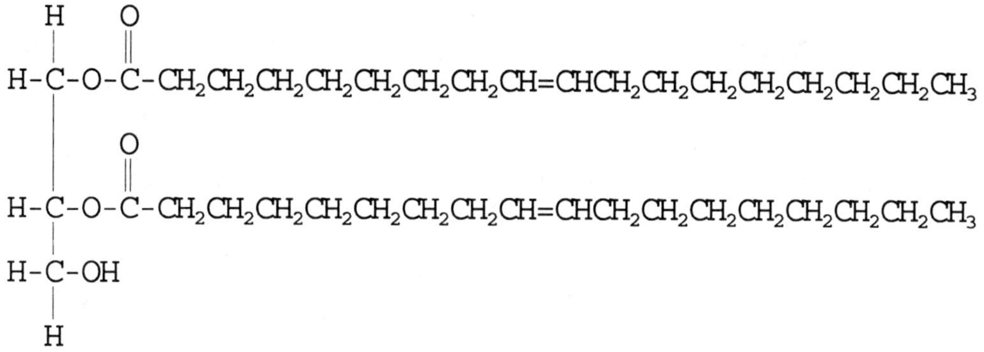

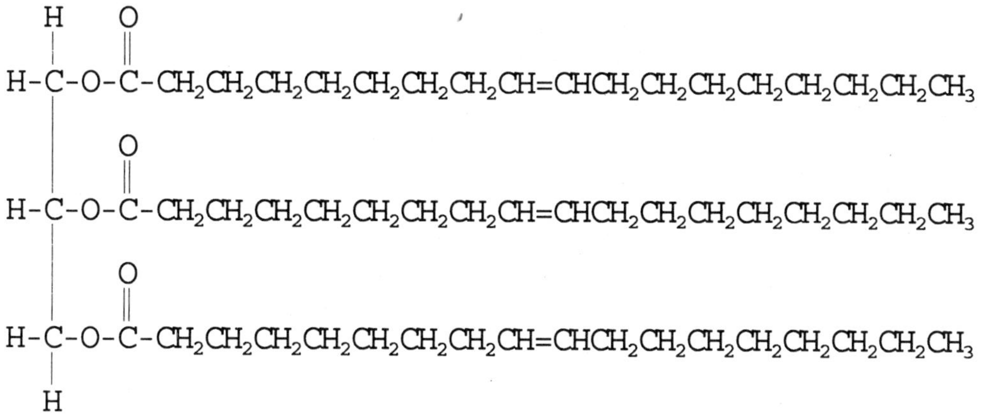

```
    H   O
    |   ||
H-C-O-C-CH₂CH₂CH₂CH₂CH₂CH₂CH₂CH=CHCH₂CH₂CH₂CH₂CH₂CH₂CH₂CH₃
    |       O
    |       ||
H-C-O-C-CH₂CH₂CH₂CH₂CH₂CH₂CH₂CH=CHCH₂CH₂CH₂CH₂CH₂CH₂CH₂CH₃
    |
H-C-OH
    |
    H
```

```
    H   O
    |   ||
H-C-O-C-CH₂CH₂CH₂CH₂CH₂CH₂CH₂CH=CHCH₂CH₂CH₂CH₂CH₂CH₂CH₂CH₃
    |       O
    |       ||
H-C-O-C-CH₂CH₂CH₂CH₂CH₂CH₂CH₂CH=CHCH₂CH₂CH₂CH₂CH₂CH₂CH₂CH₃
    |       O
    |       ||
H-C-O-C-CH₂CH₂CH₂CH₂CH₂CH₂CH₂CH=CHCH₂CH₂CH₂CH₂CH₂CH₂CH₂CH₃
    |
    H
```

b. Mono, di, and triglycerides of capric acid:

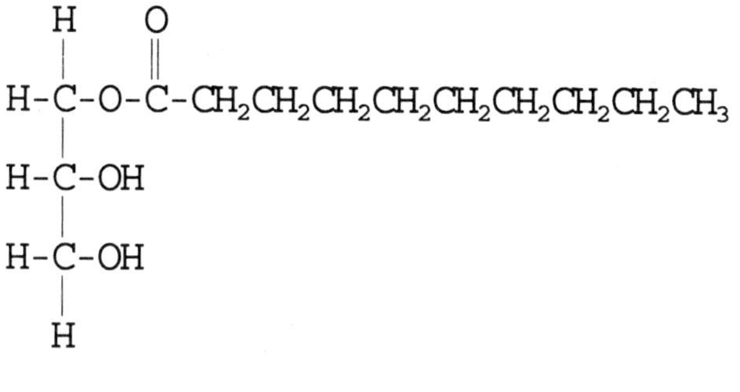

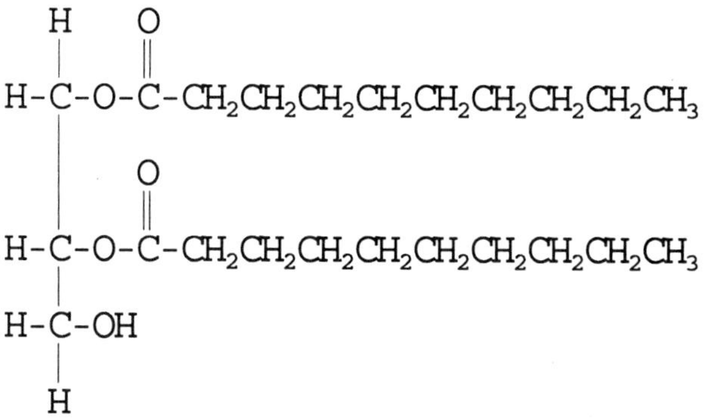

```
    H   O
    |   ||
H-C-O-C-CH₂CH₂CH₂CH₂CH₂CH₂CH₂CH₂CH₃
    |
H-C-OH
    |
H-C-OH
    |
    H
```

```
    H   O
    |   ||
H-C-O-C-CH₂CH₂CH₂CH₂CH₂CH₂CH₂CH₂CH₃
    |       O
    |       ||
H-C-O-C-CH₂CH₂CH₂CH₂CH₂CH₂CH₂CH₂CH₃
    |
H-C-OH
    |
    H
```

$$
\begin{array}{c}
\text{H} \quad\; \text{O} \\
| \quad\quad \| \\
\text{H}-\text{C}-\text{O}-\text{C}-\text{CH}_2\text{CH}_2\text{CH}_2\text{CH}_2\text{CH}_2\text{CH}_2\text{CH}_2\text{CH}_3 \\
| \\
\quad\quad\quad \text{O} \\
\quad\quad\quad \| \\
\text{H}-\text{C}-\text{O}-\text{C}-\text{CH}_2\text{CH}_2\text{CH}_2\text{CH}_2\text{CH}_2\text{CH}_2\text{CH}_2\text{CH}_2\text{CH}_3 \\
| \\
\quad\quad\quad \text{O} \\
\quad\quad\quad \| \\
\text{H}-\text{C}-\text{O}-\text{C}-\text{CH}_2\text{CH}_2\text{CH}_2\text{CH}_2\text{CH}_2\text{CH}_2\text{CH}_2\text{CH}_2\text{CH}_3 \\
| \\
\text{H}
\end{array}
$$

c. Mono, di, and triglycerides of palmitic acid:

$$
\begin{array}{c}
\text{H} \quad\; \text{O} \\
| \quad\quad \| \\
\text{H}-\text{C}-\text{O}-\text{C}-\text{CH}_2\text{CH}_2\text{CH}_2\text{CH}_2\text{CH}_2\text{CH}_2\text{CH}_2\text{CH}_2\text{CH}_2\text{CH}_2\text{CH}_2\text{CH}_2\text{CH}_2\text{CH}_2\text{CH}_3 \\
| \\
\text{H}-\text{C}-\text{OH} \\
| \\
\text{H}-\text{C}-\text{OH} \\
| \\
\text{H}
\end{array}
$$

$$
\begin{array}{c}
\text{H} \quad\; \text{O} \\
| \quad\quad \| \\
\text{H}-\text{C}-\text{O}-\text{C}-\text{CH}_2\text{CH}_2\text{CH}_2\text{CH}_2\text{CH}_2\text{CH}_2\text{CH}_2\text{CH}_2\text{CH}_2\text{CH}_2\text{CH}_2\text{CH}_2\text{CH}_2\text{CH}_2\text{CH}_3 \\
| \\
\quad\quad\quad \text{O} \\
\quad\quad\quad \| \\
\text{H}-\text{C}-\text{O}-\text{C}-\text{CH}_2\text{CH}_2\text{CH}_2\text{CH}_2\text{CH}_2\text{CH}_2\text{CH}_2\text{CH}_2\text{CH}_2\text{CH}_2\text{CH}_2\text{CH}_2\text{CH}_2\text{CH}_2\text{CH}_3 \\
| \\
\text{H}-\text{C}-\text{OH} \\
| \\
\text{H}
\end{array}
$$

$$
\begin{array}{c}
\text{H} \quad\; \text{O} \\
| \quad\quad \| \\
\text{H}-\text{C}-\text{O}-\text{C}-\text{CH}_2\text{CH}_2\text{CH}_2\text{CH}_2\text{CH}_2\text{CH}_2\text{CH}_2\text{CH}_2\text{CH}_2\text{CH}_2\text{CH}_2\text{CH}_2\text{CH}_2\text{CH}_2\text{CH}_3 \\
| \\
\quad\quad\quad \text{O} \\
\quad\quad\quad \| \\
\text{H}-\text{C}-\text{O}-\text{C}-\text{CH}_2\text{CH}_2\text{CH}_2\text{CH}_2\text{CH}_2\text{CH}_2\text{CH}_2\text{CH}_2\text{CH}_2\text{CH}_2\text{CH}_2\text{CH}_2\text{CH}_2\text{CH}_2\text{CH}_3 \\
| \\
\quad\quad\quad \text{O} \\
\quad\quad\quad \| \\
\text{H}-\text{C}-\text{O}-\text{C}-\text{CH}_2\text{CH}_2\text{CH}_2\text{CH}_2\text{CH}_2\text{CH}_2\text{CH}_2\text{CH}_2\text{CH}_2\text{CH}_2\text{CH}_2\text{CH}_2\text{CH}_2\text{CH}_3 \\
| \\
\text{H}
\end{array}
$$

430

d. Mono, di, and triglycerides of lauric acid:

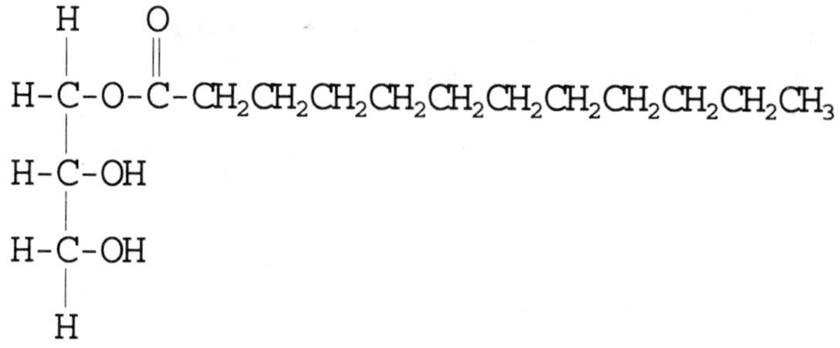

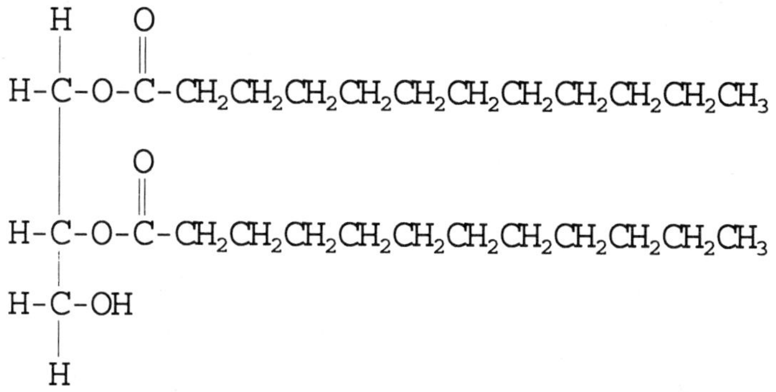

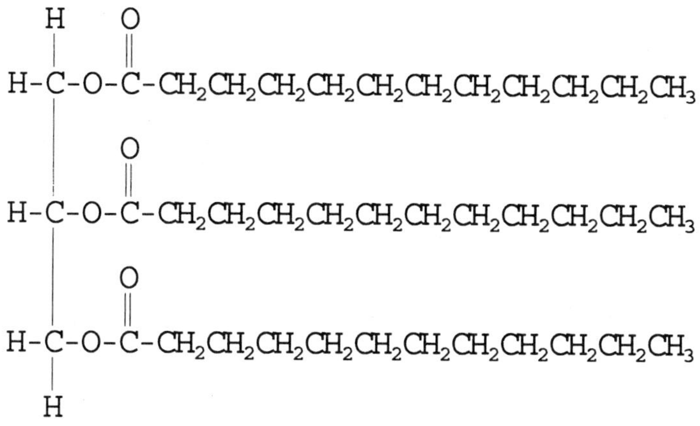

$$
\begin{array}{l}
H \quad\; O \\
| \qquad \| \\
H-C-O-C-CH_2CH_2CH_2CH_2CH_2CH_2CH_2CH_2CH_2CH_2CH_3 \\
| \\
H-C-OH \\
| \\
H-C-OH \\
| \\
H
\end{array}
$$

$$
\begin{array}{l}
H \quad\; O \\
| \qquad \| \\
H-C-O-C-CH_2CH_2CH_2CH_2CH_2CH_2CH_2CH_2CH_2CH_2CH_3 \\
| \qquad\; O \\
| \qquad\; \| \\
H-C-O-C-CH_2CH_2CH_2CH_2CH_2CH_2CH_2CH_2CH_2CH_2CH_3 \\
| \\
H-C-OH \\
| \\
H
\end{array}
$$

$$
\begin{array}{l}
H \quad\; O \\
| \qquad \| \\
H-C-O-C-CH_2CH_2CH_2CH_2CH_2CH_2CH_2CH_2CH_2CH_2CH_3 \\
| \qquad\; O \\
| \qquad\; \| \\
H-C-O-C-CH_2CH_2CH_2CH_2CH_2CH_2CH_2CH_2CH_2CH_2CH_3 \\
| \qquad\; O \\
| \qquad\; \| \\
H-C-O-C-CH_2CH_2CH_2CH_2CH_2CH_2CH_2CH_2CH_2CH_2CH_3 \\
| \\
H
\end{array}
$$

17.7 Structure of the steroid nucleus:

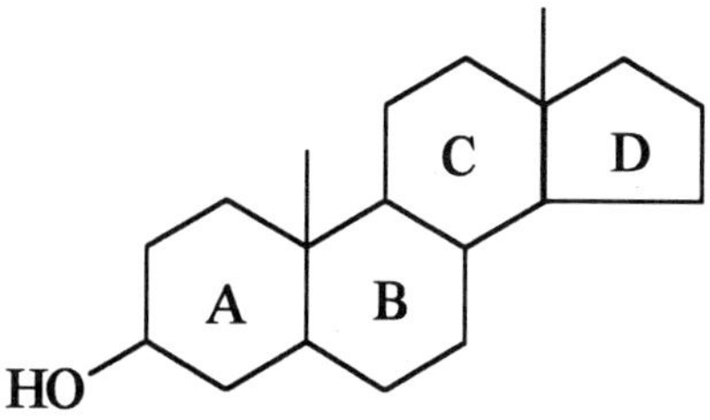

17.9   Cholesterol is carried in low density lipoprotein (LDL) particles in the plasma. These bind to specific LDL receptors within cell membranes. This binding stimulates receptor-mediated endocytosis, the invagination of the cell membrane that draws the LDL particles into the cytoplasm of the cell. The process encases the LDL within a vesicle or endosome in the cytoplasm.

17.11 Membrane transport resembles enzyme catalysis because both processes exhibit a high degree of specificity. The specificity is the result of the precise fit of the substrate into a site on the enzyme molecule or the precise fit of the molecule or ion being transported into a site on the permease protein.

****
End-of-Chapter Questions and Problems
****

17.13 The four main groups of lipids are fatty acids, glycerides, nonglyceride lipids, and complex lipids.

17.15 A saturated fatty acid is one in which the hydrocarbon tail has only carbon-to-carbon single bonds. Thus, each carbon atom is bonded to the maximum number of hydrogen atoms. An unsaturated fatty acid has at least one carbon-to-carbon double bond.

17.17 As the length of the hydrocarbon chains of fatty acids increases, the melting points increase.

17.19 a. Decanoic acid:

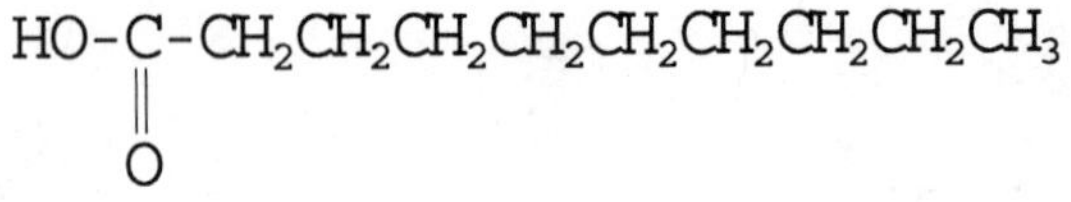

$$HO-\underset{\underset{O}{\|}}{C}-CH_2CH_2CH_2CH_2CH_2CH_2CH_2CH_2CH_3$$

b. Stearic acid:

$$HO-\underset{\underset{O}{\|}}{C}-CH_2CH_2CH_2CH_2CH_2CH_2CH_2CH_2CH_2CH_2CH_2CH_2CH_2CH_2CH_2CH_2CH_3$$

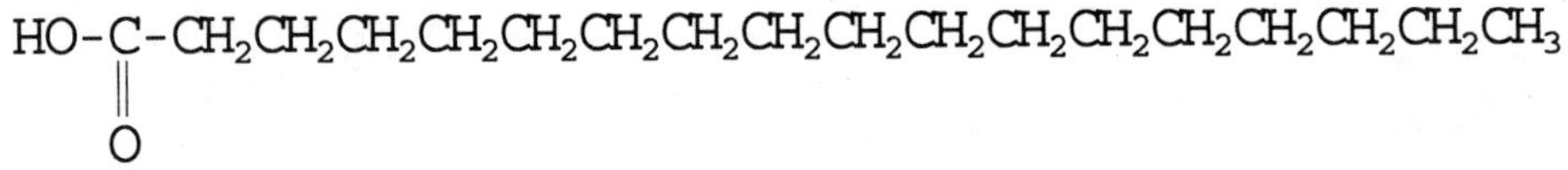

c. *trans*-5-Decenoic acid:

$$HO-\underset{\underset{O}{\|}}{C}-CH_2CH_2CH_2 \quad \underset{\underset{H}{/}}{\overset{\overset{H}{\diagdown}}{C}}=\underset{\underset{CH_2CH_2CH_2CH_3}{\diagdown}}{C}$$

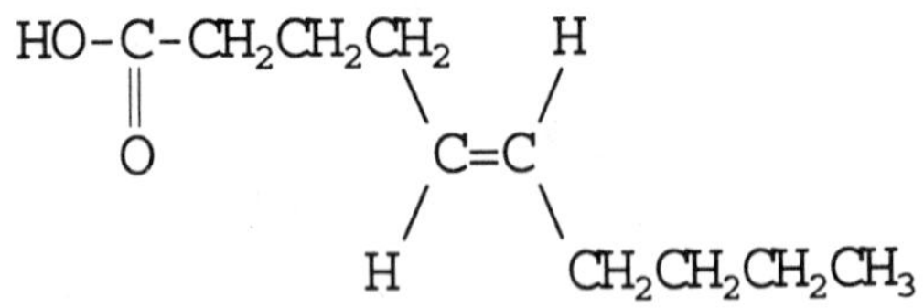

d. *cis*-5-Decenoic acid:

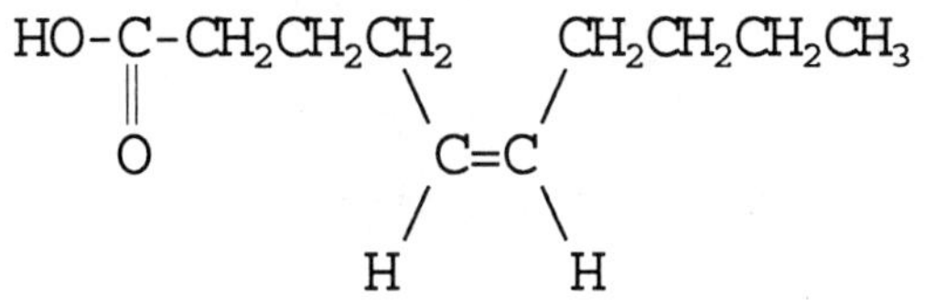

HO-C-CH$_2$CH$_2$CH$_2$ CH$_2$CH$_2$CH$_2$CH$_3$
   ‖   \   /
   O   C=C
               /   \
             H   H

17.21 a. Esterification of glycerol with three molecules of myristic acid:

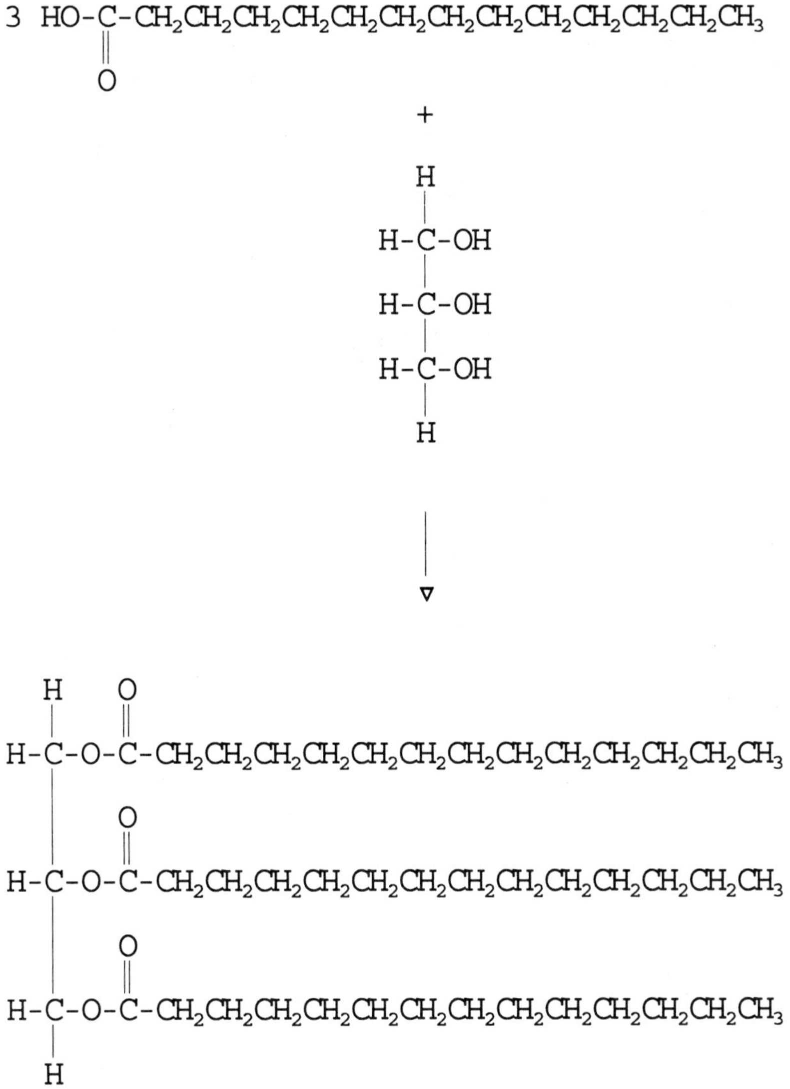

3   HO-C-CH$_2$CH$_2$CH$_2$CH$_2$CH$_2$CH$_2$CH$_2$CH$_2$CH$_2$CH$_2$CH$_2$CH$_2$CH$_3$
      ‖
      O

+

     H
     |
  H-C-OH
     |
  H-C-OH
     |
  H-C-OH
     |
     H

|
▽

   H   O
   | ‖
H-C-O-C-CH$_2$CH$_2$CH$_2$CH$_2$CH$_2$CH$_2$CH$_2$CH$_2$CH$_2$CH$_2$CH$_2$CH$_2$CH$_3$
   | O
   | ‖
H-C-O-C-CH$_2$CH$_2$CH$_2$CH$_2$CH$_2$CH$_2$CH$_2$CH$_2$CH$_2$CH$_2$CH$_2$CH$_2$CH$_3$
   | O
   | ‖
H-C-O-C-CH$_2$CH$_2$CH$_2$CH$_2$CH$_2$CH$_2$CH$_2$CH$_2$CH$_2$CH$_2$CH$_2$CH$_2$CH$_3$
   |
   H

b. Acid hydrolysis of glyceryl tristearate:

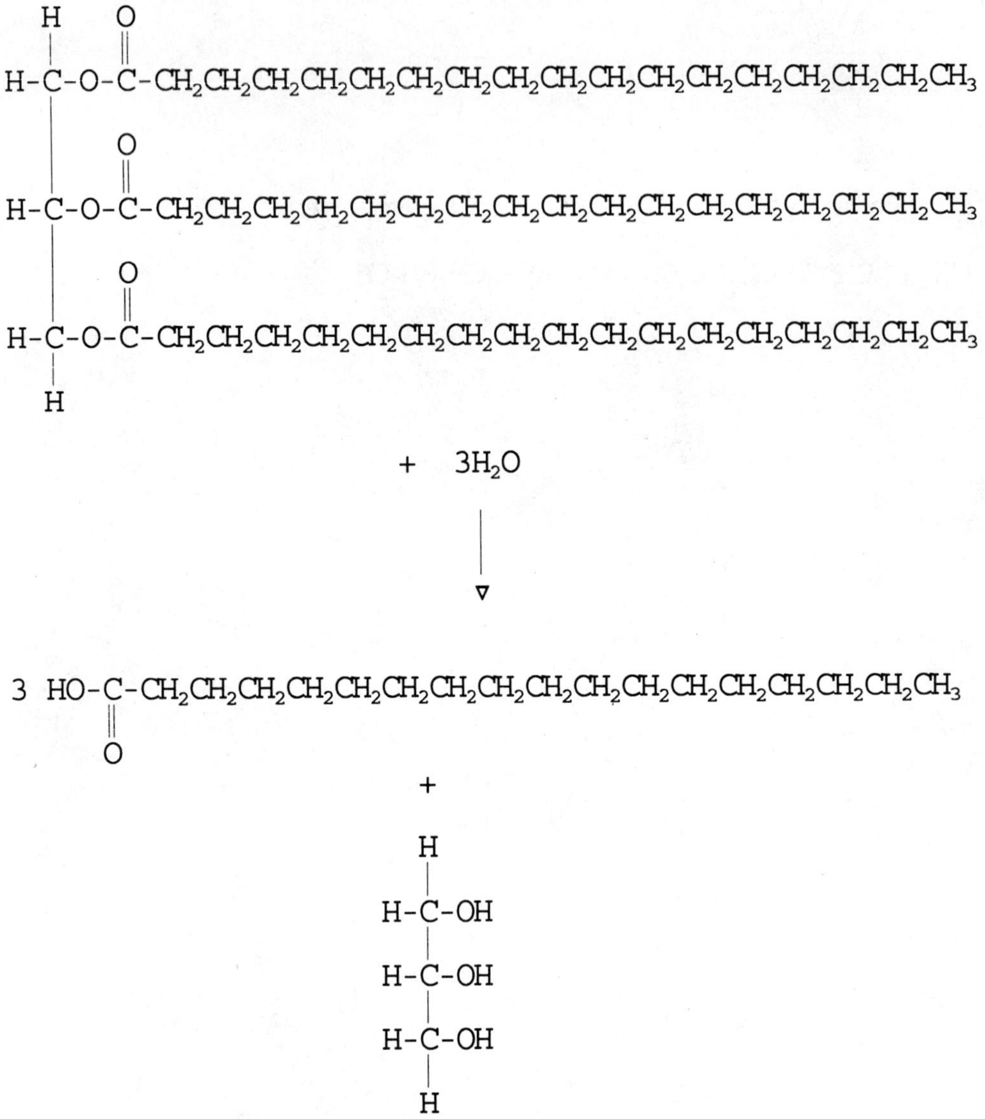

$$
\begin{array}{l}
\overset{\displaystyle H}{\underset{\displaystyle |}{}}\ \ \overset{\displaystyle O}{\underset{\displaystyle \parallel}{}} \\
H-C-O-C-CH_2CH_2CH_2CH_2CH_2CH_2CH_2CH_2CH_2CH_2CH_2CH_2CH_2CH_2CH_2CH_2CH_3 \\
\end{array}
$$

H–C–O–C–CH$_2$CH$_2$CH$_2$CH$_2$CH$_2$CH$_2$CH$_2$CH$_2$CH$_2$CH$_2$CH$_2$CH$_2$CH$_2$CH$_2$CH$_2$CH$_2$CH$_3$

H–C–O–C–CH$_2$CH$_2$CH$_2$CH$_2$CH$_2$CH$_2$CH$_2$CH$_2$CH$_2$CH$_2$CH$_2$CH$_2$CH$_2$CH$_2$CH$_2$CH$_2$CH$_3$

H

$+\quad 3H_2O$

$\triangledown$

3 HO–C–CH$_2$CH$_2$CH$_2$CH$_2$CH$_2$CH$_2$CH$_2$CH$_2$CH$_2$CH$_2$CH$_2$CH$_2$CH$_2$CH$_2$CH$_2$CH$_2$CH$_3$
$\quad\quad\parallel$
$\quad\quad O$

$+$

H

H–C–OH

H–C–OH

H–C–OH

H

c. Reaction of decanoic acid with KOH:

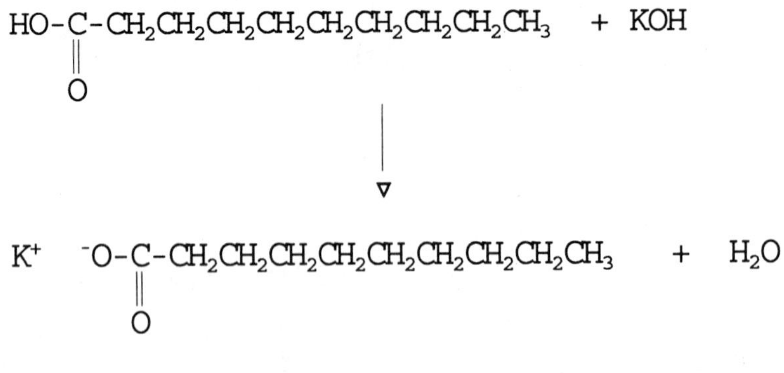

HO–C–CH$_2$CH$_2$CH$_2$CH$_2$CH$_2$CH$_2$CH$_2$CH$_2$CH$_3$ + KOH
$\quad\parallel$
$\quad O$

$\triangledown$

K$^+$ $^-$O–C–CH$_2$CH$_2$CH$_2$CH$_2$CH$_2$CH$_2$CH$_2$CH$_2$CH$_3$ + H$_2$O
$\quad\quad\parallel$
$\quad\quad O$

434

d. Hydrogenation of linoleic acid:

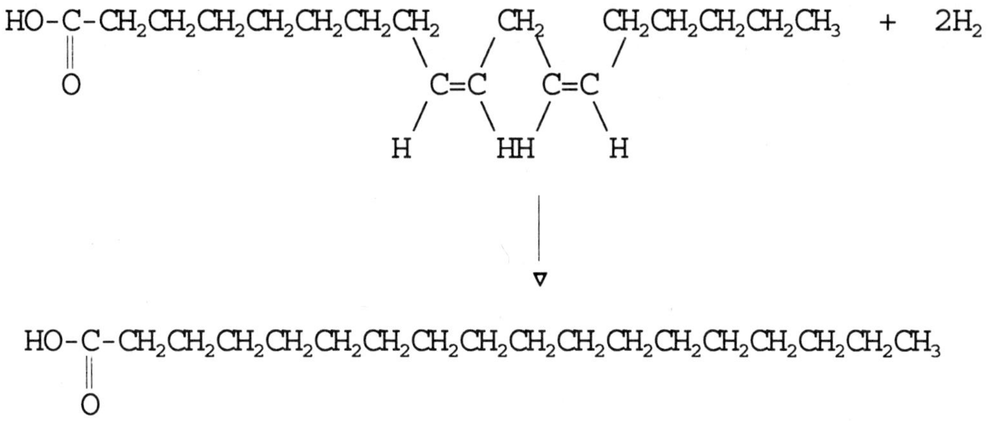

$$HO-\underset{\underset{O}{\|}}{C}-CH_2CH_2CH_2CH_2CH_2CH_2CH_2 \quad CH_2 \quad CH_2CH_2CH_2CH_2CH_3 \quad + \quad 2H_2$$

(with the C=C double bonds shown between the chains, bearing H substituents)

$$\bigtriangledown$$

$$HO-\underset{\underset{O}{\|}}{C}-CH_2CH_2CH_2CH_2CH_2CH_2CH_2CH_2CH_2CH_2CH_2CH_2CH_2CH_2CH_2CH_2CH_3$$

17.23   An essential fatty acid is one that the body cannot synthesize and thus must be supplied in the diet. The essential fatty acid linoleic acid is required for the synthesis of arachidonic acid, a precursor for the synthesis of the prostaglandins, a group of hormonelike molecules.

17.25   Aspirin effectively decreases the inflammatory response by inhibiting the synthesis of all prostaglandins. Aspirin works by inhibiting cyclooxygenase, the first enzyme in prostaglandin biosynthesis. This inhibition results from the transfer of an acetyl group from aspirin to the enzyme. Since cyclooxygenase is found in all cells, synthesis of all prostaglandins is inhibited.

17.27   Prostaglandins stimulate smooth muscle contraction, especially uterine contractions during labor. They enhance fever and swelling associated with the inflammatory response. Some prostaglandins cause bronchial dilation. Others inhibit secretion of acid into the stomach and stimulate the secretion of a mucus layer that protects the stomach lining.

17.29

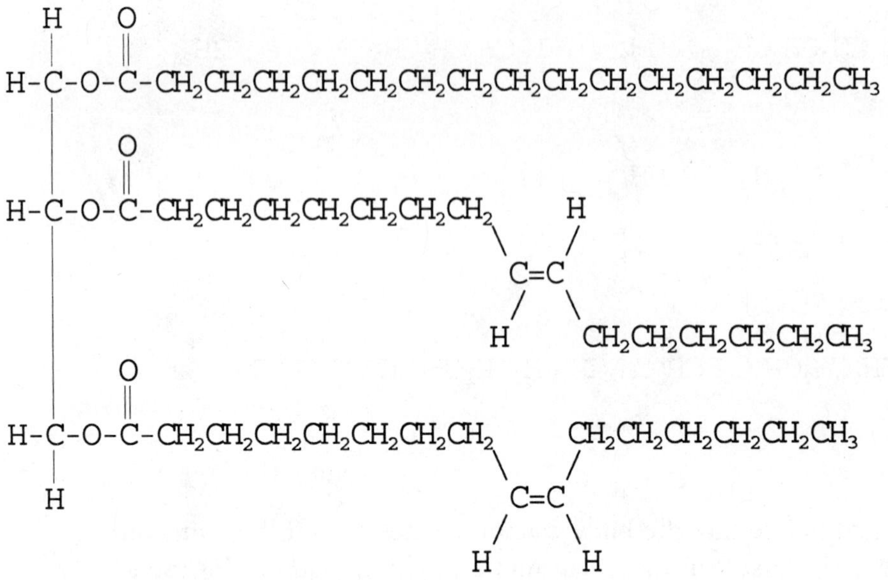

H-C-O-C-CH$_2$CH$_2$CH$_2$CH$_2$CH$_2$CH$_2$CH$_2$CH$_2$CH$_2$CH$_2$CH$_2$CH$_2$CH$_2$CH$_2$CH$_3$

H-C-O-C-CH$_2$CH$_2$CH$_2$CH$_2$CH$_2$CH$_2$CH$_2$ ... C=C ... CH$_2$CH$_2$CH$_2$CH$_2$CH$_2$CH$_3$

H-C-O-C-CH$_2$CH$_2$CH$_2$CH$_2$CH$_2$CH$_2$CH$_2$ ... C=C ... CH$_2$CH$_2$CH$_2$CH$_2$CH$_2$CH$_3$

17.31

H-C-O-C-CH$_2$CH$_2$CH$_2$CH$_2$CH$_2$CH$_2$CH$_2$CH$_2$CH$_3$

H-C-O-C-CH$_2$CH$_2$CH$_2$CH$_2$CH$_2$CH$_2$CH$_2$CH$_2$CH$_2$CH$_2$CH$_3$

$^-$O-P-O-C-H

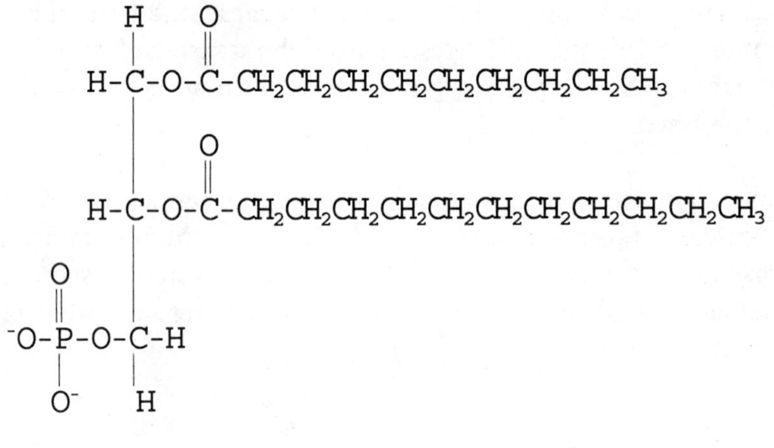

17.33 Triglycerides consist of three fatty acids esterified to the three hydroxyl groups of glycerol. In phospholipids there are only two fatty acids esterified to glycerol. A phosphoryl group is esterified (phosphoester linkage) to the third hydroxyl group.

17.35 Sphingolipids are phospholipids that are derived from sphingosine rather than glycerol. Sphingosine is a nitrogen-containing (amino) alcohol.

17.37 Cholesterol is readily soluble in the hydrophobic region of biological membranes. It is involved in regulating the fluidity of the membrane.

17.39 Progesterone is the most important hormone associated with pregnancy. It is needed for the successful initiation and completion of the pregnancy. It prepares the lining of the uterus to accept the fertilized egg, facilitates development of the fetus, and suppresses ovulation during pregnancy. Testosterone is needed for development of male secondary sexual characteristics. Estrone is required for proper development of female secondary sexual characteristics.

17.41 Cortisone is used to treat rheumatoid arthritis, asthma, gastrointestinal disorders, and many skin conditions.

17.43 Myricyl palmitate (beeswax) is made up of the fatty acid palmitic acid and the alcohol myricyl alcohol - $CH_3(CH_2)_{28}CH_2OH$.

17.45 Isoprenoids are a large, diverse collection of lipids that are synthesized from the isoprene unit:

$$
\begin{array}{c}
CH_3 \\
| \\
CH_2=C-CH=CH_2
\end{array}
$$

17.47 Some biologically important terpenes include the steroids and bile salts, lipid-soluble vitamins, certain plant hormones, and chlorophyll.

17.49 The four major types of plasma lipoproteins are chylomicrons, high density lipoproteins, low density lipoproteins, and very low density lipoproteins.

17.51 Atherosclerosis results when cholesterol and other substances coat the arteries causing a narrowing of the passageways. As the passageways become narrower, greater pressure is required to provide adequate blood flow. This results in higher blood pressure (hypertension).

17.53 If the LDL receptor is defective, it cannot function to remove cholesterol-bearing LDL particles from the blood. The excess cholesterol, along with other substances, will accumulate along the walls of the arteries, causing atherosclerosis.

17.55 If the fatty acyl tails of membrane phospholipids are converted from saturated to unsaturated, the fluidity of the membrane will increase. Each carbon-to-carbon double bond that is added will introduce a "kink" into the fatty acyl tail. As a result, the tails cannot pack together as they would if they were saturated. The result is increased fluidity.

17.57 The basic structure of a biological membrane is a bilayer of phospholipid molecules arranged so that the hydrophobic hydrocarbon tails are packed in the center and the hydrophilic head groups are exposed on the inner and outer surfaces.

17.59   A peripheral membrane protein is bound to only one surface of the membrane, either inside or outside the cell.

17.61   Cholesterol is freely soluble in the hydrophobic layer of a biological membrane. It moderates the fluidity of the membrane by disrupting the stacking of the fatty acid tails of membrane phospholipids.

17.63   Specific membrane proteins on human and mouse cells were labeled with red and green fluorescent dyes, respectively. The cells were fused into single celled hybrids and were observed using a microscope with an ultraviolet light source. The ultraviolet light caused the dyes to fluoresce. Initially the dyes were localized in regions of the membrane representing the original human or mouse cell. Within an hour, the proteins were evenly distributed throughout the membrane of the fused cell.

17.65   Both simple diffusion and facilitated diffusion are means of passive transport. Simple diffusion involves the the net movement of a solute directly across a membrane from a region of higher concentration to a region of lower concentration. Facilitated diffusion also involves movement of a substance from a region of high concentration to an area of low concentration. However, facilitated diffusion requires a channel protein or permease through which the solute must pass.

17.67   Both active transport and facilitated diffusion require a protein channel or permease through which solutes pass into or out of the cell. Active transport requires an energy input to transport molecules or ions against the gradient (from an area of lower concentration to an area of higher concentration). Facilitated diffusion is a means of passive transport in which molecules or ions pass from regions of higher concentration to regions of lower concentration through a permease protein. No energy is expended by the cell in facilitated diffusion.

17.69   An antiport transport mechanism is one in which one molecule or ion is transported into the cell while a different molecule or ion is transported out of the cell.

17.71   Each permease or channel protein has a binding site that has a shape and charge distribution that is complementary to the molecule or ion that it can bind and transport across the cell membrane. Any molecule or ion that does not fit into this binding site will not be transported by the permease.

17.73   One ATP is hydrolyzed to transport 3 Na^+ out of the cell and 2 K^+ into the cell.

17.75   Active transport is the movement of molecules or ions across a membrane against a concentration gradient (from a region of lower concentration to a region of higher concentration). Specific protein permeases are required and the cell must expend energy to accomplish this transport.

Chapter 18
Amines and Amides
Solutions to the Odd-Numbered Problems

In-Chapter Questions and Problems

18.1   a. Tertiary amine
       b. Primary amine
       c. Secondary amine

18.3

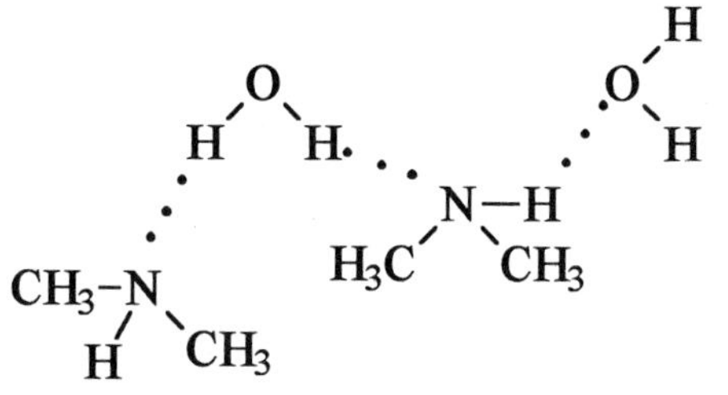

18.5   a. Methanol would have the higher boiling point than methylamine. The intermolecular hydrogen bonds between alcohol molecules will be stronger than those between two amines because because oxygen is more electronegative than nitrogen.
       b. Water would have a higher boiling point that dimethylamine. The intermolecular hydrogen bonds between water molecules will be stronger than those between two amines because because oxygen is more electronegative than nitrogen.
       c. Ethylamine will have a higher boiling point that methylamine because it has a higher molecular weight.
       d. Propylamine will have a higher boiling point that butane because propylamine molecules can form intermolecular hydrogen bonds while the nonpolar butane cannot do so.

18.7   a. *N*-Methylaniline:

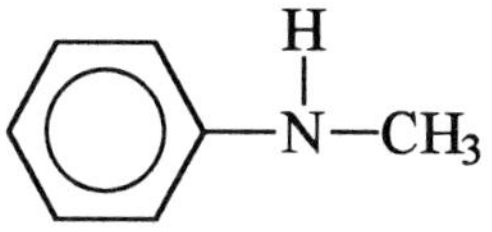

439

b. *N,N*-Dimethylaniline:

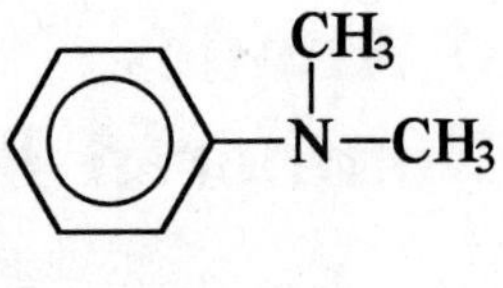

$$\text{C}_6\text{H}_5-\text{N}(\text{CH}_3)-\text{CH}_3$$

c. *N*-Ethylaniline:

$$\text{C}_6\text{H}_5-\text{N}(\text{H})-\text{CH}_2\text{CH}_3$$

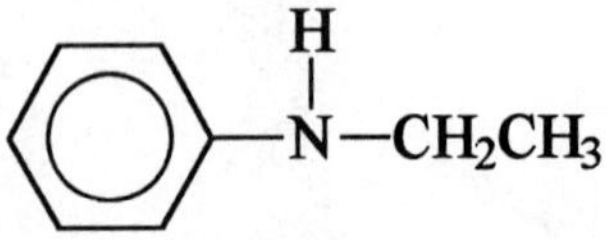

d. *N*-Isopropylaniline:

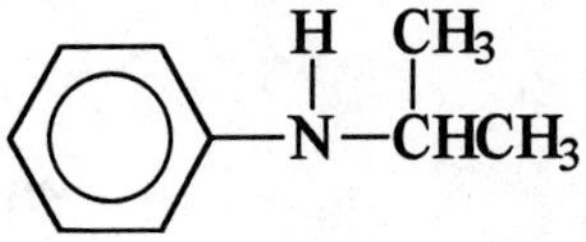

$$\text{C}_6\text{H}_5-\text{N}(\text{H})-\text{CH}(\text{CH}_3)\text{CH}_3$$

18.9   a. 2-Propanamine:

```
      H  H  H
      |  |  |
  H-C--C--C-H
      |  |  |
      H  N  H
        / \
       H   H
```

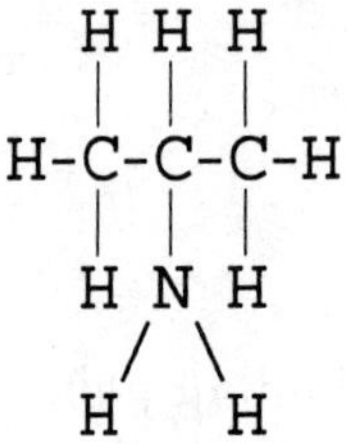

b. 3-Octanamine:

```
        H     H
         \   /
  H  H   N   H  H  H  H  H
  |  |   |   |  |  |  |  |
H-C--C---C---C--C--C--C--C-H
  |  |   |   |  |  |  |  |
  H  H   H   H  H  H  H  H
```

440

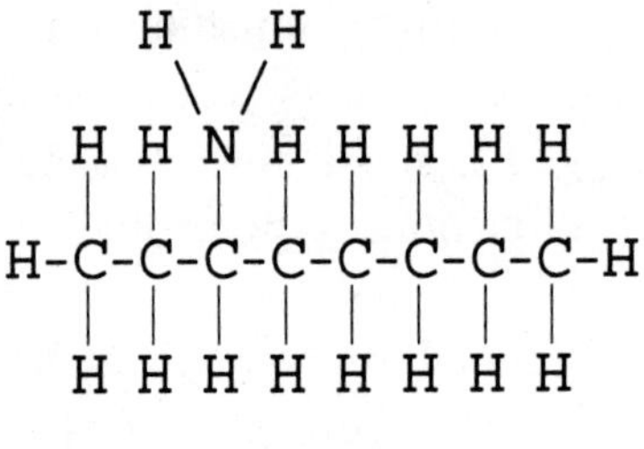

c. *N*-Ethyl-2-heptanamine:

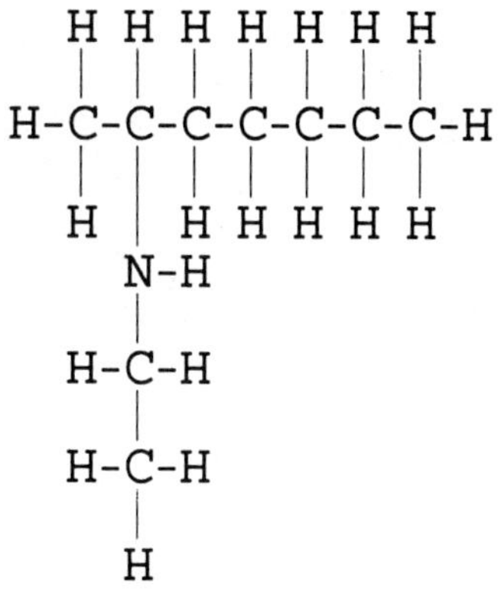

```
     H   H   H   H   H   H   H
     |   |   |   |   |   |   |
 H - C - C - C - C - C - C - C - H
     |   |   |   |   |   |   |
     H   |   H   H   H   H   H
         N - H
         |
     H - C - H
         |
     H - C - H
         |
         H
```

d. 2-Methyl-2-pentanamine:

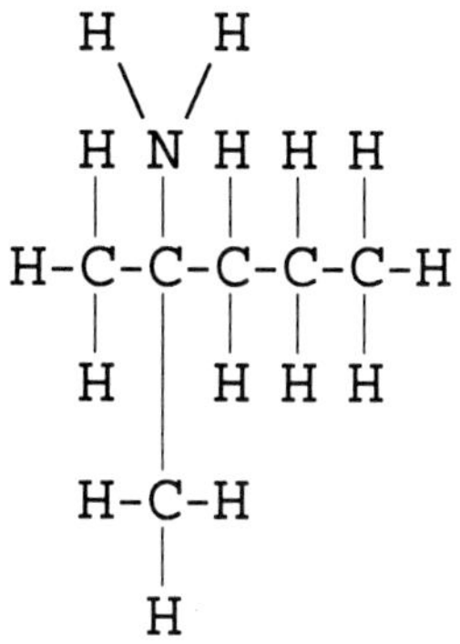

```
     H       H
      \     /
   H   N   H   H   H
   |   |   |   |   |
 H-C - C - C - C - C - H
   |   |   |   |   |
   H   |   H   H   H
   H - C - H
       |
       H
```

e. 4-Chloro-5-iodo-1-nonanamine:

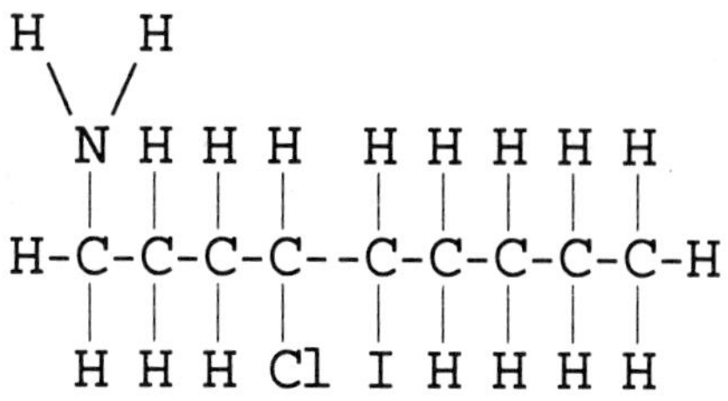

```
       H     H
        \   /
         N   H   H   H     H   H   H   H   H
         |   |   |   |     |   |   |   |   |
     H - C - C - C - C  -  C - C - C - C - C - H
         |   |   |   |     |   |   |   |   |
         H   H   H   Cl    I   H   H   H   H
```

f. *N,N*-Diethyl-1-pentanamine:

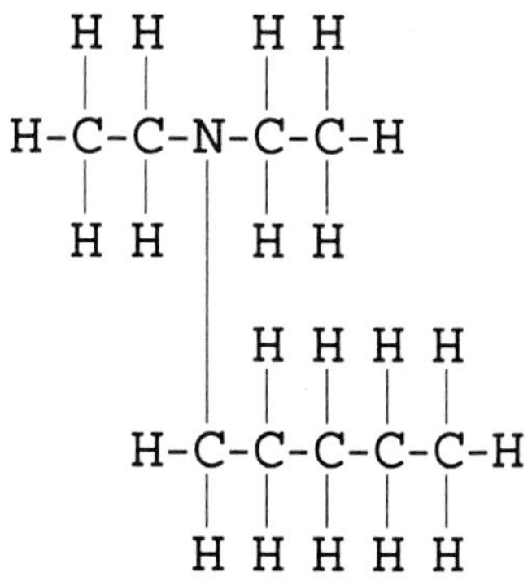

```
   H   H       H   H
   |   |       |   |
 H-C - C - N - C - C - H
   |   |   |   |   |
   H   H   |   H   H
           |
           |       H   H   H   H
           |       |   |   |   |
       H - C - C - C - C - C - H
           |   |   |   |   |
           H   H   H   H   H
```

441

18.11 a.

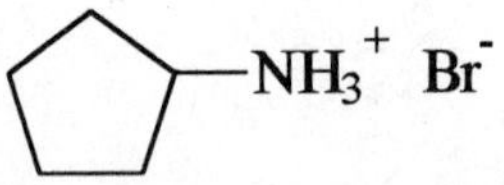

$$\text{cyclopentyl}-NH_3^+ \ \ Br^-$$

b.

$$CH_3CH_2-\overset{\overset{\displaystyle H}{|}\,+}{\underset{\underset{\displaystyle H}{|}}{N}}-CH_3 \ \ + \ \ OH^-$$

c. $CH_3-N^+H_3 \ \ + \ \ OH^-$

18.13 a. CH_3-NH_2

b. $CH_3-\overset{\overset{\displaystyle CH_3}{|}}{NH} \ \ or \ \ (CH_3)_2-NH$

****

End-of-Chapter Questions and Problems

****

18.15 a. 1-Butanamine would be more soluble in water because it has a polar amine group which can form hydrogen bonds with water molecules. Pentane is an alkane and therefore is nonpolar and water-insoluble.
b. 2-Pentanamine would be more soluble in water because it has a polar amine group which can form hydrogen bonds with water molecules. Cyclohexane is an alkane and therefore is nonpolar and water-insoluble.

18.17 Triethylamine molecules cannot form hydrogen bonds with one another, but 1-hexanamine molecules are able to do so. As a result of the greater intermolecular attraction between 1-hexanamine molecules, it has a higher boiling point.

18.19 a. 2-Butanamine
b. 3-Hexanamine
c. Cyclopentanamine
d. 2-Methyl-2-propanamine

18.21   a. Diethylamine: $CH_3CH_2-NH-CH_2CH_3$

b. Butylamine: $CH_3CH_2CH_2CH_2NH_2$

c. 3-Decanamine: $CH_3CH_2CHCH_2CH_2CH_2CH_2CH_2CH_2CH_3$
$|$
NH_2

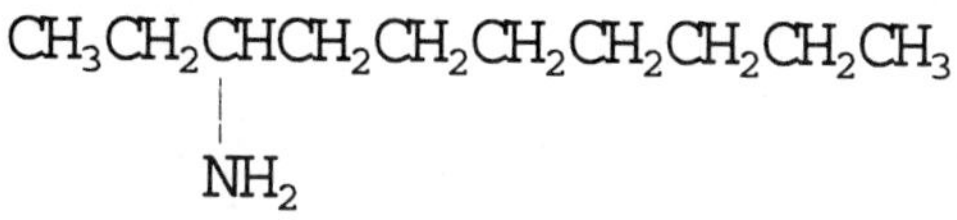

d. 3-Bromo-2-pentanamine:

Br
$|$
$CH_3CHCHCH_2CH_3$
$|$
NH_2

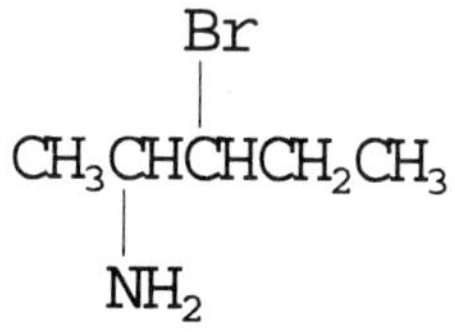

e. Triphenylamine:

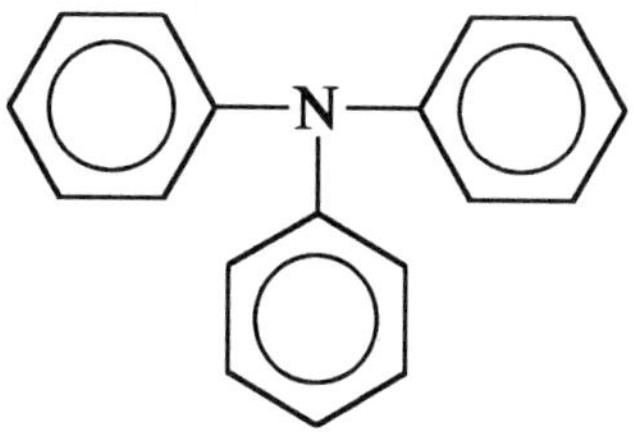

18.23   a. 2-Pentanamine:

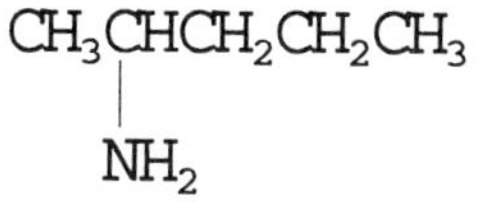

$CH_3CHCH_2CH_2CH_3$
$|$
NH_2

b. 2-Bromo-1-butanamine:

Br
$|$
$CH_3CH_2CHCH_2NH_2$

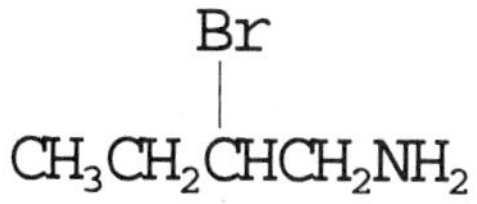

c. Ethylisopropylamine:

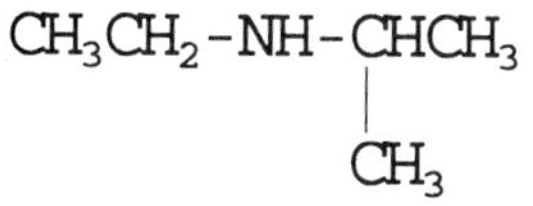

$CH_3CH_2-NH-CHCH_3$
$|$
CH_3

d. Cyclopentanamine:

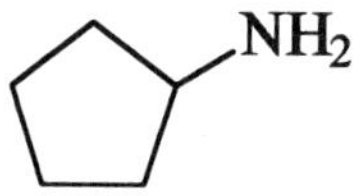

18.25

$CH_3CH_2CH_2CH_2NH_2$

1-Butanamine
(Primary amine)

$CH_3CH_2CHCH_3$
|
NH_2

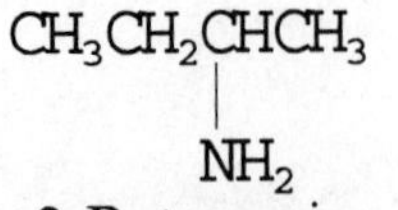

2-Butanamine
(Primary amine)

$CH_3CHCH_2NH_2$
|
CH_3

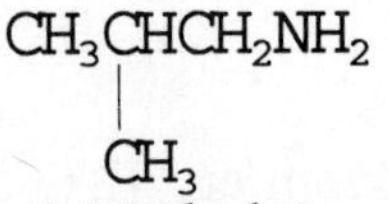

2-Methyl-1-propanamine
(Primary amine)

CH_3
|
CH_3-C-CH_3
|
NH_2

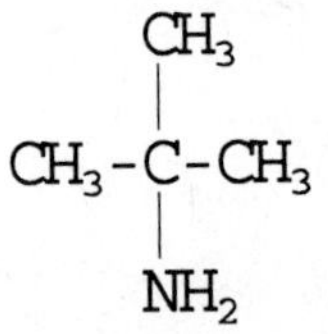

2-Methyl-2-Propanamine
(Primary amine)

CH_3
|
$CH_3CH_2-N-CH_3$

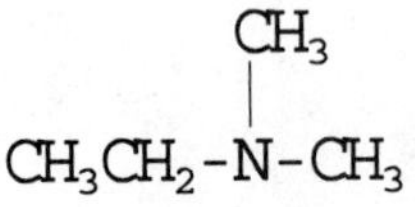

N,N-Dimethylethanamine
(Tertiary amine)

$CH_3CH_2-NH-CH_2CH_3$

N-Ethylethanamine
(Secondary amine)

CH_3CHCH_3
|
$NH-CH_3$

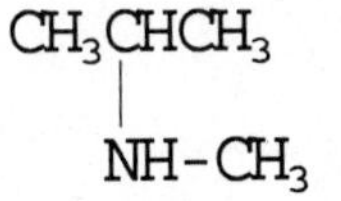

N-Methyl-2-propanamine
(Secondary amine)

$CH_3CH_2CH_2-NH-CH_3$

N-Methyl-1-propanamine
(Secondary amine)

18.27 a. Cyclohexanamine is a primary amine.
       b. Dibutylamine is a secondary amine
       c. 2-Methyl-2-heptanamine is a primary amine.
       d. Tripentylamine is a tertiary amine.

18.29

a. 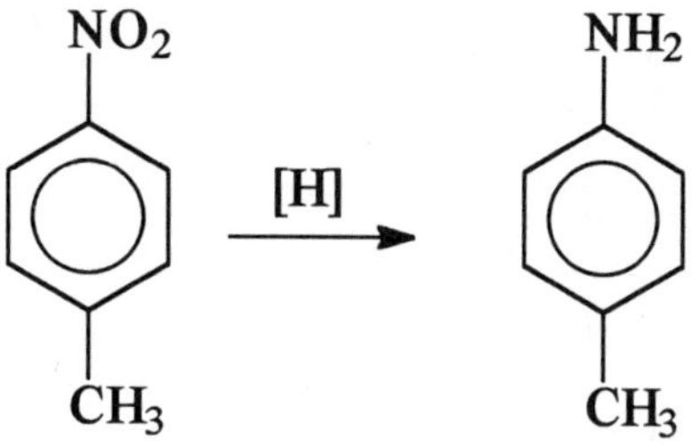 4-nitrotoluene $\xrightarrow{[H]}$ 4-methylaniline (NO$_2$ on benzene ring with CH$_3$ para, reduced to NH$_2$)

b. 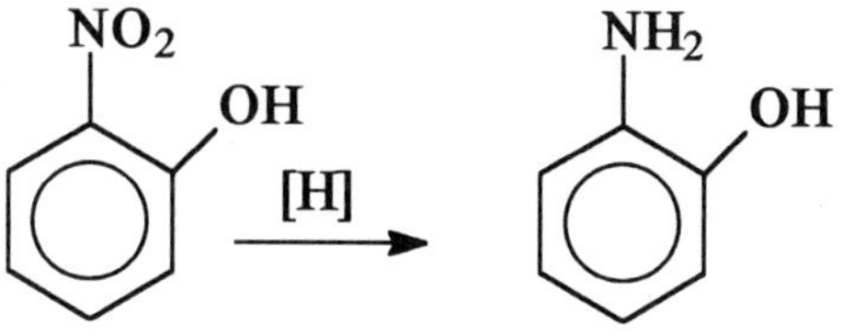 2-nitrophenol $\xrightarrow{[H]}$ 2-aminophenol (NO$_2$ with OH ortho, reduced to NH$_2$)

c. 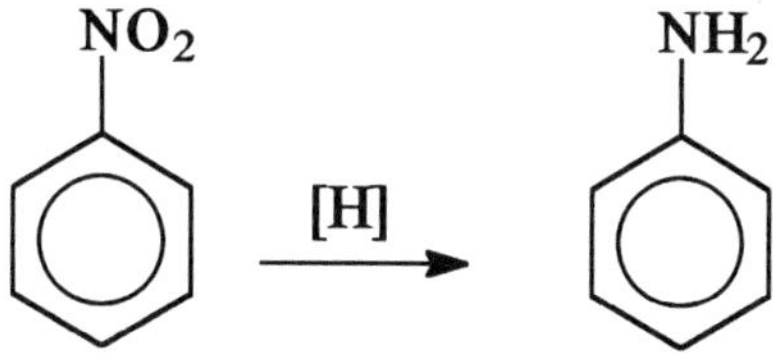 nitrobenzene $\xrightarrow{[H]}$ aniline (NO$_2$ reduced to NH$_2$)

d. 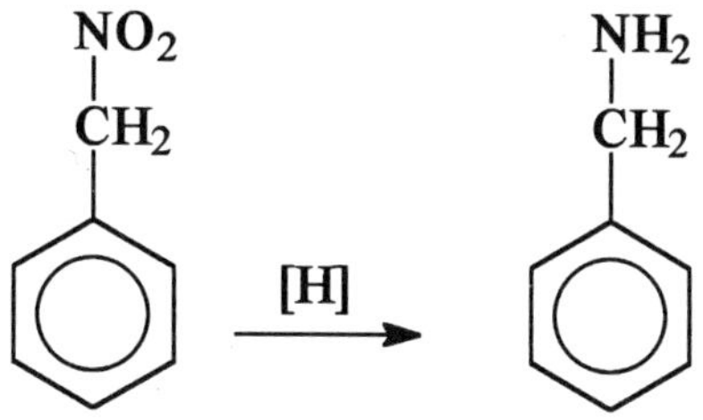 (nitromethyl)benzene $\xrightarrow{[H]}$ (aminomethyl)benzene (CH$_2$NO$_2$ reduced to CH$_2$NH$_2$)

18.31 a. H_2O

b. HBr

c. $CH_3CH_2CH_2-N^+H_3$

d. $CH_3CH_2-N^+H_2$ Cl^-

$\quad\quad\;\;\mid$

$\quad\quad CH_2CH_3$

18.33 Lower molecular weight amines are soluble in water because the N-H bond is polar and can form hydrogen bonds with water molecules. As the size of the organic substituents becomes larger, the entire molecule becomes more hydrocarbonlike and, thus, more hydrophobic overall.

18.35 Drugs containing amine groups are generally administered as ammonium salts because the salt is more soluble in water and, hence, in body fluids.

18.37 Putrescine (1,4-Diaminobutane):

$$CH_2CH_2CH_2CH_2$$
$$NH_2 \quad\quad\quad NH_2$$

Cadaverine (1,5-Diaminopentane):

$$CH_2CH_2CH_2CH_2CH_2$$
$$NH_2 \quad\quad\quad\quad NH_2$$

18.39 a.

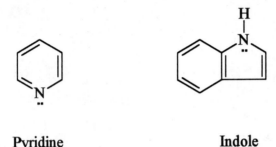

Pyridine   Indole

b. The indole ring is found in lysergic acid diethylamide, which is a hallucinogenic drug. It is also found in strychnine, which has been used as a rat poison. The pyridine ring is found in vitamin B_6, which is a water-soluble vitamin required for the synthesis and degradation of amino acids.

18.41 Morphine has been used as a pain reliever or analgesic. Codeine is used as an analgesic and cough suppressant. Cocaine has been used as an anesthetic for the sinuses and eyes. Quinine is used to treat malaria. Vitamin B_6 is a water-soluble vitamin required by the body.

18.43 a. I.U.P.A.C. name: Propanamide
          Common name: Propionamide

       b. I.U.P.A.C. name: Pentanamide
          Common name: Valeramide

       c. I.U.P.A.C. name: *N,N*-Dimethylethanamide
          Common name: *N,N*-Dimethylacetamide

18.45 a. Ethanamide:

$$\overset{\displaystyle O}{\overset{\displaystyle \|}{CH_3-C-NH_2}}$$

b. *N*-Methylpropanamide:

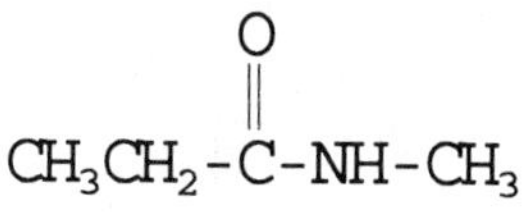

$$CH_3CH_2-\overset{\overset{\displaystyle O}{\|}}{C}-NH-CH_3$$

c. *N,N*-Diethylbenzamide:

$$C_6H_5-\overset{\overset{\displaystyle O}{\|}}{C}-\underset{\underset{\displaystyle CH_2CH_3}{|}}{N}-CH_2CH_3$$

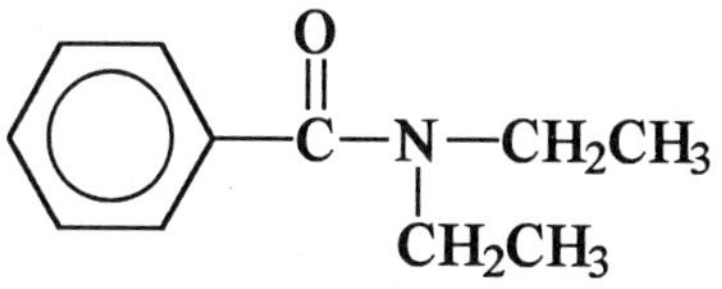

d. 3-Bromo-4-methylhexanamide:

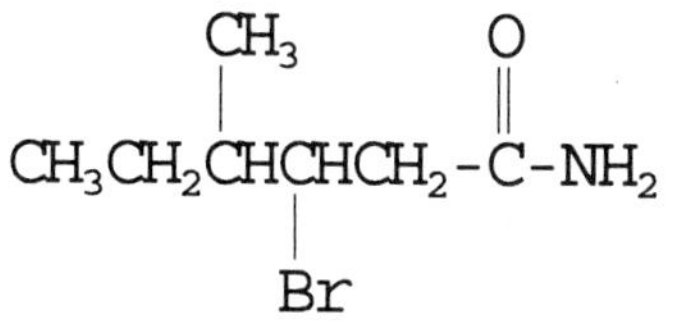

$$CH_3CH_2\underset{\underset{\displaystyle Br}{|}}{CH}\overset{\overset{\displaystyle CH_3}{|}}{CH}CH_2-\overset{\overset{\displaystyle O}{\|}}{C}-NH_2$$

e. *N,N*-Dimethylacetamide:

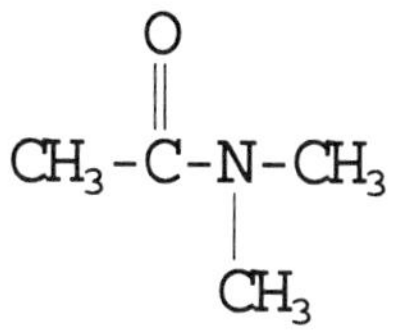

$$CH_3-\overset{\overset{\displaystyle O}{\|}}{C}-\underset{\underset{\displaystyle CH_3}{|}}{N}-CH_3$$

18.47 *N,N*-Diethyl-*m*-toluamide:

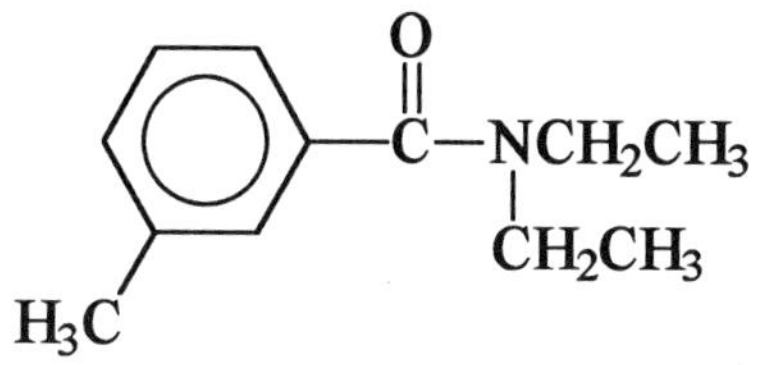

$$\overset{\overset{\displaystyle O}{\|}}{C}-\underset{\underset{\displaystyle CH_2CH_3}{|}}{N}CH_2CH_3$$

Hydrolysis of this compound would release the carboxylic acid *m*-toluic acid and the amine *N*-ethylethanamine (diethylamine).

18.49 Amides are not proton acceptors (bases) because the highly electronegative carbonyl oxygen has a strong attraction for the nitrogen lone pair of electrons. As a result they cannot "hold" a proton.

18.51

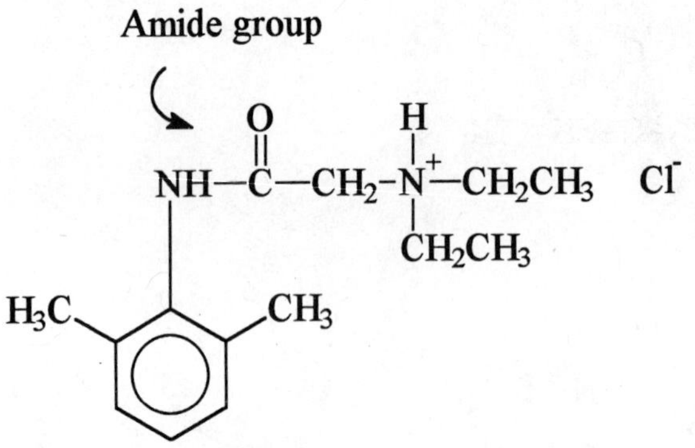

Amide group

Lidocaine hydrochloride

18.53

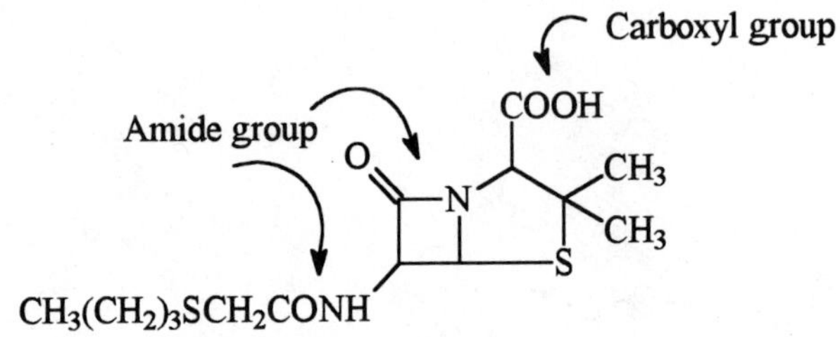

Penicillin BT

18.55 a.

$CH_3-\overset{O}{\overset{\|}{C}}-NHCH_3$ + H_3O^+ ---> CH_3COOH + $CH_3NH_3{}^+$

N-Methylethanamide   **Ethanoic acid   Methanamine**

b.

$CH_3CH_2CH_2-\overset{O}{\overset{\|}{C}}-NH-CH_3$ + H_3O^+ --> $CH_3CH_2CH_2-COOH$ + $CH_3NH_3{}^+$

***N*-Methylbutanamide**   Butanoic acid   Methanamine

c.

$$CH_3CHCH_2\text{-}\overset{\overset{\displaystyle O}{\|}}{C}\text{-}NH\text{-}CH_2CH_3 \quad + \quad H_3O^+ \quad \text{-------->}$$

with CH_3 on the second carbon

N-Ethyl-3-methylbutanamide   **Hydronium ion**
                                    (Strong acid)

$$CH_3CHCH_2\text{-}COOH \quad + \quad CH_3CH_2NH_3^+$$
with CH_3 branch

3-Methylbutanoic acid   **Ethanamine**

18.57

a. $$CH_3CH_2\text{-}\overset{\overset{\displaystyle O}{\|}}{C}\text{-}O\text{-}\overset{\overset{\displaystyle O}{\|}}{C}\text{-}CH_2CH_3 \;+\; 2\;CH_3CH_2NH_2$$

$\downarrow$

$$CH_3CH_2CH_2\text{-}NH\text{-}\overset{\overset{\displaystyle O}{\|}}{C}\text{-}CH_2CH_3 \;+\; CH_3CH_2\text{-}\overset{\overset{\displaystyle O}{\|}}{C}\text{-}O^-\overset{+}{N}H_3\text{-}CH_2CH_2CH_3$$

b. $$CH_3CH_2\text{-}\overset{\overset{\displaystyle O}{\|}}{C}\text{-}Cl \;+\; 2\,NH_3 \longrightarrow CH_3CH_2\text{-}\overset{\overset{\displaystyle O}{\|}}{C}\text{-}NH_3 \;+\; NH_4^+Cl^-$$

c. $$CH_3CH_2CH_2\text{-}\overset{\overset{\displaystyle O}{\|}}{C}\text{-}Cl \;+\; 2\;CH_3CH_2NH_2$$

$\downarrow$

$$CH_3CH_2CH_2\text{-}\overset{\overset{\displaystyle O}{\|}}{C}\text{-}NH\text{-}CH_2CH_3 \;+\; CH_3CH_2NH_3^+Cl^-$$

18.59 The following is the general structure of an amino acid:

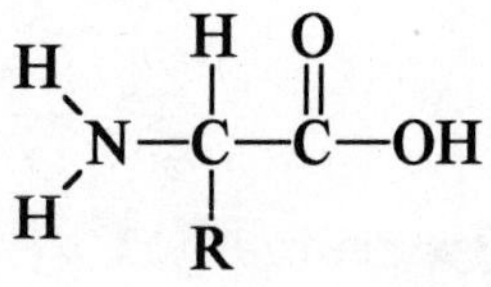

$$H_2N-\overset{\underset{\displaystyle R}{|}}{\underset{}{C}}H-\overset{\underset{}{\displaystyle O}}{\overset{\|}{C}}-OH$$

H₂N—CH(R)—C(=O)—OH

18.61 The following is the structure of the dipeptide composed of glycine and alanine:

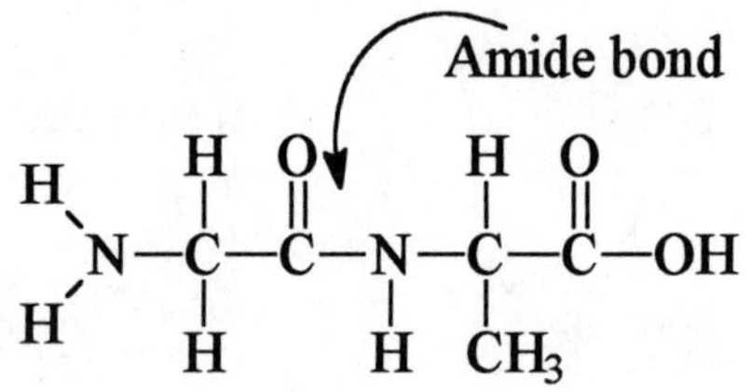

Amide bond

H₂N—CH₂—C(=O)—N(H)—CH(CH₃)—C(=O)—OH

Glycyl alanine

18.63 The following is the structure of alanine. The asterisk denotes the chiral carbon.

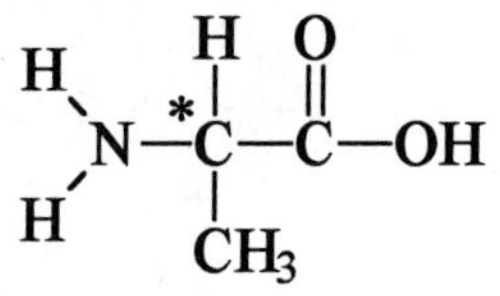

H₂N—*CH(CH₃)—C(=O)—OH

18.65 In an acyl group transfer reaction, the acyl group of an acid chloride is transferred from the Cl of the acid chloride to the N of an amine or ammonia. The product is an amide.

Chapter 19
Protein Structure and Function
Solutions to the Odd-Numbered Problems

★★
In-Chapter Questions and Problems
★★★★★★★★★★★★★★★★★★★★★★★★★★★★★★★

19.1   a. Glycine (gly):

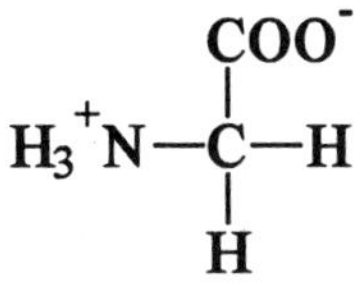

$$H_3{}^+N-\underset{\underset{H}{|}}{\overset{\overset{COO^-}{|}}{C}}-H$$

b. Proline (pro):

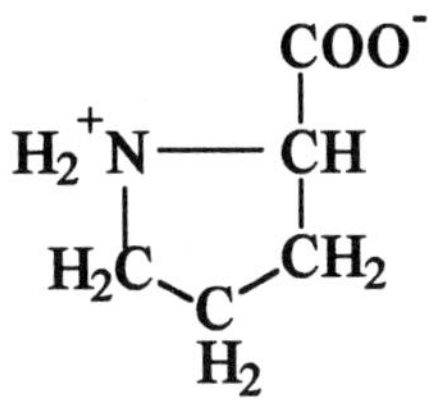

$$H_2{}^+N-\underset{H_2C}{\overset{CH}{\diagdown}}\cdots$$

c. Threonine (thr):

$$H_3{}^+N-\underset{\underset{CH_3}{|}}{\overset{\overset{COO^-}{|}}{C}}-H$$
$$H-C-OH$$

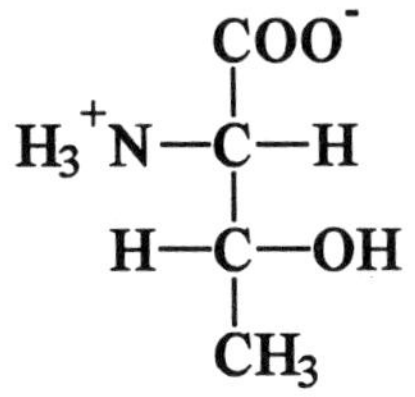

d. Aspartate (asp):

$$H_3{}^+N-\underset{\underset{COO^-}{|}}{\overset{\overset{COO^-}{|}}{C}}-H$$
$$H-C-H$$

451

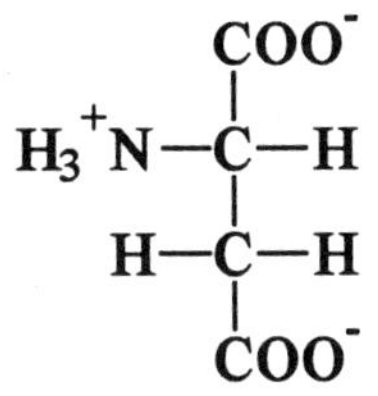

e. Lysine (lys):

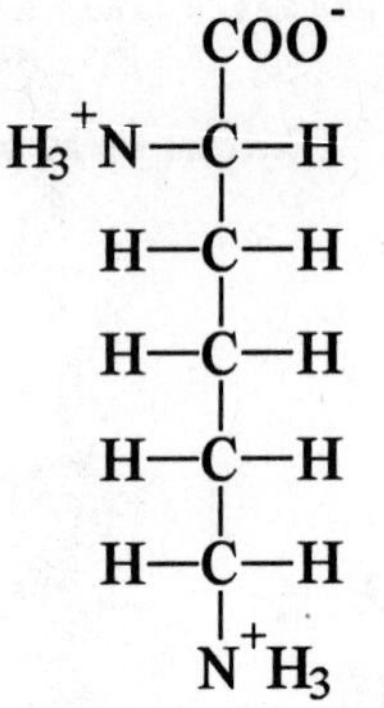

$$
\begin{array}{c}
COO^- \\
| \\
H_3{}^+N-C-H \\
| \\
H-C-H \\
| \\
H-C-H \\
| \\
H-C-H \\
| \\
H-C-H \\
| \\
N^+H_3
\end{array}
$$

19.3   a. Alanyl-phenylalanine:

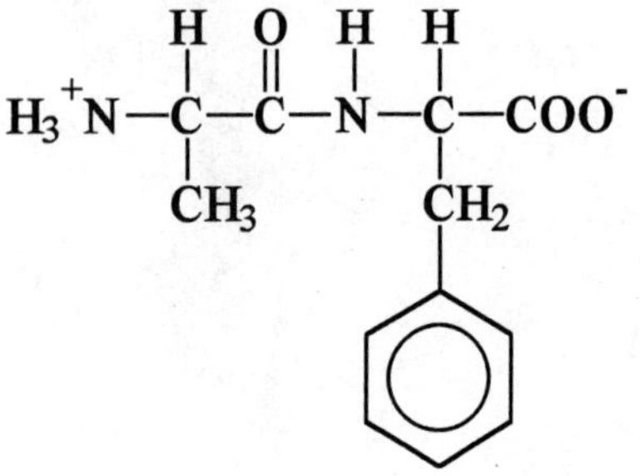

$$
\begin{array}{c}
H\quad O\quad H\quad H \\
|\quad\ \parallel\quad |\quad\ | \\
H_3{}^+N-C-C-N-C-COO^- \\
|\qquad\qquad\ | \\
CH_3\qquad\quad CH_2 \\
\qquad\qquad\qquad | \\
\qquad\qquad\qquad C_6H_5
\end{array}
$$

b. Lysyl-alanine:

$$
\begin{array}{c}
H\quad O\quad H\quad H \\
|\quad\ \parallel\quad |\quad\ | \\
H_3{}^+N-C-C-N-C-COO^- \\
|\qquad\qquad\ | \\
CH_2\qquad\quad CH_3 \\
| \\
CH_2 \\
| \\
CH_2 \\
| \\
CH_2 \\
| \\
N^+H_3
\end{array}
$$

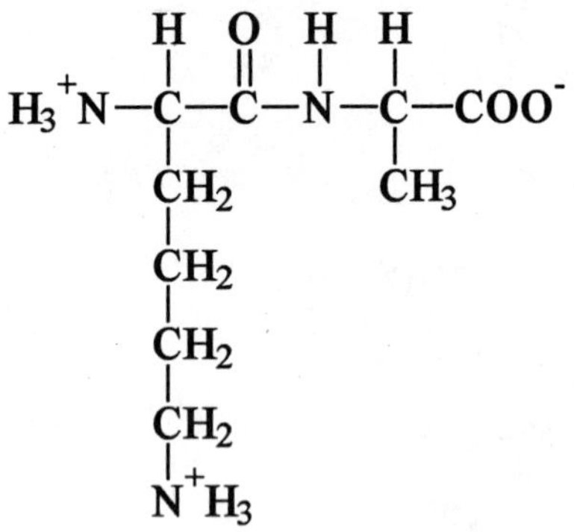

c. Phenylalanyl-tyrosyl-leucine:

$$H_3{}^+N-\overset{\overset{\displaystyle H}{|}}{\underset{\underset{\displaystyle CH_2}{|}}{C}}-\overset{\overset{\displaystyle O}{||}}{C}-\overset{\overset{\displaystyle H}{|}}{N}-\overset{\overset{\displaystyle H}{|}}{\underset{\underset{\displaystyle CH_2}{|}}{C}}-\overset{\overset{\displaystyle O}{||}}{C}-\overset{\overset{\displaystyle H}{|}}{N}-\overset{\overset{\displaystyle H}{|}}{\underset{\underset{\underset{\underset{\displaystyle CH_3}{|}}{\displaystyle CHCH_3}}{|}}{\underset{\underset{\displaystyle CH_2}{|}}{C}}}-COO^-$$

19.5   The primary structure of a protein is the amino acid sequence of the protein chain. Regular, repeating folding of the peptide chain caused by hydrogen bonding between the amide nitrogens and carbonyl oxygens of the peptide bond is the secondary structure of a protein. The two most common types of secondary structure are the α-helix and the β-pleated sheet. Tertiary structure is the further folding of the regions of α-helix and β-pleated sheet into a compact, spherical structure. Formation and maintenance of the tertiary structure results from weak attractions between amino acid R groups. The binding of two or more peptides to produce a functional protein defines the quaternary structure.

19.7   Oxygen is efficiently transferred from hemoglobin in the blood to myoglobin in the muscle because myoglobin has a greater affinity for oxygen.

19.9   High temperature disrupts the hydrogen bonds and other weak interactions that maintain protein structure. As a result, the protein loses its characteristic three-dimensional shape and becomes denatured.

19.11   Vegetables vary in amino acid composition. No single vegetable can provide all of the amino acid requirements of the body. By eating a variety of different vegetables, all the amino acid requirements of the human body can be met.

****

End-of-Chapter Questions and Problems

****

19.13   Five of the biological functions carried out by proteins include serving as enzymes to speed up biochemical reactions, acting as antibodies to protect the body against disease, transport of materials throughout the body and into and out of cells, regulation of cellular function, and serving as structural support for animals.

19.15   The general structure of an amino acid:

$$H_3\overset{+}{N}-\underset{\underset{R}{|}}{\overset{\overset{H}{|}}{C}}-\overset{\overset{O}{\|}}{C}-O^-$$

19.17   Interactions between the R groups of the amino acids in a polypeptide chain are important for the formation and maintenance of the tertiary and quaternary structures of proteins.

19.19 The following are the structures of the amino acids that have hydrophobic R groups:

Glycine

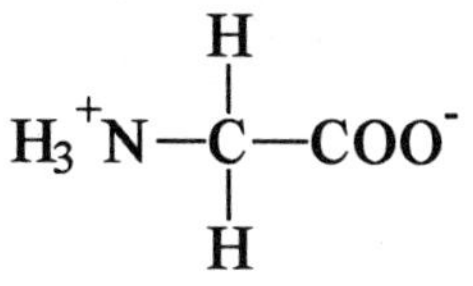

$$H_3^+N-\underset{\underset{H}{|}}{\overset{\overset{H}{|}}{C}}-COO^-$$

Alanine

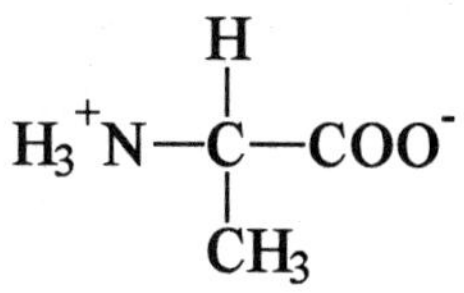

$$H_3^+N-\underset{\underset{CH_3}{|}}{\overset{\overset{H}{|}}{C}}-COO^-$$

Valine

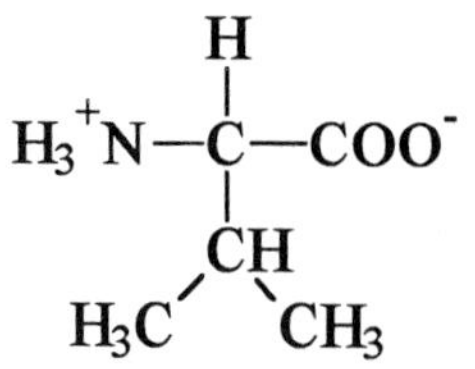

$$H_3^+N-\underset{\underset{CH}{|}}{\overset{\overset{H}{|}}{C}}-COO^-$$
$$H_3C \quad CH_3$$

Leucine

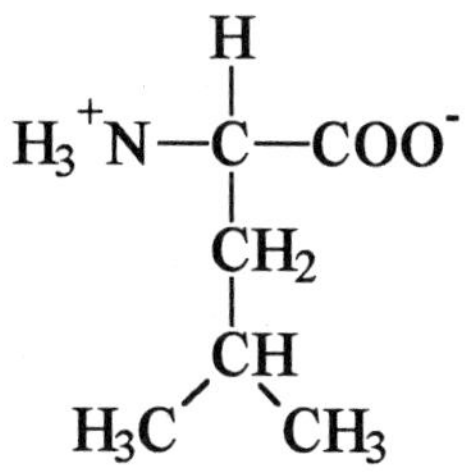

$$H_3^+N-\underset{\underset{CH_2}{|}}{\overset{\overset{H}{|}}{C}}-COO^-$$
$$CH$$
$$H_3C \quad CH_3$$

Isoleucine

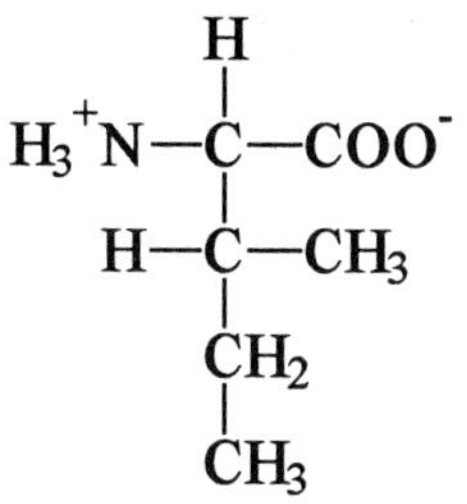

$$H_3^+N-\underset{\underset{CH_2}{|}}{\overset{\overset{H}{|}}{C}}-COO^-$$
$$H-C-CH_3$$
$$CH_3$$

Phenylalanine

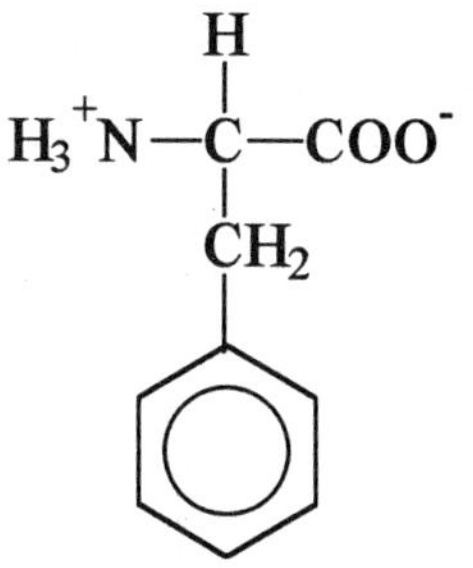

$$H_3^+N-\underset{\underset{CH_2}{|}}{\overset{\overset{H}{|}}{C}}-COO^-$$

Proline

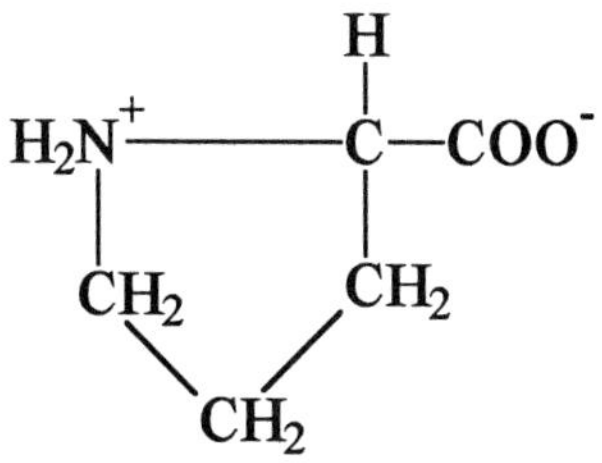

$$H_2^+N-\underset{\underset{CH_2}{|}}{\overset{\overset{H}{|}}{C}}-COO^-$$
$$CH_2 \qquad CH_2$$
$$CH_2$$

Tryptophan

$$H_3^+N-\underset{\underset{CH_2}{|}}{\overset{\overset{H}{|}}{C}}-COO^-$$

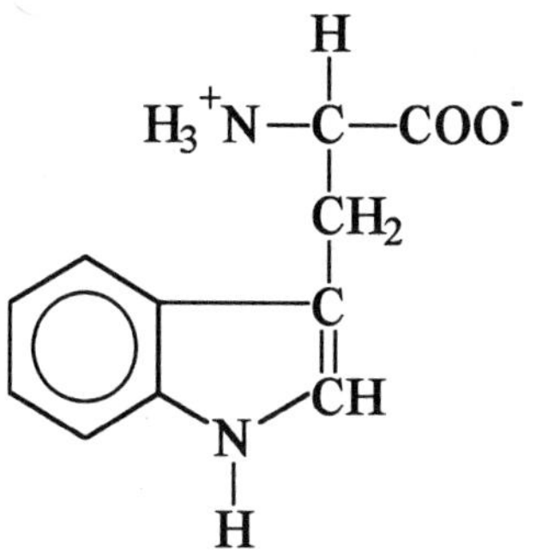

Methionine

$$H_3^+N-\underset{\underset{CH_2}{|}}{\overset{\overset{H}{|}}{C}}-COO^-$$
$$CH_2$$
$$S$$
$$CH_3$$

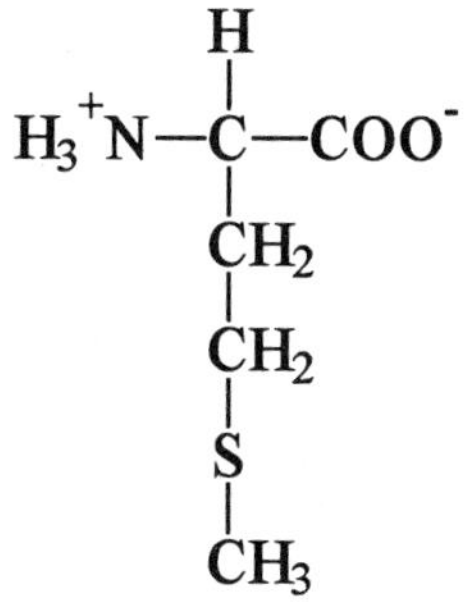

19.21

a. His-trp-cys:

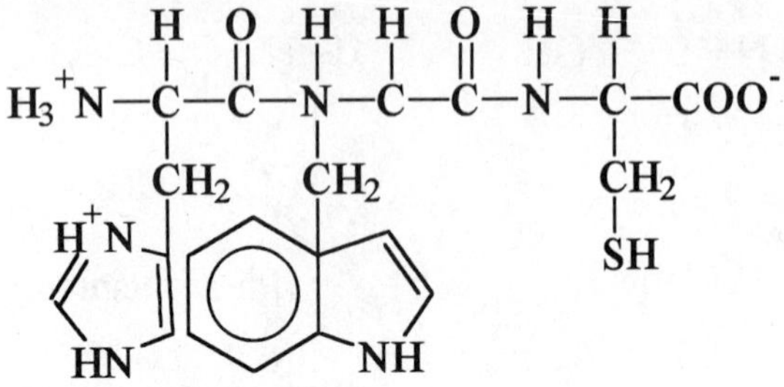

H_3N^+—C—C—N—C—C—N—C—COO^- (peptide backbone with side chains: CH_2-imidazolium (His), CH_2-indole (Trp), CH_2-SH (Cys))

b. Gly-leu-ser:

H_3N^+—C—C—N—C—C—N—C—COO^- (side chains: H (Gly), CH_2-CH(CH_3)-CH_3 (Leu), H-C-OH/H (Ser))

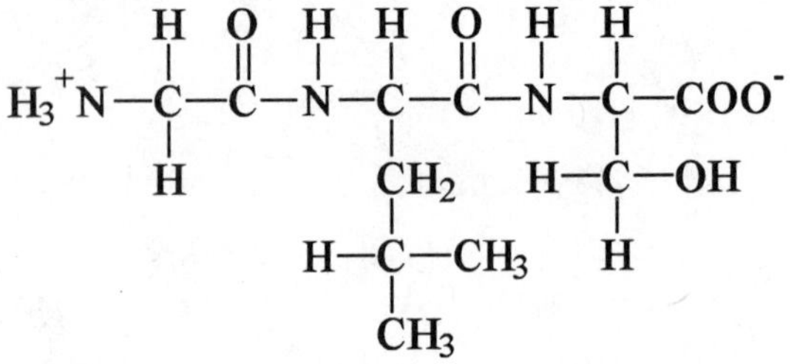

c. Arg-ile-val:

H_3N^+—C—C—N—C—C—N—C—COO^- (side chains: (CH_2)_3-NH-C(=N^+H_2)-NH_2 (Arg), CH(CH_3)-CH_2-CH_3 (Ile), CH(CH_3)-CH_3 (Val))

456

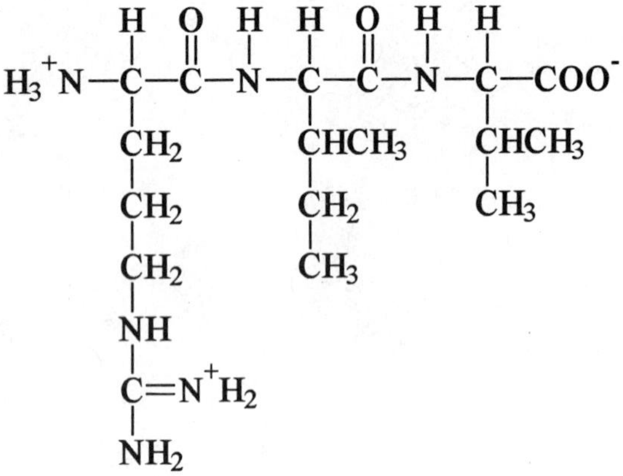

19.23 The peptide bond consists of an amide group. There is no free rotation around the peptide bond because the lone pair of electrons of the nitrogen atom interacts with the carbon and oxygen of the carbonyl group. This results in a resonance structure with a partially double-bonded character:

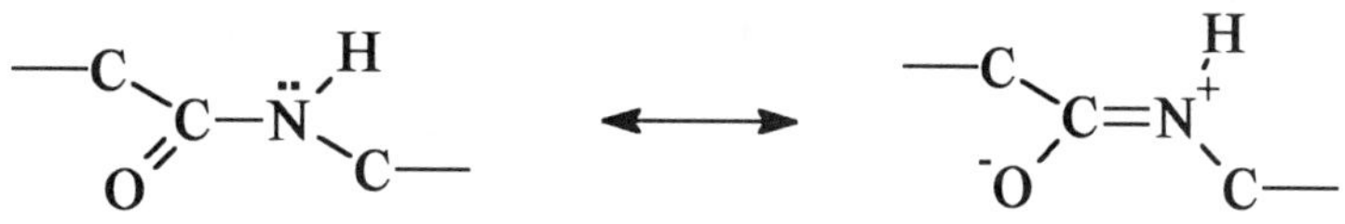

19.25 The genetic information in the DNA dictates the order in which amino acids will be added to the protein chain. The order of the amino acids is the primary structure of the protein.

19.27 The primary structure of a protein is the linear arrangement of amino acids joined to one another by peptide bonds.

19.29 The secondary structure of a protein is the folding of the primary structure into an α-helix or β-pleated sheet. These are maintained by hydrogen bonds between the amide hydrogen and carbonyl oxygen of the peptide bond.

19.31 a. α-Helix secondary structure is characteristic of the α-keratins.
    b. β-Pleated sheet secondary structure is characteristic of silk fibroin.

19.33 A fibrous protein is one that is composed of peptides arranged in long sheets or fibers. Silk fibroin, α-keratins, and collagens are fibrous proteins. Typically, they are insoluble and provide mechanical strength.

19.35 A parallel β-pleated sheet is one in which the hydrogen bonded peptide chains have their amino-termini aligned head-to-head.

19.37 The tertiary structure of a protein is the globular, three-dimensional structure of a protein that results from folding the regions of secondary structure. This folding occurs spontaneously as a result of interactions between the side chain R groups of the amino acids.

19.39

The oxidation of two cysteine molecules to produce cystine:

$$H_3{}^+N-\underset{\underset{SH}{|}}{\underset{\underset{CH_2}{|}}{\overset{\overset{H}{|}}{C}}}-COO^- \quad + \quad H_3{}^+N-\underset{\underset{SH}{|}}{\underset{\underset{CH_2}{|}}{\overset{\overset{H}{|}}{C}}}-COO^- \quad \longrightarrow \quad H_3{}^+N-\underset{\underset{\underset{\underset{H_3{}^+N-\underset{\underset{H}{|}}{C}-COO^-}{|}}{\underset{CH_2}{|}}}{\underset{\underset{S}{|}}{\underset{\underset{S}{|}}{\underset{\underset{CH_2}{|}}{\overset{\overset{H}{|}}{C}}}}}-COO^-$$

19.41   The tertiary structure is a level of folding of a protein chain that has already undergone secondary folding. The regions of α-helix and β-pleated sheet are folded into a globular structure.

19.43   Quaternary protein structure is the aggregation of two or more folded peptide chains to produce a functional protein. Amino acid R group interactions produce and maintain the quaternary structure.

19.45   A conjugated protein is a protein that requires an additional nonprotein group in order to be functional.

19.47   The function of hemoglobin is to carry oxygen from the lungs to oxygen demanding tissues throughout the body. Hemoglobin is found in red blood cells.

19.49   Hemoglobin is a protein composed of four subunits - 2 α-globin and 2 β-globin subunits. Each subunit holds a heme group, which in turn, carries an Fe^{2+} ion.

19.51   The function of the heme group in hemoglobin and myoglobin is to bind to molecular oxygen.

19.53   Because carbon monoxide binds tightly to the heme groups of hemoglobin, it is not easily removed or replaced by oxygen. As a result, the effects of oxygen deprivation (suffocation) occur.

19.55   When sickle cell hemoglobin (HbS) is deoxygenated, the amino acid valine fits into a hydrophobic pocket on the surface of another HbS molecule. Many such sickle cell hemoglobin molecules polymerize into long rods that cause the red blood cell to sickle. In normal hemoglobin, glutamic acid is found in the place of the valine. This negatively charged amino acid will not "fit" into the hydrophobic pocket.

19.57 When individuals have one copy of the sickle cell gene and one copy of the normal gene, they are said to carry the *sickle cell trait*. These individuals will not suffer serious side-effects, but may pass the trait to their offspring. Individuals with two copies of the sickle cell globin gene exhibit all the symptoms of the disease and are said to have *sickle cell anemia*.

19.59 A single collagen strand is a left-handed helix. About 25% of the amino acids in collagen are 5-hydroxyproline and 5-hydroxylysine. Glycine makes up another one-third of the amino acids in collagen.

19.61 The following are the structures of 4-hydroxyproline and 5-hydroxylysine:

4-Hydroxyproline   **5-Hydroxylysine**

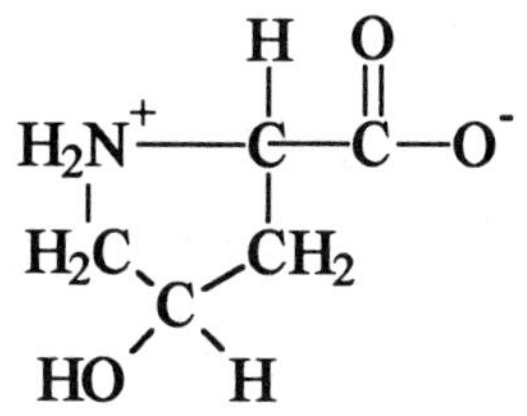

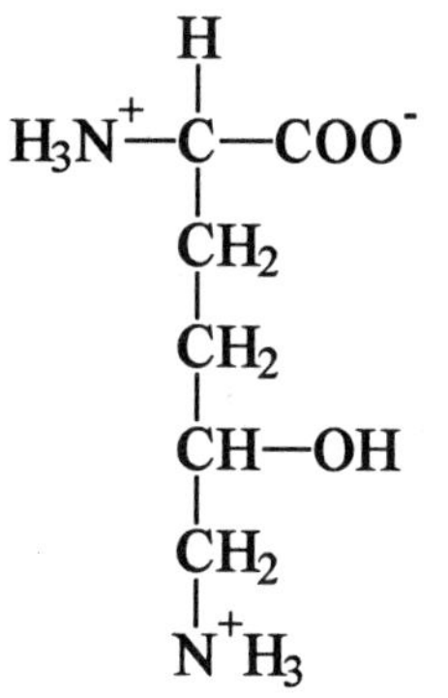

19.63 Vitamin C is a cofactor for the enzymes that catalyze the hydroxylation of proline (prolyl hydroxylase) and lysine (lysyl hydroxylase) in the structure of collagen.

19.65 Hydrogen bonding maintains the secondary structure of a protein and contributes to the stability of the tertiary and quaternary levels of structure.

19.67 The resonance structure of the amide (peptide) bond result in a partially double bonded character. This causes the rigidity of the peptide bond.

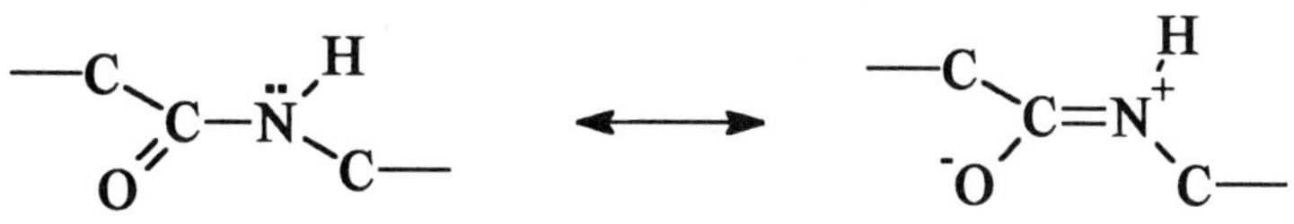

19.69 The code for the primary structure of a protein is carried in the genetic information (DNA). This information will be translated into the linear sequence of amino acids during the process of protein synthesis (translation). The sequence of amino acids will dictate the sequence of R groups along the peptide chain. This, in turn, will determine the three dimensional structure of the final protein.

19.71 *Denaturation* is the process by which the organized structure of a protein is disrupted, resulting in a completely disorganized, nonfunctional form of the protein.

19.73 Heat is an effective means of sterilization because it destroys the proteins of microbial life-forms, including fungi, bacteria, and viruses.

19.75 If the pH of the blood were to become too acidic or too basic, blood proteins and enzymes and the proteins of blood cells would be denatured. Blood proteins could no longer carry out important transport and communication functions. Hemoglobin would no longer carry oxygen. Death would result. Thus, buffering mechanisms in the blood are essential.

19.77 Proteins become polycations at low pH because the additional protons will protonate the carboxylate groups. As these negative charges are neutralized, the charge on the proteins will be contributed only by the protonated amino groups ($-N^+H_3$).

19.79 The low pH of the yogurt denatures the proteins of microbial contaminants, inhibiting their growth.

19.81 In a vegetarian diet, vegetables are the only source of dietary protein. Since individual vegetable sources do not provide all the needed amino acids, vegetables must be mixed to provide all the essential and nonessential amino acids in the amounts required for biosynthesis.

19.83 Nonessential amino acids can be synthesized by the body and are, therefore, not required in the diet. Essential amino acids cannot be synthesized by the body and must be provided by the diet.

19.85 Synthesis of digestive enzymes must be carefully controlled because the active enzyme would digest and destroy the cell that produces it. As a result, these enzymes are produced in an inactive form that is "activated" in the gastrointestinal tract.

Chapter 20
Enzymes
Solutions to the Odd-Numbered Problems

****
In-Chapter Questions and Problems
****

20.1   a. The substrate of sucrase is sucrose.
       b. The substrate of pyruvate decarboxylase is pyruvate.
       c. The substrate of succinate dehydrogenase is succinate.

20.13   a. Pyruvate kinase is a transferase.
       b. RNA ligase is a ligase.
       c. Triose isomerase is an isomerase.
       d. Pyruvate dehydrogenase is an oxidoreductase.
       e. Phosphoglucoisomerase is an isomerase

20.5   The induced fit model assumes that the enzyme is flexible. Both the enzyme and the
       substrate are able to change shape to form the enzyme-substrate complex. The lock-
       and-key model assumes that the enzyme is inflexible (the lock) and the substrate (the
       key) fits into a specific rigid site (the active site) on the enzyme to form the enzyme-
       substrate complex.

20.7   An enzyme might put pressure on a bond, thereby catalyzing bond breakage. An
       enzyme could bring two reactants into close proximity and in the proper orientation for
       the reaction to occur. Finally, an enzyme could alter the pH of the microenvironment
       of the active site, thereby serving as a transient donor or acceptor of H^+.

20.9   Water-soluble vitamins are required by the body for the synthesis of coenzymes that
       are required for the function of a variety of enzymes.

20.11   A decrease in pH will change the degree of ionization of the R groups within a
       peptide chain. This disturbs the weak interactions that maintain the structure of an
       enzyme, which may denature the enzyme. Less drastic alterations in the charge of R
       groups in the active site of the enzyme can inhibit enzyme-substrate binding or destroy
       the catalytic ability of the active site.

20.13   Irreversible inhibitors bind very tightly, sometimes even covalently, to an R group in
       enzyme active sites. They generally inhibit many different enzymes. The loss of
       enzyme activity impairs normal cellular metabolism, resulting in death of the cell or
       the individual.

20.15 A structural analog is a molecule that has a structure and charge distribution very similar to that of the natural substrate of an enzyme. Generally they are able to bind to the enzyme active site. This inhibits enzyme activity because the normal substrate must compete with the structural analog to form an enzyme-substrate complex.

20.17

a. 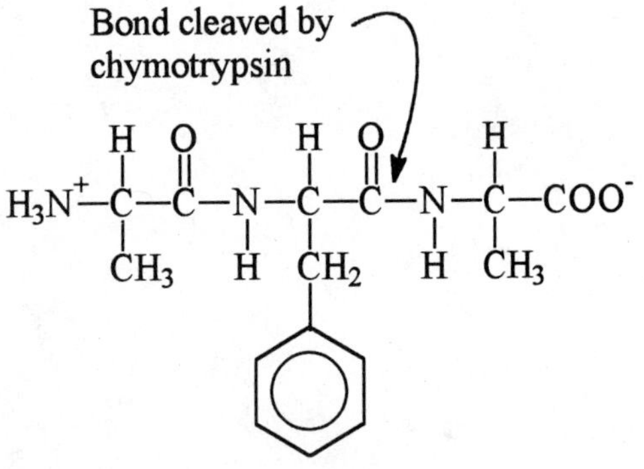 ala-phe-ala

b. 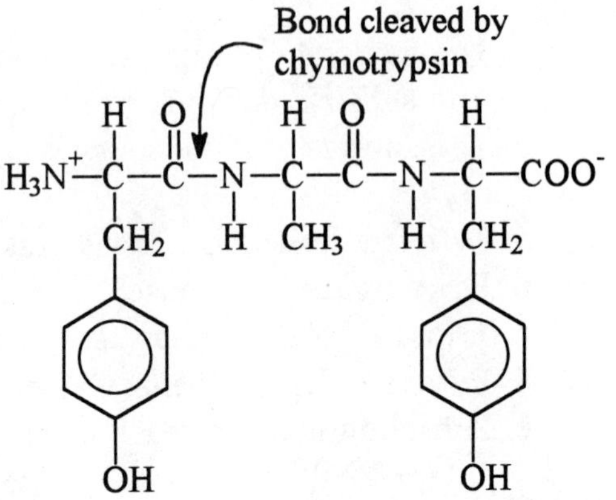 tyr-ala-tyr

c. 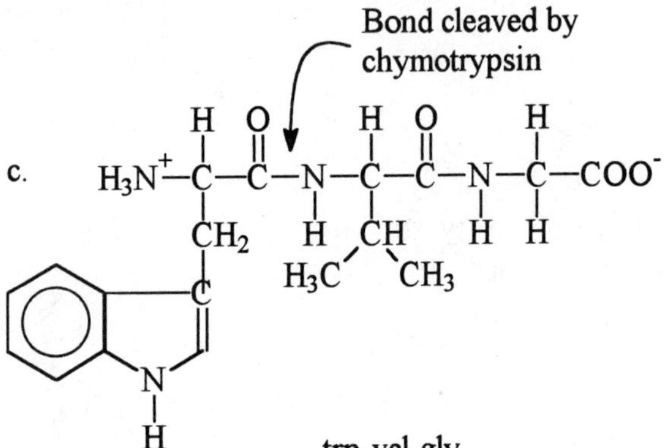 trp-val-gly

d. 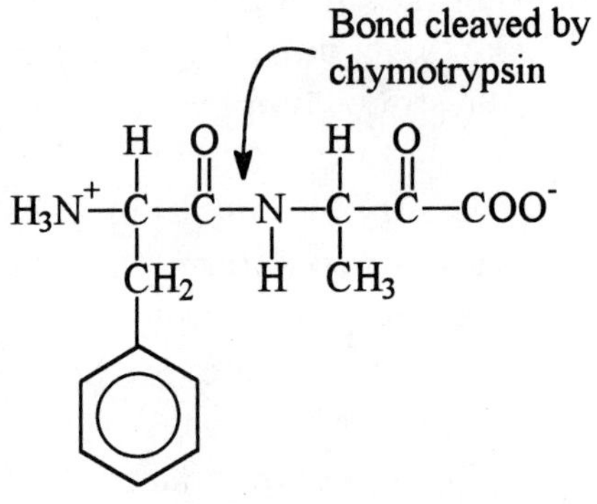 phe-ala-pro

20.19 Hemophilia A is the most common form of hemophilia. It results from the production of an abnormal factor VIII that cannot promote clotting. This is an X-linked recessive trait most commonly seen in males. It is characterized by spontaneous hemorrhages and serious bleeding after even minor wounds. Hemophilia B results from a deficiency of factor IX and is also X-linked. Hemophilia C is a mild form of the disease caused by a deficiency of factor XI.

20.21 Vitamin K is produced by our intestinal bacteria. In the diet it is obtained from green leafy vegetables and liver.

****
End-of-Chapter Questions and Problems
****

20.23   **Substrate**   **Enzyme**
           1. urea   e. urease
           2. hydrogen peroxide   c. peroxidase
           3. lipid   a. lipase
           4. aspartic acid   f. aspartase
           5. glucose-6-phosphate   b. glucose-6-phosphatase
           6. sucrose   d. sucrase

20.25   a. Citrate decarboxylase catalyzes the cleavage of a carboxyl group from citrate.
         b. Adenosine diphosphate phosphorylase catalyzes the addition of a phosphate group to ADP.
         c. Oxalate reductase catalyzes the reduction of oxalate.
         d. Nitrite oxidase catalyzes the oxidation of nitrite.
         e. *cis-trans* Isomerase catalyzes interconversion of *cis* and *trans* isomers.

20.27   The activation energy of a reaction is the energy required for the reaction to occur.

20.29   The equilibrium constant for a chemical reaction is a reflection of the difference in energy of the reactants and products. Consider the following reaction:

$$aA + bB \rightarrow cC + dD$$

The equilibrium constant for this reaction is:

$$K_{eq} = [D]^d[C]^c/[A]^a[B]^b = [products]/[reactants]$$

Since the difference in energy between reactants and products is the same regardless of what path the reaction takes, an enzyme does not alter the equilibrium constant of a reaction.

20.31   The rate of an uncatalyzed chemical reaction typically doubles every time the substrate concentration is doubled.

20.33

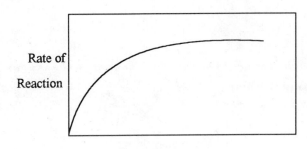

Substrate Concentration

20.35 Enzyme active sites are pockets in the surface of an enzyme that include R groups involved in binding and R groups involved in catalysis. The shape of the active site is complementary to the shape and charge distribution of the substrate. Thus, the conformation of the active site determines the specificity of the enzyme. Enzyme-substrate binding involves weak, noncovalent interactions.

20.37 The lock-and-key model of enzyme-substrate binding was proposed by Emil Fischer in 1894. He thought that the active site was a rigid region of the enzyme into which the substrate fit perfectly. Thus, the model purports that the substrate simply snaps into place within the active site, like two pieces of a jigsaw puzzle fitting together.

20.39 *Absolute specificity* - an enzyme catalyzes the reaction of only one substrate.
Group specificity - an enzyme catalyzes processes involving similar molecules having the same functional group.
Linkage specificity - an enzyme catalyzes the formation or breakage of only one type of bond.
Stereochemical specificity - an enzyme distinguishes one enantiomer from another.

20.41 The first step of an enzyme-catalyzed reaction is the formation of the enzyme-substrate complex. In the second step, the transition state is formed. This is the state in which the substrate assumes a form intermediate between the original substrate and the product. In step 3 the substrate is converted to product and the enzyme-product complex is formed. Step 4 involves the release of the product and regeneration of the enzyme in its original form.

20.43 In a reaction involving bond breaking, the enzyme might put pressure on a bond, producing a transition state in which the bond is stressed. An enzyme could bring two reactants into close proximity and in the proper orientation for the reaction to occur, producing a transition state in which the proximity of the reactants facilitates bond formation. Finally, an enzyme could alter the pH of the microenvironment of the active site, thereby serving as a transient donor or acceptor of H^+.

20.45 A cofactor helps maintain the shape of the active site of an enzyme.

20.47 NAD$^+$ (nicotinamide adenine dinucleotide) is a coenzyme that serves as a donor or acceptor of hydride anions in biochemical reactions. Since these would be oxidation-reduction reactions, NAD$^+$ serves as a coenzyme for oxidoreductases.

20.49 Changes in pH or temperature affect the activity of enzymes, as can changes in the concentration of substrate and the concentrations of certain ions.

20.51 A drastic change in pH above or below the pH optimum for an enzyme will denature the protein. Since a change in the conformation of the protein will drastically alter its active site, it will no longer be able to bind the substrate and catalyze the reaction.

20.53 High temperature denatures bacterial enzymes and structural proteins. Since the life of the cell is dependent on the function of these proteins, the cell dies. High temperatures also destroy cell membranes, as well.

20.55 A lysosome is a membrane-bound vesicle in the cytoplasm of cells which contains approximately 50 hydrolytic enzymes. These enzymes degrade large molecules into smaller molecules that can be used in cellular energy-harvesting reactions.

20.57 Enzymes used for clinical assays in hospitals are typically stored at refrigerator temperatures to ensure that they are not denatured by heat. In this way they retain their activity for long periods of time.

20.59 Cells regulate the level of enzyme activity to conserve energy. It is a waste of cellular energy to produce an enzyme if its substrate is not present or if its product is in excess. Production of proteolytic digestive enzymes must be carefully controlled because the active enzyme could destroy the cell that produces it. Thus, they are produced in an inactive form in the cell and are only activated at the site where they carry out digestion.

20.61 In positive allosterism, binding of the effector molecule turns the enzyme on. In negative allosterism, binding of the effector molecule turns the enzyme off. The effect on enzyme activity is mediated by shape changes in the active site of the enzyme as a result of effector binding. In positive allosterism, effector binding causes the active site to take on an active configuration. In negative allosterism, effector binding causes the active site to assume an inactive configuration.

20.63 A zymogen is the inactive form of an enzyme that is produced. It is converted to the active form of the enzyme, generally by proteolysis, at the site of its activity. An example is the production of the zymogen pepsinogen in cells lining the stomach. It is converted to the active form, pepsin, in the stomach.

20.65 *Competitive enzyme inhibition* occurs when a structural analog of the normal substrate occupies the enzyme active site so that the reaction cannot occur. The structural analog and the normal substrate compete for the active site because they have the same general shape, size, and charge distribution. Thus, the rate of the reaction will depend on the relative concentrations of the two molecules.

20.67 A structural analog is a molecule that can mimic the substrate of an enzyme because of the structural similarity between the two molecules. It has the same general size, shape, and charge distribution as the normal substrate for an enzyme.

20.69 Irreversible inhibitors bind tightly to an enzyme and eliminate catalysis. This disrupts the metabolic pathway and destroys cell function. The binding of the inhibitor is often tighter than the binding of the normal substrate. In some cases an irreversible inhibitor may bind covalently to the enzyme.

20.71 The compound would be a competitive inhibitor of the enzyme.

20.73 The structural similarities among chymotrypsin, trypsin, and elastase suggest that these enzymes evolved from a single ancestral gene that was duplicated. Each copy then evolved independently.

20.75 Chymotrypsin would cleave the peptide bond on the carbonyl side of tyrosine.

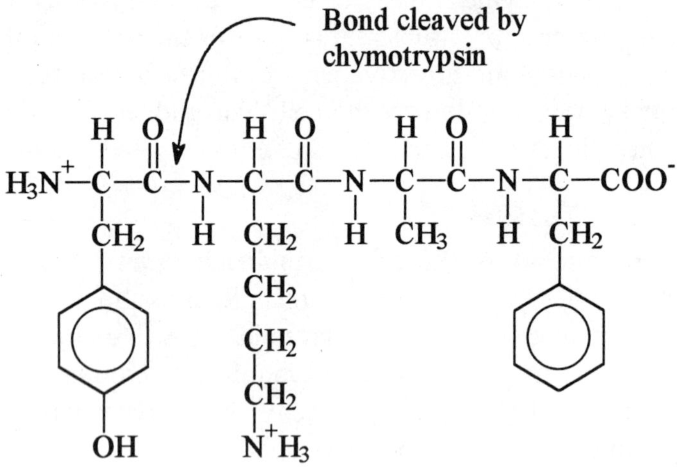

Bond cleaved by chymotrypsin

tyr-lys-ala-phe

20.77 Elastase will cleave the peptide bonds on the carbonyl side of alanine and glycine. Trypsin will cleave the peptide bonds on the carbonyl side of lysine and arginine. Chymotrypsin will cleave the peptide bonds on the carbonyl side of tryptophan and phenylalanine.

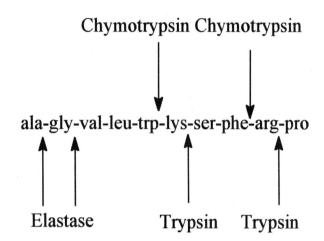

20.79 Prothrombin is a zymogen that is proteolytically cleaved to produce thrombin. Thrombin is a protease involved in the blood clotting cascade. It cleaves fibrinogen to produce fibrin which forms a meshwork of polymerized fibrin threads that, along with blood cells and plasma proteins, forms a clot. Fibrinogen is a soluble protein found in the blood. Fibrin, produced from fibrinogen by proteolytic cleavage, is an insoluble protein that forms the meshwork that becomes a clot.

20.81 Structural analogs of vitamin K act as competitive inhibitors. Competitive inhibition blocks the formation of prothrombin and blood clotting factors VII, IX, and X.

20.83 Acetylcholine is a neurotransmitter that is released from the nerve cell at a neuromuscular junction. Acetylcholine binds to its receptor on the muscle cell and this stimulates a muscle contraction. Acetylcholinesterase hydrolyzes the acetylcholine within the neuromuscular junction to stop continued stimulation and muscle contraction.

20.85 Organofluorophosphates covalently bind to the active site of acetylcholinesterase and act as irreversible, noncompetitive inhibitors. With acetylcholinesterase no longer available, nerve transmission continues and muscle spasms result. In humans, death may result from laryngeal spasms.

20.87 Creatine phosphokinase (CPK), lactate dehydrogenase (LDH), and aspartate aminotransferase (AST/SGOT) levels are elevated in blood serum following a myocardial infarction.

Chapter 21
Carbohydrate Metabolism
Solutions to the Odd-Numbered Problems

****
In-Chapter Questions and Problems
****

21.1 ATP is called the universal energy currency because it is the major molecule used by all organisms to store energy. Hydrolysis of the high energy phosphoanhydride bonds releases energy that is used for cellular work.

21.3 The first stage of catabolism is the digestion (hydrolysis) of dietary macromolecules in the stomach and intestine. Polysaccharides are hydrolyzed to monosaccharides; proteins are degraded to amino acids; and triglycerides are broken down into glycerol and fatty acids. The small molecules produced by digestion are taken into the cells lining the intestine by active or passive transport.

   In the second stage of catabolism, monosaccharides, amino acids, fatty acids, and glycerol are converted by metabolic reactions into molecules that can be completely oxidized. Often they are converted into acetyl CoA.

   In the third stage of catabolism, the two carbon acetyl group of acetyl CoA is completely oxidized by the reactions of the citric acid cycle. The energy of the electrons harvested in these oxidation reactions is used to make ATP.

21.5 Substrate level phosphorylation is one way the cell can make ATP. In this reaction, a high energy phosphoryl group of a substrate in the reaction is transferred to ADP to produce ATP. Substrate level phosphorylation can be summarized as follows:

$$\text{Substrate} \sim P + ADP \rightarrow \text{Product} + ATP$$

21.7 Glycolysis is a pathway involving nine reactions. In reactions 1-3 energy is invested in the beginning substrate, glucose. This is done by transferring high energy phosphoryl groups from ATP to the intermediates in the pathway. The product is fructose-1,6-bisphosphate. In the energy harvesting reactions of glycolysis, fructose-1,6-bisphosphate is split into two three-carbon molecules that begin a series of rearrangement, oxidation-reduction, and substrate level phosphorylation reactions that produce 4 ATP (net yield or 2 ATP), 2 NADH, and 2 pyruvate molecules.

21.9 Both the alcohol and lactate fermentations are anaerobic reactions that use the pyruvate and re-oxidize the NADH produced in glycolysis. In the alcohol fermentation, pyruvate is first decarboxylated to produce acetaldehyde. The acetaldehyde is then reduced as NADH is oxidized. The products are CO_2, ethanol, and NAD^+. In the lactate fermentation pyruvate is reduced to lactate and NADH is oxidized to NAD^+.

21.11 Gluconeogenesis (synthesis of glucose from noncarbohydrate sources) appears to be the reverse of glycolysis (the first stage of carbohydrate degradation) because the intermediates in the two pathways are the same. However, reactions 1, 3, and 9 of glycolysis are not reversible reactions. Thus, the reverse reactions must be carried out by different enzymes. Reaction 1 of glycolysis, the transfer of a high energy phosphoryl group from ATP to glucose, is carried out by the enzyme hexokinase. The reverse reaction in gluconeogenesis is catalyzed by glucose-6-phosphatase. Reaction 3 of glycolysis, the transfer of a high energy phosphoryl group from ATP to fructose-6-phosphate, is catalyzed by phosphofructokinase. The reverse reaction of gluconeogenesis is carried out by fructose bisphosphatase. The final reaction of glycolysis, the transfer of a high energy phosphoryl group from phosphoenolpyruvate to ADP, is catalyzed by pyruvate kinase. This is reversed in gluconeogenesis by the action of two enzymes. Pyruvate carboxylase adds CO_2 to pyruvate to produce oxaloacetate and phosphoenolpyruvate carboxykinase removes the CO_2 and transfers a high energy phosphoryl group from GTP to produce phosphoenolpyruvate.

21.13 The enzyme glycogen phosphorylase catalyzes the phosphorolysis of a glucose unit at one end of a glycogen molecule. The reaction involves the displacement of the glucose by a phosphate group. The products are glucose-1-phosphate and a glycogen molecule that is one glucose unit shorter.

21.15 Glucokinase traps glucose within the liver cell by phosphorylating it. Because the product, glucose-6-phosphate is charged, it cannot be exported from the cell.

21.17 Glucagon indirectly stimulates glycogen phosphorylase, the first enzyme of glycogenolysis. This speeds up glycogen degradation. Glucagon also inhibits glycogen synthase, the first enzyme in glycogenesis. This inhibits glycogen synthesis.

21.19 Pyruvate is converted to acetyl CoA by the pyruvate dehydrogenase complex. This huge enzyme complex requires four coenzymes, each of which is made from a different vitamin. The four coenzymes are thiamine pyrophosphate (made from thiamine), FAD (made from riboflavin), NAD^+ (made from niacin), and coenzyme A (made from the vitamin pantothenic acid). The coenzyme lipoamide is also involved in this reaction.

****

End-of-Chapter Questions and Problems

****

21.21 ATP is the molecule that is primarily responsible for conserving the energy released in catabolism.

21.23

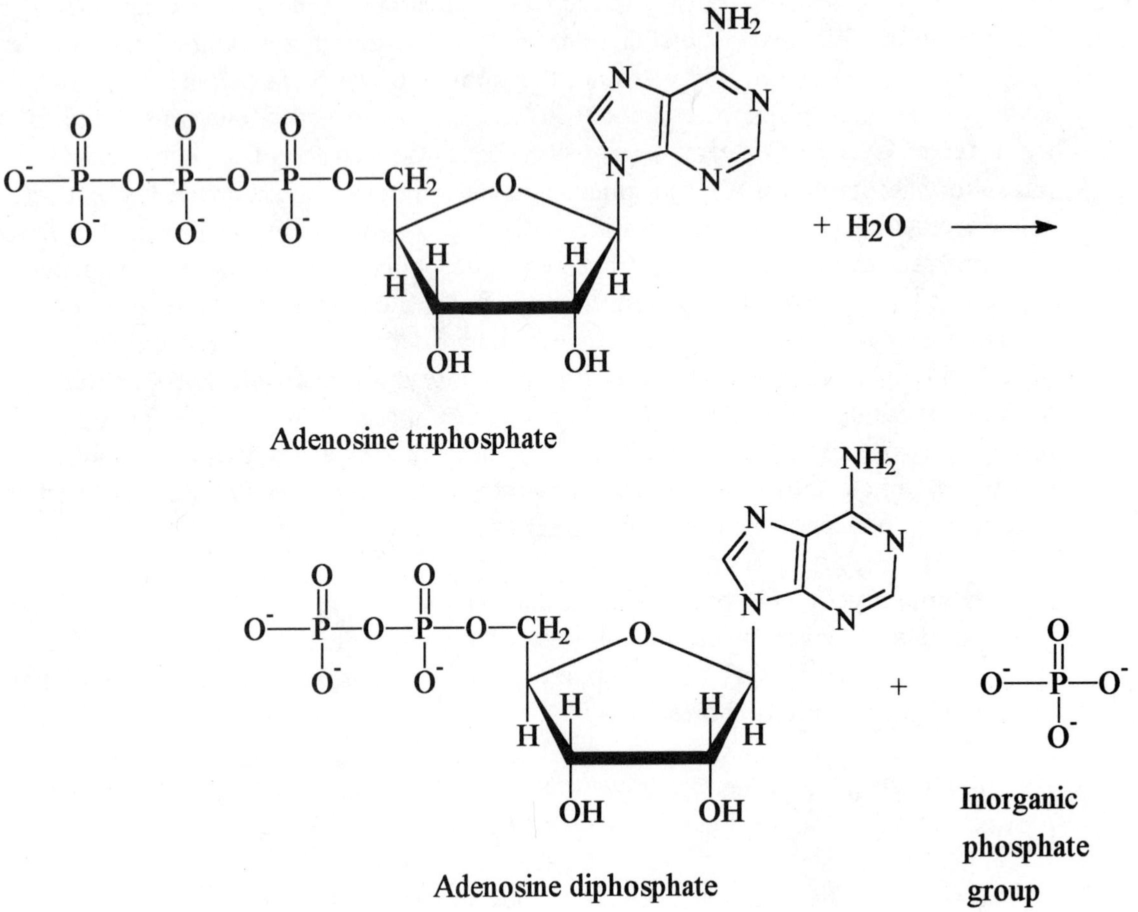

Adenosine triphosphate

+ H2O ⟶

Adenosine diphosphate

+

Inorganic
phosphate
group

21.25 Glycolysis requires NAD^+ for reaction 5 in which glyceraldehyde-3-phosphate dehydrogenase catalyzes the oxidation of glyceraldehyde-3-phosphate. NAD^+ is reduced and, thus, serves as the hydride anion acceptor in this reaction. If NAD^+ were not available, this reaction would not occur, and glycolysis, and therefore ATP synthesis, would stop.

21.27 The net ATP yield of glycolysis is 2 ATP molecules per glucose molecule.

21.29 Although muscle cells only have enough ATP stored for a few seconds of activity, glycolysis speeds up dramatically when there is a demand for more energy. If the cells have a sufficient supply of oxygen, aerobic respiration (the citric acid cycle and oxidative phosphorylation) will contribute large amounts of ATP. If oxygen is limited, the lactate fermentation will speed up. This will use up the pyruvate and re-oxidize the NADH produced by glycolysis and allow continued synthesis of ATP for muscle contraction.

21.31

$$C_6H_{12}O_6 + 2ADP + 2P_i + 2NAD^+ \rightarrow 2\ C_3H_3O_3 + 2ATP + 2NADH + 2H_2O$$

Glucose   Pyruvate

21.33   a. Hexokinase catalyzes the phosphorylation of glucose in the first step of glycolysis. The product is glucose-6-phosphate.
b. Pyruvate kinase catalyzes the transfer of a phosphoryl group from phosphoenolpyruvate to ADP. The products of this substrate level phosphorylation reaction are pyruvate and ATP.
c. Phosphoglyceromutase catalyzes the isomerization reaction that converts 3-phosphoglycerate to 2-phosphoglycerate.
d. Glyceraldehyde-3-phosphate dehydrogenase catalyzes the oxidation and phosphorylation of glyceraldehyde-3-phosphate and the reduction of NAD^+ to NADH.

21.35   The enzyme alcohol dehydrogenase catalyzes the conversion of acetaldehyde to ethanol.

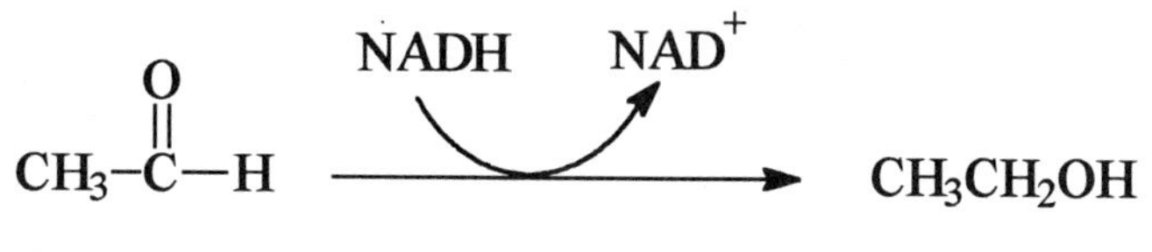

Acetaldehyde   Ethanol

21.37   The lactate fermentation produces lactate from pyruvate when the amount of oxygen is limiting. Over time, lactate will accumulate in the muscle.

21.39   The tangy flavor of yogurt and some cheeses is the result of lactate produced by the lactate fermentation. It is the pH decrease caused by the lactate build-up that causes milk protein to coagulate. This coagulation produces the soft curd of yogurt and the hard curd of some cheeses.

21.41   Lactate dehydrogenase catalyzes the reduction of pyruvate to lactate.

21.43   This child must have the enzymes to carry out the alcohol fermentation. When the child exercised hard, there was not enough oxygen in the cells to maintain aerobic respiration. As a result, glycolysis and the alcohol fermentation were responsible for the majority of the ATP production by the child. The accumulation of alcohol (ethanol) in the child caused the symptoms of drunkenness.

21.45 The first stage of the pentose phosphate pathway is an oxidative stage in which glucose-6-phosphate is converted to ribulose-5-phosphate. Two NADPH molecules and one CO_2 molecule are also produced in these reactions. The second stage of the pentose phosphate pathway involves isomerization reactions that convert ribulose-5-phosphate into other five-carbon sugars, ribose-5-phosphate and xylulose-5-phosphate. The third stage of the pathway involves a complex series of rearrangement reactions that result in the production of two fructose-6-phosphate and one glyceraldehyde-3-phosphate molecules from three molecules of pentose phosphate.

21.47 The ribose-5-phosphate produced in the pentose phosphate pathway is used for the biosynthesis of nucleotides, such as ATP. The erythrose-4-phosphate is used for the biosynthesis of aromatic amino acids, such as phenylalanine, tyrosine, and tryptophan.

21.49 The liver is primarily responsible for gluconeogenesis.

21.51 Lactate is first converted to pyruvate. Pyruvate is the starting substrate for gluconeogenesis.

21.53 Because steps 1, 3, and 9 of glycolysis are irreversible, gluconeogenesis is not simply the reverse of glycolysis. The reverse reactions must be carried out by different enzymes. Reaction 1 of glycolysis, the transfer of a high energy phosphoryl group from ATP to glucose, is carried out by the enzyme hexokinase. The reverse reaction in gluconeogenesis is catalyzed by glucose-6-phosphatase. Reaction 3 of glycolysis, the transfer of a high energy phosphoryl group from ATP to fructose-6-phosphate, is catalyzed by phosphofructokinase. The reverse reaction of gluconeogenesis is carried out by fructose bisphosphatase. The final reaction of glycolysis, the transfer of a high energy phosphoryl group from phosphoenolpyruvate to ADP, is catalyzed by pyruvate kinase. This is reversed in gluconeogenesis by the action of two enzymes. Pyruvate carboxylase adds CO_2 to pyruvate to produce oxaloacetate and phosphoenolpyruvate carboxykinase removes the CO_2 and transfers a high energy phosphoryl group from GTP to produce phosphoenolpyruvate.

21.55 Steps 1, 3, and 9 of glycolysis are irreversible. Step 1 is the transfer of a phosphoryl group from ATP to carbon-6 of glucose and is catalyzed by hexokinase. Step 3 is the transfer of a phosphoryl group from ATP to carbon-1 of fructose-6-phosphate and is catalyzed by phosphofructokinase. Step 9 is the substrate level phosphorylation in which a phosphoryl group is transferred from phosphoenolpyruvate to ADP and is catalyzed by pyruvate kinase.

21.57 The liver is the primary organ involved in maintaining proper blood glucose levels.

21.59 *Hypoglycemia* is the condition in which blood glucose levels are too low.

21.61 Insulin stimulates glycogen synthase, the first enzyme in glycogen synthesis. It also stimulates uptake of glucose from the bloodstream into cells and phosphorylation of glucose by the enzyme glucokinase. This traps glucose within liver cells, increases the storage of glucose in the form of glycogen, and decreases blood glucose levels.

21.63 Any defect in the enzymes required to degrade glycogen or export it from liver cells will result in a reduced ability of the liver to provide glucose at times when blood glucose levels are low. This will cause hypoglycemia.

21.65 Under aerobic conditions pyruvate is converted to acetyl CoA.

21.67 The coenzymes NAD^+, FAD, thiamine pyrophosphate, and coenzyme A are required by the pyruvate dehydrogenase complex for the conversion of pyruvate to acetyl CoA. These coenzymes are synthesized from the vitamins niacin, riboflavin, thiamine, and pantothenic acid, respectively. If the vitamins are not available, the coenzymes will not be available and pyruvate cannot be converted to acetyl CoA. Since the complete oxidation of the acetyl group of acetyl CoA produces the vast majority of the ATP for the body, ATP production would be severely inhibited by a deficiency of any of these vitamins.

Chapter 22
Aerobic Respiration and Energy Production
Solutions to the Odd-Numbered Problems

****
In-Chapter Questions and Problems
****

22.1   Mitochondria are the organelles responsible for aerobic respiration. They have enzymes that carry out the final oxidations of carbohydrates, amino acids, and fatty acids. They produce the majority of the ATP for the cell.

22.3

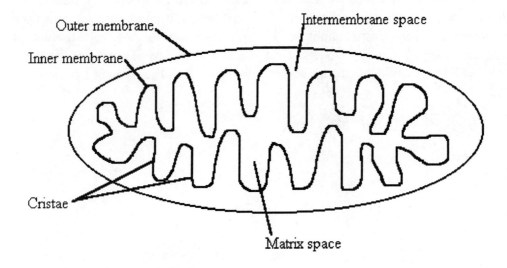

22.5   *Oxidative phosphorylation* is the process by which the energy of electrons harvested from oxidation of a fuel molecule is used to phosphorylate ADP to produce ATP.

22.7 $NAD^+ + H:^- \rightarrow NADH$

22.9   Pyridoxal phosphate is a coenzyme required by all transaminases. During transamination reactions, the α-amino group is transferred to pyridoxal phosphate. In the last part of the reaction, the α-amino group is transferred from pyridoxal phosphate to an α-keto acid.

22.11   The purpose of the urea cycle is to convert toxic ammonium ions to urea which is excreted in the urine of land animals. This keeps toxic ammonium ions out of the bloodstream.

22.13 An amphibolic pathway is a metabolic pathway that functions both in anabolism and catabolism. The citric acid cycle is amphibolic because it has a catabolic function - it completely oxidizes the acetyl group carried by acetyl CoA to provide electrons for ATP synthesis. Because citric acid cycle intermediates are precursors for the biosynthesis of many other molecules, it also serves a function in anabolism.

****

End-of-Chapter Questions and Problems
****

22.15 The intermembrane compartment is the location of the high energy proton (H^+) reservoir produced by the electron transport system. The energy of this H^+ reservoir is used to make ATP.

22.17 The outer mitochondrial membrane is freely permeable to substances of molecular weight less than 10,000. The inner mitochondrial membrane is highly impermeable. Only certain fuel molecules and H^+ are transported through it. Embedded within the inner mitochondrial membrane are the electron carriers of the electron transport system and ATP synthase, the multisubunit enzyme that makes ATP.

22.19 a. False
     b. False
     c. True
     d. True

22.21 The acetyl group of acetyl CoA is transferred to oxaloacetate to produce citrate.

22.23 Three NAD^+ are reduced to 3 NADH in one turn of the citric acid cycle.

22.25 The net yield of ATP for anaerobic glycolysis is 2 ATP per glucose.

22.27 The function of acetyl CoA in the citric acid cycle is to bring the two carbon remnant (acetyl group) of pyruvate from glycolysis and transfer it to oxaloacetate. In this way the acetyl group enters the citric acid cycle for the final stages of oxidation.

22.29 GTP is produced in the citric acid cycle. The high energy phosphoryl group of the GTP is transferred to ADP to produce ATP. This reaction is catalyzed by the enzyme dinucleotide diphosphokinase.

22.31 The oxidation of NADH via oxidative phosphorylation yields 3 ATP.

22.33 The oxidation of a variety of fuel molecules, including carbohydrates, the carbon skeletons of amino acids, and fatty acids provides the electrons. The energy of these electrons is used to produce an H^+ reservoir. The energy of this proton reservoir is used for ATP synthesis.

22.35 The electron transport system passes electrons harvested during oxidation of fuel molecules to molecular oxygen. At three sites protons are pumped from the mitochondrial matrix into the intermembrane compartment. Thus, the electron transport system builds the high energy H^+ reservoir that provides energy for ATP synthesis.

22.37 Two ATP per glucose (net yield) are produced in glycolysis, while the complete oxidation of glucose in aerobic respiration (glycolysis, the citric acid cycle, and oxidative phosphorylation) results in the production of 36 ATP per glucose. Thus, aerobic respiration harvests nearly 40% of the potential energy of glucose, while anaerobic glycolysis harvests only about 2% of the potential energy of glucose.

22.39 Energy-harvesting pathways, such as the citric acid cycle, must be responsive to the energy needs of the cell. If the energy requirements are high, as during exercise, the reactions must speed up. If energy demands are low and ATP is in excess, the reactions of the pathway slow down.

22.41 ADP serves as a signal to increase the rate of the reactions of the citric acid cycle. If ADP is found in high concentration, then ATP levels must be low. This is a signal that the cell requires more ATP energy. The enzyme isocitrate dehydrogenase is an allosteric enzyme of the citric acid cycle that is stimulated by ADP.

22.43 Aminotransferases (transaminases) transfer amino groups from amino acids to ketoacids.

22.45 The glutamate family of transaminases is very important because the ketoacid corresponding to glutamate is α-ketoglutarate, one of the citric acid cycle intermediates. This provides a link between the citric acid cycle and amino acid metabolism. These transaminases provide amino groups for amino acid synthesis and collect amino groups during catabolism of amino acids.

22.47 Alanine loses the α-amino group and becomes pyruvate. The complete oxidation of pyruvate is summarized below:

Pyruvate --> acetyl CoA:

1 NADH x 3 ATP/NADH	3 ATP

Citric acid cycle:

3 NADH x 3 ATP/NADH	9 ATP
1 FADH$_2$ x 2 ATP/FADH$_2$	2 ATP
1 GTP x 1 ATP/GTP	1 ATP
	15 ATP

22.49 The synthesis of glutamate catalyzed by glutamate dehydrogenase can be represented by the following equation:

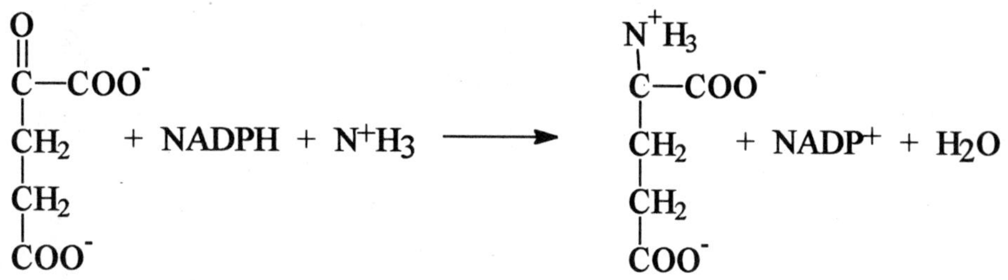

$$\underset{\alpha\text{-Ketoglutarate}}{\begin{array}{c} O \\ \parallel \\ C-COO^- \\ | \\ CH_2 \\ | \\ CH_2 \\ | \\ COO^- \end{array}} + NADPH + N^+H_3 \longrightarrow \underset{\text{Glutamate}}{\begin{array}{c} N^+H_3 \\ | \\ C-COO^- \\ | \\ CH_2 \\ | \\ CH_2 \\ | \\ COO^- \end{array}} + NADP^+ + H_2O$$

α-Ketoglutarate   Ammonia   Glutamate

22.51 Hyperammonemia, an elevation of the concentration of ammonium ions in the body, results when the urea cycle is not functioning. A complete deficiency of a urea cycle enzyme results in death in early infancy. A partial deficiency causes retardation, convulsions, and vomiting and can be treated with a low protein diet.

22.53 a. The source of one amino group of urea is the ammonium ion and the source of the other is the α-amino group of the amino acid aspartate.
b. The carbonyl group of urea is derived from CO_2.

22.55 The amino acid glutamate is synthesized from the citric acid cycle intermediate α-ketoglutarate.

22.57 Citric acid cycle intermediates are the starting materials for the biosynthesis of many biological molecules.

22.59 An essential amino acid is one that cannot be synthesized by the body and must be provided in the diet.

22.61

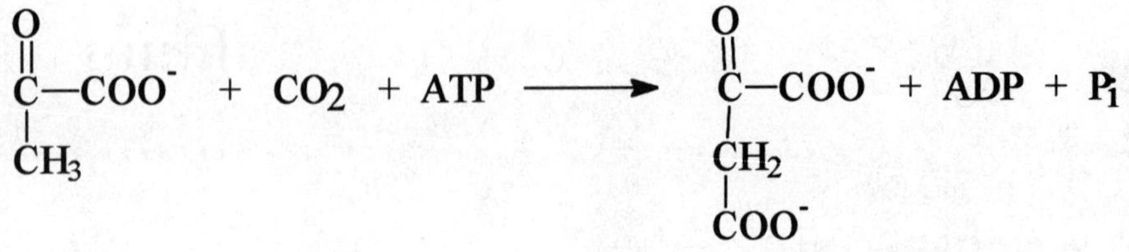

$$\underset{\text{Pyruvate}}{\overset{\text{O}}{\underset{\text{CH}_3}{\overset{\|}{\text{C}}}}-\text{COO}^-} + \text{CO}_2 + \text{ATP} \longrightarrow \underset{\text{Oxaloacetate}}{\overset{\text{O}}{\underset{\substack{\text{CH}_2 \\ \text{COO}^-}}{\overset{\|}{\text{C}}}}-\text{COO}^-} + \text{ADP} + \text{P}_\text{i}$$

Chapter 23
Fatty Acid Metabolism
Solutions to the Odd-Numbered Problems

****

In-Chapter Questions and Problems

****

23.1   Because dietary lipids are hydrophobic, they arrive in the small intestine as large fat globules. The bile salts emulsify these fat globules into tiny fat droplets. This greatly increases the surface area of the lipids, allowing them to be more accessible to pancreatic lipases and thus more easily digested.

23.3   a. The products of the β-oxidation of 9-phenylnonanoic acid are 4 acetyl CoA, 1 benzoate, 4 NADH, and 4 FADH$_2$.
b. The products of the β-oxidation of 8-phenyloctanoic acid are 3 acetyl CoA, 1 phenyl acetate, 3 NADH, and 3 FADH$_2$.
c. The products of the β-oxidation of 7-phenylheptanoic acid are 3 acetyl CoA, 1 benzoate, 3 NADH, and 3 FADH$_2$.
d. The products of the β-oxidation of 12-phenyldodecanoic acid are 5 acetyl CoA, 1 phenyl acetate, 5 NADH, and 5 FADH$_2$.

23.5

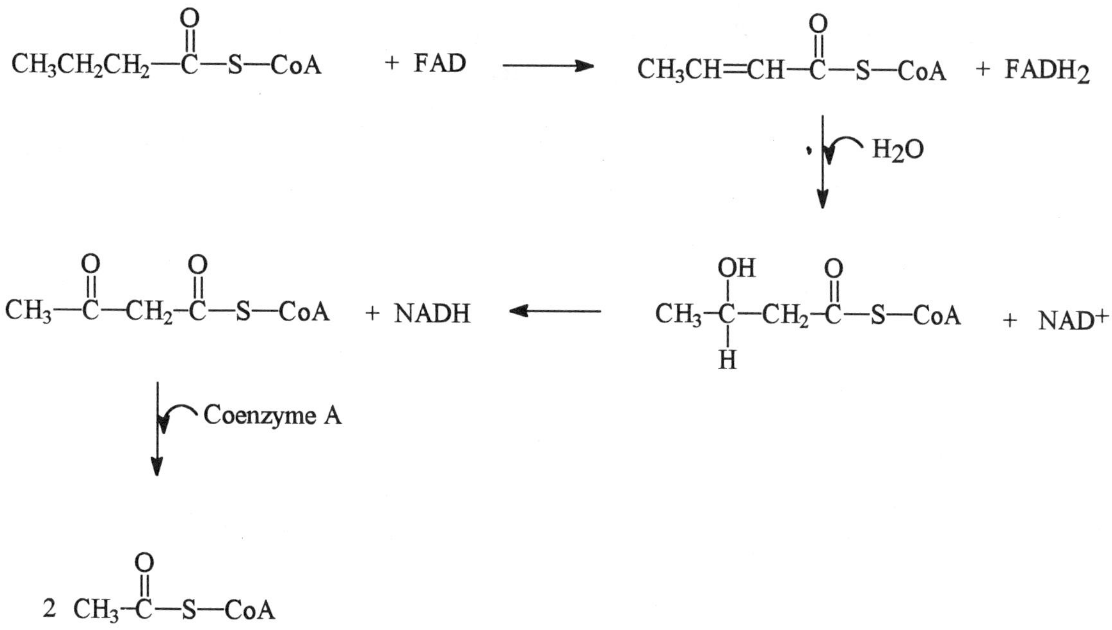

23.7   Starvation, a diet low in carbohydrates, and diabetes mellitus are conditions that lead to the production of ketone bodies. Lack of carbohydrates causes a decreased amount of oxaloacetate. This slows the citric acid cycle so that less acetyl CoA is oxidized. The excess acetyl CoA is converted to ketone bodies to recover coenzyme A.

23.9   The following are four differences between fatty acid biosynthesis and β-oxidation: (1) Fatty acid biosynthesis occurs in the cytoplasm while β-oxidation occurs in the mitochondria.
(2) The acyl group carrier in fatty acid biosynthesis is acyl carrier protein while the acyl group carrier in β-oxidation is coenzyme A.
(3) The seven enzymes of fatty acid biosynthesis are associated as a multienzyme complex called fatty acid synthase. The enzymes involved in β-oxidation are not physically associated with one another.
(4) NADPH is the reducing agent used in fatty acid biosynthesis. NADH and FADH$_2$ are produced by β-oxidation.

23.11   The liver regulates blood glucose levels under the control of the hormones insulin and glucagon. When blood glucose levels are too high, insulin stimulates the uptake of glucose by liver cells and the storage of the glucose in glycogen polymers. When blood glucose levels are too low, the hormone glucagon stimulates the breakdown of glycogen and release of glucose into the bloodstream. Glucagon also stimulates the liver to produce glucose for export into the bloodstream by the process of gluconeogenesis.

23.13   Insulin stimulates uptake of glucose and amino acids by cells, glycogen and protein synthesis, and storage of lipids. It inhibits glycogenolysis, gluconeogenesis, breakdown of stored triglycerides, and ketogenesis.

****
End-of-Chapter Questions and Problems
****

23.15   Triglycerides in adipose tissue are the major storage form of lipids.

23.17   The most outstanding feature of an adipocyte is the large fat globule that takes up nearly the entire cytoplasm.

23.19 Lipases catalyze the hydrolysis of the ester bonds of triglycerides, as seen in the following equation:

$$
\begin{array}{l}
\text{H} \quad\quad \text{O} \\
| \quad\quad \| \\
\text{H—C—O—C—R}_1 \\
\quad\quad\quad \text{O} \\
\quad\quad\quad \| \\
\text{H—C—O—C—R}_2 \quad + 3\,H_2O \longrightarrow \\
\quad\quad\quad \text{O} \\
\quad\quad\quad \| \\
\text{H—C—O—C—R}_3 \\
| \\
\text{H}
\end{array}
\qquad
\begin{array}{l}
\text{H} \\
| \\
\text{H—C—OH} \\
| \\
\text{H—C—OH} \quad + \\
| \\
\text{H—C—OH} \\
| \\
\text{H}
\end{array}
\qquad
\begin{array}{l}
\quad\quad \text{O} \\
\quad\quad \| \\
\text{HO—C—R}_1 \\
\quad\quad \text{O} \\
\quad\quad \| \\
\text{HO—C—R}_2 \\
\quad\quad \text{O} \\
\quad\quad \| \\
\text{HO—C—R}_3
\end{array}
$$

23.21 Acetyl CoA is the precursor for fatty acids, several amino acids, cholesterol, and other steroids.

23.23 Chylomicrons are plasma lipoproteins (aggregates of protein and triglycerides) that carry dietary triglycerides from the intestine to all tissues via the bloodstream.

23.25 Bile salts serve as detergents. Fat globules stimulate their release from the gall bladder. The bile salts then emulsify the lipids, increasing the surface area and making them more accessible to digestive enzymes (pancreatic lipases).

23.27 When dietary lipids in the form of fat globules reach the duodenum, they are emulsified by bile salts. The triglycerides in the resulting tiny fat droplets are hydrolyzed into monoglycerides and fatty acids by the action of pancreatic lipases, assisted by colipase. The monoglycerides and fatty acids are absorbed by cells lining the intestine. Within intestinal cells triglycerides are reassembled and are packaged into chylomicrons (lipoprotein particles made up of protein and dietary triglycerides). Chylomicrons are secreted into the lymphatic vessels and eventually reach the bloodstream. In the bloodstream the triglycerides are hydrolyzed once again and the products (glycerol and free fatty acids) are absorbed by the cells of the body.

23.29 The β-oxidation of 14-phenyltetradecanoic acid yields 6 acetyl CoA, 1 phenyl acetate, 6 NADH, and 6 $FADH_2$.

23.31 The following diagram summarizes the energy harvested by the β-oxidation of tetradecanoic acid:

Step 1 (Activation)   -2ATP

Steps 2-6 (repeated six times):

6 FADH$_2$ x 2 ATP/FADH$_2$	12 ATP
6 NADH x 3 ATP/NADH	18 ATP

7 acetyl CoA to citric acid cycle:

7 x 1 GTP x 1 ATP/GTP	7 ATP
7 x 3 NADH x 3 ATP/NADH	63 ATP
7 x 1 FADH$_2$ x 2 ATP/FADH$_2$	14 ATP
	112 ATP

23.33 Two molecules of ATP are produced for each FADH$_2$ produced by β-oxidation.

23.35 The acetyl CoA produced by β-oxidation will enter the citric acid cycle unless there is too little oxaloacetate available. If that is the case, the acetyl CoA will be used in ketogenesis.

23.37

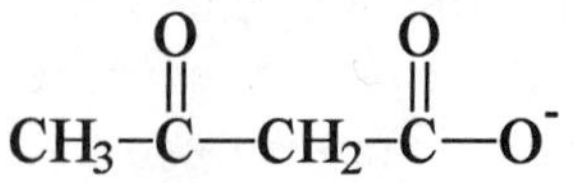

$$CH_3-\overset{\overset{\displaystyle O}{\|}}{C}-CH_2-\overset{\overset{\displaystyle O}{\|}}{C}-O^- \qquad\qquad CH_3-\underset{\underset{\displaystyle OH}{|}}{CH}CH_2-\overset{\overset{\displaystyle O}{\|}}{C}-O^-$$

         Acetoacetate 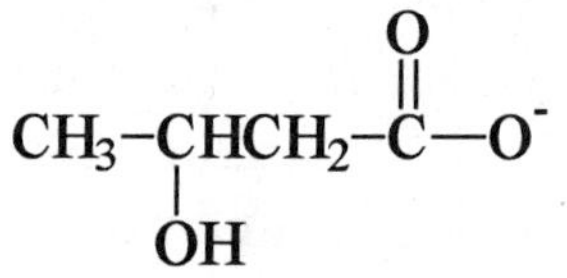   β-Hydroxybutyrate

23.39 In those suffering from uncontrolled diabetes, the glucose in the blood cannot get into the cells of the body. The excess glucose is excreted in the urine. Body cells degrade fatty acids because glucose is not available. β-Oxidation of fatty acids yields enormous quantities of acetyl CoA, so much acetyl CoA, in fact, that it cannot all enter the citric acid cycle because there is not enough oxaloacetate available.. Excess acetyl CoA is used for ketogenesis.

23.41 Ketone bodies are the preferred energy source of the heart.

23.43 The phosphopantetheine group allows formation of a high energy thioester bond with a fatty acid. It is derived from the vitamin pantothenic acid.

23.45 Fatty acid synthase is a huge multienzyme complex consisting of the seven enzymes involved in fatty acid synthesis. It is found in the cell cytoplasm. The enzymes involved in β-oxidation are not physically associated with one another. They are free in the mitochondrial matrix space.

23.47 The major metabolic function of the liver is to regulate blood glucose levels. It also produces triglycerides for storage when excess calories are consumed. During fasting or starvation the liver converts fatty acids to ketone bodies which are then exported to other organs.

23.49 Ketone bodies are the major fuel for the heart. Glucose is the major energy source of the brain and the liver obtains most of its energy from the oxidation of amino acid carbon skeletons.

23.51 Fatty acids are absorbed from the bloodstream by adipocytes. Using glucose-3-phosphate, produced as a by-product of glycolysis, triglycerides are synthesized. Triglycerides are constantly being hydrolyzed and resynthesized in adipocytes. The rates of hydrolysis and synthesis are determined by lipases that are under hormonal control.

23.53 Insulin is produced in the β-cells of the islets of Langerhans in the pancreas.

23.55 Insulin stimulates the uptake of glucose from the blood into cells. It enhances glucose storage by stimulating glycogenesis and inhibiting glycogen degradation and gluconeogenesis.

23.57 Insulin stimulates synthesis and storage of triglycerides.

23.59 Untreated diabetes mellitus is starvation in the midst of plenty because blood glucose levels are very high. However, in the absence of insulin, blood glucose can't be taken up into cells. The excess glucose is excreted into the urine while the cells of the body are starved for energy.

Chapter 24
Introduction to Molecular Genetics
Solutions to the Odd-Numbered Problems

****
In-Chapter Questions and Problems
****

24.1 a. Adenosine diphosphate:

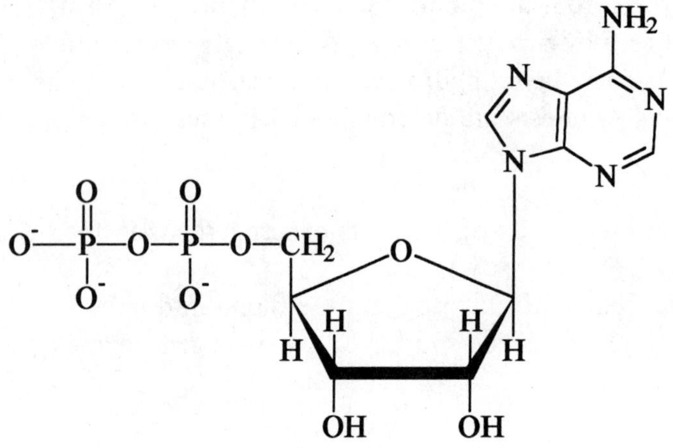

b. Deoxyguanosine triphosphate:

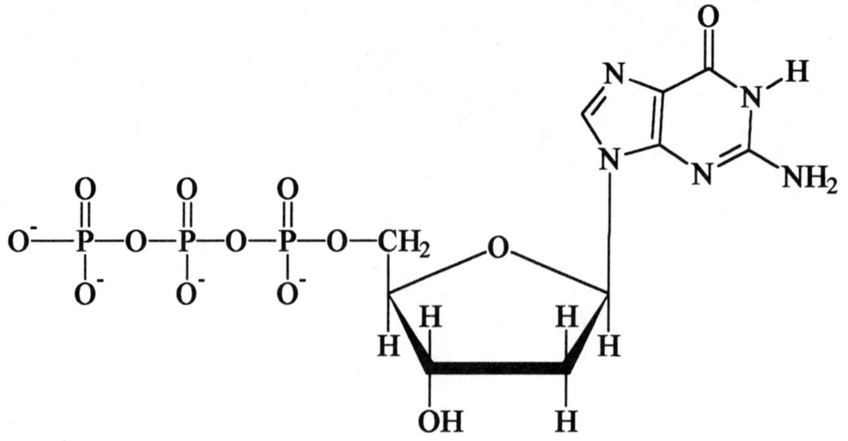

24.3   The deoxyribonucleotides of guanine are:

Deoxyguanosine monophosphate (dGMP)
Deoxyguanosine diphosphate (dGDP)
Deoxyguanosine triphosphate (dGTP)

The ribonucleotides of guanine are:

Guanosine monophosphate (GMP)
Guanosine diphosphate (GDP)
Guanosine triphosphate (GTP)

24.5   The RNA polymerase recognizes the promoter site for a gene, separates the strands of DNA, and catalyzes the polymerization of an RNA strand complementary to the DNA strand that carries the genetic code for a protein. It recognizes a termination site at the end of the gene and releases the RNA molecule.

24.7   The genetic code is said to be degenerate because several different triplet codons may serve as code words for a single amino acid.

24.9   The nitrogenous bases of the codons are complementary to those of the anticodons. As a result they are able to hydrogen bond to one another according to the base pairing rules.

24.11   The ribosomal P-site holds the peptidyl tRNA during protein synthesis. The peptidyl tRNA is the tRNA carrying the growing peptide chain. The only exception to this is during initiation of translation when the P-site holds the initiator tRNA.

24.13   The normal mRNA sequence, AUG-CCC-GAC-UUU, would encode the peptide sequence, methionine-proline-aspartate-phenylalanine. The mutant mRNA sequence, AUG-CGC-GAC-UUU, would encode the mutant peptide sequence, methionine-arginine-aspartate-phenylalanine. This would not be a silent mutation because a hydrophobic amino acid (proline) has been replaced by a positively charged amino acid (arginine).

****

End-of-Chapter Questions and Problems

****

24.15 It is the N-9 of the purine that forms the N-glycosidic bond with C-1 of the five carbon sugar. The general structure of the purine ring is shown below:

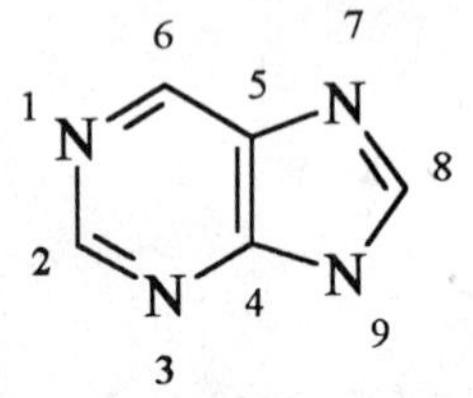

24.17 The ATP nucleotide is composed of the five-carbon sugar ribose, the purine adenine, and a triphosphate group.

24.19 The two strands of DNA in the double helix are said to be *antiparallel* because they run in opposite directions. One strand progresses in the 5'→ 3' direction, while the opposite strand progresses in the 3' → 5' direction.

24.21 Two hydrogen bonds link the adenine-thymine base pair.

24.23

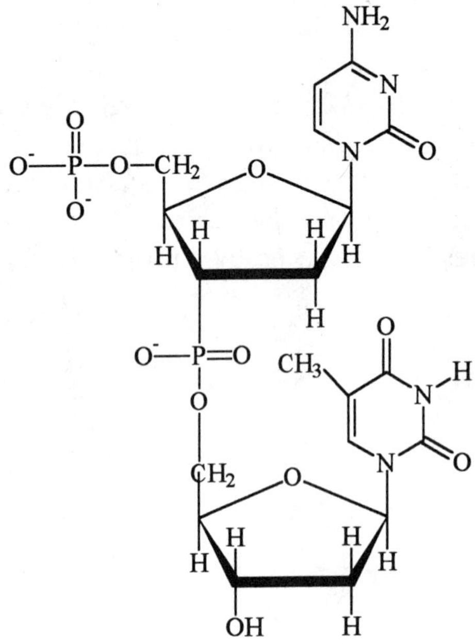

24.25 The term semiconservative DNA replication refers to the fact that each parental DNA strand serves as the template for the synthesis of a daughter strand. As a result, each of the daughter DNA molecules is made up of one strand of the original parental DNA and one strand of newly synthesized DNA.

24.27 The two primary functions of DNA polymerase III are to read a template DNA strand and catalyze the polymerization of a new daughter strand and to proofread the newly synthesized strand and correct any errors by removing the incorrectly inserted nucleotide and adding the proper one.

24.29 If the parental DNA strand had the following nucleotide sequence: 5'-ATGCGGCTAGAATATTCCA-3', the sequence of the complementary daughter strand would be 3'-TACGCCGATCTTATAAGGT-5'.

24.31 The *replication origin* of a DNA molecule is the unique sequence on the DNA molecule where DNA replication begins.

24.33 The central dogma of molecular biology states that information flow in cellular biological systems is unidirectional: DNA → RNA → Protein. The DNA carries the genetic information; RNA molecules carry out the expression of the genetic information for proteins; the final products are proteins which carry out the work of the cell and serve as cellular structural components.

24.35 Anticodons are found on transfer RNA molecules.

24.37 If a gene had the sequence, 5'-TACCTAGCTCTGGTCATTAAGGCAGTA-3', the mRNA would have the sequence, 3'-AUGGAUCGAGACCAGUAAUUCCGUCAU-5'.

24.39 *RNA splicing* is the process by which the noncoding sequences (introns) of the primary transcript of a eukaryotic mRNA are removed and the protein coding sequences (exons) are spliced together.

24.41 The three classes of RNA molecules are messenger RNA (mRNA), transfer RNA (tRNA), and ribosomal RNA (rRNA).

24.43 In the first transesterification reaction of RNA splicing, the sugar-phosphate backbone of the RNA is cut on the 5' side of the sequence GpU, which is always found at the 5' intron boundary. Then the 5' phosphoryl group of the guanosine forms a 2'-5' phosphodiester bond with an adenosine found in the sequence CUPuAPy within the intron. In the second transesterification reaction the 3'-OH at the end of the liberated exon forms a phosphodiester bond with the 5' phosphoryl group of the nucleotide at the 5' end of the next exon. This releases the intron as a lariat structure.

24.45 The *poly(A) tail* is a stretch of 100-200 adenosine nucleotides polymerized onto the 3' end of a mRNA by the enzyme poly(A) polymerase.

24.47 The *cap structure* is made up of the nucleotide 7-methylguanosine attached to the 5' end of a mRNA by a 5'-5' triphosphate bridge. Generally the first two nucleotides of the mRNA are also methylated. The cap structure is required for efficient translation of the mRNA.

24.49 There are 64 codons in the genetic code.

24.51 The reading frame of a gene is the sequential set of triplet codons that carries the genetic code for the primary structure of a protein. Each triplet specifies the addition of a particular amino acid to the growing peptide chain.

24.53 Methionine (AUG) and tryptophan (UGG) are encoded by only one codon.

24.55 The codon 5'-UUU-3' encodes the amino acid phenylalanine. The mutant codon 5'-UUA-3' encodes the amino acid leucine. Both leucine and phenylalanine are hydrophobic amino acids, however, leucine has a smaller R group. It is possible that the smaller R group would disrupt the structure of the protein.

24.57 The ribosomes serve as a platform on which protein synthesis can occur. They also carry the enzymatic activity that forms peptide bonds.

24.59 In the initiation of translation, initiation factors, methionyl tRNA (the initiator tRNA), the mRNA, and the small and large ribosomal subunits form the initiation complex. During the elongation stage of translation an aminoacyl tRNA binds to the A-site of the ribosome. Peptidyl transferase catalyzes the formation of a peptide bond and the peptide chain is transferred to the tRNA in the A-site. Translocation shifts the peptidyl tRNA from the A-site into the P-site, leaving the A-site available for the next aminoacyl tRNA. In the termination stage of translation, a termination codon is encountered. A release factor binds to the empty A-site and peptidyl transferase catalyzes the hydrolysis of the bond between the peptidyl tRNA and the completed peptide chain.

24.61 The bond between an amino acid and a tRNA is an ester bond formed between the carboxylate group of the amino acid and the 3'-OH of the sugar ribose in the tRNA molecule.

24.63 UV light causes the formation of thymine dimers, the covalent bonding of two adjacent thymine bases. Mutations occur when the UV damage repair system makes an error during the repair process. This causes a change in the nucleotide sequence of the DNA.

24.65 A *carcinogen* is a compound that causes cancer. Cancers are caused by mutations in the genes responsible for controlling cell division. Carcinogens cause DNA damage that results in changes in the nucleotide sequence of the gene. Thus, carcinogens are also mutagens.

24.67 A *restriction enzyme* is a bacterial enzyme that "cuts" the sugar-phosphate backbone of DNA molecules at a specific nucleotide sequence.

24.69 Nucleic acid hybridization is based on the fact that complementary DNA or RNA sequences will hydrogen bond to one another according to the base pairing rules.

24.71 Human insulin, interferon, human growth hormone, and human blood clotting factor VIII are protein products of recombinant DNA technology that are of great value in the field of medicine.